# Selected Titles in This Series

18 **Serge Dubuc and Gilles Deslauriers, Editors,** Spline functions and the theory of wavelets, 1999

17 **Olga Kharlampovich, Editor,** Summer school in group theory in Banff, 1996, 1998

16 **Alain Vincent, Editor,** Numerical methods in fluid mechanics, 1998

15 **François Lalonde, Editor,** Geometry, topology, and dynamics, 1998

14 **John Harnad and Alex Kasman, Editors,** The bispectral problem, 1998

13 **Michel C. Delfour, Editor,** Boundaries, interfaces, and transitions, 1998

12 **Peter C. Greiner, Victor Ivrii, Luis A. Seco, and Catherine Sulem, Editors,** Partial differential equations and their applications, 1997

11 **Luc Vinet, Editor,** Advances in mathematical sciences: CRM's 25 years, 1997

10 **Donald E. Knuth,** Stable marriage and its relation to other combinatorial problems: An introduction to the mathematical analysis of algorithms, 1997

9 **D. Levi, L. Vinet, and P. Winternitz, Editors,** Symmetries and integrability of difference equations, 1996

8 **J. Feldman, R. Froese, and L. M. Rosen, Editors,** Mathematical quantum theory II: Schrödinger operators, 1995

7 **J. Feldman, R. Froese, and L. M. Rosen, Editors,** Mathematical quantum theory I: Field theory and many-body theory, 1994

6 **Guido Mislin, Editor,** The Hilton Symposium 1993: Topics in topology and group theory, 1994

5 **D. A. Dawson, Editor,** Measure-valued processes, stochastic partial differential equations, and interacting systems, 1994

4 **Hershy Kisilevsky and M. Ram Murty, Editors,** Elliptic curves and related topics, 1994

3 **Rémi Vaillancourt and Andrei L. Smirnov, Editors,** Asymptotic methods in mechanics, 1993

2 **Philip D. Loewen,** Optimal control via nonsmooth analysis, 1993

1 **M. Ram Murty, Editor,** Theta functions: From the classical to the modern, 1993

Volume 18

# CRM PROCEEDINGS & LECTURE NOTES

Centre de Recherches Mathématiques
Université de Montréal

# Spline Functions and the Theory of Wavelets

Serge Dubuc
Gilles Deslauriers
*Editors*

The Centre de Recherches Mathématiques (CRM) of the Université de Montréal was created in 1968 to promote research in pure and applied mathematics and related disciplines. Among its activities are special theme years, summer schools, workshops, postdoctoral programs, and publishing. The CRM is supported by the Université de Montréal, the Province of Québec (FCAR), and the Natural Sciences and Engineering Research Council of Canada. It is affiliated with the Institut des Sciences Mathématiques (ISM) of Montréal, whose constituent members are Concordia University, McGill University, the Université de Montréal, the Université du Québec à Montréal, and the Ecole Polytechnique. The CRM may be reached on the Web at www.crm.umontreal.ca.

**American Mathematical Society**
Providence, Rhode Island USA

The production of this volume was supported in part by the Fonds pour la Formation de Chercheurs et l'Aide à la Recherche (Fonds FCAR) and the Natural Sciences and Engineering Research Council of Canada (NSERC).

1991 *Mathematics Subject Classification.* Primary 65D07, 41A30;
Secondary 41A15, 42C15, 94A12, 65D15, 81V45, 62G07.

---

**Library of Congress Cataloging-in-Publication Data**

Spline functions and the theory of wavelets / Serge Dubuc, Giles Deslauriers, editors.
p. cm. — (CRM proceedings & lecture notes, ISSN 1065-8580 ; v. 18)
Includes bibliographical references.
ISBN 0-8218-0875-3
1. Spline theory. 2. Wavelets (Mathematics) I. Dubuc, Serge, 1939– . II. Deslauriers, Gilles, 1941– . III. Series.
QA224.S6525 1999
511′.42—dc21 98-54441
CIP

---



∞ The paper used in this book is acid-free and falls within the guidelines
established to ensure permanence and durability.
This volume was typeset using $\mathcal{A}\mathcal{M}\mathcal{S}$-TEX,
the American Mathematical Society's TEX macro system,
and submitted to the American Mathematical Society in camera ready
form by the Centre de Recherches Mathématiques.
Visit the AMS home page at URL: http://www.ams.org/

10 9 8 7 6 5 4 3 2 1 04 03 02 01 00 99

# Contents

# Preface

Quel or ondé en tresses s'allongeant
Frapoit ce jour sa gorge nouvelette,
Et sus son col, ainsi qu'une ondelette
Flotte aux zephyrs, au vent alloit nageant

*Les Amours*, Pierre de Ronsard

Spline functions emerged from a 1946 article of Schoenberg and owe much of their development to de Boor while the theory of wavelets was born in the work of Morley and Grossmann in 1980 and took off as a result of a number of authors among which Meyer and Daubechies are the most reknowned.

Among the historical bench marks in the development of the theory of wavelets is the discovery by Meyer of the first orthogonal wavelet basis. This occurred in 1985 in France, in a manner that Meyer himself described as mysterious and accidental. Shortly thereafter Mallat introduced multiresolutional analysis which explained some of the mysteries of Meyer's wavelet. Then on a cold, snowy February day in 1987 in Montreal, Daubechies found a wavelet with compact support. Next came biorthogonal wavelets which were related to spline functions. In Paris in May 1985, Deslauriers and Dubuc used dyadic interpolation to explain multiresolution, a term which by then was familiar to many. Since these developments, alot of water has gone under the bridge (much of it French) and we might also say that many waves and wavelets have also gone under that bridge. Within a few years, a multitude of books and other publications on the theory of wavelets appeared, several conferences were held, and congresses were organized.

Was this the birth of a theory? During this same period, related subjects, such as multiresolutional analysis, subdivision schemes, dilatation equations, to cite just a few, were developped.

As a result of this brew of ideas there were applications to physics, signal processing and imaging, the theory of fractals, statistics, and computer science. These interactions allowed us to clarify the theory. Meanwhile the development of the theory of splines proceeded in a parallel fashion.

From all this, emerged the idea of holding a series of thematic workshops relating the theory of wavelets and the theory of splines without losing sight of related applications which were of increasing importance. These workshops took place at the Centre de recherches mathématiques in 1996. Ten themes were treated, each for one week: Geometric Modelling with Splines, Splines in Approximation and in Differential Equations, Splines and Wavelets, Wavelets and Approximation, Multiresolution Analysis and Subdivision Operators, Wavelets and Differential Equations, Wavelets in Signal Processing and Image Analysis, Wavelets and Fractals, Wavelets in Physics and finally Splines and Wavelets in Statistics. The contributions of Professor Yves Meyer who gave a series of lectures as Aisenstadt chairholder, were most appreciated. They resulted in a book entitled Wavelets, Vibrations and Scalings published in the CRM Monograph Series (Volume 9) by the American Mathematical Society.

The proceedings of these workshops consist of 27 contributions which were grouped in four parts entitled Spline Functions, Theory of Wavelets, Wavelets in Physics, and Splines and Wavelets in Statistics. Each section begins with a presentation due respectively to Brunner, Jaffard, Arnéodo, and MacGibbon.

We wish to thank all those who contributed to the success of these workshops, the support staff at the Centre de recherches mathématiques: Josée Laferrière, Louise Letendre, André Montpetit, and Louis Pelletier, those who helped with the organizing: Martin Bilodeau, Véronique Hussin, Jean-Marc Lina, Brenda MacGibbon, Marc Moore, and Sherman Riemenschneider. We also note the consciencious work of the members of the scientific committee for these proceedings: Alain Arnéodo, John J. Benedetto, Hermann Brunner, Stéphane Jaffard, Alain Le Méhauté, Brenda MacGibbon, Sherman Riemenschneider and Philippe Tchamitchian. Finally we express our appreciation for the moral support of Luc Vinet, director of the CRM, and the financial support of NSERC (Canada).

Gilles Deslauriers
Serge Dubuc

# Part 1

# Spline Functions

# Introduction and Summary

The papers collected in this section convey a lively picture of the broad spectrum of current research in the theory and applications of spline functions: here, the theory ranges from classical univariate spline approximation (optimal error bounds) to an abstract framework for multivariate spline interpolation on nonregular data, while applications deal, for example, with scattered data interpolation in $\mathbb{R}^n$ and algorithms for implementing blossoming techniques in CAGD, as well as with the use of various kinds of (deficient) spline functions in the numerical solution of initial-value problems for (ordinary and delay) differential and integral equations.

The first of the two papers by Dubeau and Savoie presents a review of results on best constants in bounds for the interpolation error for a class of odd/even degree deficient splines, both for arbitrary partitions and (in the case of periodic splines) for uniform partitions. Their second paper is concerned with cardinal spline interpolation of a function at shifted nodes: the techniques employed for deriving the optimal bounds for the error and its $n$-th derivative, and non-optimal error bounds for intermediate derivatives, exhibit a nice blend of classical approximation results and modern spline theory.

Contrasting the above paper on refinements of classical univariate spline theory is the contribution by Le Méhauté and Bouhamidi: it treats the problem of interpolation with respect to a finite number of linear, continuous functionals defined on a semi-Hilbert space and studies the question of existence and uniqueness of the interpolating $(m, \ell, s)$-spline. The results have interesting applications in, e.g., thin plate spline and multiquadric interpolation.

Scattered vertex data interpolation in $\mathbb{R}^n$ is the topic of the paper by Bos and Holland: here, the basic idea consists in solving a sequence of univariate problems whose solutions (a system of interpolating "curves") is then extended to the whole space. The main purpose of the paper is to derive a method that can easily be implemented in any dimension; the approach is illustrated by illuminating concrete examples.

The modelling of static flexible objects is considered by Dussault and Pfister. Using tensor-product B-splines, objects are represented by some control polygonal meshes which are progressively refined; after each refinement, a merit function (representing the object's energy) is minimized. The resulting algorithm is then applied to the problem of modelizing cloth and tire-like objects.

There are numerous techniques for analyzing and unifying properties and algorithms related to polynomial curves and surfaces in Computer Aided Geometric Design (CAGD). The use of so-called blossoming techniques has turned out to be particularly attractive, and they are discussed in the contribution by Fiorot and Jeannin. Background information and new results are discussed in some detail.

Sablonnière gives a survey of recent results by himself and his co-workers on existence and various properties of families of piecewise polynomials with bounded support (B-splines) on the two classical uniform triangulations of the plane. These B-splines are obtained by suitable repeated convolution of simple B-splines, with classical box-splines as important examples. Sections on quasi-interpolants and Lagrange interpolation on rectangular domains conclude this paper.

The above papers all deal with univariate or multivariate spline approximation techniques for various types of data. During the last 30 years, classical and deficient splines ("piecewise polynomials") have also played a significant role in the numerical solution of differential and integral equations. The survey paper by Brunner looks at the use of such spline functions in collocation methods for ordinary and delay differential equations, differential-algebraic equations, and Volterra integral and integro-differential equations. An extensive list of representative references complements this review.

H. Brunner

Centre de Recherches Mathématiques
CRM Proceedings and Lecture Notes
Volume **18**, 1999

# Radial Extensions of Vertex Data

L. P. Bos and D. Holland

Abstract. A new method to interpolate vertex data in $\mathbb{R}^n$ is developed with an emphasis on relative simplicity and ease of implementation. The vertices are first (simplicially) triangulated and then interpolating curves along edges in the triangulation are defined. Edge curves are then transfinitely extended, at each vertex $\mathbf{v}_i$, to the interior of surrounding simplices by means of local radial extensions. These extensions are defined in terms of a system of generalized spherical coordinates which describe the local simplex geometries surrounding $\mathbf{v}_i$. By building the local geometry directly into the interpolant, further sets of conditions to enforce cross-boundary smoothness are thus avoided. An application of this method in conjunction with the technique of Network splines (Minimum Norm Networks) yields good results. Some simple examples in $\mathbb{R}^3$ and $\mathbb{R}^4$ are presented.

## 1. Introduction

In this work, we address the problem of scattered data interpolation in $\mathbb{R}^n$, namely:

> Given nodes $\mathbf{v}_i \in \Omega \subset \mathbb{R}^n$ $i = 1, 2, \ldots, N$ and values $\{z_i^{(\alpha)}\}_{i=1,2,\ldots,N}^{|\alpha|\le k}$ in $\mathbb{R}$, find $F \in C^k(\Omega)$ such that $D^\alpha F(\mathbf{v}_i) = z_i^{(\alpha)}$ for $|\alpha| \le k$ and $i = 1, 2, \ldots, N$.

Now suppose that from the nodes (or vertices) we have constructed a simplicial triangulation, say $\mathcal{T}$. Then, restricted to each edge $\mathbf{e}_{ij}$ (joining $\mathbf{v}_i$ and $\mathbf{v}_j$) we have a univariate Hermite interpolation with (directional) derivatives up to order $k$ at the endpoints determined by the $z_i^{(\alpha)}$ and $z_j^{(\alpha)}$. Each of these univariate problems may be solved, for example, by a polynomial of degree $2k+1$ and thus we immediately and easily arrive at a system of interpolating "curves" defined over the edges of $\mathcal{T}$. The problem is how to extend these curves to all of $\mathbb{R}^n$.

A number of methods, such as the side-vertex method, have been discussed in the literature (see **[9]** for example), but these either work well only in two dimensions, or become increasingly complex as the dimension grows. In this work we describe a simple method that can easily be implemented in any number of dimensions and yet seems to remain effective. The idea is as follows. First consider each

1991 *Mathematics Subject Classification.* Primary: 41A63; Secondary: 41A05.
This is the final form of the paper.

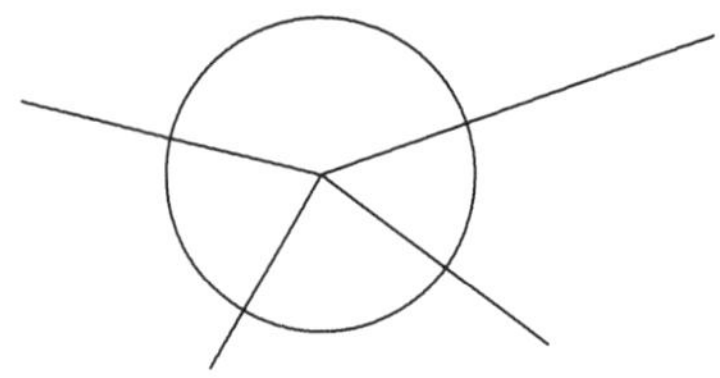

FIGURE 1.

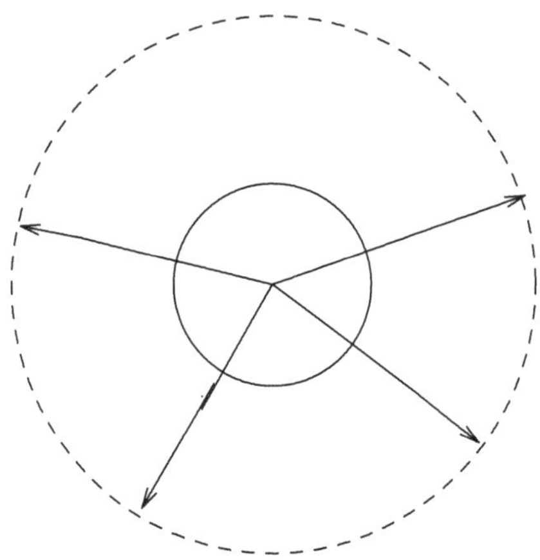

FIGURE 2.

vertex $v_i$ locally. A vertex and the edges emanating from it might typically look as in Figure 1 (in 2-d).

Along each edge is defined a univariate interpolant of the vertex data at each end. We may extrapolate the edges *and the corresponding interpolants* to $\infty$ to obtain a picture like in Figure 2.

We then regard the central vertex $\mathbf{v}_i$ as the origin of a spherical coordinate system with a certain number of given functions $f_{ij}$ along (infinite) rays emanating from $v_i$ and by means of the radial extension explained later, construct a function $F_i \in C^k(\mathbb{R}^n)$ which agrees with this data. In this way, we obtain a set of *globally* defined vertex functions $F_i$ which are "correct" on the edges emanating from $\mathbf{v}_i$. Notice that $F_i$ and $F_j$ agree everywhere along the edge $\mathbf{e}_{ij}$.

The next step is to "blend" these vertex functions together to form a global interpolant. Although this may be done in many ways, we have chosen a method of the following simple form. Suppose we wish to evaluate the global function at a point $\mathbf{x}_0$ somewhere in the convex hull of the vertices. This point $\mathbf{x}_0$ must lie in one of the simplices of $\mathcal{T}$, say $T$. All we do is take a certain "weighted average" or "blend" of the vertex functions $F_i$ corresponding to the vertices of $T$. The details will be given below.

## 2. Notation

Again, let $\mathcal{T}$ represent a non-degenerate simplicial triangulation of a given set of vertices $\{\mathbf{v}_i\}_{i=1}^N$ in $\mathbb{R}^n$ and let $S^{n-1}$ denote the unit sphere in $\mathbb{R}^n$, i.e., $S^{n-1} = \{x \in \mathbb{R}^n : r := |x| = 1\}$.

The (univariate) edges joining $\mathbf{v}_i$ and $\mathbf{v}_j$ in $\mathcal{T}$ will be denoted by $\mathbf{e}_{ij}$ and we define the graph $G$ to consist of all edges in $\mathcal{T}$. The set of indices of all edges (listed once) in $G$ is $N_G := \{ij : \mathbf{e}_{ij} \text{ is in } G\}$ and the set of indices of those edges emanating from the vertex $\mathbf{v}_i$ is $N_i := \{ij : \mathbf{e}_{ij} \text{ is in } G\}$ ($i$ is fixed).

We make use of the following elementary fact in the sequel. Suppose that $g \in C^k(S^{n-1})$ and $f(x) := g(x/r)$. Then there is a function $\tilde{g} \in C^k(\mathbb{R}^n)$ such that $g(x) = \tilde{g}(x)$ on $S^{n-1}$ so that $f(x) = \tilde{g}(x/r)$. Hence

$$\begin{aligned} \frac{\partial f}{\partial x_i} &= -\frac{x_i}{r^3}\sum_{j=1}^{n}\frac{\partial \tilde{g}}{\partial x_j}\left(\frac{x}{r}\right)x_j + \frac{1}{r}\frac{\partial \tilde{g}}{\partial x_i}\left(\frac{x}{r}\right) \\ &= \frac{1}{r}\left\{-\frac{x_i}{r}\sum_{j=1}^{n}\frac{\partial \tilde{g}}{\partial x_j}\left(\frac{x}{r}\right)\frac{x_j}{r} + \frac{\partial \tilde{g}}{\partial x_i}\left(\frac{x}{r}\right)\right\} \end{aligned}$$

so that there are functions $h_i \in C^{k-1}(\mathbb{R}^n)$, specifically,

$$h_i(x) := -x_i \sum_{j=1}^{n} \frac{\partial \tilde{g}}{\partial x_j}(x) x_j + \partial\tilde{g}\partial x_i(x),$$

such that

$$\frac{\partial f}{\partial x_i} = \frac{1}{r} h_i\left(\frac{x}{r}\right). \tag{1}$$

## 3. Extensions

Given vertex data $\{D^\alpha F(\mathbf{v}_i)\}_{|\alpha|\le k}$ $i = 1, 2, \ldots, N$, along each edge $\mathbf{e}_{ij}$ a smooth interpolating univariate function $f_{ij}$ (parametrized from 0 to 1) can be fixed by various means. For example, a univariate polynomial of degree $2k+1$ which interpolates position and directional derivatives as follows would suffice (for $p = 0, 1, \ldots k$):

$$\begin{cases} f_{ij}^{(p)}(0) &= \sum_{|\alpha|=p} \frac{|\alpha!|}{\alpha!}(\mathbf{v}_j - \mathbf{v}_i)^\alpha D^\alpha F(\mathbf{v}_i) \\ f_{ij}^{(p)}(1) &= \sum_{|\alpha|=p} \frac{|\alpha|!}{\alpha!}(\mathbf{v}_j - \mathbf{v}_i)^\alpha D^\alpha F(\mathbf{v}_j) \end{cases}. \tag{2}$$

In order to describe the geometry of the edges emanating from a vertex $\mathbf{v}_i$, introducc a local system of generalized spherical coordinates with origin at $\mathbf{v}_i$. Every edge direction is thus associated with a point $\boldsymbol{\theta}_{ij}$ on the unit sphere $S^{n-1}$ i.e., $\boldsymbol{\theta}_{ij} := (\mathbf{v}_j - \mathbf{v}_i)/||\mathbf{v}_j - \mathbf{v}_i||$. Let $\{l_{ij}(\boldsymbol{\theta})\}_{ij\in N_i}$ be a set of cardinal functions contained in $C^k(S^{n-1})$ which satisfy $l_{ij}(\boldsymbol{\theta}_{kp}) = \delta_{ij,kp}$ for $ij, kp \in N_i$. These curves will be used to transfinitely extend the edge curves into the interior of the simplex. For vertices in $\mathbb{R}^2$, one simple choice for $l_{ij}(\boldsymbol{\theta})$ would be a (weighted) univariate spline with periodic end conditions (see for example [**2**]). In higher dimensions, this idea could be extended using tensor products of univariate splines i.e.

$$l_{ij}(\boldsymbol{\theta}) := \prod_{p=1}^{n-1} l_{ij}(\theta^p) \text{ where } \boldsymbol{\theta} = (\theta^1, \theta^2, \ldots, \theta^{n-1}) \text{ and } \{l_{ij}(\theta^p)\}_{ij\in N_i}$$

is a set of univariate functions cardinal with respect to the distinct values of $\{\theta^p_{ij}\}_{ij\in N_i}$. Other natural choices of $l_{ij}(\boldsymbol{\theta})$ include, for example, multivariate functions like Spherical Multiquadrics (see [**6**]).

Having fixed the edge curves and radial extensions, a vertex function $F_i$ is defined as :

$$F_i(r,\boldsymbol{\theta}) := \left\{ \sum_{ij\in N_i} \left(f_{ij}(r) - p_{ij}(r)\right) \cdot l_{ij}(\boldsymbol{\theta}) \right\} + P_i(r,\boldsymbol{\theta})$$

where $P_i$ is the Taylor polynomial defined by the vertex data $\{D^\alpha F(\mathbf{v}_i)\}_{|\alpha|\le k}$ and $p_{ij} := P_i|_{\mathbf{e}_{ij}}$. The term $\left(f_{ij}(r) - p_{ij}(r)\right)$ normalizes the emanating edge curves to have a zero of order $k$ at $\mathbf{v}_i$, $l_{ij}(\boldsymbol{\theta})$ is used to effect a radial extension and finally $P_i(r,\boldsymbol{\theta})$ restores the original vertex data present before the normalization and extension. It should be noted here that due to the unequal lengths of the edges emanating from $\mathbf{v}_i$, some degree of extrapolation may occur.

CLAIM. The vertex function $F_i(r,\boldsymbol{\theta})$ possesses the following basic properties:

(1) $F_i(r,\boldsymbol{\theta})$ interpolates the curves $f_{ij}(r)$ $\forall ij \in N_i$.
(2) $F_i(r,\boldsymbol{\theta})$ reproduces $P_k(\Omega)$.
(3) $F_i(r,\boldsymbol{\theta})$ is a $C^k(\Omega)$ function.

PROOF. Property 1 follows from the cardinality of the functions $l_{ij}(\boldsymbol{\theta})$ and property 2 is clear as the summation vanishes when $f_{ij}(r) = p_{ij}(r)$ $\forall ij \in N_i$. Finally, smoothness away from $\mathbf{v}_i$ $(r > 0)$ is a consequence of the smoothness of the functions involved. All that remains is the question of smoothness as $r \to 0^+$ and this follows from the form of $F_i$ and the following theorem. □

THEOREM. *Suppose that* $g \in C^k(S^{n-1})$ *and* $f \in C^k[0,\infty)$ *has the property that* $f(0) = f'(0) = \cdots = f^{(k)}(0) = 0$. *Then*

$$F(x) := g(\theta)f(r) \in C^k(\mathbb{R}^n)$$

*where* $r := |x|$ *and* $\theta := x/r$.

PROOF. The only point in question is the origin. We first show that the derivatives of $F$ have a special structure from which the theorem easily follows. More specifically, we claim that for $|\alpha| = i \le k$ and $x \ne 0$,

$$D^\alpha F(x) = \sum_{j=0}^{i} g_{j,\alpha}(\theta)\left(\frac{f^{(j)}(r)}{r^{i-j}}\right) \tag{3}$$

for some functions $g_{j,\alpha} \in C^{k-i}(S^{n-1})$. To show this, we proceed by induction on $i$. The case $i = 0$ is merely our assumption on $F$ and so assume that the claim is true for all $|\alpha| \le i$ and consider a multi-index $\alpha'$ with $|\alpha'| = i+1$. We may, without loss of generality, assume that $\alpha' = \alpha + e_1$ where $|\alpha| = i$ and $e_1 := \langle 1, 0, \ldots, 0\rangle$. Then, for $x \neq 0$,

$$D^{\alpha'} F(x) = \frac{\partial}{\partial x_1} D^{\alpha} F(x) = \frac{\partial}{\partial x_1} \sum_{j=0}^{i} g_{j,\alpha}(\theta) \frac{f^{(j)}(r)}{r^{i-j}}$$

$$= \sum_{j=0}^{i} \left\{ \left( \frac{\partial}{\partial x_1} g_{j,\alpha}(\theta) \right) \frac{f^{(j)}(r)}{r^{i-j}} + g_{j,\alpha}(\theta) \left[ \frac{f^{(j+1)}(r)}{r^{i-j}} \frac{x_1}{r} - \frac{(i-j)}{r^{i-j+1}} \frac{x_1}{r} f^{(j)}(r) \right] \right\}$$

$$= \sum_{j=0}^{i} \left\{ \left( \frac{\partial}{\partial x_1} g_{j,\alpha} \right) \frac{f^{(j)}(r)}{r^{i-j}} + g_{j,\alpha}(\theta) \left( \frac{x_1}{r} \right) \left[ \frac{f^{(j+1)}(r)}{r^{i+1-(j+1)}} - (i-j) \frac{f^{(j)}(r)}{r^{i+1-j}} \right] \right\}.$$

But, by (1), there are functions $h_{j,\alpha} \in C^{(k-i)-1}(S^{n-1}) = C^{k-(i+1)}(S^{n-1})$ such that

$$\frac{\partial}{\partial x_1} g_{j,\alpha}(\theta) = \frac{1}{r} h_{j,\alpha}(\theta)$$

so that

$$\left( \frac{\partial}{\partial x_1} g_{j,\alpha}(\theta) \right) \frac{f^{(j)}(r)}{r^{i-j}} = h_{j,\alpha}(\theta) \frac{f^{(j)}(r)}{r^{i+1-j}}.$$

Moreover, the functions

$$g_{j,\alpha}(\theta) \frac{x_1}{r} \in C^{k-i}(S^{n-1}) \subset C^{k-(i+1)}(S^{n-1})$$

and our claim follows.

To complete the proof of the theorem, note that, for $0 \leq j \leq i \leq k$,

$$\lim_{r \to 0^+} \frac{f^{(j)}(r)}{r^{i-j}} = \lim_{r \to 0^+} \frac{f^{(j+1)}}{(i-j) r^{i-j-1}} = \cdots = \lim_{r \to 0^+} \frac{f^{(i)}(r)}{(i-j)!} = 0.$$

Further, the functions, $g_{j,\alpha}(\theta)$, being continuous functions on the sphere, remain bounded as $r \to 0^+$. Hence, for $|\alpha| \leq k$,

$$\lim_{x \to 0} D^{\alpha} F(x) = 0. \quad \square$$

The final interpolant, evaluated at a point $\mathbf{x} \leftrightarrow (r, \boldsymbol{\theta})$ in a simplex $T$, is taken to be:

$$F(\mathbf{x}) := \sum_{i} \Phi_i(\mathbf{x}) F_i(\mathbf{x})$$

where the summation is taken over those $i$ such that $\mathbf{v}_i$ is a vertex of $T$ and $\Phi_i(\mathbf{x})$ are appropriate blending functions (see for example [**1**]).

Of course, in many applications only position data $\{F(\mathbf{v}_i)\}_{i=1}^{N}$ may be available and higher order data must be generated in some fashion. One effective method to accomplish this is provided by the technique of Network spline (or Minimum Norm Networks) in $\mathbb{R}^n$ (see [**3**], [**4**], [**7**], [**10**], [**11**], [**12**]). These Network splines (subject to certain unisolvency conditions) provide vertex data and smooth curves defined over the edges in a triangulation and optimal with respect to a certain "minimal energy" principle.

TABLE 1. Data

| X | Y | Z | X | Y | Z | X | Y | Z |
|---|---|---|---|---|---|---|---|---|
| 0.0 | 0.0 | 0.0 | 0.5 | 0.0 | 0.0 | 1.0 | 0.0 | 0.0 |
| 0.0 | 0.5 | 1.0 | 0.5 | 0.5 | 1.0 | 1.0 | 0.5 | 1.0 |
| 0.0 | 1.0 | 1.0 | 0.5 | 1.0 | 1.0 | 1.0 | 1.0 | 1.0 |

## 4. Examples

In the examples presented here, a $C^1$ Network spline interpolating given data points was first constructed and then the resulting curve network was extended to the interior of the triangulation using a variety of radial extensions.

It turns out that a useful property of Network splines is the ability to deal with rapidly varying data by introducing a modified energy functional into the definition of optimality. More specifically, we consider a weight function on each edge,

$$w_{ij} := \frac{1}{(1+m_{ij}^2)^{5/2}} \text{ where } m_{ij} := \frac{F(\mathbf{v}_j) - F(\mathbf{v}_i)}{\|\mathbf{v}_j - \mathbf{v}_i\|_2}.$$

The functional to be minimized then contains terms of the form $\int_0^1 (f_{ij}'')^2\ (1 + m_{ij}^2)^{5/2}\, dx$ which yields an approximation to the true strain energy of a (homogeneous) thin elastic beam (see [**8**]). The geometric effect is to allow the curve to bend more sharply where $w_{ij}$ is small ($|m_{ij}|$ is large) and be held rigid otherwise. Univariate splines can be similarly weighted and in what follows the terms *weighted* and *unweighted* will be taken to mean $w_{ij} := 1/(1+m_{ij}^2)^{5/2}$ or $w_{ij} := 1$ respectively.

Figures 3–4: We begin by illustrating the general construction with a very simple example corresponding to the data given in Table 1.

Figures 3a–3b shows the Network spline curves and also the radial extensions at the middle vertex, $\mathbf{v}_5$, for four discrete values $r = 0.1$, $r = 0.2$, $r = 0.3$ and $r = 0.4$. In Figure 1a, the Network spline is unweighted and the radial extensions $l_{ij}(\theta)$ were taken to be (unweighted) univariate cubic splines with $C^1$ periodic end conditions. In Figure 1b the Network spline and radial extensions were weighted. The effect of the weighting is evident with a flattening of the surface and the creation of a "ramp" effect. To produce a flatter effect, splines under tension, which are composed of polynomial and exponential basis functions, can be employed.

Figures 4a–4b show a single vertex function $F_5(\mathbf{x})$ with the curves from Figure 3 superimposed. Again, Figure 4a shows the unweighted case while Figure 4b illustrates the weighted case. Note the single vertex function $F_5(\mathbf{x})$ interpolates the Network spline curves emanating from the middle vertex but does not interpolate the Network spline curves around the boundary of the figure. The final interpolant (not shown) which consists of a blend of the vertex functions does in fact interpolate all of the Network spline curves.

Figure 5: In this example, we consider 4-D data provided by sampling the trivariate function $F(x, y, z) = x^2 + y^2 - z^2$ over a regular $5 \times 5 \times 5$ grid. A $C^1$ Network spline was constructed over the edges in a Delauney triangulation and then extended radially using Spherical Multiquadrics i.e., $l_{ij}(\boldsymbol{\theta})$ has the form

$$l_{ij}(\boldsymbol{\theta}) = \sum_p A_{ip} G_{ip}(\boldsymbol{\theta}) \text{ where } G_{ip}(\boldsymbol{\theta}) = \sqrt{1 + R^2 - 2R(\theta \cdot \theta_{ip})} \text{ and } R = 1 - \pi/4.$$

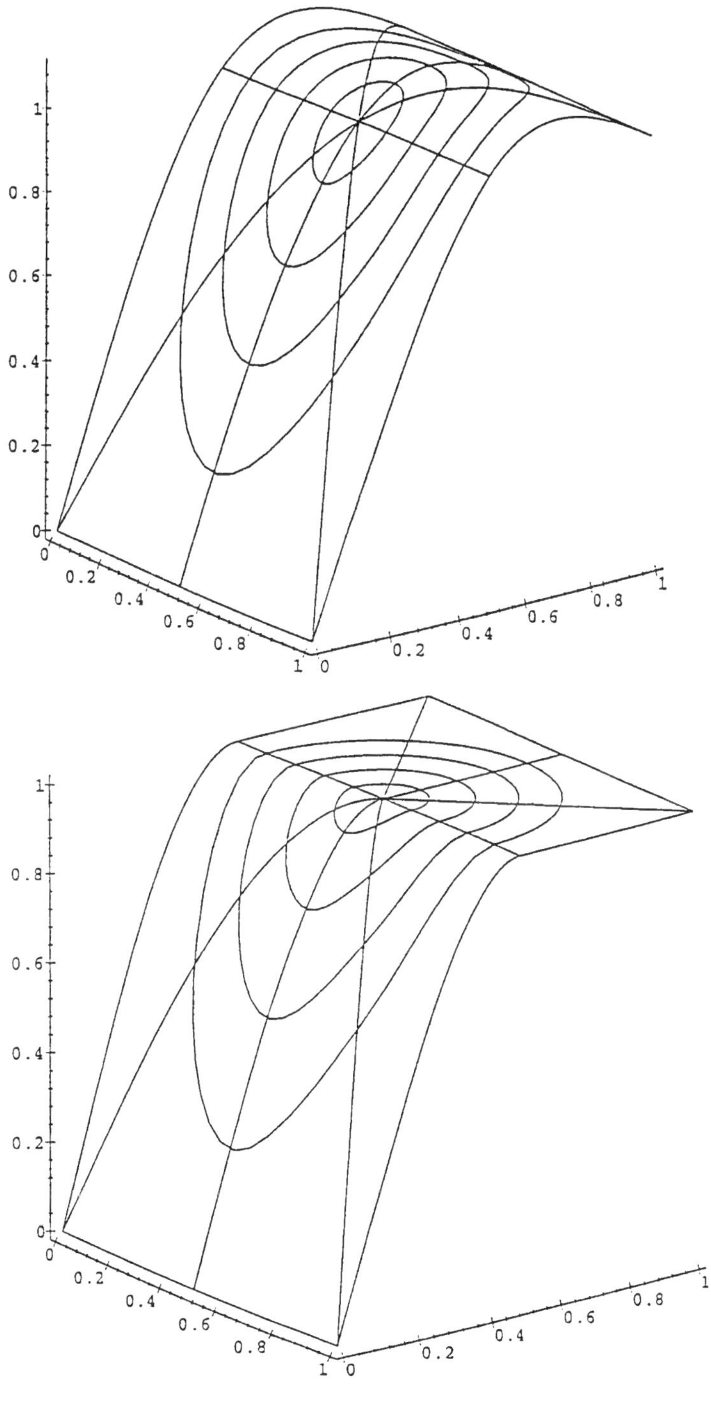


FIGURE 3. Ramp data. a) Unweighted network spline curves and radial extensions $r = 0.1, 0.2, 0.3, 0.4$. b) Weighted network spline curves and radial extensions $r = 0.1, 0.2, 0.3, 0.4$.

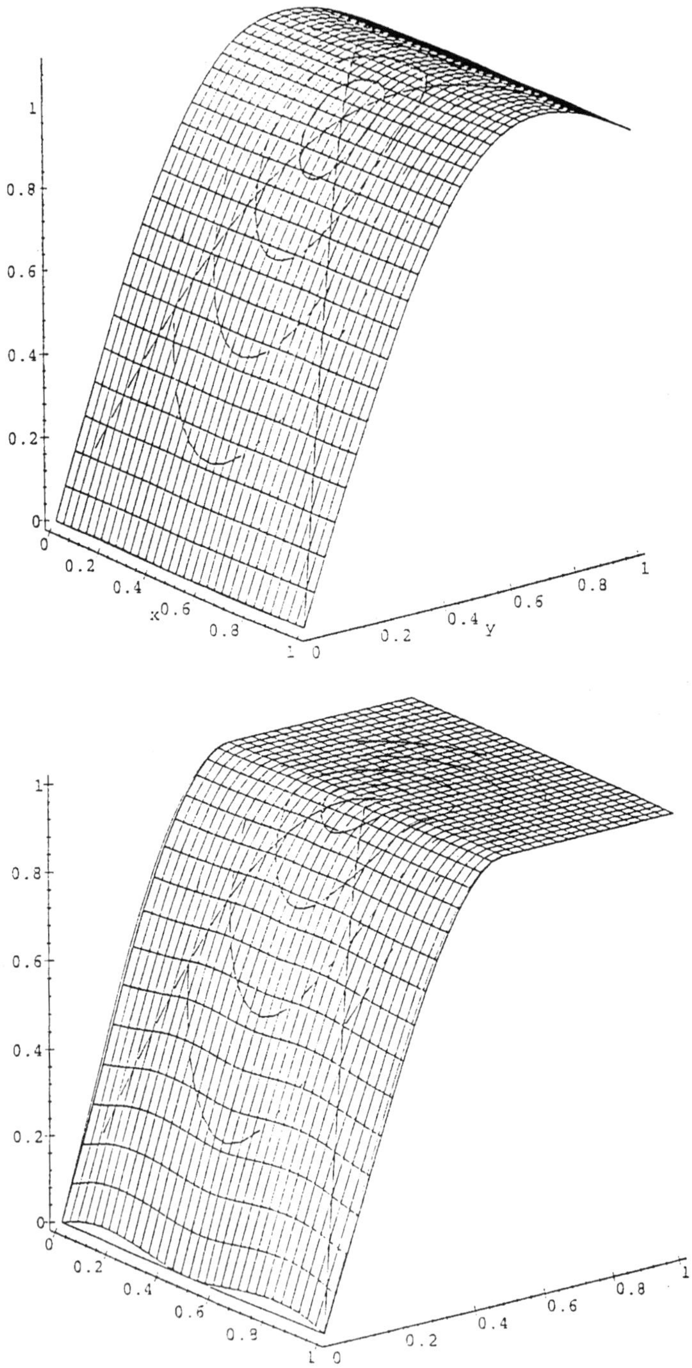


FIGURE 4. Vertex function $F_5(\mathbf{x})$ for ramp data. a) Unweighted vertex function. b) Weighted vertex function.

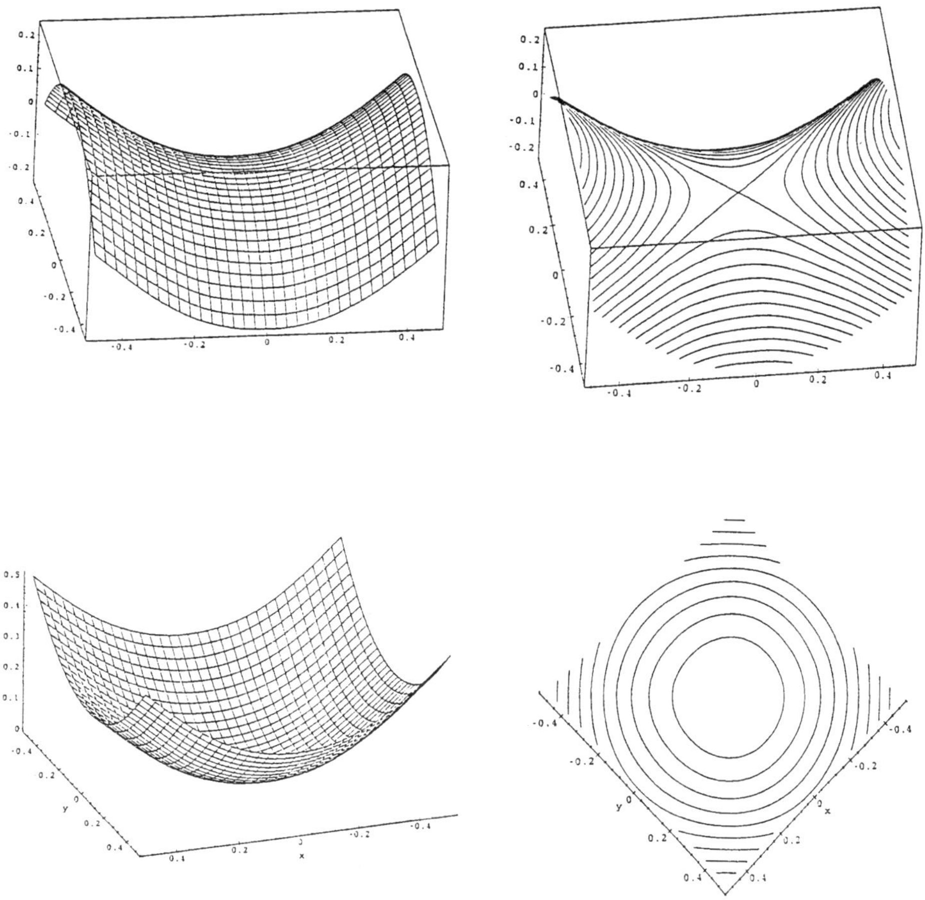

FIGURE 5. Unweighted interpolant of $F(x, y, z) = x^2 + y^2 - z^2$ test function. a)–b) The $x = 0$ cross-section and contours. c)–d) The $z = 0$ cross-section and contours.

Figures 5a–5b show a cross-section of the interpolant obtained by fixing $y = 0$ and the associated contours of this cross-section. Similarly, Figures 3c–3d show the cross-section obtained by fixing $z = 0$ and corresponding contours. The cross-sections would appear to approximate the cross-sections of the original test function very accurately i.e., $F(x, 0, z) = x^2 - y^2$ and $F(x, y, 0) = x^2 + y^2$. The contours would seem to verify that the interpolant is in fact smooth.

Further examples, including rapidly varying data in $\mathbb{R}^4$ can be found in [**4**].

## References

1. P. Alfeld, *Multivariate perpendicular interpolation*, SIAM J. Numer. Anal. **22** (1985), no. 1, 96–106.
2. L. P. Bos and K. Salkauskas, *Weighted splines based on piecewise polynomial weight functions*, Curve and Surface Design (Tempe, 1989), Geom. Des. Publ., SIAM, Philadelphia, PA, 1992, pp. 87–98.
3. L. P. Bos and D. Holland, *Bivariate weighted network splines*, (1996) (submitted).

4. ______, *Network splines*, Results Math. **30** (1996), nos. 3–4, 228–258.
5. T. A. Foley, *Local control of interval tension using weighted splines*, Comput. Aided Geom. Design **3** (1986), no. 4, 281–294.
6. R. L. Hardy and W. M. Goepfert, *Least squares predictions of gravity anomalies, geoidal undulations and deflections of the vertical with multiquadric harmonic functions*, Geophys. Res. Lett. **2** (1975), 423–426.
7. D. Holland, *Generalized minimum norm networks*, Ph.D. thesis, University of Calgary, 1994.
8. M. A. Malcolm, *On the computation of nonlinear spline functions*, SIAM J. Numer. Anal. **14** (1977), no. 2, 254–282.
9. G. Nielson, *The side-vertex method for interpolation in triangles*, J. Approx. Theory **25** (1979), no. 4, 318–336.
10. ______, *A method for interpolating scattered data based upon a minimum norm network*, Math. Comp. **40** (1983), no. 161, 253–271.
11. H. Pottmann, *Scattered data interpolation based upon generalized minimum norm networks*, Constr. Approx. **7** (1991), no. 2, 247–256.
12. ______, *Interpolation on surfaces using minimum norm networks*, Comput. Aided Geom. Design **9** (1992), no. 1, 51–67.
13. K. Salkauskas, *$C^1$ splines for interpolation of rapidly varying data*, Rocky Mountain J. of Math. **14** (1984), no. 1, 239–250.

Department of Mathematics and Statistics, University of Calgary, Calgary, Alberta, Canada, T2N 1N4

*E-mail address*: lpbos@math.ucalgary.ca

Department of Mathematics and Statistics, University of Calgary, Calgary, Alberta, Canada, T2N 1N4

*E-mail address*: dholland@math.ucalgary.ca

Centre de Recherches Mathématiques
CRM Proceedings and Lecture Notes
Volume **18**, 1999

# The Use of Splines in the Numerical Solution of Differential and Volterra Integral Equations

Hermann Brunner

ABSTRACT. In the numerical analysis of initial-value problems for ordinary differential equations (ODEs), three principal questions have to be answered:

(i) Does the numerical method *converge* ("$h \to 0$, $nh = \text{const.}$");

(ii) does the numerical method mimic the *stability properties* of the given problem ("$h > 0$ fixed, $n \to \infty$"); and (ii) what is the *optimal order of convergence* (globally, on the prescribed interval; or locally, at the mesh points) of the method? Analogous questions arise for integral equations and integro-differential equations of Volterra type (functional equations with memory terms), since they may be viewed as generalized initial-value problems.

In this survey we consider the above questions for collocation methods in various spaces of (polynomial) spline functions. After a brief survey of the use of spline collocation methods for initial-value problems in ODEs, the discussion will then focus on Volterra integral equations and on how the memory term (smooth versus weakly singular kernel; kernel with delay arguments; etc.) governs the order of convergence and the stability of collocation methods. It turns out that, as in ODEs, splines with high regularity lead to divergent methods. However, in contrast to ODEs, the problem of what degrees of regularity yield feasible methods has not yet been completely solved.

We also look briefly at related functional equations (differential-algebraic equations and Volterra integro-differential equations), and describe a number of open problems, particularly for Volterra equations with variable delays, Volterra integral equations of the first kind, and wavelet-based (spatial) semidiscretization of parabolic integro-differential equations.

## 1. Introduction

**1.1. Splines spaces as solution spaces.** Let $I := [0, T]$ be the interval on which the given initial-value problem is to be solved (compare Section 1.2 for a description of the problems discussed in this talk), and let $\Pi_N : 0 = t_0 < t_1 < \cdots <$

1991 *Mathematics Subject Classification.* Primary: 65L05, 65L60, 65R20, 65D07; Secondary: 45D05, 45J05, 41A25.

This work has been supported by the Natural Sciences and Engineering Research Council of Canada (NSERC Research Grant No. OGP0009406)

This is the final form of the paper.

$t_N = T$ be a (not necessarily uniform) mesh for $I$. We set

$$h_n := t_{n+1} - t_n, \quad h := \max_{0\le n\le N-1} \{h_n\}, \quad I_n := [t_n, t_{n+1}] \ (n = 0, \dots, N-1),$$

and we denote by

$$S_\mu^{(d)}(\Pi_N) := \{u \in C^d(I) \colon u\,|_{I_n} \in \pi_\mu \ (0 \le n \le N-1)\}$$

the space of (real) piecewise polynomials (or splines) of degree at most $\mu$ that have continuous derivatives of order $d$ on $I$, with $-1 \le d < \mu$. In the case where $d = -1$ the elements of $S_\mu^{(-1)}(\Pi_N)$ will in general have (finite) jump discontinuities at the interior mesh points; we then set $I_n = (t_n, t_{n+1}]$ $(n = 1, \dots, N-1)$.

LEMMA 1.1.

$$\dim S_\mu^{(d)}(\Pi_N) = N(\mu - d) + (d+1).$$

*In particular, if* $\mu = m + d$ $(m \ge 1)$ *we have*

$$\dim S_{m+d}^{(d)}(\Pi_N) = Nm + (d+1) \quad (d \ge -1). \tag{1.1}$$

*The "classical" spline spaces are given by* $d = \mu - 1$; *thus,*

$$\dim S_\mu^{(\mu-1)}(\Pi_N) = N + \mu.$$

The idea behind a *collocation method* in the present setting is to find an element $u \in S_\mu^{(d)}(\Pi_N)$ which satisfies the given functional equation on a suitable finite set of *collocation points* in $I$. The above lemma suggests an obvious way of placing these collocation points if they are of *multiplicity one*: each of the $N$ subintervals $I_n$ will contain $\mu - d$ collocation points (or, in the case of the space $S_{m+d}^{(d)}(\Pi_N)$, $m$ such points for all $d \ge -1$). We will denote the points in $I_n$ by $t_n + c_i h_n$, and we set

$$X_N := \{t_n + c_i h_n \colon 0 \le c_1 < \cdots < c_{\mu-d} \le 1 \ (n = 0, \dots, N-1)\}. \tag{1.2}$$

(The case of *multiple collocation parameters* $\{c_i\}$ will not be considered in this talk; the reader is referred to, e.g., [**59, 88, 89, 90, 91, 93, 77**].) Observe that for $\mu - d \ge 2$ the special choice $c_1 = 0$ and $c_{\mu-d} = 1$ increases the regularity of the spline space by one: for problems with continuous data the collocation solution will then lie in $S_\mu^{(d)}(\Pi_N) \cap C^{d+1}(I) =: S_\mu^{(d+1)}(\Pi_N)$.

We will focus on the following questions:

(1) Which choices of $\mu$ and $d$ in the approximating space $S_\mu^{(d)}(\Pi_N)$ lead to convergent/divergent methods? Are there classes of equations where, for given $\mu$ and $d$, the collocation solution is convergent only for certain choices of the collocation parameters $\{c_i\}$ but diverges for others?
(2) If the collocation method is convergent and has global order $p$ on $I$, are there sets $\{c_i\}$ for which an order $p^* > p$ is obtained at the mesh points ("local superconvergence of order $p^*$ on $\Pi_N$")?
(3) For which choices of the $\{c_i\}$ does the collocation solution have the same qualitative (asymptotic) stability properties as the exact solution of the given problem?

A closely related question concerns the choice of the *"natural"* solution (spline) space for a given class of problems—natural in the sense that the collocation solution is determined completely from the *given data* (the equation and its initial conditions), and no further data (additional "artificial" initial values; differentiated form of the equation; etc.) are used. Lemma 1.1 provides the answer to this question in the case of *simple* collocation parameters $\{c_i\}$: the space $S_{m+d}^{(d)}(\Pi_N)$ is the natural solution space if $d$ is chosen such that $d+1$ (the second term on the right-hand side of (1.1)) equals the number of given initial conditions. Thus, for integral equations the natural space is $S_{m-1}^{(-1)}(\Pi_N)$ $(d=-1)$; for first-order problems (ODEs, DDEs, VIDEs), it is $S_m^{(0)}(\Pi_N)$ $(d=0)$; etc. It turns out, not surprisingly, that due to the *recursive* nature of the approximation methods, spline spaces with too much regularity (to be made precise below) will lead to divergent methods.

**1.2. Initial-value problems.** This survey will centre on spline collocation methods for initial-value problems in ordinary differential equations (ODEs), differential-algebraic equations (DAEs), and Volterra integral equations of the first and second kind (Sections 2, 3, and 4), with brief excursions to the use of these methods in the numerical solution of delay differential and delay integral equations (Section 5), and Volterra integro-differential equations (VIDEs) (Section 6). We also add some remarks on ongoing and future work, particularly on the spatial (semi-) discretization of parabolic VIDEs using wavelet-based approximations of elliptic operators and its implication on the attainable order in time collocation methods. The intended role of the *list of references* is to provide a representative (rather than comprehensive) selection of relevant papers and monographs.

In the remainder of this section we briefly describe the equations that will be considered in the following.

*Initial-value problems for ODEs*:

$$y'(t) = f(t, y(t)), \quad t \in I := [0, T]; \quad y(0) = y_0. \tag{1.3}$$

In the list of references, items labelled [DE*] refer to works on collocation and related methods for (1.3).

*Differential-algebraic equations*: Consider the (system of) DAEs $F(t, w, w') = 0$ $(t \in I)$. The *index* of this equation is the minimum number of times that the DAE (or part of it) must be differentiated with respect to $t$ in order to determine $w'$ as a continuous function of $w$ and $t$.

A particular class of *Index* 1 DAEs is given by

$$y' = f(t, y, z), \quad 0 = g(t, y, z) \quad (t \in I), \tag{1.4}$$

where $g_z$ has a bounded inverse in a neighborhood of the exact solution.

The system

$$y' = f(t, y, z), \quad 0 = g(t, y) \quad (t \in I), \tag{1.5}$$

where $g_y f_z$ is invertible for all $(t, y, z)$ in a neighborhood of the solution, forms an important subclass of *Index* 2 DAEs.

The system

$$y' = f(t, y, z), \quad z' = h(t, y, z, v), \quad 0 = g(y), \tag{1.6}$$

is a DAE of *Index* 3 provided $g_y f_z h_v$ has a bounded inverse in a neighborhood of the solution.

The monographs [**13, 58, 60**] contain good introductions to the theory and numerical analysis of DAEs (including equations of higher index). A sample of additional papers may be found under [DA*] in the list of references.

*Volterra integral equations of the second kind*:

$$y(t) = g(t) + \int_0^t k\big(t, s, y(s)\big)\, ds, \quad t \in I. \tag{1.7}$$

*Volterra integral equations of the first kind*:

$$\int_0^t k\big(t, s, y(s)\big)\, ds = g(t), \quad t \in I. \tag{1.8}$$

Publications relating to (1.7) and (1.8) are listed under [VE*].

*Delay differential equations*:

$$y'(t) = f\big(t, y(t), y(\theta(t))\big), \quad t \in I; \quad \theta(t) < t. \tag{1.9}$$

We shall be concerned in particular with the cases $\theta(t) = t - \tau$ $(\tau > 0)$ and $\theta(t) = qt$ $(0 < q < 1)$. References related to (1.6) are labelled [DD*].

*Volterra integral equations with delay*:

$$y(t) = g(t) + \int_0^t k\big(t, s, y(s), y(\theta(s))\big)\, ds, \quad t \in I. \tag{1.10}$$

The delay function $\theta(t)$ will be as in (1.9). Relevant papers in the list of references are labelled [DV*].

*Volterra integro-differential equations*:

$$y'(t) = f\big(t, y(t)\big) + \int_0^t k\big(t, s, y(s)\big)\, ds, \quad t \in I; \quad y(0) = y_0. \tag{1.11}$$

A selection of references on collocation and related methods for (1.11), as well as for VIDEs with delay arguments, is listed under [VD*].

**1.3. Optimal order of convergence on $\Pi_N$.** While a collocation method yields a global approximation on $I$ whose order of convergence, $p$, is essentially given by the approximation power of the underlying piecewise polynomial space, there arises the question as to whether a judicious choice of the collocation parameters $\{c_i\}$ may lead to a higher order of convergence, $p^* > p$, at the mesh points $\Pi_N$ (*local superconvergence*).

Consider, as an illustration, the class of $m$-stage implicit Runge-Kutta methods (cf. Section 2.1) for the ODE (1.3). A Runge-Kutta method is said to be of order $p$ (at the mesh points $\Pi_N$) if the RK approximation $u_1$ at $t = t_1 = h$ satisfies

$$\big|y(t_1) - u_1\big| \leq Ch^{p+1} \tag{1.12}$$

for all sufficiently regular problems. It was shown in the early 1960s (see [**79, 39**]; compare also Section 2.1 below) that Runge-Kutta methods based on the $\{c_i\}$ given by the *Gauss points*, i.e. the zeros of the (shifted) Legendre polynomial $P_m(2s-1)$, possess the (optimal) order $p^* = 2m$. A few years later it was recognized that these Runge-Kutta methods can in fact be interpreted as *collocation methods* in the spline space $S_m^{(0)}(\Pi_N)$ [**54, 105, 59**]: their global order (on $I$) is $p = m$, and they are locally superconvergent of order $p^* = 2m$.

Does collocation using the $m$ Gauss points (or other special sets $\{c_i\}$) always lead to local superconvergence of order $m < p^* \leq 2m$? Moreover, is there a general way of deciding when, and for which of the functional equations listed in Section 1.2, local superconvergence on $\Pi_N$ is possible? For ODEs (1.3) it was shown in [**54**] (see also [**105, 59**]) that the attainable order of local superconvergence on $\Pi_N$ is related to the degree of precision of the (interpolatory) $m$-point quadrature formula having the $\{c_i\}$ as abscissas: if the collocation parameters satisfy

$$\int_0^1 s^{k-1}M(s)\,ds = 0 \text{ for } k = 1, \ldots, r, \tag{1.13}$$

with $M(s) := \prod_{i=1}^{m}(s - c_i)$, and if $u$ is the corresponding collocation solution in $S_m^{(0)}(\Pi_N)$, then

$$\max_{1\le n\le N}|y(t_n) - u(t_n)| \le Ch^{m+r} \quad (r \le m). \tag{1.14}$$

This result is also valid when collocation in $S_{m+1}^{(1)}(\Pi_N)$ is applied to second-order ODEs [**77, 59, 21**], and analogous results hold for two-point BVPs of arbitrary order [**95, 12**]. Does the above statement ("(1.13) → (1.14)") remain valid for collocation solutions corresponding to the other equations listed in Section 1.2? As we will see in the following, the answer is "yes" in some cases (e.g. equations with constant delay and integro-differential equations); a qualified "yes" in the case of DAEs; a "no" for integral equations of the first and second kind; and a (big) "?" for equations with certain variable or state-dependent delays.

## 2. Initial-Value Problems for ODEs

**2.1. A brief history of spline collocation methods.** Spline collocation methods for the initial-value problem (1.3) were introduced in the late 1960s by Loscalzo [**81, 82**]: the underlying spline space was $S_m^{(m-1)}(\Pi_N)$ $(m \ge 2)$ (see Lemma 1.1), with $c_1 = 1$ (collocation at $t_{n+1}$). It was shown that the resulting method is convergent for $m = 2, 3$, and *diverges* for $m \ge 4$. However, Hermite-type collocation in $S_{2d}^{(d)}(\Pi_N)$ $(d \ge 2)$ yields convergent methods.

Hung [**68**] derived analogous results for second-order IVPs: collocation in $S_m^{(m-1)}(\Pi_N)$ $(m \ge 3)$, again with $c_1 = 1$, converges for $m = 3, 4$ but is divergent for $m \ge 5$. Hermite-type collocation in $S_5^{(3)}(\Pi_N)$ leads to a convergent method.

Callender [**40**] used the space $S_m^{(1)}(\Pi_N)$ (whose dimension is $N(m-1)+2$) and equidistant collocation parameters $c_i = i/(m-1)$ $(i = 1, \ldots, m-1)$ to obtain convergent methods for all $m \ge 2$. This approach requires one additional initial value, $u'(0) = y'(0) = f(0, y_0)$.

While these papers are concerned with the *global order* of collocation approximations, Guillou and Soulé [**54**] (see also [**105, 93, 59**]) had shown in 1969 that collocation in $S_m^{(0)}(\Pi_N)$ yields an $m$-stage implicit Runge-Kutta method: since $u$ is to satisfy (1.3) on $X_N$ (cf. (1.2)) we have, setting $U_{n,i} := u'(t_n + c_ih_n)$ and $y_n := u(t_n)$ (note that on each $I_n$, $u$ is a polynomial of degree at most $m$),

$$u(t_n + sh_n) = y_n + h_n \sum_{i=1}^{m} \beta_i(s)U_{n,i}, \quad s \in [0, 1],$$

with

$$U_{n,i} = f\left(t_n + c_ih_n, y_n + h_n \sum_{j=1}^{m} a_{i,j}U_{n,j}\right) \quad (i = 1, \ldots, m).$$

Here, $a_{i,j} := \int_0^{c_i} L_j(s)\,ds$ and $\beta_i(s) := \int_0^s L_i(v)\,dv$; the $L_i$ denote the Lagrange canonical polynomials with respect to the set $\{c_i\}$. The *local order* of the method at the mesh points satisfies $p^* \le 2m$, with equality iff the $\{c_i\}$ are the Gauss points.

Local orders $p^* = 2m - 1$ and $p^* = 2m - 2$ are obtained for the Radau II points (the zeros of $P_m(2s-1) - P_{m-1}(2s-1)$) and the Lobatto points (the zeros of $s(s-1)P'_{m-1}(2s-1)$), respectively. Similar results were derived by Hulme [**66, 67**] for (discrete) Galerkin methods in the same spline space, and by Delfour et al. [**45, 46**] for discrete Galerkin methods in the discontinuous space $S_m^{(-1)}(\Pi_N)$. A complete analysis of collocation methods (including methods using multiple collocation points) was given by Mülthei [**88, 89, 90, 91**] in the early 1980s. Theoretical and practical aspects of *nonlinear spline methods* are discussed in [**102, 103**] and [**104**]; the latter also presents more elementary proofs for a number of Mülthei's divergence results.

Collocation methods for *second-order* initial-value problems were studied in [**86, 87, 77**] and [**64**]; compare also [**21**] where local superconvergence results were established for second-order Volterra integro-differential equations (which include second-order ODEs as a special case).

Except for the paper by Guillou and Soulé [**54**] (which was largely overlooked at the time), much of the interest in local superconvergence was triggered by the papers by Russell and Shampine [**95**] and, especially, by de Boor and Swartz [**12**] in the early 1970s: here, collocation approximations for $r$th-order boundary-value problems in the space $S_{m+d}^{(d)}(\Pi_N)$ (with $d = r - 1$) using the Lobatto points [**95**] and the Gauss points [**12**] were analyzed. Compare also [**3, 4, 94, 78**] and the survey [**53**] for related results, and consult [**96**] for a overview of superconvergence results for Fredholm integral equations.

**2.2. Local superconvergence on $\Pi_N$.** We briefly summarize the principal local superconvergence results for (1.3) [**54, 59, 26, 24**]. The general principle underlying the proof of Theorem 2.1 (resolvent representation of the collocation error; [**54, 59, 34**]) can be used for many of the other problems under consideration.

THEOREM 2.1. *Let $u \in S_m^{(0)}(\Pi_N)$ be the collocation solution to the initial-value problem* (1.3), *and assume that the collocation parameters $\{c_i\}$ satisfy the orthogonality condition* (1.13).

(a) *For all sufficiently regular problems and all sufficiently small $h > 0$ we have*

$$\max_{1 \le n \le N} |y(t_n) - u(t_n)| \le Ch^{m+r}.$$

(b) *A necessary condition for $u$ and $u'$ to possess the same order of local superconvergence at the mesh points is given by $c_m = 1$. If the collocation parameters are the $m$ Radau* II *points ($r = m - 1$), or the Lobatto points ($r = m - 2$), then $u$ and $u'$ both have the order of local superconvergence $p^* = 2m - 1$ ($m \ge 2$), and $p^* = 2m - 2$ ($m \ge 3$), respectively, on $\Pi_N$.*

(c) *For the Gauss points ($r = m$), however, we only have*

$$\max_{1 \le n \le N} |y'(t_n) - u'(t_n)| \le Ch^m,$$

*while $u$ itself has the local order $p^* = 2m$.*

For DAEs some of the superconvergence results of Theorem 3.1 are no longer valid: due to the algebraic component in a DAE there results a reduction in the order. Most of the work on the attainable local order of spline collocation methods for DAEs in $S_m^{(0)}(\Pi_N)$ dates from the last six years. The monographs [**13**] (Brenan, Campbell and Petzold), [**58**] (Hairer, Lubich and Roche), and the survey paper

[**56**] contain comprehensive descriptions of these results. In addition we refer to [**2**, problems of Indexes 1 and 2] , [**80**, Index 1 problems], [**60**, Index 2 problems], and [**71**, DAEs of Index 3]. Part of this analysis has been extended to differential-algebraic problems in VIDEs [**74**].

For DAEs of the special forms (1.4), (1.5), and (1.6) the (local) order of collocation methods (or their implicit Runge-Kutta forms) will in general be different for the individual components $y$, $z$ and $v$. Moreover, collocation at the *Gauss points* will often lead to *order reduction*: for example, in the Index 2 problem (1.5) the local order for the $y$-component drops from $2m$ to either $m+1$ (for $m$ odd) or $m$ (for $m$ even); compare [**58**].

Order reduction can be avoided either by employing collocation parameters $\{c_i\}$ with $c_m = 1$ (leading to so-called "stiffly stable" RK methods)—here, the Radau II points are the recommended points—or by using *projected* Gauss methods [**5**].

## 3. Second-Kind Volterra Integral Equations

**3.1. Early collocation methods.** In 1939 Huber [**65**] discussed the numerical solution of Volterra integral equations in $S_1^{(0)}(\Pi_N)$; his method was shown to be a collocation method (with $c_1 = 1$) by Kadner [**72**]. Hung [**68**] then proved that collocation in $S_m^{(m-1)}(\Pi_N)$ with $c_1 = 1$ is convergent for $m = 1, 2$ but diverges for $m \geq 3$ (compare also [**92**] for a convergent modification of Hung's $S_3^{(2)}$-method). Similar, but more general, results were obtained by El Tom [**50, 49**]: for example, collocation (at equally spaced points) in $S_m^{(m-k)}(\Pi_N)$ is divergent if $m \geq k+2$ ($k = 1, 2, 3$) and is conjectured to be divergent whenever $k \geq m-1$ ($m \geq 3$). Compare also Micula [**85**] for related results. Esser [**52**] showed that Hermite-type collocation in the space $S_{2m-1}^{(m-1)}(\Pi_N)$ leads to high-order methods, also for certain delay VIEs.

These papers, and [**32**] (collocation in $S_m^{(0)}(\Pi_N)$), discuss the *global* convergence of spline collocation methods (for equally spaced collocation points).

**3.2. Local superconvergence on $\Pi_N$.** The question of possible local superconvergence on $\Pi_N$ appears to have been raised first by Axelsson [**6**]: he considered the integrated form of the IVP (1.3) (which is a special case of (1.4)),

$$y(t) = y_0 + \int_0^t f\big(s, y(s)\big)\, ds, \quad t \in I,$$

and applied collocation in $S_m^{(0)}(\Pi_N)$ under the assumption $0 \leq c_1 < \cdots < c_m = 1$ (which excludes the Gauss points). He showed that the local superconvergence result (1.14) remains valid (with $r = m-1$ and $r = m-2$, respectively) for the Radau II and the Lobatto points.

De Hoog and Weiss [**61**] extended these results to general second-kind Volterra integral equations (1.7) (still assuming $c_m = 1$); their methods were shown to be particular discretizations of collocation in $S_{m-1}^{(-1)}(\Pi_N)$,

$$u(t) = g(t) + \int_0^t k\big(t, s, u(s)\big)\, ds, \quad t \in X_N$$

(see [**30**]). The analysis for the Gauss points (where $c_m < 1$) remained elusive.

Collocation at the *Gauss points* does *not* lead to local superconvergence on $\Pi_N$ ([**27, 36, 26, 23, 11**]; see also [**34**] and the survey [**20**]); however, the corresponding *iterated collocation solution*,

$$u_{it}(t) := g(t) + \int_0^t k\big(t, s, u(s)\big)\, ds, \quad t \in I,$$

exhibits local superconvergence of order $p^* = 2m$ on $\Pi_N$.

The results summarized in Theorem 3.1 may be found in, e.g., [**27, 36, 23, 34, 11**]; compare also [**106**]. Note that we have $u_{it}(t) = u(t)$ whenever $t \in X_N$.

THEOREM 3.1. *Let $u \in S_{m-1}^{(-1)}(\Pi_N)$ be the collocation solution for the VIE* (1.7).

(a) *For all sufficiently regular Volterra integral equations* (1.7) *and all sufficiently small $h > 0$, the iterated collocation solution $u_{it}$ corresponding to $u \in S_{m-1}^{(-1)}(\Pi_N)$ and the $m$ Gauss points $\{c_i\}$ satisfies*

$$\max_{1\le n\le N} \big|y(t_n) - u_{it}(t_n)\big| \le Ch^{2m},$$

*while the local order of $u$ itself on $\Pi_N$ coincides with its global order $m$.*

(b) *If collocation is at the Radau* II *points, then*

$$\max_{1\le n\le N} \big|y(t_n) - u(t_n)\big| \le Ch^{2m-1}.$$

(c) *The local order in the* continuous *spline space $S_{m-1}^{(0)}(\Pi_N)$ (which corresponds to setting $c_1 = 0$, $c_m = 1$; $m \ge 2$) cannot exceed $2m - 2$; it is attained for the Lobatto points.*

We recall in passing that in the case of *Fredholm* integral equations of the second kind, $u_{it}$ based on collocation at the Gauss points exhibits *global superconvergence* on $I$ of order $2m$; see [**42, 23, 20, 96**] for details. It has recently been shown in [**37**] that for Volterra integral equations global superconvergence of order $m+1$ in $u_{it}$ is possible.

**3.3. Weakly singular kernels.** The superconvergence results described in Section 3.2 are valid only if the given data $g$ and $k$ are smooth. If the kernel $k(t,s,y)$ in (1.7) is weakly singular, e.g., $k(t,s,y) = (t-s)^{-\alpha}G(t,s,y)$, with smooth $G$ and $0 < \alpha < 1$, then the attainable order of local superconvergence for $u_{it}$ reduces to $\mathcal{O}(N^{-(m+1-\alpha)})$, provided the mesh $\Pi_N$ is *graded*: $t_n = (n/N)^q T$, with $q = m/(1-\alpha)$ [**22, 34**]. For *uniform meshes* it becomes $\mathcal{O}(h^{1-\alpha})$ for $u$ and $\mathcal{O}(h^{2(1-\alpha)})$ for $u_{it}$ (Gauss points), regardless of $m$. Higher-order convergence on uniform meshes can be obtained by working in *nonpolynomial spline spaces* [**25**].

## 4. Volterra Integral Equations of the First Kind

**4.1. Singularly perturbed Volterra equations.** Consider the singularly perturbed VIE

$$\varepsilon y(t) = g(t) - \int_0^t K(t,s)y(s)\, ds \quad (0 < \varepsilon \ll 1), \quad t \in I. \tag{4.1}$$

For fixed positive $\varepsilon$, (4.1) is a Volterra integral equation of the second kind, and hence the local superconvergence results of Theorem 3.1 hold. The "reduced equation" corresponding to $\varepsilon = 0$ is the first-kind Volterra equation (4.2), the linear version of (1.8).

Do local superconvergence results hold for such equations, and how does a positive $\varepsilon \ll 1$ affect the local orders of Theorem 3.1? The answer to the first question is (essentially) "no" and made precise in the following section; the answer to the second question is yet to be found (but see [**74, 73**] for related problems).

**4.2. Convergence of collocation methods.** Assume that $K$ and $g$ in

$$\int_0^t K(t,s)y(s)\,ds = g(t), \quad t \in I, \tag{4.2}$$

are sufficiently smooth and satisfy $K(t,t) \neq 0$ $(t \in I)$ and $g(0) = 0$. Hung [**68**] showed that collocation in $S_m^{(m-1)}(\Pi_N)$ with $c_1 = 1$ diverges for $m \geq 2$ but converges for $m = 0, 1$. Extensions of these results were found by El Tom (see e.g. [**49**]). The papers [**32, 31**] deal with the global convergence of collocation (using uniformly spaced collocation points) in $S_m^{(0)}(\Pi_N)$.

The results of Theorem 4.1 were established in [**29, 76**]; compare also [**34**]. As shown in [**30**], the high-order (and fully discrete) methods of de Hoog and Weiss [**62, 63**] may be interpreted as special discretizations of spline collocation methods in $S_{m+d}^{(d)}(\Pi_N)$ with $d = -1$ and $d = 0$ for which results analogous to those given below were proved (assuming $c_m = 1$). Theorem 4.2 can be found in [**76**]; a weaker form of part (a) occurs in [**34**, page 302].

THEOREM 4.1. *Assume that* $0 < c_1 < \cdots < c_m \leq 1$, *and let* $u \in S_{m-1}^{(-1)}(\Pi_N)$ *be the collocation solution for* (4.2). *Then* $\|y - u\|_\infty \to 0$ *as* $h \to 0$ ($Nh = \text{const.}$) *for all* $m \geq 1$ *iff*

$$-1 \leq \rho_m := (-1)^m \prod_{i=1}^{m} \frac{1 - c_i}{c_i} \leq 1, \tag{4.3}$$

*and we have*

$$\|y - u\|_\infty = \begin{cases} \mathcal{O}(h^m) & \text{if } -1 \leq \rho_m < 1 \\ \mathcal{O}(h^{m-1}) & \text{if } \rho_m = 1. \end{cases}$$

THEOREM 4.2. *Let* $u \in S_m^{(0)}(\Pi_N)$ *be the collocation solution for* (4.2), *and assume* $0 < c_1 < \cdots < c_m \leq 1$. *The required initial value is given by* $y(0) = g'(0)/K(0,0)$.

(a) *If* $c_m = 1$, *then* $u$ *is convergent iff*

$$-1 \leq \rho_{m-1} := (-1)^m \prod_{i=1}^{m-1} \frac{1 - c_i}{c_i} \leq 1. \tag{4.4}$$

*In this case we have*

$$\|y - u\|_\infty = \begin{cases} \mathcal{O}(h^{m+1}) & \text{if } -1 \leq \rho_{m-1} < 1 \\ \mathcal{O}(h^m) & \text{if } \rho_{m-1} = 1. \end{cases}$$

(b) *Assume that* $c_m < 1$. *If the collocation parameters are symmetric, i.e. if* $c_i = 1 - c_{m+1-i}$ $(i = 1, \ldots, m)$, *then* $u$ *does not converge to* $y$ *as* $h \to 0$ ($Nh = \text{const.}$).

For nonsymmetric $\{c_i\}$ with $c_m < 1$ there are convergent methods: the collocation parameters have to be such that the spectral radius of a certain rank-2 matrix associated with the $\{c_i\}$ does not exceed one (compare [**76**] for details). For the space $S_{m+1}^{(1)}(\Pi_N)$ with $m \geq 2$ the question regarding convergence of collocation methods remains unanswered (for $m = 1$, Hung [**68**] showed divergence in the case $c_1 = 1$); the same is true for the spaces $S_\mu^{(d)}(\Pi_N)$ with $\mu \geq d + 2$ $(d \geq 0)$.

**4.3. Local superconvergence.** For first-kind Volterra integral equations of the forms (1.8), (4.2), there are no collocation parameters with $0 < c_1 < \cdots < c_m \leq 1$ for which collocation in $S_{m+d}^{(d)}(\Pi_N)$, with $d = -1$ or $d = 0$, yields local superconvergence on $\Pi_N$ [**28, 34, 76**]. However, in the case $d = -1$ local superconvergence of order $p^* = m+1$ (compared to the global order $p = m$) is possible if the collocation parameters are the $m$ positive Lobatto points in $(0,1]$ (i.e. the zeros of $(s-1)P'_{m-1}(2s-1)$). The local order $p^*$ is then attained at the Gauss points. An analogous result is true for the space $S_m^{(0)}(\Pi_N)$ [**76**].

**4.4. Equations with weakly singular kernels.** Let $0 < \alpha < 1$ and assume that the first-kind Volterra (-Abel) integral equation,

$$\int_0^t (t-s)^{-\alpha} K(t,s) y(s)\, ds = g(t), \quad t \in I, \tag{4.5}$$

is solved by collocation in $S_{m+d}^{(d)}(\Pi_N)$ $(d = -1, 0)$. For which collocation parameters satisfying $0 < c_1 < \cdots < c_m \leq 1$ is the collocation solution $u$ convergent?

Except for the pairs $(m,d) = (1,-1)$ and $(m,d) = (1,0)$ this problem remains open (compare [**34, 48, 15**]). Numerical evidence suggests that the following *conjecture* is true:

- If the collocation solution $u$ is convergent for (4.5) with $\alpha = 0$, i.e. if (4.3) $(d = -1)$ or (4.4) $(d = 0)$ holds, then the same is true for all $\alpha \in (0,1)$. (Compare also [**48**] for a discussion of this conjecture.)

We mention in passing that the convergence of Galerkin methods in $S_{m-1}^{(-1)}(\Pi_N)$ for (4.5) was studied in [**47**]; the paper also contains a numerical comparison of Galerkin and collocation methods in this space. The question on the convergence of Galerkin methods in the continuous spline space $S_m^{(0)}(\Pi_N)$ has not yet been answered.

The regularity of the exact solution of (4.5) is governed by the regularity of the function

$$G_\alpha(t) := \int_0^t (t-s)^{\alpha-1} g(s)\, ds, \quad t \in I.$$

If $g^{(j)}(0) = 0$ $(j = 0, \ldots, q)$, then $G_\alpha$ may be written as

$$G_\alpha(t) = \frac{1}{(\alpha)_{q+1}} \int_0^t (t-s)^{\alpha+q} g^{(q+1)}(s)\, ds,$$

with $(\alpha)_{q+1} := \alpha(\alpha+1)\cdots(\alpha+q)$. This implies that the solution $y$ of (4.5) is in $C^q(I)$ but $y^{(q+1)}(t)$ behaves like $t^{\alpha-1}$ near $t = 0$. Thus, optimal error estimates for the collocation solution hold only (under the assumptions on $K$ and $g$ stated at the beginning of Section 4.2) on suitably graded meshes, as the following theorem shows.

THEOREM 4.3. *Let $g \in C^{m+1}(I)$ satisfy $g^{(j)}(0) = 0$ for $j = 0, \ldots, q$ $(q \leq m)$. If the collocation parameters $\{c_i\}$ are such that the collocation solution $u \in S_{m-1}^{(-1)}(\Pi_N)$ for (4.5) is uniformly convergent on $I$, and if $\Pi_N$ is the graded mesh given by*

$$t_n = (n/N)^r T \quad (n = 0, \ldots, N), \quad \text{with } r = m/(\alpha+q),$$

*then*

$$\|y - u\|_\infty = \mathcal{O}(N^{-m}). \tag{4.6}$$

We conjecture that if the algebraic singularity $(t-s)^{-\alpha}$ $(0<\alpha<1)$ in (4.5) is replaced by $\log(t-s)$, then the optimal estimate (4.6) holds if $r=m/(1+q)$.

## 5. ODEs and VIEs with Delay Arguments

**5.1. Equations with constant delays.** It has been known for some ten years (Bellen [**8**]; also [**100**]) that for differential equations with *constant delay* $\tau>0$ collocation at the Gauss points, using the space $S_m^{(0)}(\Pi_N)$, leads to the same local superconvergence results $(p^*=2m)$ as for ODEs, provided the underlying mesh $\Pi_N$ is *constrained*; that is, $h_n=h=\tau/r$ for some positive integer $r$.

The same is true for Volterra integral equations (1.10) with constant delay [**7, 16**]: the results of Theorem 3.1 remain valid for such equations and for constrained $\Pi_N$.

However, "all is not well" with Gauss collocation for equations with constant delays. Consider the two DDEs

$$y'(t)=ay(t)+by(t-\tau)\quad(\tau>0); \tag{5.1}$$

with constant $a$ and $b$, and

$$y'(t)=a(t)y(t)+b(t)y(t-\tau), \tag{5.2}$$

where $a(t)$ and $b(t)$ are continuous functions. In (5.1),

$$|b|<-\mathrm{Re}(a) \text{ implies } y(t)\to 0, \text{ as } t\to\infty,$$

for any initial function $\phi\in C[-\tau,0]$, while for (5.2) the condition

$$\big|b(t)\big|\le -\mathrm{Re}\big(a(t)\big)\ (t\ge 0) \text{ implies } \big|y(t)\big|\le \max_{-\tau\le t\le 0}\big|\phi(t)\big| \text{ for } t\ge 0.$$

On constrained meshes $\Pi_N$ the Gauss collocation solution for (5.1) (constant coefficients) inherits the stability property of the analytical solution (*P-stability*), but for (5.2) (variable coefficients) this is no longer true (Torelli [**98**]; see also [**108, 107**]): Gauss collocation is *not PN-stable*—the collocation solution may become unbounded, even on constrained meshes.

Related stability results appear to hold for Volterra integral equations with delay [**99**].

**5.2. Variable delays.** Let $0<q<1$ and consider the simple delay differential equation

$$y'(t)=ay(t)+by(qt),\quad t\ge 0;\quad y(0)=y_0\ne 0, \tag{5.3}$$

together with the equivalent delay integral equation

$$y(t)=y_0+\int_0^t ay(s)\,ds+\int_0^{qt}(b/q)y(s)\,ds. \tag{5.4}$$

The paper [**69**] by Iserles contains a lucid exposition of the various—and often surprising—properties of the analytical solution of (5.3). The theory for general VIEs with variable delays goes back much further, to a 1914 paper by Andreoli [**1**]; see also [**41**].

It is well known [**34, 7**] that, for $b=0$, the collocation solutions $u\in S_m^{(0)}(\Pi_N)$ for (5.3) and $v\in S_{m-1}^{(-1)}(\Pi_N)$ for (5.4) are related by

$$v(t_n)\ne u(t_n) \text{ but } v_{it}(t_n)=u(t_n)\quad(n\ge 1). \tag{5.5}$$

Moreover, if collocation is at the Gauss points, then

$$u(h) = v_{it}(h) = R_{m,m}(z)y_0, \quad h = t_1 \quad (z := ah), \tag{5.6}$$

where $R_{m,m}(z)$ is the $(m, m)$-Padé approximant to $\exp(z)$. Thus,

$$y(h) - u(h) = y(h) - v_{it}(h) = \mathcal{O}(h^{2m+1}). \tag{5.7}$$

The statements (5.5)–(5.7) are no longer true if $b \neq 0$ and $0 < q < 1$ [**14**]. The general convergence analysis is still incomplete.

Also, little is presently known about the stability properties of collocation solutions to (5.3) and (5.4), except for the case $q = 1/2$ and the space $S_1^{(0)}(\Pi_N)$ [**38**].

## 6. Volterra Integro-Differential Equations

Spline collocation methods for the VIDE (1.11) date back to [**68**] and [**55**]. Since for $k \equiv 0$, (1.11) reduces to the ODE (1.3), the divergence results for collocation in the classical spline space $S_m^{(m-1)}(\Pi_N)$ parallel those of Loscalzo: $m \geq 4$ leads to divergent methods, while Hermite-type collocation in $S_4^{(2)}(\Pi_N)$ is convergent. The global order of convergence of fully discretized collocation methods was analyzed in [**84**] (like the method of de Hoog and Weiss [**61**] for second-kind VIEs, this block method may be interpreted as a discretized version of the collocation method in $S_m^{(0)}(\Pi_N)$).

The local superconvergence results for ODEs (Theorem 2.1) carry over to VIDEs of the form (1.11) ([**24, 34, 18**]; see also [**106**]). They remain valid also for higher-order, neutral VIDEs (where the highest derivative of the unknown solution occurs in the integrand, too; [**21, 19**]), and for VIDEs with constant delay [**17**]. However, if the kernel of (1.11) contains a weak singularity, then superconvergence on $\Pi_N$ is lost, except if the mesh is suitably graded ([**97**] and its references). In this case, marginal local superconvergence is possible (see also Section 3.3).

In view of the remarks stated in Section 5.2 it is not surprising that little is known yet about local superconvergence or stability of spline collocation methods for VIDEs with variable delay $\theta(t) = qt\,(0 < q < 1)$. This problem is currently being studied by a Ph.D. student at Memorial University, W. Zhang.

A major source of (systems of) VIDEs is the *semidiscretization in space* of *parabolic* VIDEs,

$$\begin{gathered} v_t(x,t) + A(x,t)v(x,t) = f(t) + \int_0^t B(x;t,\tau)v(x,\tau)\,d\tau, \\ \hfill x \in \Omega \subset \mathbb{R}^d,\ t > 0; \\ v(x,0) = v_0(x), \quad x \in \Omega; \\ v(x,t) = 0, \quad x \in \partial\Omega,\ t > 0. \end{gathered} \tag{6.1}$$

Here, $A$ is a second-order elliptic and positive definite spatial partial differential operator ($A = -\triangle$ is an obvious example), and $B$ is a general (spatial) partial differential operator of order at most two. Spatial discretization based on finite difference, finite element, or collocation techniques (see, e.g., the surveys [**53, 101**]) yields a (generally large) system of VIDEs,

$$y'(t) + A_h(t)y(t) = f_h(t) + \int_0^t B_h(t,\tau)y(\tau)\,d\tau, \quad t > 0;\ y(0) = y_0 \tag{6.2}$$

for $y(t) \in \mathbb{R}^M$, where the size of $M$ is determined by the spatial mesh $\Omega_h$ of $\Omega$ and the corresponding spatial discretization method. If (6.2) is solved by collocation or collocation-based Volterra-Runge-Kutta methods, then the (unbounded) "stiffness" of $A_h(t)$ will in general lead to a reduction in the (classical) order of the time discretization method ([**35**]; see also [**75**]).

Since the early 1990s, a number of authors have studied *wavelet-Galerkin methods* for elliptic problems, using multiresolution analysis to construct suitable wavelet bases (see, e.g., [**70, 44**] and their lists of references, and [**83, 51**] for time-dependent problems). It has been shown that there exists a (diagonal) preconditioner for the matrix corresponding to the discretized elliptic operator such that the preconditioned matrix has a *bounded condition number* regardless of the size of the underlying spatial mesh (e.g. [**70**, page 968]). Thus, it will be of interest to understand this result in the context of parabolic VIDEs (6.1): can wavelet-based spatial discretization be used to reduce, or even eliminate, the amount of order reduction in the subsequent time discretization by implicit (spline collocation) Volterra-Runge-Kutta methods of high-order, such as studied in [**35**]?

## References


1. G. Andreoli, *Sulle equazioni integrali*, Rend. Palermo, **37** (1914), 76–112.
2. U. Ascher, *On numerical differential algebraic problems with application to semiconductor device simulation*, SIAM J. Numer. Anal. **26** (1989), 517–538.
3. ———, *Collocation for two-point boundary value problems revisited*, SIAM J. Numer. Anal. **23** (1986), 596–609.
4. U. Ascher and G. Bader, *Stability of collocation at Gaussian points*, SIAM J. Numer. Anal. **23** (1986), 412–422.
5. U. Ascher and L. R. Petzold, *Projected implicit Runge-Kutta methods for differential-algebraic equations*, SIAM J. Numer. Anal. **28** (1991), 1097–1120.
6. O. Axelsson, *A class of A-stable methods*, BIT **9** (1969), 185–199.
7. N. Baddour and H. Brunner, *Continuous Volterra-Runge-Kutta methods for integral equations with pure delay*, Computing **50** (1993), 213–227.
8. A. Bellen, *One-step collocation for delay differential equations*, J. Comput. Appl. Math. **10** (1984), 275–283.
9. A. Bellen, Z. Jackiewicz, R. Vermiglio, and M. Zennaro, *Stability analysis of Runge-Kutta methods for Volterra integral equations of the second kind*, IMA J. Numer. Anal. **10** (1990), 103–118.
10. L. Blank, *Stability of collocation for weakly singular Volterra equations*, IMA J. Numer. Anal. **15** (1995), 357–375.
11. J. G. Blom and H. Brunner, *The numerical solution of nonlinear Volterra integral equations of the second kind by collocation and iterated collocation methods*, SIAM J. Sci. Statist. Comput. **8** (1987), 806–830.
12. C. De Boor and B. Swartz, *Collocation at Gaussian points*, SIAM J. Numer. Anal. **10** (1973), 582–606.
13. K. E. Brenan, S. L. Campbell, and L. R. Petzold, *Numerical solution of initial-value problems in differential-algebraic equations*, Elsevier, New York, 1989. (Revised and corrected reprint of the 1989 original), Classics in Appl. Math., vol. 14, Society for Industrial and Applied Mathematics (SIAM), Philadelphia, PA, 1996.
14. H. Brunner, *On the discretization of differential and Volterra integral equations with variable delay*, BIT **37** (1997), 1–12.
15. ———, *Open problems in the discretization of Volterra integral equations*, Numer. Funct. Anal. Optim. **17** (1996), 717–736.
16. ———, *Iterated collocation methods for Volterra integral equations with delay arguments*, Math. Comp. **62** (1994), 581–599.
17. ———, *The numerical solution of neutral Volterra integro-differential equations with delay arguments*, Ann. Numer. Math. **1** (1994), 309–322.

18. ———, *Collocation methods for nonlinear Volterra integro-differential equations with infinite delay*, Math. Comp. **53** (1989), 571–587.
19. ———, *The numerical solution of initial value problems for integro-differential equations*, Numerical Analysis 1987 (Dundee, 1987), Pitman Res. Notes Math. Ser., vol. 170, Longman Sci. Tech., Harlow, 1988, pp. 18–38.
20. ———, *Collocation methods for one-dimensional Fredholm and Volterra integral equations*, The State of the Art in Numerical Analysis (A. Iserles and M. J. D. Powell, eds.), Clarendon Press, Oxford, 1987, pp. 563–600.
21. ———, *Implicit Runge-Kutta-Nyström methods for general second-order Volterra integro-differential equations*, Comput. Math. Appl. **14** (1987), 549–559.
22. ———, *The numerical solution of weakly singular Volterra integral equations by collocation on graded meshes*, Math. Comp. **45** (1985), 417–437.
23. ———, *Iterated collocation methods and their discretizations for Volterra integral equations*, SIAM J. Numer. Anal. **21** (1984), 1132–1145.
24. ———, *Implicit Runge-Kutta of optimal order for Volterra integro-differential equations*, Math. Comp. **42** (1984), 95–109.
25. ———, *Nonpolynomial spline collocation for Volterra integral equations with weakly singular kernels*, SIAM J. Numer. Anal. **20** (1983), 1106–1119.
26. ———, *The application of the variation of constants formula in the numerical analysis of integral and integro-differential equations*, Utilitas Math. **19** (1981), 255–290.
27. ———, *Superconvergence in collocation and implicit Runge-Kutta methods for Volterra-type integral equations of the second kind*, Numerical Treatment of Integral Equations (J. Albrecht and L. Collatz, eds.), ISNM, vol. 53, Birkhäuser Verlag, Basel, 1980, pp. 54–72.
28. ———, *Superconvergence of collocation methods for Volterra integral equations of the first kind*, Computing **21** (1979), 151–157.
29. ———, *Discretization of Volterra integral equations of the first kind* II., Numer. Math. **30** (1978), 117–136.
30. ———, *Discretization of Volterra integral equations of the first kind*, Math. Comp. **31** (1977), 708–716.
31. ———, *The solution of Volterra integral equations of the first kind by piecewise polynomials*, J. Inst. Math. Appl. **12** (1973), 295–302.
32. ———, *The solution of nonlinear Volterra integral equations by piecewise polynomials*, Proc. First Manitoba Confer. Numer. Math. (R. S. Thomas and H. C. Williams, eds.), Univ. Manitoba, Winnipeg, 1971, pp. 65–78.
33. H. Brunner, M. R. Crisci, E. Russo, and A. Vecchio, *A family of methods for Abel integral equations of the second kind*, J. Comput. Appl. Math. **34** (1991), 211–219.
34. H. Brunner and P. J. Van Der Houwen, *The numerical solution of Volterra equations*, CWI Monograph, vol. 3, North-Holland Publ. Co., Amsterdam–New York, 1986.
35. H. Brunner, J.-P. Kauthen, and A. Ostermann, *Runge-Kutta time discretizations of parabolic Volterra integro-differential equations*, J. Integral Equations Appl. **7** (1995), 1–16.
36. H. Brunner and S. P. Nörsett, *Superconvergence of collocation methods for Volterra and Abel integral equations of the second kind*, Numer. Math. **36** (1981), 347–358.
37. H. Brunner and N. Yan, *On global superconvergence of iterated collocation solutions to linear second-kind Volterra integral equations*, J. Comput. Appl. Math. **69** (1996), 185–189.
38. M. D. Buhmann, A. Iserles, and S. P. Nörsett, *Runge-Kutta methods for neutral delay differential equations*, Contributions in Numerical Analysis (R. P. Agarwal, ed.), World Scientific, Singapore, 1993, pp. 85–98.
39. J. C. Butcher, *Implicit Runge-Kutta processes*, Math. Comp. **18** (1964), 50–64.
40. E. D. Callender, *Single step methods and low order splines for solutions of ordinary differential equations*, SIAM J. Numer. Anal. **8** (1971), 61–66.
41. L. G. Chambers, *Some properties of the functional equation* $\phi(x) = f(x)+\int_0^{\lambda x} g(x, y, \phi(y))\, dy$, Internat. J. Math. Math. Sci. **14** (1991), 27–44.
42. F. Chatelin and R. Lebbar, *The iterated projection solution for the Fredholm integral equation of the second kind*, J. Austral. Math. Soc. Ser. B **22** (1981), 439–451.
43. M. R. Crisci, E. Russo, and A. Vecchio, *Stability of collocation methods for Volterra integro-differential equations*, J. Integral Equations Appl. **4** (1992), 491–507.
44. W. Dahmen, A. Kunoth, and K. Urban, *A wavelet-Galerkin method for the Stokes equations*, Computing **56** (1996), 259–301.

45. M. C. Delfour and F. Dubeau, *Discontinuous polynomial approximations in the theory of one-step, hybrid and multistep methods for nonlinear ordinary differential equations*, Math. Comp. **47** (1986), 169–189.
46. M. C. Delfour, W. Hager, and F. Trochu, *Discontinuous Galerkin methods for ordinary differential equations*, Math. Comp. **36** (1981), 455–473.
47. P. P. B. Eggermont, *On Galerkin methods for Abel-type integral equations*, SIAM J. Numer. Anal. **25** (1988), 1093–1117.
48. ———, *Stability and robustness of collocation methods for Abel-type integral equations*, Numer. Math. **45** (1984), 431–445.
49. M. E. A. El Tom, *Application of spline functions to systems of Volterra integral equations of the first and second kinds*, J. Inst. Math. Appl. **17** (1976), 295–310.
50. ———, *Numerical solution of Volterra integral equations by spline functions*, BIT **13** (1973), 1–7.
51. B. Engquist, S. Osher, and S. Zhong, *Fast wavelet based algorithms for linear evolution equations*, SIAM J. Sci. Comput. **15** (1994), 755–775.
52. R. Esser, *Numerische Behandlung einer Volterraschen Integralgleichung*, Computing **10** (1978), 269–284.
53. G. Fairweather and D. Meade, *A survey of spline collocation methods for the numerical solution of differential equations*, Mathematics for Large Scale Computing (J. C. Díaz, ed.), Lecture Notes in Pure and Appl. Math., vol. 120, Marcel Dekker, New York, 1989, pp. 297–341.
54. A. Guillou and J. L. Soulé, *La résolution numérique des problèmes différentiels aux conditions initiales par des méthodes de collocation*, R.I.R.I. **3** (1969), 17–44.
55. J. A. Guzek and G. A. Kemper, *A new error analysis for a cubic spline approximate solution of a class of Volterra integro-differential equations*, Math. Comp. **27** (1973), 563–570.
56. E. Hairer and L. Jay, *Implicit Runge-Kutta methods for higher index differential-algebraic systems*, Contributions in Numerical Mathematics (R. P. Agarwal, ed.), World Scientific, Singapore, 1993, pp. 213–224.
57. E. Hairer and C. Lubich, *On the stability of Volterra-Runge-Kutta methods*, SIAM J. Numer. Anal. **21** (1984), 123–135.
58. E. Hairer, C. Lubich, and M. Roche, *The numerical solution of differential-algebraic systems by Runge-Kutta methods*, Lecture Notes in Math., vol. 1409, Springer-Verlag, Berlin-Heidelberg-New York, 1989.
59. E. Hairer, S. P. Nörsett and G. Wanner, *Solving ordinary differential equations* I. *Nonstiff problems*, 2nd ed., Springer-Verlag, Berlin-Heidelberg, 1993.
60. E. Hairer and G. Wanner, *Solving ordinary differential equations.* II. *Stiff and differential-algebraic problems*, Springer Ser. Comput. Math. **8**. Springer-Verlag, Berlin, 1993.
61. F. De Hoog and R. Weiss, *Implicit Runge-Kutta methods for second kind Volterra integral equations*, Numer. Math. **23** (1975), 199–213.
62. ———, *On the solution of Volterra integral equations of the first kind*, Numer. Math. **21** (1973), 22–32.
63. ———, *High order methods for Volterra integral equations of the first kind*, SIAM J. Numer. Anal. **10** (1973), 647–664.
64. P. J. Van Der Houwen, B. P. Sommeijer, and N. H. Cong, *Stability of collocation-based Runge-Kutta-Nyström methods*, BIT **31** (1991), 469–481.
65. A. Huber, *Eine Näherungsmethode zur Auflösung Volterrascher Integralgleichungen*, Monatsh. Math. Phys. **47** (1939), 240–246.
66. B. L. Hulme, *One-step piecewise polynomial Galerkin methods for initial value problems*, Math. Comp. **26** (1972), 415–426.
67. ———, *Discrete Galerkin and related one-step methods for ordinary differential equations*, Math. Comp. **26** (1972), 881–891.
68. H.-S. Hung, *The numerical solution of differential and integral equations by spline functions*, MRC Techn. Summ. Report 1053, Univ. Wisconsin, Madison, 1970.
69. A. Iserles, *On the generalized pantograph functional differential equation*, Europ. J. Appl. Math. **4** (1993), 1–38.
70. S. Jaffard, *Wavelet methods for fast resolution of elliptic problems*, SIAM J. Numer. Anal. **29** (1992), 965–986.
71. L. Jay, *Collocation methods for differential-algebraic equations of Index* 3, Numer. Math. **65** (1993), 407–421.

72. H. Kadner, *Die numerische Behandlung von Integralgleichungen nach der Kollokationsmethode*, Numer. Math. **10** (1957), 241–260.
73. J.-P. Kauthen, *Implicit Runge-Kutta methods for singularly perturbed integro-differential systems*, Appl. Numer. Math. **18** (1995), 201–210.
74. ______, *Implicit Runge-Kutta methods for some integrodifferential-algebraic equations*, Appl. Numer. Math. **13** (1993), 125–134.
75. ______, *The method of lines for parabolic partial integro-differential equations*, J. Integral Equations Appl. **4** (1992), 69–81.
76. J.-P. Kauthen and H. Brunner, *Continuous collocation approximations of solutions of first-kind Volterra equations*, Math. Comp. **66** (1997), 1441–1459.
77. L. Kramarz, *Global approximations to solutions of initial value problems*, Math. Comp. **32** (1978), 35–59.
78. M. Křížek and P. Neittaanmäki, *On superconvergence techniques*, Acta Appl. Math. **9** (1987), 175–198.
79. J. Kuntzmann, *Neuere Entwicklungen der Methode von Runge-Kutta*, Z. Angew. Math. Mech. (ZAMM) **41** (1961), 28–31.
80. L. Lopez, *One-step collocation methods for differential-algebraic systems of Index* 1, J. Comput. Appl. Math. **29** (1990), 145–159.
81. F. R. Loscalzo, *An introduction to the application of spline functions to initial value problems*, Theory and Applications of Spline Functions (T.N.E. Greville, ed.), Academic Press, New York, 1969, pp. 37–64.
82. F. R. Loscalzo and T. D. Talbot, *Spline function approximations for solutions of ordinary differential equations*, SIAM J. Numer. Anal. **4** (1967), 433–445.
83. Y. Maday and J.-C. Ravel, *Adaptivité par ondelettes: conditions aux limites et dimensions supérieures*, C. R. Acad. Sci. Paris Sér. I Math. **315** (1992), 85–90.
84. A. Makroglou, *Convergence of a block-by-block method for nonlinear Volterra integro-differential equations*, Math. Comp. **35** (1980), 783–796.
85. G. Micula, *Numerische Behandlung der Volterra-Integralgleichungen mit Splines*, Studia Univ. Babeş-Bolyai Math. **24** (1979), 46–54.
86. ______, *Approximate solution of the differential equation $y'' = f(x,y)$ with spline functions*, Math. Comp. **27** (1973), 807–816; erratum: Math. Comp. **29** (1975), 673–674.
87. ______, *Spline functions and applications* (Romanian), Editura Technica, Bucarest, 1978.
88. H. N. Mülthei, *Splineapproximationen von beliebigem Defekt zur numerischen Lösung von gewöhnlichen Differentialgleichungen.* I. II. III., Numer. Math. **32** (1979), 147–157; 343–358; **34** (1980), 143–154.
89. ______, *Numerische Lösung gewöhnlicher Differentialgleichungen mit Splinefunktionen*, Computing **25** (1980), 317–335.
90. ______, *Maximale Konvergenzordnung bei der numerischen Lösung von Anfangswertproblemen mit Splines*, Numer. Math. **39** (1982), 449–463.
91. ______, *A-stabile Kollokationsverfahren mit mehrfachen Knoten*, Computing **29** (1982), 317–335.
92. A. N. Netravali, *Spline approximation to the solution of the Volterra integral equation of the second kind*, Math. Comp. **27** (1973), 99–106.
93. S. P. Nörsett, *Splines and collocation for ordinary initial value problems*, Approximation Theory and Spline Functions (S. P. Singh, ed.), (St. John's, Nfld., 1983), NATO Adv. Sci. Inst. Ser. C Math. Phys. Sci. **136**, Reidel, Dordrecht-Boston, Mass., 1984, pp. 397-417.
94. G. W. Reddien, *Projection methods for two-point boundary value problems*, SIAM Rev. **22** (1980), 156–171.
95. R. D. Russell and L. F. Shampine, *A collocation method of boundary value problems*, Numer. Math. **19** (1972), 1–28.
96. I. H. Sloan, *Superconvergence*, Numerical Solution of Integral Equations (M. A. Golberg, ed.), Plenum Press, New York, 1990, pp. 35–70.
97. T. Tang, *Superconvergence of numerical solutions to weakly singular Volterra integro-differential equations*, Numer. Math. **61** (1992), 373–382.
98. L. Torelli, *Stability of numerical methods for delay differential equations*, J. Comput. Appl. Math. **25** (1989), 15–26.
99. R. Vermiglio, *On the stability of Runge-Kutta methods for delay integral equations*, Numer. Math. **61** (1992), 561–577.

100. ———, *A one-step subregion method for delay differential equations*, Calcolo **22** (1985), 429–455.
101. L. B. Wahlbin, *Finite element methods for evolution equations*, Advances in Numerical Analysis, vol. I (Lancaster, 1990), Oxford Sci. Publ., Oxford Univ. Press, New York, 11–33.
102. H. Werner, *The development of nonlinear splines and their application*, Approximation Theory III (E. W. Cheney, ed.), Academic Press, New York, 1980, pp. 125–150.
103. ———, *Calculation of singularities for solutions of algebraic differential equations*, Computational Aspects of Complex Analysis (H. Werner *et al.*, eds.), D. Reidel, Dordrecht-Boston, 1983, pp. 325–360.
104. H. Werner and H. Hilgers, *Lösung von Differentialgleichungen mit Splinefunktionen. Eine Störungstheorie. I. Divergenzaussagen*, Numer. Math. **48** (1986), 323–336.
105. K. Wright, *Some relationships between implicit Runge-Kutta, collocation and Lanczos $\tau$ methods, and their stability properties*, BIT **10** (1970), 217–227.
106. W. Yuan and T. Tang, *The numerical analysis of implicit Runge-Kutta methods for a certain nonlinear integro-differential equation*, Math. Comp. **54** (1990), 155–168.
107. M. Zennaro, *Delay differential equations: theory and numerics*, Theory and Numerics of Ordinary and Partial Differential Equations (Leicester, 1994) (M. Ainsworth *et al.*, eds.), Adv. Numer. Anal., vol. IV, Oxford Univ. Press, New York, 1995, pp. 291–333.
108. ———, *On the stability of one-step collocation for delay differential equations*, Delay Equations, Approximation and Application (G. Meinardus and G. Nürnberger, eds.), ISNM, vol. 74, Birkhäuser Verlag, Basel, 1985, pp. 334–343.

Department of Mathematics and Statistics, Memorial University of Newfoundland St. John's, Newfoundland, Canada A1C 5S7

*E-mail address*: hbrunner@morgan.ucs.mun.ca

Centre de Recherches Mathématiques
CRM Proceedings and Lecture Notes
Volume **18**, 1999

# On Best Error Bounds for Deficient Splines

François Dubeau and Jean Savoie

## 1. Introduction

The purpose of this paper is to present a brief review (not exhaustive) of results concerning the best constants in error bounds for a class of odd and even degree deficient splines. After introducing the definitions and notation in Section 2, results for deficient splines on an arbitrary partition are presented in Section 3, and Section 4 contains results for periodic splines on a uniform partitions.

## 2. Definitions and Notation

Let $\Pi = \{x_i\}_{i=0}^{N}$ be a mesh or a partition of the interval $[a,b]$, $a,b \in \mathbb{R}$ and $a = x_0 < \cdots < x_i < \cdots < x_N = b$. We define

$$\begin{aligned} \Delta x_i &= x_i - x_{i-1} \quad (i = 1, \dots, N), \\ h &= \max_i \Delta x_i = \text{mesh gauge}, \\ \beta &= \frac{h}{\min_i \Delta x_i} = \text{mesh ratio}. \end{aligned}$$

The $x_i$'s are the knots of the partition. Let $\{z_i\}_{i=1}^{N}$ be the sequence of midpoints $z_i = (x_{i-1} + x_i)/2$ $(i = 1, \dots, N)$.

A function $s$ is said to be a spline of degree $n$ with deficiency $m$ defined on $\Pi$ if $s$ is a polynomial of degree at most $n$ on each subinterval $[x_{i-1}, x_i]$ $(i = 1, \dots, N)$ and $s \in C^{n-m}\big([a,b];\mathbb{R}\big)$.

In this paper we consider the following function spaces

$$AC^{n+1}\big([a,b];\mathbb{R}\big) = \left\{ f \in C^n\big([a,b];\mathbb{R}\big) \;\middle|\; \begin{array}{ll} \text{(i)} & f^{(n+1)} \in L^\infty\big([a,b];\mathbb{R}\big) \\ \text{(ii)} & f^{(n)}(\beta) - f^{(n)}(\alpha) = \int_\alpha^\beta f^{(n+1)}(\xi)\, d\xi \\ & \text{for any } \alpha, \beta \in [a,b] \end{array} \right\},$$

1991 *Mathematics Subject Classification.* Primary: 41A44, 41A05; Secondary: 65D05, 65D07.

This work has been supported in part by the Natural Sciences and Engineering Research Council of Canada and by the "ministère de l'Éducation du Québec".

This is the final form of the paper.

and

$$AC_p^{n+1}\big([a,b];\mathbb{R}\big) = \big\{f \in AC^{n+1}\big([a,b];\mathbb{R}\big) \mid f^{(k)}(a) = f^{(k)}(b) \text{ for } k = 0,\dots,n\big\}.$$

## 3. Deficient Splines on an Arbitrary Partition

This section contains a review of best interpolating error bounds for deficient splines on an arbitrary partition.

**3.1. Cubic splines.** In [**1**] and [**9**] error bounds were derived for cubic spline interpolation over arbitrary partitions. In [**10**] the optimality of the bounds was investigated and the following result was obtained.

THEOREM 3.1 (Hall and Meyer [**10**]). *Assume $f \in AC^4\big([a,b];\mathbb{R}\big)$. Let $s$ be the unique cubic spline interpolant defined on $\pi$, $s \in C^2\big([a,b];\mathbb{R}\big)$ (spline of degree 3 and deficiency 1), such that $s(x_i) = f(x_i)$ $(i = 0,\dots,N)$, and*

**Type I:** $s^{(1)}(x_j) = f^{(1)}(x_j) \quad (j = 0, N)$;
**Type II:** $s^{(2)}(x_j) = f^{(2)}(x_j) \quad (j = 0, N)$.

*Then*

$$\big\|f^{(k)} - s^{(k)}\big\|_\infty \le C_k h^{4-k} \|f^{(4)}\|_\infty \quad (k = 0,1,2,3)$$

*where*

$$C_0 = \frac{5}{384}, \quad C_1 = \frac{1}{24}, \quad C_2 = \frac{3}{8}, \text{ and } C_3 = \frac{\beta + \beta^{-1}}{2}.$$

*Moreover $C_0$ and $C_1$ are the best possible constants.*

To prove this result, Hall and Meyer [**10**] used Hermite interpolation to be able to consider only error at the nodes. Bounds for these errors were obtained from appropriate kernels explicitely constructed using two fundamental splines.

The main idea for a general proof were included in this paper but the proof itself has been done only recently (see Section 3.4 and [**6**]). The first step in this direction was to include the periodic case in Theorem 3.1.

THEOREM 3.2 (Dubeau and Savoie [**4**]). *Assume $f \in AC_p^4\big([a,b];\mathbb{R}\big)$. Let $s$ be the unique periodic cubic spline interpolant defined on $\Pi$, $s \in C^2\big([a,b];\mathbb{R}\big)$ (spline of degree 3 and deficiency 1), such that $s(x_i) = f(x_i)$ $(i = 0,\dots,N)$ and*

**Type III:** $s^{(k)}(a) = s^{(k)}(b) \quad (k = 1,2)$.

*Then*

$$\big\|f^{(k)} - s^{(k)}\big\|_\infty \le C_k h^{4-k} \|f^{(4)}\|_\infty \quad (k = 0,1,2,3)$$

*where*

$$C_0 = \frac{5}{384}, \quad C_1 = \frac{1}{24}, \quad C_2 = \frac{3}{8}, \text{ and } C_3 = \frac{\beta + \beta^{-1}}{2}.$$

*Moreover, $C_0$ and $C_1$ are the best possible constants.*

**3.2. Quartic Splines.** Using the method of Hall and Meyer, Howell and Varma [**11**] were able to obtain the following result for deficient quartic splines.

THEOREM 3.3 (Howell and Varma [**11**]). *Assume* $f \in AC^5([a,b];\mathbb{R})$. *Let* $s$ *be the unique quartic spline interpolant defined on* $\Pi$, $s \in C^2([a,b];\mathbb{R})$ *(spline of degree 4 with deficiency 2), such that*

$$\begin{aligned} s(x_i) &= f(x_i) && (i = 0,\ldots,N),\\ s(z_i) &= f(z_i) && (i = 1,\ldots,N),\\ s^{(1)}(x_j) &= f^{(1)}(x_j) && (j = 0,N). \end{aligned}$$

*Then*

$$\|f - s\|_\infty \le C_0 \frac{h^5}{5!}\|f^{(5)}\|_\infty$$

*where*

$$C_0 = \left(\frac{1}{30} + \frac{1}{3\sqrt{30}}\right)\left(\frac{1}{4} - \frac{1}{\sqrt{30}}\right)^{1/2} = \max_{0\le t\le 1}|c(t)|$$

$$c(t) = \frac{t(1-t)(1-2t)[1+3t(1-t)]}{6}$$

*and* $C_0$ *is optimal. Also*

$$\left|f^{(1)}(x_i) - s^{(1)}(x_i)\right| \le \frac{h^4}{6!}\|f^{(5)}\|_\infty \quad (i = 1,\ldots,N-1)$$

*which cannot be improved. Finally, there exists a constant* $C_1$ *such that*

$$\|f^{(1)} - s^{(1)}\|_\infty \le C_1 \frac{h^4}{6!}\|f^{(5)}\|_\infty.$$

**3.3. Quadratic splines.** In a first attempt to obtain results similar to Theorems 3.1 and 3.2 for even degree splines, four classes of quadratic splines already known in the literature have been considered. These quadratic splines defined on $\Pi$ are described below.

**Type I:** (Marsden[**13**]): $f$ and $s$ are periodic and

$$\begin{aligned} s(z_i) &= f(z_i) && (i = 1,\ldots,N),\\ s^{(k)}(a) &= s^{(k)}(b) && (k = 0,1). \end{aligned}$$

**Type II:** (Kammerer et al [**12**], Demko [**2**], Xie [**16**]):

$$\begin{aligned} s(z_i) &= f(z_i) && (i = 1,\ldots,N),\\ s(x_j) &- f(x_j) && (j = 0,N). \end{aligned}$$

**Type III:** (Xie [**16**]):

$$\begin{aligned} s(z_i) &= f(z_i) && (i = 1,\ldots,N),\\ s^{(1)}(x_j) &= f^{(1)}(x_j) && (j = 0,N). \end{aligned}$$

**Type IV:** (Xie [**16**]):

$$\begin{aligned} s(z_i) &= f(z_i) && (i = 2,\ldots,N-1),\\ s^{(k)}(x_j) &= f^{(k)}(x_j) && (j = 0,N; k = 0,1). \end{aligned}$$

For these types of quadratic splines the following result holds.

THEOREM 3.4 (Dubeau and Savoie [**5**]). *Assume $f \in AC^3([a,b];\mathbb{R})$ or, if necessary, $f \in AC_p^3([a,b];\mathbb{R})$. Let $s$ be the quadratic spline interpolant defined on $\Pi$ of Type I, II, III or IV and $s \in C^1([a,b];\mathbb{R})$ (spline of degree 2 with deficiency 1). Then*

$$\|f^{(k)} - s^{(k)}\|_\infty \le C_k h^{3-k} \|f^{(3)}\|_\infty \quad (k = 0,1,2)$$

*where*

$$C_0 = \begin{cases} \frac{1}{24} & \text{for Types I, II, and III,} \\ \frac{1}{12} & \text{for Type IV,} \end{cases}$$

$$C_1 = \frac{7}{24},$$

$$C_2 = \begin{cases} \frac{5}{6} & \text{if } 1 \le \beta \le \frac{3}{2}, \\ \frac{\beta+\beta^{-1}}{2} & \text{if } \beta \ge \frac{3}{2}. \end{cases}$$

*Moreover $C_0$ is optimal.*

**3.4. Odd and even degree deficient splines.** For the general deficient spline as defined in Section 2, the next two theorems generalize the preceding results for $\|f - s\|_\infty$. These results can be obtained by an appropriate construction of two fundamental splines and an analysis of their properties.

THEOREM 3.5 (Odd case (Dubeau and Savoie [**6**])). *Assume $m \ge 2$ and $f \in AC^{2m}([a,b];\mathbb{R})$. Let $s$ be the unique spline interpolant of degree $2m-1$ with deficiency $m-1$ defined on $\Pi$ and associated with $f$ such that $s^{(k)}(x_i) = f^{(k)}(x_i)$ $(i = 0,\dots,N$ and $k = 0,\dots,m-2)$ and*

**Type I:** $s^{(m-1)}(x_i) = f^{(m-1)}(x_i) \quad (i = 0, N)$,
**Type II:** $s^{(m)}(x_i) = f^{(m)}(x_i) \quad (i = 0, N)$,
**Type III:** $s^{(k)}(a) = s^{(k)}(b) \quad (k = m-1, m)$ *and $f$ periodic.*

*Then*

$$\|f - s\|_\infty \le \frac{(m+3)2^{-2m}}{(m-1)(2m)!} h^{2m} \|f^{(2m)}\|_\infty$$

*and the bound is optimal. Moreover*

$$\left|f^{(m-1)}(x_i) - s^{(m-1)}(x_i)\right| \le \frac{(m-2)!}{(2m)!} h^{m+1} \|f^{(2m)}\|_\infty \qquad (i = 0,\dots,N)$$

$$\left|f^{(m)}(x_i) - s^{(m)}(x_i)\right| \le \frac{(m+1)!}{(2m)!(m-1)} h^{m} \|f^{(2m)}\|_\infty \quad (i = 0,\dots,N)$$

*and the first inequality cannot be improved.*

Let us observe that the statement of the first part of this result for Type I was given in [**10**] with no proof.

THEOREM 3.6 (Even Case (Dubeau and Savoie [**6**])). *Assume $m \ge 1$ and $f \in AC^{2m+1}([a,b];\mathbb{R})$. Let $s$ be the unique spline interpolant of degree $2m$ with deficiency $m$ defined on $\Pi$ and associated with $f$ such that*

$$s(z_i) = f(z_i) \qquad (i = 1,\dots,N)$$
$$s^{(k)}(x_i) = f^{(k)}(x_i) \quad (i = 0,\dots,N;\ k = 0,\dots,m-2),$$

*and*

**Type I:** $s^{(m-1)}(x_i) = f^{(m-1)}(x_i) \quad (i = 0, N)$,
**Type II:** $s^{(m)}(x_i) = f^{(m)}(x_i) \quad (i = 0, N)$,
**Type III:** $s^{(k)}(a) = s^{(k)}(b) \quad (k = m-1, m)$ *and* $f$ *periodic.*

*Then*

$$\|f - s\|_\infty \le C_m \frac{2^{-2m}}{(2m+2)!} h^{2m+1} \|f^{(2m+1)}\|_\infty$$

*and the bound is optimal.* $C_m$ *is given by*

$$C_m = \max_{t\in[0,1]} c(t) = c(\tau_m)$$

*where*

$$c(t) = \big[(m+1)(1-t^2) + 4\big] t(1-t^2)^{m-1},$$

*and* $1 - \tau_m^2 = \xi_m$ *where* $\xi_m$ *is the unique root in* $[0,1]$ *of the quadratic equation*

$$(2m+1)(m+1)\xi^2 - 2(m-2)(m-1)\xi - 8(m-1) = 0.$$

*Moreover*

$$\big|f^{(m-1)}(x_i) - s^{(m-1)}(x_i)\big| \le \frac{(m-1)!}{(2m+2)!} h^{m+2} \qquad (i = 0, \ldots, N)$$

$$\big|f^{(m)}(x_i) - s^{(m)}(x_i)\big| \le \frac{m!(m+3)}{(2m+2)!} h^{m+1} \qquad (i = 0, \ldots, N)$$

*and the first inequality cannot be improved.*

Let us point out that interpolation at midknots for even degree splines has been introduced by Subbotin [**15**].

## 4. Periodic Splines on a Uniform Partition

For a uniform partition in the periodic cases (quadratic and cubic splines), Miroshnichenho [**14**] obtained best bounds that depend on the regularity of the functions $f$ (see also references in [**14**]). These results are based on an analysis of kernels obtained by inverting a circulant matrix.

THEOREM 4.1 (V. L. Miroshnichenko [**14**]). *Assume* $f \in AC_p^r\big([a,b];\mathbb{R}\big)$ *and* $r = 1, 2, 3$. *Let* $s$ *be the unique periodic quadratic spline interpolant of* $f$ *defined on a uniform partition* $\Pi$ *such that*

$$s(z_i) = f(z_i) \quad (i = 0, \ldots, N).$$

*Then*

$$\|f^{(k)} - s^{(k)}\|_\infty \le K_{kr} h^{r-k} \|f^{(r)}\|_\infty \ (r = 1,2,3;\ k = 0, \ldots, r-1),$$

*where*

| $K_{kr}$ | $r$ | | |
|---|---|---|---|
| $k$ | 1 | 2 | 3 |
| 0 | $\frac{\sqrt{2}}{2}$ | $\frac{1+\sqrt{2}}{6}$ | $\frac{1}{24}$ |
| 1 | | $\frac{3\sqrt{2}}{8}$ | $\frac{1}{8}$ |
| 2 | | | $\frac{1+2\sqrt{2}}{6}$ |

*Moreover these constants are optimal.*

THEOREM 4.2 (V. L. Miroschnichenko [**14**]). *Assume $f \in AC_p^r([a,b];\mathbb{R})$ and $r = 1,2,3,4$. Let $s$ be the unique periodic cubic spline interpolant of $f$ defined on a uniform partition of $[a,b]$ such that*

$$s(x_i) = f(x_i) \quad (i = 0,\dots,N).$$

*Then*

$$\|f^{(k)} - s^{(k)}\|_\infty \le K_{kr} h^{r-k} \|f^{(r)}\|_\infty \quad (r = 1,2,3,4;\ k = 0,\dots,r-1)$$

*where*

| $K_{kr}$ | $r$ | | | |
|---|---|---|---|---|
| $k$ | 1 | 2 | 3 | 4 |
| 0 | $\frac{1+3\sqrt{3}}{8}$ | $\frac{3\sqrt{3}}{32}$ | $\frac{(3-\sqrt{3})(7/4+\sqrt{3}+\sqrt{2})}{144}$ | $\frac{5}{384}$ |
| 1 | | $\frac{\sqrt{3}}{3}$ | $\frac{1+\sqrt{2}}{18}$ | $\frac{1}{24}$ |
| 2 | | | $\frac{21+5\sqrt{3}+4\sqrt{2}(3-\sqrt{3})}{72}$ | $\frac{\sqrt{3}}{12}$ |
| 3 | | | | $\frac{1+\sqrt{3}}{4}$ |

*Moreover these constants are optimal.*

It is expected that some of these results, or similar results, could be obtained in the more general setting of uniformly translated nodes. Peano kernels in that case can be obtained from [**3**], and an analysis can be done as in [**7, 8**].

## References


1. G. Birkhoff and C. de Boor, *Error bounds for spline interpolation*, J. Math. Mech. **13** (1964), 827–835.
2. S. Demko, *Interpolation by quadratic splines*, J. Approx. Theory **23** (1978), 392–400.
3. F. Dubeau and J. Savoie, *Splines périodiques avec partage uniforme de la droite réelle*, Utilitas Math. **32** (1987), 111–120.
4. ______, *A note on optimal error bounds for periodic cubic spline interpolation*, techn. report, C.M.R. St-Jean, 1992.
5. ______, *Optimal error bounds for quadratic spline interpolation*, J. Math. Anal. Appl. **198** (1996), 49–63.
6. ______, *Best error bounds for odd and even degree deficient splines*, SIAM J. Numer. Anal. **34** (1997), no. 3, 1167–1184.
7. ______, *Explicit error bounds for spline interpolation on a uniform partition*, J. Approx. Theory **82** (1995), 1–14.
8. ______, *On optimal error bounds for spline interpolation on a uniform partition*, Approximation Theory VIII, vol. 1 (College Station, TX, 1995), Ser. Approx. Decompos. **6**, World Sci. Publishing, River Edge, NJ, 1995, pp. 485–490.
9. C. Hall, *On error bounds for cubic spline interpolation*, J. Approx. Theory **1** (1968), 209–218.
10. C. Hall and W. Meyer, *Optimal error bounds for cubic spline interpolation*, J. Approx. Theory **16** (1976), 105–122.
11. G. Howell and A. K. Varma, *Best error bounds for quartic spline interpolation*, J. Approx. Theory **58** (1989), 58–67.
12. W. J. Kammerer, G. W. Reddien, and R. S. Varga, *Quadratic interpolatory splines*, Numer. Math. **22** (1974), 241–259.
13. M. J. Marsden, *Quadratic spline interpolation*, Bull. Amer. Math. Soc. (N.S.) **80** (1974), 903–906.
14. V. L. Miroshnichenko, *Exact error bounds for the periodic cubic and parabolic spline interpolation on the uniform mesh*, Math. Balkanica (N.S.), **2** (1988), 210–221.
15. Y. N. Subbotin, *On piecewise polynomial (spline) interpolation*, Mat. Zametki **1** (1967), 63–67; English transl. in Math. Notes **1** (1967), 41–46.

16. S. Xie, *Quadratic spline interpolation*, J. Approx. Theory **40** (1984), 66–90.

Département de mathématiques et d'informatique, Faculté des sciences, Université de Sherbrooke, 2500, boulevard Université, Sherbrooke, Québec J1K 2R1, Canada
*E-mail address*: francois.dubeau@dmi.usherb.ca

Centre de recherches pour la défense, 2459 boul. Pie-XI Nord, Val-Bélair, Québec, Canada, G0A 1R0
*E-mail address*: jean.savoie@drev.dnd.ca

Centre de Recherches Mathématiques
CRM Proceedings and Lecture Notes
Volume **18**, 1999

# Optimal Error Bounds for Spline Interpolation on a Uniform Partition

François Dubeau and Jean Savoie

ABSTRACT. In this paper we are concerned with the error for the cardinal spline interpolation of a function at shifted nodes. Optimal bounds are presented for the error and its $n$-th order derivative. For the intermediate derivatives ($1 \leq k \leq n-1$) nice error bounds are also obtained. The bounds are given in terms of Euler splines and Euler-Frobenius polynomials.

## 1. Introduction

An $n$-degree cardinal spline $s$ is a function $s \in C^{n-1}(\mathbb{R})$ such that $s$ restricted to $[l, l+1]$ is an algebraic polynomial of degree at most $n$ for any $l \in \mathbb{Z}$ . A cardinal spline $s$ is said to be an interpolating spline of $f$ if for a given $v \in [0,1]$ we have $s(l+v) = f(l+v)$ for all $l \in \mathbb{Z}$.

Let the Euler-Frobenius polynomial be defined by $p_n(t,z) = \sum_{j=0}^{n} Q_n(n+t-j)z^j$ for $t \in [0,1]$ and $z \in \mathbb{R}$, where $Q_n$ is the cardinal B-spline of degree $n$. The main properties of $p_n(t,z)$ can be found in [**1,3,7,8**] and [**15**] for example.

Using the linear dependence relationships of Ter Morsche [**15,16**] for an $n$-degree cardinal spline $s$, we obtain

$$p_n(v,E)s^{(k)}(l+u) - p_n^k(u,E)s(l+v) = 0 \tag{1.1}$$

for any $l \in \mathbb{Z}$, $u$, $v \in [0,1]$, $k = 0, \ldots, n$, where $E$ is the shift operator $Ef(t) = f(t+1)$, and $p_n^k(u,E) = p_{n-k}(u,E)(E-I)^k$.

A function $f$ is said to be of polynomial growth on $\mathbb{R}$ if there exists an integer $\nu \geq 0$ such that $f(x) = O(|x|^\nu)$ for $|x| \to +\infty$. As a consequence of (1.1) and the properties of $p_n(t,z)$, there exists a unique $n$-degree interpolating spline $s$ for any function $f$ of polynomial growth when $p_n(v,-1) \neq 0$ [**5,11,13,15**]. Moreover, we have

$$s^{(k)}(l+u) = p_n(v,E)^{-1}p_n^k(u,E)s(l+v) \tag{1.2}$$

---

1991 *Mathematics Subject Classification.* Primary: 41A44, 41A05; Secondary: 65D05,65D07.

This work has been supported in part by the "Ministère de l'Éducation du Québec" (FCAR grant), by the Natural Sciences and Engineering Research Council of Canada, and by the "Centre de recherches pour la défense de Valcartier".

This is the final form of the paper.

for $u \in [0,1]$, where $s(l+v) = f(l+v)$.

Let us consider the following function spaces

$$L^1_{\text{loc}}(\mathbb{R}) = \left\{ f:\mathbb{R} \to \mathbb{R} \;\middle|\; \int_a^b |f(x)|\,dx < \infty \text{ for any interval } [a,b] \subset \mathbb{R} \right\}$$

and

$$AC^{n+1}_{\text{loc}}(\mathbb{R}) = \left\{ f \in C^n(\mathbb{R}) \;\middle|\; \begin{array}{l} \text{(i) } f^{(n+1)} \in L^1_{\text{loc}}(\mathbb{R}) \\ \text{(ii) for all } [a,b] \subset \mathbb{R},\ f^{(n)}(x)\,\big|_a^b = \displaystyle\int_a^b f^{(n+1)}(x)\,dx \end{array} \right\}.$$

For any $f \in AC^{n+1}_{\text{loc}}(\mathbb{R})$, using its Taylor expansion and (1.1) we obtain

$$p_n(v,E)f^{(k)}(l+u) - p_n^k(u,E)f(l+v) = \int_0^{n+1} \frac{K_n^k(u,v,\theta)}{n!} f^{(n+1)}(l+\theta)\,d\theta \tag{1.3}$$

where

$$\frac{K_n^k(u,v,\theta)}{n!} = p_n(v,E)\frac{(u-\theta)_+^{n-k}}{(n-k)!} - p_n^k(u,E)\frac{(v-\theta)_+^n}{n!}$$

for any $l \in \mathbb{Z}$, $u \in [0,1]$, and $k = 0,\ldots,n$. This is a direct consequence of the Peano Kernel Theorem [**2,4,5**]; the Peano Kernel $K_n^k$ was analysed in [**6**]. It follows from (1.1) and (1.3) that

$$p_n(v,E)e^{(k)}(l+u) = \int_0^{n+1} \frac{K_n^k(u,v,\theta)}{n!} f^{(n+1)}(l+\theta)\,d\theta$$

where $e^{(k)} = f^{(k)} - s^{(k)}$ $(k = 0,\ldots,n)$. Moreover, if $f^{(n+1)}$ is of polynomial growth, we have

$$e^{(k)}(l+u) = p_n(v,E)^{-1} \int_0^{n+1} \frac{K_n^k(u,v,\theta)}{n!} f^{(n+1)}(l+\theta)\,d\theta, \tag{1.4}$$

and this last expression can be transformed into

$$e^{(k)}(l+u) = \int_{-\infty}^{\infty} N_n^k(u,v,\theta) f^{(n+1)}(l+\theta)\,d\theta \tag{1.5}$$

where

$$N_n^k(u,v,\theta) = \frac{(u-\theta)_+^{n-k}}{(n-k)!} - p_n(v,E)^{-1}p_n^k(u,E)\frac{(v-\theta)_+^n}{n!}. \tag{1.6}$$

If $f^{(n+1)} \in L^\infty(\mathbb{R})$, we therefore obtain

$$\left|e^{(k)}(l+u)\right| \le A_n^k(u,v)\|f^{(n+1)}\|_\infty \tag{1.7}$$

for any $u \in [0,1]$, and

$$\|e^{(k)}\|_\infty \le A_n^k(v)\|f^{(n+1)}\|_\infty \tag{1.8}$$

where

$$A_n^k(u,v) = \int_{-\infty}^{\infty} \left|N_n^k(u,v,\theta)\right|\,d\theta, \tag{1.9}$$

and

$$A_n^k(v) = \sup_{u \in [0,1]} A_n^k(u, v) \tag{1.10}$$

are the best constants in (1.7) and (1.8) for any function $f \in \mathcal{C} = \{f \in AC_{\text{loc}}^{n+1}(\mathbb{R}) \mid f^{(n+1)} \in L^\infty(\mathbb{R})\}$.

In Section 2 we present exact values for $A_n^0(v)$ which are the smallest possible for $v = v_n$ ($v_n = 1/2$ for $n$ even, $v_n = 0$ or 1 for $n$ odd). We also present some upper bounds for $A_n^k(v)$ ($k = 1, \ldots, n$). We note that the value of $A_n^k(v)$ for $v = v_n$ and $k = 0$ was first obtained in [**1**]. In Section 3, properties of the kernels $N_n^k(u, v, \theta)$ are given. In the next sections we consider only the case $v = v_n$. In Section 4, the exact value for $A_n^n(v_n)$ is given, and in Section 5 near optimal bounds for $A_n^k(v_n)$ are presented for $1 \leq k \leq n-1$. The last section contains bounds for quadratic and cubic cases. The proofs of the results presented here are contained in [**6**] and [**9**].

## 2. The Case $k = 0$ and the Case $k > 0$

Let $p_n(v, -1) \neq 0$. For $k = 0$, using the properties of $K_n^k(u, v, \theta)$, it was obtained in [**6**] that

$$A_n^0(u, v) = \frac{1}{2^{n+1}} \left| \mathcal{E}_{n+2}(u) - \mathcal{E}_{n+1}(u) \frac{\mathcal{E}_{n+2}(v)}{\mathcal{E}_{n+1}(v)} \right| \tag{2.1}$$

and

$$A_n^0(v) = \frac{1}{2^{n+1}} \max_{u \in [0,1]} \left( |\mathcal{E}_{n+2}(u)| + |\mathcal{E}_{n+1}(u)| \frac{|\mathcal{E}_{n+2}(v)|}{|\mathcal{E}_{n+1}(v)|} \right) \tag{2.2}$$

where $\mathcal{E}_{n+1}$ is the Euler spline of degree $n$ [**14**]. Moreover,

$$\min_{0 \leq v \leq 1} A_n^0(v) = A_n^0(v_n) = \frac{|\mathcal{E}_{n+2}(v_{n+1})|}{2^{n+1}} = \frac{|p_{n+1}(v_{n+1}, -1)|}{2^{n+1}} \tag{2.3}$$

where

$$v_n = \begin{cases} \frac{1}{2} & \text{for } n \text{ even,} \\ 0 & \text{for } n \text{ odd.} \end{cases}$$

For $k > 0$, it was also obtained in [**6**] that

$$A_n^k(v) \leq \max_{0 \leq u \leq 1} \frac{(\mathcal{E}_{n-k+2}(u))^2 - \mathcal{E}_{n-k+1}(u)\mathcal{E}_{n-k+3}(u)}{2^{n-k+1}|\mathcal{E}_{n+1}(v)|} \tag{2.4}$$

and

$$A_n^k(v) \leq \frac{(\pi^2/2)^{k-n}}{2|p_n(v, -1)|}. \tag{2.5}$$

Moreover, the smallest value of the righthand side of (2.5) is for $v = v_n$. It followed from these results, for $v = v_n$ and $k = 0, \ldots, n$, that

$$\|e^{(k)}\|_\infty \leq B_n^k \|f^{(n+1)}\|_\infty \tag{2.6}$$

where

$$B_n^k = \frac{(\pi^2/2)^k}{2\pi^n}. \tag{2.7}$$

The last bound is good for low values of $k$, but can be improved for large values of $k$.

## 3. Peano Kernels

Let

$$p_n(t,z) = \sum_{j=-\infty}^{\infty} Q_n(n+t-j)z^j$$

be the extension for $t \in \mathbb{R}$ of the Euler-Frobenius polynomial $p_n(t,z)$. If we use the notation

$$R_n^k(u,v,\theta) = \frac{1}{2\pi i}\int_{C_{1-\varepsilon}} \frac{p_{n-k}(1-u,z)}{p_n(1-v,z)} \frac{p_n(\theta-v,z)}{(1-z)^{n-k+1}}\,dz$$

and

$$L_n^k(u,v,\theta) = \frac{1}{2\pi i}\int_{C_{1-\varepsilon}} \frac{p_{n-k}(u,z)}{p_n(v,z)} \frac{p_n(v-\theta,z)}{(z-1)^{n-k+1}}\,dz$$

where $\varepsilon > 0$ is sufficiently small and $C_r$ is the counterclockwise oriented circle of radius $r$ centered at 0, we obtain the following results related to $N_n^k$.

THEOREM 3.1. *If either $k = n$ and $\theta \neq u$ or $0 \le k < n$, we have the representations*

$$\begin{aligned} N_n^k(u,v,\theta) &= \frac{(u-\theta)_+^{n-k}}{(n-k)!} - R_n^k(u,v,\theta) \\ &= (-1)^{n-k+1}\frac{(\theta-u)_+^{n-k}}{(n-k)!} - L_n^k(u,v,\theta). \end{aligned}$$

COROLLARY 3.2.
(i) *If either $n = k$ and $u \in (0,1)$ or $0 \le k < n$ , then $N_n^k(u,v_n,j) = 0$ for $j \in \mathbb{Z}$.*
(ii) *If $n = k$ and $u \in \{0,1\}$, then $N_n^k(u,v_n,j) = 0$ for $j \in \mathbb{Z} - \{u\}$. Moreover, we have $N_n^n(1,v,1^+) = 0$ and $N_n^n(0,v,0^-) = 0$.*

As a consequence of the Residue Theorem, we obtain from Theorem 3.1 the following representation.

COROLLARY 3.3. *Let*

$$\mathcal{U}_{n,\alpha}(\theta;v) = \frac{p_n(v-\theta,\alpha)}{\partial/(\partial z)p_n(v,\alpha)(1-\alpha)}.$$

*If $p_n(v,-1) \neq 0$, then*

$$N_n^k(u,v,\theta) = \begin{cases} \frac{(u-\theta)_+^{n-k}}{(n-k)!} - \sum_{\alpha\in\mathcal{A}_n(1-v)} \frac{p_{n-k}(1-u,\alpha)}{(1-\alpha)^{n-k}}\mathcal{U}_{n,\alpha}(1-\theta;1-v) & \text{if } \theta > v \\ (-1)^{n-k+1}\frac{(\theta-u)_+^{n-k}}{(n-k)!} + \sum_{\alpha\in\mathcal{A}_n(v)} \frac{p_{n-k}(u,\alpha)}{(\alpha-1)^{n-k}}\mathcal{U}_{n,\alpha}(\theta;v) & \text{if } \theta < v \end{cases}$$

*where $\mathcal{A}_n(v)$ is the set of roots of $p_n(v,\cdot)$ inside the unit circle. Moreover, for $\alpha \in \mathcal{A}_n(v)$ we have*
(i) $\mathcal{U}_{n,\alpha}(\theta-j;v) = \alpha^j\mathcal{U}_{n,\alpha}(\theta;v)$,
(ii) $\int_0^1 \mathcal{U}_{n,\alpha}(\theta;v)\,d\theta = 1/(n+1)$,
(iii) $\mathcal{U}_{n,\alpha}(\theta;v) \ge 0$ *for* $\theta \in [0,1]$.

The next theorem gives some symmetry properties of the kernels $N_n^k$.

THEOREM 3.4. *We have*

$$N_n^k(u,v,\theta) = (-1)^{n-k+1} N_n^k(1-u,1-v,1-\theta),$$
$$N_n^k(u,v_n,\theta) = (-1)^{n-k+1} N_n^k(1-u,v_n,1-\theta),$$

*and if* $k < n$

$$N_n^k(0,v_n,\theta) = (-1)^{n-k+1} N_n^k(0,v_n,-\theta).$$

Let $\alpha(v) = \min\{\alpha \mid \alpha \in \mathcal{A}_n(v)\}$ and $\alpha_n = \alpha(v_n)$. Using integration by parts and Corollary 3.2, we obtain (1.5) and the next theorem.

THEOREM 3.5. *If* $p_n(v,-1) \neq 0$, *then there exists a constant* $C$ *such that*

$$\left|N_n^k(u,v,\theta)\right| \le C\big(|\gamma|\big)^{|\theta|}$$

*where* $\gamma = \min\{\alpha_n(v), \alpha_n(1-v)\}$. *Also, if* $s$ *is the interpolating spline of degree* $n$ *of* $f \in AC_{\mathrm{loc}}^{n+1}(\mathbb{R})$ *such that* $s(l+v) = f(l+v)$ *for* $l \in \mathbb{Z}$, *then*

$$e^{(k)}(l+u) = \int_{-\infty}^{\infty} N_n^k(u,v,\theta) f^{(n+1)}(l+\theta)\,d\theta.$$

Using Corollary 3.3, we obtain the next theorem.

THEOREM 3.6. *Let* $j \in \mathbb{Z}$ *and*

$$G_j(u) = \int_j^{j+1} N_n^k(u,v_n,\theta)\,d\theta.$$

*Then for* $j \ge 0$,

$$G_j(u) = \left(\frac{u^{n-k+1}}{(n-k+1)!} - \frac{u^{n-k}(1-v_n)}{(n-k)!(n+1)}\right)\delta_{0,j} - \sum_{\alpha \in \mathcal{A}_n(v_n)} \frac{p_{n-k}(1-u,\alpha)\alpha^j}{(n+1)(1-\alpha)^{n-k}}$$

*and for* $j < 0$,

$$G_j(u) = (-1)^{n-k} G_{-j}(1-u).$$

## 4. Application to the Case $k = n$

In this section we obtain the value of the bounds $A_n^n(v_n)$ in terms of the roots of $p_n(v_n,z)$. The main result is the following theorem.

THEOREM 4.1. *We have*

$$A_n^n(v_n) = A_n^n(0,v_n) = \frac{1}{2} - \frac{2}{n+1} \sum_{\alpha \in \mathcal{A}_n(v_n)} \frac{\alpha}{1+\alpha}. \tag{4.1}$$

The proof of this result is based on the analysis of the sign property of the kernel $N_n^n(u,v,\theta)$. Using the Budan-Fourier Theorem for the derivative $\partial_\theta N_n^n(u,v,\theta)$, we obtain the desired result. The expression (4.1) follows from Corollary 3.3 and Theorem 3.6. We note here that the sign property of $N_n^n$ should not be confused with sign property of $\partial_\theta^n N_n^0(u,v,\theta)$ obtained in [**1**] even though Budan-Fourier Theorem is used.

Using the bound

$$\sum_{\alpha \in \mathcal{A}_n(v_n)} \left(\frac{1-\alpha}{1+\alpha}\right)^2 \le \frac{8}{\pi^3}(n+1)(n+2)$$

obtained in [**8**], we get the next result.

THEOREM 4.2.

$$A_n^n(v_n) \leq \frac{(1-v_n)}{(n+1)} + \frac{2}{\pi^{3/2}}\sqrt{(n+1)}. \tag{4.2}$$

## 5. The Case $1 \leq k \leq n-1$

This section is concerned with finding a good upper approximation for the optimal bound $A_n^k(v_n)$. Since we consider only $v = v_n$ in this section, we will use $N_n^k(u,\theta)$ for $N_n^k(u, v_n, \theta)$. Our method to obtain a good approximation is based on the determination of the sign of the kernel $N_n^k(u,\cdot)$, which is directly related to the nature of its zeros. In order to do that, we consider the kernel $N_n^k(u,\cdot)$ as a $HB$-spline and use corresponding Budan-Fourier Theorem [**10**].

We already know from Section 3 that $N_n^k(u,l) = 0$ for $l \in \mathbb{Z}$. The next theorem indicates that $N_n^k(u,\cdot)$ has at most one more zero which depends on $u$. It will be denoted by $\psi_n^k(u)$. This last result is based in part on the fact that

$$-1 < \cdots < \alpha_n < \alpha_{n-1} < \ldots \alpha_2 < 0$$

which is proved in [**8**] (see also [**7**]).

THEOREM 5.1. *For every $k$, $1 \leq k \leq n-1$, there exists a unique $u_n^k \in [0,1]$ such that $p_{n-k}(u_n^k, \alpha_n) = 0$ if $n-k$ is even, and $p_{n-k}(1-u_n^k, \alpha_n) = 0$ if $n-k$ is odd. Moreover, $u_n^k \in (0, 1/2)$.*

*Also, if*

$$\mathcal{I}_n^k = \begin{cases} (u_n^k, 1-u_n^k) & \text{if } n-k \text{ is even} \\ [0, u_n^k) \cup (1-u_n^k, 1] & \text{if } n-k \text{ is odd,} \end{cases}$$

*and if $\mathcal{J}_n^k \subset [0,1]$ is such that*

$$\begin{gathered} \mathcal{I}_n^k \cap \mathcal{J}_n^k = \emptyset, \\ [0,1] - \mathcal{I}_n^k \cup \mathcal{J}_n^k = \{u_n^k, 1-u_n^k\}, \quad \text{for } n = 2 \text{ or } 3, \\ \mathcal{I}_n^k \cup \mathcal{J}_n^k = [0,1], \quad \text{for } n \geq 4, \end{gathered}$$

*then the zeros of $N_n^k(u,\cdot)$ are related to the intervals $\mathcal{I}_n^k$ and $\mathcal{J}_n^k$ in the following way.*

(i) *If $u \in \mathcal{J}_n^k$, then $\mathbb{Z}$ is the set of zeros of $N_n^k(u,\cdot)$, and these zeros are simple and isolated.*

(ii) *If $u \in \mathcal{I}_n^k$, then the set of zeros of $N_n^k(u,\cdot)$ is $\mathbb{Z} \cup \{\psi_n^k(u)\}$. If $\psi_n^k(u) \notin \mathbb{Z}$, then all the zeros are simple and isolated. If $\psi_n^k(u) \in \mathbb{Z}$, $\psi_n^k(u)$ is a double zero and all other zeros are simple. Also,*

$$\psi_n^k(u) + \psi_n^k(1-u) = 1.$$

(iii) *If $n = 2$ or $3$, and $u \in \{u_n^k, 1-u_n^k\}$ then $N_n^k(u,\theta) = 0$ on $\mathbb{Z} \cup I$ where $I$ is of the form $(-\infty, v_n]$ or $[1-v_n, \infty)$ and $\mathbb{Z} - I$ is the set of isolated simple zeros.*

LEMMA 5.2. *The Kernels $N_n^k(u,\cdot)$ have the following special properties:*
(i) *$N_n^k(0,\cdot)$ is an even function for $n-k$ odd and an odd function for $n-k$ even. It follows that $\theta=0$ is a double zero of $N_n^k(0,\cdot)$ for $n-k$ odd.*
(ii) *$N_n^k(1/2,1/2+\cdot)$ is an even function for $n-k$ odd and an odd function for $n-k$ even. It follows that $\theta=1/2$ is a simple zero of $N_n^k(1/2,\cdot)$ for $n-k$ even.*

The next result states sign properties of $N_n^k$ which are useful to analyse $\psi_n^k$.

THEOREM 5.3. *Let $\tau(\theta)=(-1)^l$ if $\theta\in(l,l+1)$. If $u\in\mathcal{I}_n^k$ and $\theta\notin\mathbb{Z}$,*

$$\operatorname{sign} N_n^k(u,\theta)=\begin{cases} -\tau(\theta)\operatorname{sign} p_{n-k}(1-u,\alpha_n) & \text{if } \theta>\psi_n^k(u),\\ (-1)^{n-k}\tau(\theta)\operatorname{sign} p_{n-k}(u,\alpha_n) & \text{if } \theta<\psi_n^k(u),\end{cases}$$

*and if $u\in\mathcal{J}_n^k$,*

$$\operatorname{sign} N_n^k(u,\theta)=-\tau(\theta)\operatorname{sign} p_{n-k}(1-u,\alpha_n)=(-1)^{n-k}\tau(\theta)\operatorname{sign} p_{n-k}(u,\alpha_n).$$

*Furthermore,*
(i) *if $n-k$ is odd,*

$$\operatorname{sign} p_{n-k}(u,\alpha_n)=\begin{cases} \sigma_n^k=(-1)^{(n-k+1/2)} & \text{for } 1-u_n^k<u\le 1,\\ -\sigma_n^k=(-1)^{(n-k-1/2)} & \text{for } 0\le u<1-u_n^k,\end{cases}$$

(ii) *if $n-k$ is even,*

$$\operatorname{sign} p_{n-k}(u,\alpha_n)=\begin{cases} \sigma_n^k=(-1)^{(n-k/2)} & \text{for } u_n^k<u\le 1,\\ -\sigma_n^k=(-1)^{(n-k/2)-1} & \text{for } 0\le u<u_n^k,\end{cases}$$

*where $\sigma_n^k=p_{n-k}(1,\alpha_n)$.*

To establish monotonicity and differentiability properties of $\psi_n^k$, we analyse its inverse function $\mu_n^k$.

THEOREM 5.4a. *Let $n-k$ be even. There exists a unique increasing continuously differentiable function $\mu_n^k\colon[1/2,\infty)\to[1/2,1-u_n^k]$ such that $\mu_n^k(\psi_n^k(u))=u$ for $u\in[1/2,1-u_n^k)$. Moreover, $\mu_n^k(1/2^+)=1/2$, and*
(i) *if $n=3$, $\mu_3^1$ is strictly increasing on $[1/2,1]$ and $\mu_3^1(\theta)=1-u_3^1$ for $\theta\ge 1$;*
(ii) *if $n\ge 4$, $\mu_n^k$ is strictly increasing on $[1/2,\infty)$ and $\lim_{\theta\to\infty}\mu_n^k(\theta)=1-u_n^k$.*
*We can extend $\mu_n^k$ on $(-\infty,1/2]$ such that $\mu_n^k(1/2-\theta)=1-\mu_n^k(1/2+\theta)$.*

THEOREM 5.4b. *Let $n-k$ be odd. There exists a unique increasing continuously differentiable function $\mu_n^k\colon[0,\infty)\to[0,u_n^k]$ such that $\mu_n^k(\psi_n^k(u))=u$ for $u\in[0,u_n^k)$. Moreover, $\mu_n^k(0^+)=0$, and*
(i) *if $n=2$, $\mu_2^1$ is strictly increasing on $[0,1/2]$ and $\mu_2^1(\theta)=u_2^1$ for $\theta\ge 1/2$;*
(ii) *if $n=3$, $\mu_3^2$ is strictly increasing on $[0,1]$ and $\mu_3^2(\theta)=u_3^2$ for $\theta\ge 1$;*
(iii) *if $n\ge 4$, $\mu_n^k$ is strictly increasing on $[0,\infty)$ and $\lim_{\theta\to\infty}\mu_n^k(\theta)=u_n^k$.*
*We have also a second function $\bar\mu_n^k\colon(-\infty,1]\to[1-u_n^k,1]$ such that $\bar\mu_n^k(\theta)=1-\mu_n^k(1-\theta)$ for $\theta\in(-\infty,1]$.*

THEOREM 5.5. *The functions $\psi_n^k$ are continuously differentiable, strictly increasing on $\mathcal{I}_n^k$, and $D\psi_n^k(u)>0$.*

From preceding results we obtain the following bounds.

THEOREM 5.6. *Let $u \notin \mathcal{I}_n^k$, then*

$$A_n^k(u, v_n) = \left|\mathcal{E}_{n-k+2}(u)\right|/2^{n-k+1}.$$

*It follows that*

$$\max_{u \notin \mathcal{I}_n^k} A_n^k(u, v_n) = A_n^k(v_{n-k+1}, v_n)$$

*since $v_{n-k+1} \notin \mathcal{I}_n^k$.*

THEOREM 5.7. *For $u \in \mathcal{I}_n^k$, we have*

$$A_n^k(u, v_n) \leq A_n^k(v_{n-k+1}, v_n) + A_n^k(v_{n-k}, v_n).$$

THEOREM 5.8. *For $k = 1, \ldots, n-1$*

$$\max\left\{A_n^k(v_{n-k+1}, v_n), A_n^k(v_{n-k}, v_n)\right\} \leq A_n^k(v_n) \leq A_n^k(v_{n-k+1}, v_n) + A_n^k(v_{n-k}, v_n)$$

*where*

$$A_n^k(v_{n-k+1}, v_n) = \frac{|\mathcal{E}_{n-k+2}(v_{n-k+1})|}{2^{n-k+1}},$$

*and for $n-k$ odd*

$$A_n^k(v_{n-k}, v_n) = A_n^k(0, v_n) = \frac{2}{n+1}\left|\sum_{\alpha \in \mathcal{A}_n(v_n)} \frac{p_{n-k}(1, \alpha)}{(1+\alpha)(1-\alpha)^{n-k}}\right|,$$

*for $n-k$ even*

$$A_n^k(v_{n-k}, v_n) = A_n^k(\tfrac{1}{2}, v_n) = \frac{2}{n+1}\left|\sum_{\alpha \in \mathcal{A}_n(v_n)} \frac{\alpha p_{n-k}(1/2, \alpha)}{(1-\alpha)^{n-k}}\left[\frac{1}{1+\alpha} + \frac{p_{n+1}(1-v_n, \alpha) - p_{n+1}(1/2 - v_n, \alpha)}{(1-\alpha)p_{n+1}(v_n, \alpha)}\right]\right|.$$

THEOREM 5.9. *For $1 \leq k \leq n-1$, we have*

(i) *for $n-k$ odd:*

$$\frac{1}{\pi^{n-k+1}} \leq A_n^k(v_n) \leq \left[\frac{1}{2} + \frac{\sqrt{n+1}}{\pi^{1/2}}\right]\frac{1}{\pi^{n-k}} \leq \frac{\sqrt{n+1}}{\pi^{n-k}};$$

(ii) *for $n-k$ even:*

$$\frac{1}{\pi^{n-k+1}} \leq A_n^k(v_n) \leq \left[\frac{2+\pi}{4} + \frac{\pi^{1/2}}{2}\sqrt{n+1}\right]\frac{1}{\pi^{n-k}} \leq 2\frac{\sqrt{n+1}}{\pi^{n-k}}.$$

The next result, which depends on a relation between $u_n^k$ and $u_n^{k+1}$, is appropriate for low values of $k$.

THEOREM 5.10. *Let $1 \leq k \leq n-1$. If either*

(i) *$n-k$ is odd and $u_n^k < u_n^{k+1}$, or*

(ii) *$n-k$ is even and $u_n^{k+1} < u_n^k$,*

*then*

$$A_n^k(v_n) \leq \frac{\mathcal{E}_{n-k+2}(v_{n-k+1})}{2^{n-k}} \leq \frac{1}{\pi^{n-k}}.$$

## 6. The Cases $n = 2$ and $n = 3$

From the results of the preceding sections we obtain the following optimal values for quadratic and cubic splines.

COROLLARY 6.1. *Let $n = 2$ and $v = v_2 = 1/2$. Then*

$$\begin{aligned} A_2^0(\tfrac{1}{2}) &= A_2^0(0, \tfrac{1}{2}) = \tfrac{1}{24}, \\ A_2^1(\tfrac{1}{2}) &= A_2^1(\tfrac{1}{2}, \tfrac{1}{2}) = \tfrac{1}{8}, \\ A_2^2(\tfrac{1}{2}) &= A_2^2(0, \tfrac{1}{2}) = \tfrac{1+2\sqrt{2}}{6}. \end{aligned}$$

COROLLARY 6.2. *Let $n = 3$ and $v = v_3 = 0$. Then*

$$\begin{aligned} A_3^0(0) &= A_3^0(\tfrac{1}{2}, 0) = \tfrac{5}{384}, \\ A_3^1(0) &= A_3^1(0, 0) = \tfrac{1}{24}, \\ A_3^2(0) &= A_3^2(0, 0) = \tfrac{\sqrt{3}}{12}, \\ A_3^3(0) &= A_3^3(0, 0) = \tfrac{1+\sqrt{3}}{4}. \end{aligned}$$

Let us observe that these values have also been obtained in [**12**] for periodic quadratic and cubic splines defined on a uniform partition of a given interval.

## References


1. C. de Boor and I. J. Schoenberg, *Cardinal interpolation and spline functions* VIII. *The Budan-Fourier theorem for splines and applications*, Spline Functions (Karlsruhe, 1975) (K. Böhmer, G. Meinardus, and W. Schempp, eds.), Lecture Notes in Math, vol. 501, Springer, Berlin, 1976, pp. 1–77.
2. P. J. Davis, *Interpolation and approximation*, Dover, New York, 1963.
3. F. Dubeau and J. Savoie, *Généralisation de la série géométrique et applications*, Ann. Sci. Math. Québec **11** (1987), 305–320.
4. ———, *Splines périodiques avec partage uniforme de la droite réelle*, Utilitas Math. **32** (1987), 111–120.
5. ———, *Développements asymptotiques de fonctions splines avec partage uniforme de la droite réelle*, SIAM J. Numer. Anal. **2** (1989), 468–479.
6. ———, *Explicit error bounds for spline interpolation on a uniform partition*, J. Approx. Theory **82** (1995), 1–14.
7. ———, *On interlacing properties of the roots of orthogonal and Euler-Frobenius polynomials*, Approximation Theory VIII, (College Station, TX, 1995) (C. K. Chui and L. L. Schumaker, eds.), vol. 1, World Sci. Publishing, River Edge, NJ, 1995, pp. 185–191.
8. ———, *More on the roots of Euler-Frobenius polynomials*, tech. report CRM-2350, Université de Montréal, 1996.
9. ———, *On optimal error bounds for derivatives of interpolating splines on a uniform partition*, Approximation Theory VIII, (College Station, TX, 1995), vol. 1, World Sci. Publishing, River Edge, NJ, 1995, pp. 485–490.
10. T. N. T. Goodmann and S. S. Lee, *The Budan-Fourier theorem and Hermite-Birkhoff spline interpolation*, Trans. Amer. Math. Soc. **271** (1982), 451–467.
11. C. A. Micchelli, *Cardinal $\mathcal{L}$-splines*, Studies in Spline Functions and Approximation Theory (S. Karlin, C. A. Micchelli, A. Pinkus, and I. J. Schoenberg, eds.), Academic Press, New York, 1976, pp. 203–250.
12. V. L. Miroshnichenko, *Exact error bounds for the periodic cubic and parabolic spline interpolation on a uniform mesh*, Math. Balkanica (N.S.) **2** (1988), 210–221.
13. I. J. Schoenberg, *On Micchelli's theory of cardinal $\mathcal{L}$-splines*, Studies in Spline Functions and Approximation Theory (S. Karlin, C. A. Micchelli, A. Pinkus, and I. J. Schoenberg, eds.), Academic Press, New York, 1976, pp. 251–276.
14. L. L. Schumaker, *Spline functions: Basic theory* (1981), John Wiley & Sons, New York.

15. H. ter Morsche, *On the existence and convergence of periodic spline functions of arbitrary degree*, Spline-Funktionen, (Oberwolfach, 1973) (K. Böhmer, G. Meinardus, and W. Schempp, eds.), Bibliographisches Inst., Mannhein, 1974, pp. 197–214.
16. ———, *On the relations between finite differences and derivatives of cardinal spline function*, Lecture Notes in Math. **501** (1976), Springer, Berlin, 210–219.

Département de mathématiques et d'informatique, Facultés des sciences, Université de Sherbrooke, 2500 boul. Université, Sherbrooke, Québec, Canada, J1K 2R1

*E-mail address*: `francois.dubeau@dmi.usherb.ca.`

Centre de recherches pour la défense, 2459 boul. Pie-XI Nord, Val-Bélair, Québec, Canada, G0A 1R0

*E-mail address*: `jean.savoie@drev.dnd.ca`

Centre de Recherches Mathématiques
CRM Proceedings and Lecture Notes
Volume **18**, 1999

# Modelization of Flexible Objects Using Constrained Optimization and B-Spline Surfaces

Jean-Pierre Dussault and Nicolas Pfister

Abstract. A general framework to modelize stationary flexible objects is presented. The technique uses two concepts. The first is the representation of the object using a polygonal control mesh together with a subdivision scheme, such as rectangular meshes of control points for a Bézier or B-spline bicubic surface with the de Casteljau or Oslo subdivision schemes. The second is the minimization (subject to non-penetration constraints between objects) of a merit function representing the object's energy; a minimizing point then describes the stationary solution for the flexible model.

The method consists of alternately minimizing the merit function, and refining the mesh. Because of the minimization process, the bounding volumes associated with the subdivision scheme must be adapted in order to provide an efficient ray/object intersection procedure.

As examples, the modelization of cloth and tire like objects using B-splines bicubic surfaces is presented.

## Introduction

We consider the modelization of flexible objects in *stationary state.* That is, our method computes the position and deformation of an object in a static scene as the minimum point of some energy function. In contrast with [**16**], we discuss in details the optimization models and algorithms and the representation (control meshes) of the objects; however, our method only considers stationary solutions. In [**16**], the optimization is used to introduce constraints in a *dynamic* differential equations model solved by standard differential equations solvers.

The merit function represents the object's energy. Stationary states of the object correspond to constrained local minima of the energy function. The minimization process is constrained in order to avoid objects inter penetration. Since it is usually very difficult to identify the exact energy function which corresponds with an object, we propose a general scheme based on the penalization of *structural constraints.* The structural constraint specify some properties of the object at rest, such as its length/area/volume, shape, etc. Those constraints usually involve

1991 *Mathematics Subject Classification.* Primary: 65C20; Secondary: 65D05.

This research was partially supported by NSERC grant OGP0005491.

This is the final form of the paper.

distances between mesh points, or angles between mesh edges. The structural constraints may be physically based (springs, elastics), or, as in the cloth example in Section 4 empirical, reflecting macroscopic properties of the object. In this paper we limit ourselves to tensor product B-spline surfaces and length constraints.

Objects are represented by some control polygonal meshes which are progressively refined using some subdivision algorithm. After each refinement, the objective function is further minimized. The limiting object is thus the product of a subdivision algorithm in which perturbations (by the optimization process) are interleaved with the subdivisions.

The required precision at a given subdivision level may be varied in order to produce deformations of different scales. If the coarse grid is optimized very accurately, large deformations will appear, progressively fine-tuned as the subdivision progresses. On the other hand, if the coarse grid is optimized loosely, much progress in the optimization process is left for finer grids, producing comparatively smaller scaled deformations. Unless the object is of fractal nature, there will be a subdivision level from which the optimization process will cease to progress significantly. When this situation is detected, we may claim that the optimal point is found.

We present in Section 1 the mathematical optimization model that we consider. We discuss in general terms its applicability. In Section 2, we briefly discuss the solution algorithm.

In order to allow efficient ray tracing of the objects, we must adapt the bounding volume to take into account the optimization process. The solution we present in Section 3 consists in computing a scale factor which enlarge the usual bounding volume (convex hull) in order to genuinely contain the perturbed refined meshes.

As an example in Section 4, we present two models based on B-splines: one for cloth objects and the other for tire like objects. Both are represented by rectangular meshes of control points for a bicubic B-spline surface. The convex hull property allows to use $4 \times 4$ sub-meshes as bounding volumes for the (curved) surface patches. However, because of the perturbation induced by the optimization step, the bounding volumes need to be perturbed as well.

## 1. The Optimization Model

We consider an optimization model of the form

$$\min_p f(p) \tag{1}$$

$$\text{subject to } c(p) \leq 0. \tag{2}$$

The variable $p$ of the model consists of the points $p_i$ of the grid $G$.

### 1.1. The objective function.

The merit function is usually composed of a gravity term, the sum of the height of the points, $p_{i3}$ plus the penalization of some structural constraints $s_j(p) = 0, j \in J$ :

$$f(p) = \sum_{i \in G} p_{i3} + \sum_{j \in J} \frac{1}{2r_j} s_j(p)^2.$$

By varying the weight $r_j$ of the penalty, a given structural constraint is more or less *encouraged* to be satisfied, but not enforced. The remaining constraints are collision constraints $c$.

**1.2. Formulation of structural constraints.** Structural constraints involve relation between points of the grid, such as distances between neighbors, angles between adjacent edges or facets, etc. Since we consider flexible objects, those constraints are better described by their penalization. For instance, if the distance between $p_i$ and $p_j$ must be say $\alpha$, then the constraint takes the form $\|p_i - p_j\| - \alpha = 0$. However, for a flexible model, this length is subject to (small) variations, depending on other factors: the flexible object may be squeezed, reducing the length, etc. Therefore, it is natural to represent the constraint by incorporating it into the objective function via a penalty term:

$$\frac{1}{2r}\left(\|p_i - p_j\| - \alpha\right)^2 \quad \text{with } r > 0.$$

The factor $1/2r$ is the weight. If $r$ is large, the constraint's residual will probably be large while if $r$ is small, the residual will be small. Actually, under mild assumptions, one shows that the residual is a differentiable function of $r$, so that it decreases to zero at the same speed as $r$.

**1.3. Collision constraints.** Collision constraints forbid object inter penetration. If a set of implicit equations is known for a given object, as $\{p : F(p) \leq 0\}$, then the function $F$ defines the collision constraint for other objects. This situation allows to treat quadric surfaces, etc. For example, a unit sphere is represented by the function $F(p) = \|p\| - 1 \leq 0$, and so the constraint forbidding penetration into the sphere is $\|p\| \geq 1$, or $c(p) = 1 - \|p\| \leq 0$.

For objects defined themselves using control polygon meshes and polynomial or subdivision schemes, one has to resort to the hierarchical bounding volumes to produce suitable equations for the constraints. Then, however, to improve the efficiency of the method, such constraints are generated only when needed, i.e. when inter penetration is observed.

## 2. The Optimization Algorithm

Since we already consider penalization terms in the model of the structural constraints, it is natural to use a penalty algorithm to solve the program (1)–(2). However, since the collision constraints must be satisfied, here, their weight factor, denoted $\rho$ to distinguish them from the "soft" structural constraints, must be driven to zero. In [**16**], the authors explain that $\rho$ need to be fixed at a suitably small value, and they further use an augmented Lagrangian term in hope that this value is not too small. However, as they discuss, the smaller this value is chosen (their $c$ gets larger), the stiffer the resulting equations become, making the numerical solution difficult. Moreover, even within the augmented Lagrangian scheme, it is impossible to know in advance what *small enough* value will ensure convergence, so that there is some need for trial and error. In contrast, our algorithm possesses the global convergence property, that is its accumulation points are stationary, regardless the starting point.

**2.1. Sequences of unconstrained minimization subproblems.** Since we are interested in stationary behavior, we use a decreasing sequence of values for $\rho$. This allows to prove that the algorithm to be presented is efficient. Moreover, since the non-penetration constraints often involve inequalities, we use the so-called exponential penalty function $\rho e^{c_i(p)/\rho}$ [**5**].

As discussed in the introduction, optimization is performed after each refinement of the grid. Let us label $G^k$ the grid after $k$ subdivision steps. The (unconstrained) minimization problem for this grid is written as

$$\min_{p\in G^k} f(p) = \sum_{i\in G^k} p_{i3} + \sum_{j\in J} \frac{1}{2r_j^k} s_j(p)^2 + \rho^k \sum_l e^{c_l(p)/\rho^k}.$$

The first order optimality conditions for this last problem is written as $\nabla f(p) = 0$, where

$$\nabla f(p) = e_{i3} + \sum_{j\in J} \frac{s_j(p)}{r_j^k} \nabla s_j(p) + \sum_l e^{c_l(p)/\rho^k} \nabla c_l(p).$$

However, there is no need to solve those equations exactly, and convergence may be proven if the residual converges to zero [**6**]. Therefore, we require only that $\|\nabla f(p)\| \le \varepsilon^k$.

By denoting $p^k$ the approximate minimum of the $k^{\text{th}}$ unconstrained subproblem, one may interpret the algorithm as minimizing a functional in some function space (Hilbert) by an iterative method in which the iterates are approximated by some finite dimensional approximation $p^k$.

**2.2. Sequences of weight factors.** The resulting algorithm involves three sequences: $\rho^k \searrow 0$, $\varepsilon^k \searrow 0$ and $r^k \ge 0$.

The sequence $r^k$ allows to weight the structural constraints, and is part of the model. The sequence $\rho^k$ asymptotically enforces the collision constraints. The sequence $\varepsilon^k$ ensures the global convergence of the whole algorithm while allowing savings in computational costs for each unconstrained subproblem. Those sequences are part of the algorithm, and both must converge to zero. However, the speed at which they converge to zero may influence the appearance of the object, and thus are more or less part of the model.

**2.3. Solution of the unconstrained subproblems.** Each subproblem may be solved by any unconstrained minimization algorithm. We use a modified Newton's method. The linear systems within Newton's method are solved using a conjugate gradient iterative method.

Newton's method requires second derivatives (Hessian matrices) of the objective function. Fortunately, the conjugate gradient only requires that the product of the Hessian with some vector is computed. This allows to use a finite difference scheme to evaluate the second order information.

**2.4. The complete algorithm.** The complete algorithm is sketched below. It is expressed using assignment statements and computer variables, so that $p$ contains $p^k$, etc.

A few more remarks are in order at this point.

The algorithmic scheme is quite robust, and it can be shown that whatever the initial point $p^0$, if the gradients of the constraints are linearly independent at an accumulation point, then the accumulation point is stationary. One well known flaw of penalty algorithms is the fact that the subproblems become ill-conditioned as $\rho$ approaches zero, as remarked above. If very precise solution were needed, this difficulty would become a serious concern. However, in our application, the required precision is quite loose, so that the ill-conditioning is never severe and even if it was, recent work have proposed ways of removing it [**9, 2, 3, 5, 6**].

Algorithm 1: Penalty algorithm.

```
Penalty
  { Given: p^0, sequences {r^i}, {ρ^i}, {ε^i} }
  r ← r^0
  ρ ← ρ^0
  ε ← ε^0
  p ← p^0
  while (ρ > Precision )
    { Newton corrections }
    while (‖∇f(p)‖ > ε )
      d_N ← Newton_Direction(p)
      θ ← Armijo_Stepsize(p, d_N)
      p ← p + θ d_N
    { p is the k^th approximation }
    r ←  next(r)
    ρ ←  next(ρ)
    ε ←  next(ε)
    p ←  subdivide(p)
```

Similarly, the asymptotic convergence order of penalty methods may be improved by using an extrapolation strategy. Indeed, the points $p^k$ define, under mild assumptions, a differentiable trajectory reaching the optimal solution. However, for our application, we stop the algorithm well before being in the vicinity of the solution, $10^{-3}$ being a small enough precision to witness a realistic image.

Because of all those observations, we were very careful in our implementation to improve the efficiency in the early iterations.

## 3. Bounding Volumes

In order to ray trace objects defined with a polygonal mesh refined using some subdivision scheme, hierarchical bounding volumes are used. For instance, B-splines or Bézier tensor product surfaces are enclosed in the convex hull of the control mesh. However, since the optimization process modifies the refined grids, the bounded volumes need to be modified accordingly. We use a technique developed in [**7**] to compute scale factors of the bounding volumes while the optimization/subdivision algorithm is executed. When ray-tracing, we use the scaled volumes instead of the usual volumes.

## 4. Two Examples: A Cloth and a Tire

Modelization of cloth and other flexible object has attracted much attention recently. Previous models include static and dynamic ones that are either based on physical properties of cloth, geometric aspects or some combination of both. For a complete and comprehensive survey of cloth modeling techniques one should referred to the paper of Ng and Grimsdale [**13**]. We will give here only a short summary of the works that are most closely related to ours. Weil [**22**] used catenaries to simulate hanging threads obtaining an initial position of the grid before making the points to satisfy some length constraints. The idea of assigning an energy function to the cloth and finding its minimum using a multi-grid scheme was introduced by Feynman [**8**] and further developed by Pfister [**15**] and by Ng *et al.* [**14**]. In Pfister however the optimization is done on the whole grid at once avoiding the bias introduced when searching a minimum point by point. In one of the most versatile approach, Terzopoulos *et al.* [**19**] modeled the differential equations of motion of the object obtaining as a solution its motion trough time. This idea led to the simulation of inelastic deformation in Terzopoulos and Fleischer [**18**], the use of dynamic constraint methods to define the differential equations as in Platt and Barr [**16**]. The Thalmann team [**4, 10, 20**] and [**12**] in what may be the most advanced software on the subject, also used the equations of motion to develop a complex and complete model simulating the behavior of clothing according to the movement of their owner. In a more empirical manner Provot [**17**] approximated cloth by a network of springs dampened by a correction factor. Rejecting the continuum approach, Breen *et al.* [**1**] and Eberhardt *et al.* [**21**] successfully investigated the possibility of using interacting particles to model cloth objects.

Most physically based models have to be modified to avoid artifacts probably difficult to take into account given the complexity of cloth objects [**11**]. Therefore, no genuine theoretical model has given convincing results.

Our approach for flexible object is easily applied to cloth or tire like objects. After having chosen a bicubic B-spline representation, all we need is to specify some structural constraints, their weight within the merit function and a starting grid of control points representing the original position of the object. We then let the optimization/subdivision algorithm converge until a satisfying image is obtained. Both cloth and tire are represented as rectangular grids of points. However what turns the rectangular grid into a tire like object is the cyclic nature of the structural constraints. For the tire the points on opposite sides of the cloth are related with each other by the constraints. The grid then becomes a "closed" grid, which is not the case with the cloth where the grid remains "open". Another difference between those two objects is the number of control points usually required to obtain a realistic image. Cloth must be optimized on a much finer grid because it needs to have more creases and wrinkles to look realistic.

Since for now our model is only using bicubic B-splines representation for all our examples, the variety of flexible shapes we can actually work with is somewhat limited to the topology of a deformed rectangle (cylinder, torus, etc.). For example, it would be difficult to simulate a flexible sphere using this representation. However by using triangular meshes within the butterfly subdivision scheme, we could extend our model to include a large variety of differently shaped objects

**4.1. Structural constraints.** We use an empirical observation of flexible objects to develop distinct macroscopic constraints.The final appearance of the object will depend on the obstacles the scene contains (floor, wall, fixed points, etc.) and the structural properties of the object itself. For the two flexible objects represented below we simulate their structural properties by some combination of the three following constraints

- Length constraint;
- Shear constraint;
- Curvature constraint.

All these constraints in our examples are in fact distance constraints between different control points of the grid.

4.1.1. *Length constraint.* A mathematically exact approach to establish length constraints is based on evaluating the length of every B-spline pieces. However as each of these pieces depends on four control points of the grid, the handling of the function and its derivative prove to be complicated and expensive in terms of calculations. Moreover, the number of variables in the cloth model is artificially increased because of the need to triple the border points of the grid in order for the B-spline surface to interpolate these points. To avoid these kind of difficulties, we propose to work directly on the distance between the control points on the grid. The visual aspect of the result is not affected by this modification and in fact because the grid will converge toward the B-spline surface as the subdivision progresses, the distance between the control points will also converge toward the length of the B-spline piece.

In both cases: cloth and tire, the grid is rectangular as we have said and the length constraint simulating the elasticity of the fibers is obtain after specifying the distances between each points along both directions: the warp and the weft directions (the names are a reference to the fibers of cloth). We then sum the square of the differences between the actual and those desired distances: $\alpha_{\text{warp}_{ij}}$ and $\alpha_{\text{weft}_{ij}}$. The expression of the constraint is of the form

$$\sum_i \sum_j \left[ \frac{1}{2r_{\text{warp}}} \left( \|p_{i+1j} - p_{ij}\| - \alpha_{\text{warp}_{ij}} \right)^2 \frac{1}{2r_{\text{weft}}} \left( \|p_{ij+1} - p_{ij}\| - \alpha_{\text{weft}_{ij}} \right)^2 \right].$$

where $1/2r_{\text{warp}}$ and $1/2r_{\text{weft}}$ are the weights, allowing a difference between the elasticity of the fibers in the two directions.

4.1.2. *Shear constraint.* Shearing represent a deformation caused either by a force applied in a direction not perpendicular to the side of the object being deformed or by forces that cause two contiguous parts of the object to slide relatively to each other.

In a cloth, shearing occur mainly when the direction of elongation is not parallel to the threads axis. It represents the capacity of the fibers to move relative to one another. To simulate this behavior, we restrain the diagonal distance between the control points of the grid to be of some specified length. Similarly to the length constraint, the shearing constraint is obtain by summing the squared difference between the actual and specified distance $\alpha_{\text{diag}_{ij}}$. In the case of the cloth this distance is simply given by $\sqrt{\alpha^2_{\text{warp}_{ij}} + \alpha^2_{\text{weft}_{ij}}}$, but for the tire it also depends on the relative position of the control points because unlike cloth, a tire at rest is not (always) flat. The constraint is then

$$\sum_i \sum_j \left[ \frac{1}{2r_{\text{diag}}} \left( \|p_{i+1j+1} - p_{ij}\| - \alpha_{\text{diag}_{ij}} \right)^2 \right].$$

where $1/2r_{\text{diag}}$ represent the weight of the shearing constraint. For example, a loosely knitted cloth, like a wool sweater, $1/2r_{\text{diag}}$ should be small, enabling more movement between the fibers. We can observe that in the case of the cloth it is possible to save some computations by restraining only one diagonal for each square of four points since elongation of a diagonal necessarily shortens the other.

For a structurally three dimensional object, such as tire, the two diagonals must be constrained to avoid spiraling effects in the resulting image.

Figure 1: Hanging.

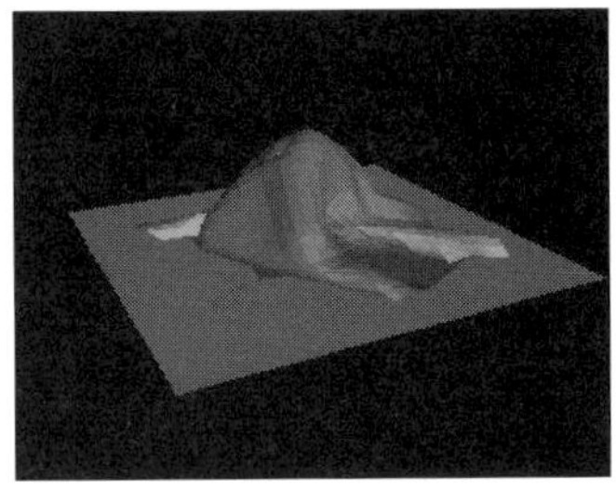

Figure 2: Dropped.

4.1.3. *Curvature constraint.* In our first attempt the curvature constraint was obtained by computing the cosine of the angle made by every triplet of consecutive points on the grid. But since there is an important discrepancy between minimizing a difference in distances (length and shear constraints) and minimizing an angle, the optimization algorithm was not performing as expected. The constraint is now obtained by using the distance between every second neighboring point of the grid. It is thus quite similar to the other constraints for length and shear. The curvature constraint is of the form

$$\sum_i \sum_j \frac{1}{2r_{\text{curv}}} \left[ \left( \|p_{i+2j} - p_{ij}\| - 2\alpha_{\text{warp}} \right)^2 + \left( \|p_{ij+2} - p_{ij}\| - 2\alpha_{\text{weft}} \right)^2 \right].$$

where $1/2r_{\text{curv}}$ is the weight of the constraint.

4.1.4. *Self collision detection.* Self collision happens when one part of the object penetrate another part of that same object. Actually this is not a truly structural constraint but more like a dynamic one. It has some effect only when a collision takes place. The more flexible an object is, the more chance there is the situation will occur. Although this constraint has not been implemented yet, the problem of self collision has occurred only in few cases. The combination of curvature constraint and stopping the optimization procedure soon enough seems to prevent its appearance. However cloth which has not much resistance to curvature, should be quite sensitive to that problem. For example, to simulate a piece of clothing crumpled on the floor, there must be some protection against self collision. In the two first figures, we can see a square shaped piece of cloth hanging by its middle point (Figures 1) which is then dropped on the floor (Figure 2). If we look closely at Figure 2, self-penetration of the cloth is apparent on both the right and the left side. The cloth has turned around to go through itself. In this example, the grid is initially horizontal and contains $11 \times 11$ points. The subdivision process was done only once to obtain a final $21 \times 21$ representation.

Since the curvature constraint is taking care of the distance between every second point on the grid, it is sufficient to check the distance between every fifth point before each iteration and to consider in the merit function only those points whose distance between them is under a certain threshold $\gamma$. We then add the constraint

$$\sum_i \sum_j \left[ e^{\gamma - \|p_{ij} - p_{i+5j}\|/\rho} + e^{\gamma - \|p_{ij} - p_{ij+5}\|/\rho} \right].$$

Figure 3: Cotton.

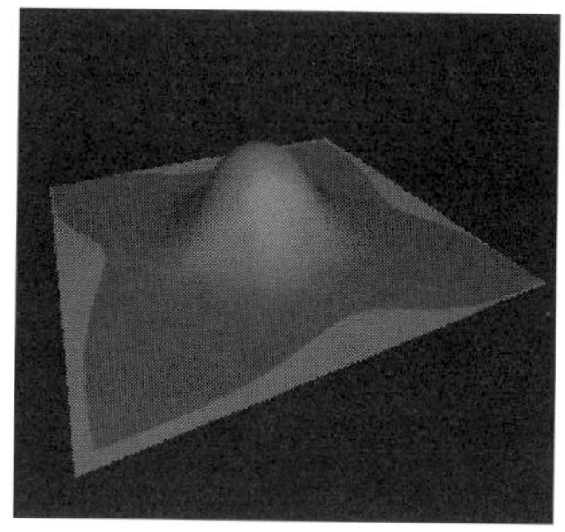

Figure 4: Lycra.

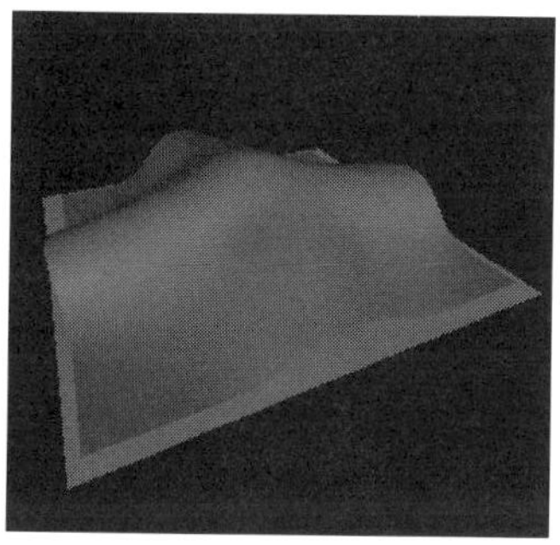

Figure 5: Rubber.

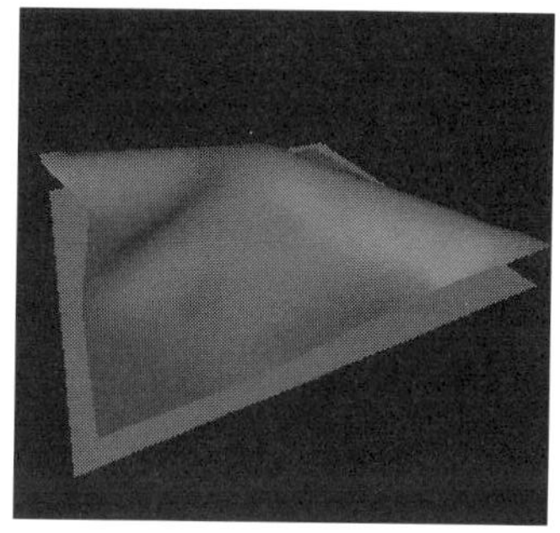

Figure 6: Paper.

Table 1: Weights of the constraints.

| Fig. | $\frac{1}{2r_{\text{warp}}}$ | $\frac{1}{2r_{\text{weft}}}$ | $\frac{1}{2r_{\text{diag}}}$ | $\frac{1}{2r_{\text{curv}}}$ |
|---|---|---|---|---|
| 3 | 1 | 1 | 1 | 0 |
| 4 | 1 | 1 | 0,01 | 0,01 |
| 5 | 2 | 2 | 2 | 1 |
| 6 | 2 | 2 | 20 | 0,2 |

We can observe that most of the displacement of the points happens during the first stages of the subdivision process. This is when self collision is most likely to occur. But since the grid does not contain that many points on such a coarse grid, the calculations should remain reasonable.

**4.2. Effect of varying the weights of penalized structural constraints.** Unlike collision and self collision constraints, the weights of the structural constraints remain constant during the whole optimization procedure. It is possible however that generating a particular sequence of weights for one or more constraints would accelerate the convergence of the algorithm.

By setting the weights $1/2r_{\text{warp}}$, $1/2r_{\text{weft}}$, $1/2r_{\text{diag}}$ and $1/2r_{\text{curv}}$ to different values relatively to each others, we are able to simulate more or less elastic, more or less rigid and more or less tightly knitted pieces of cloth. To illustrate that, we

Figure 7: Inflated tubes.

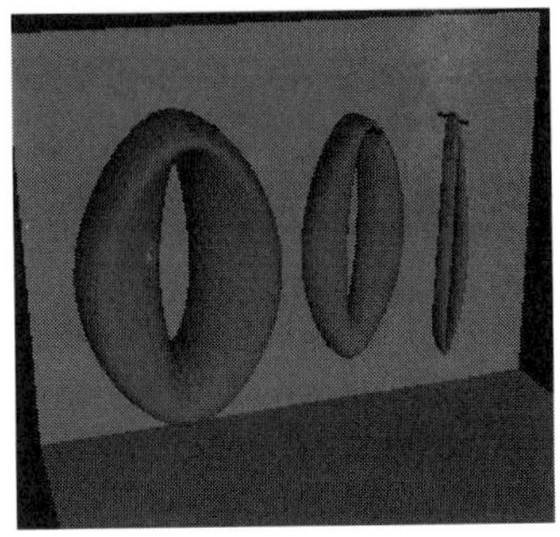

Figure 8: Hanging tubes.

have dropped a square shaped piece of cloth on a sphere resting on the floor using different weights for the constraints. We can see that is possible to simulate a wide variety of material ranging from paper to rubber (Figures 3, 4, 5 and 6). In all these examples the dimension of the starting grid is $8 \times 8$ which is then subdivided twice to give a $29 \times 29$ final representation. The initial grid is always horizontal and placed just above the top of the sphere.The values for the weights are summarized in the Table 1 below.

The effect of varying the weights can also be seen in Figure 7 where we have simulated three differently inflated tubes. When the weights for the length and the shear constraints are given bigger values, the tube is more rigid and appear to be more inflated.

Our second example shows that similar objects of different sizes will have different behavior under the same optimization procedure and using the same weights for the constraints. Figure 8 shows three tubes hanging on a nail by a wall. As expected, the bicycle's tube is much more deformed than the one for the truck. In those two last examples the initial grid was a $4 \times 4$ square shaped torus with the top side placed horizontally. The torus was then dropped, being held only by one corner.

**4.3. Effect of varying the sequence of accuracy factors.** By varying the precision required at each subdivision level we will obtain, as said earlier, deformation of different scales. To illustrate this, we have simulated a hanging drape using three different optimization schemes. The three simulations are obtained using an horizontal $8 \times 8$ grid which is subdivided a first time to obtain a $15 \times 15$ grid and a second time resulting in a final $29 \times 29$ grid. In the first scheme (Figure 9), optimization was performed using a precision of $10^{-5}$ on the coarser grid, $10^{-3}$ for the next one and $10^{-2}$ for the finer one.

For the second scheme (Figure 10), optimization was performed on the two finer grids only, using a precision of $10^{-3}$ for the first and $10^{-2}$ for the second.

In the last scheme (Figure 11), only the final grid is optimized to a precision of $10^{-2}$.

We can see that as the optimization procedure is delayed to be started on finer grid, the size of the deformations tend to decrease. In the last case (Figure 11), there are more wrinkles and they are relatively smaller than in the first or second cases. This results from the fact that initial solutions for the two first cases (Figures 9

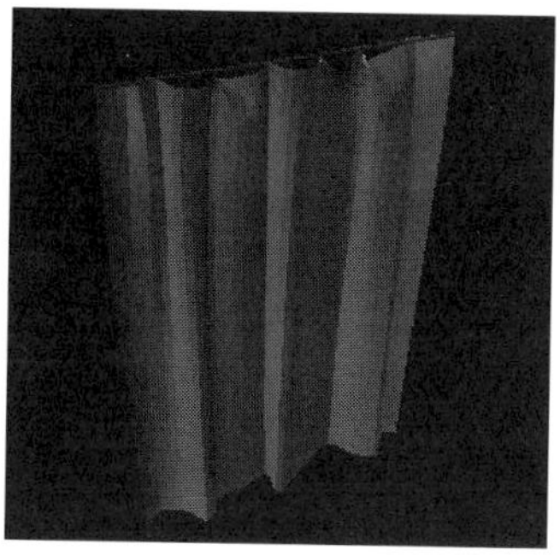

Figure 9: Three steps optimization.

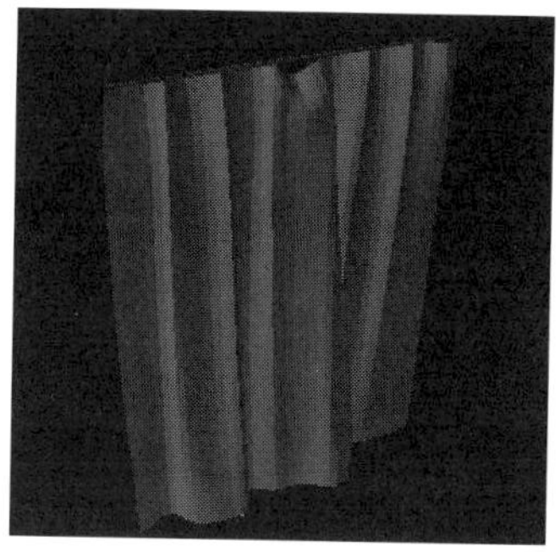

Figure 10: Two steps optimization.

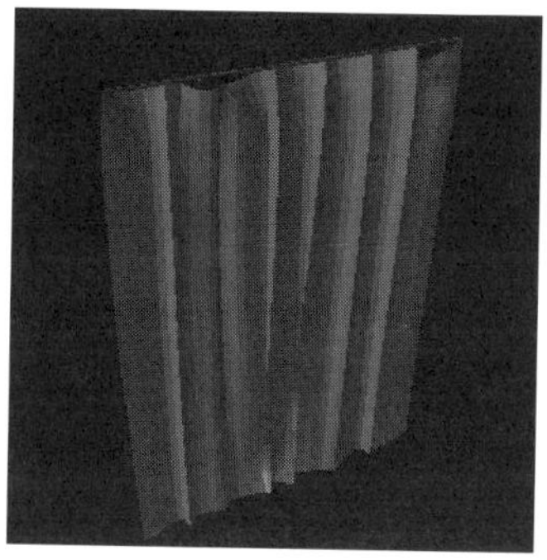

Figure 11: One step optimization.

Table 2: Computation time.

| Figure | Time |
|---|---|
| 3 | 48 sec |
| 4 | 60 sec |
| 5 | 2 mn 27 sec |
| 6 | 2 mn 2 sec |
| 8 | 1 mn 45 sec |
| 9 | 1 mn 14 sec |
| 10 | 2 mn 59 sec |
| 11 | 11 mn 02 sec |
| 17 | 35 sec |

and 10) were obtained on a coarse grid, creating large deformations in the drape that were not much displaced in the later optimization steps. It is important to note that the computation time needed to reach the final solution in the third scheme was much longer than in the first one. The optimization procedure has to work harder when started on a fine grid far away from the solution. The advantage of using a subdivision process can be seen here: we are able to optimize a coarse grid first to get a rough approximation of the solution at a relatively low cost. A

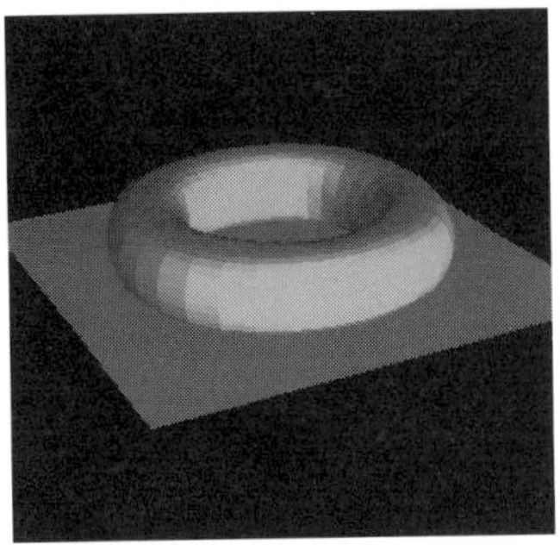

Figure 12: Iteration no. 6.

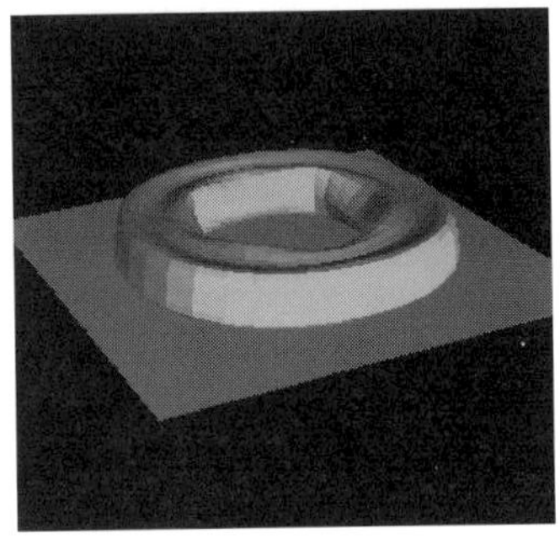

Figure 13: Iteration no. 12.

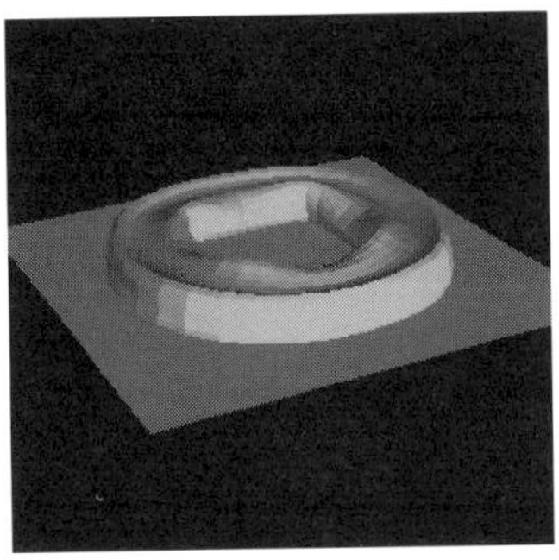

Figure 14: Iteration no. 18.

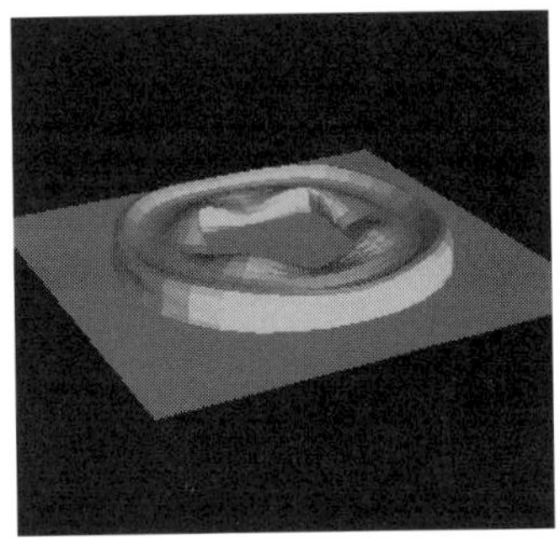

Figure 15: Iteration no. 24.

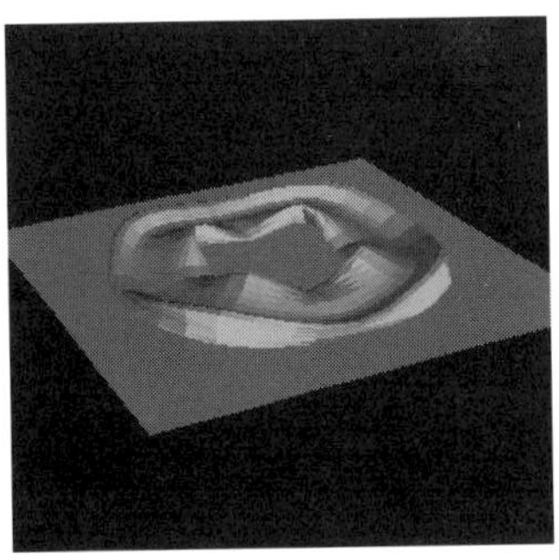

Figure 16: Iteration no. 30.

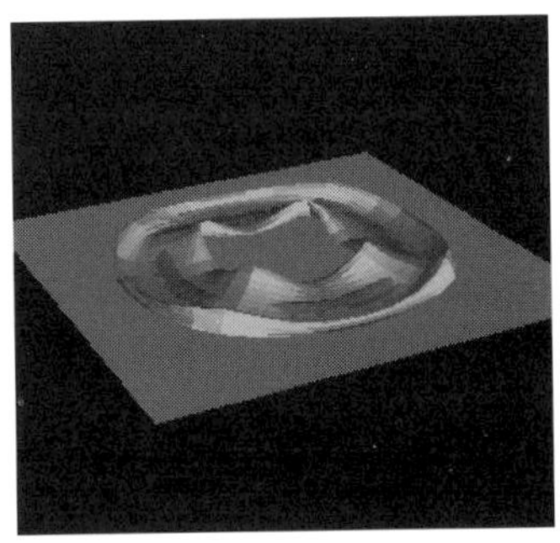

Figure 17: Iteration no. 36.

summary of the time needed to calculate the different images is given in Table 2. They should only be taken as a general indication of how long it takes to obtain the images: the code being far away from being optimized. All the computations where done on a Sun SPARCserver 20/712.

## 5. Concluding Remarks

The empirical method we propose to model flexible objects was originally intended to obtain stationary representations. But it could also used to generate sequences of animation. As the algorithm converges toward a solution, it produces an intermediate image at each iteration representing the transformations the flexible object goes through between its starting point and its final position. Figures 12–17 represent every sixth image generated simulating a deflating tire. We confess that without cheating, it is a bit more complex to have the tire inflated! it would be necessary to have a constraint on the volume of the object.

Our next step is to be able to model a larger variety of flexible objects using for instance triangular meshes within the butterfly subdivision scheme. We also want to implement the constraint for self-collision detection and a procedure for collision detection between flexible objects.

## References


1. David E. Breen, Donald H. House, and Michael J. Wozny, *Predicting the drape of woven cloth using interacting particles*, SIGGRAPH'94, Comput. Graphics **28** (1994), no. 3, 365–371.
2. C. G. Broyden and N. F. Attia, *A smooth sequential penalty function method for solving nonlinear programming problems*, System Modelling and Optimization (P. Thoft-Christenses, ed.) (Copenhagen, 1983), Lecture Notes in Control and Inform. Sci. **59**, Springer, Berlin-New York, 1984, pp. 237–245.
3. ______, *Penalty functions, Newton's method, and quadratic programming*, J. Optim. Theory Appl. **58** (1988), no. 3, 377–385.
4. Michel Carignan, Ying Yang, Nadia Magnenat Thalmann, and Daniel Thalmann, *Dressing animated synthetic actors with complex deformable clothes*, SIGGRAPH'92, Comput. Graphics **26** (1992), no. 2, 99–104.
5. Roberto Cominetti and Jean-Pierre Dussault, *Stable exponential-penalty algorithm with superlinear convergence*, J. Optim. Theory Appl. **83** (1994), no. 2, 285–309.
6. Jean-Pierre Dussault, *Numerical stability and efficiency of penalty algorithms*, SIAM J. Numer. Anal. **32** (1995), no. 1, 296–317.
7. Richard Egli, *Interpolation de surface par la technique de subdivision butterfly*, Master's thesis, Univ. Sherbrooke, 1994.
8. Carl Richard Feynman, *Modeling the appearance of cloth*, Master's thesis, MIT, 1986.
9. Nicolas Ian Mark Gould, *On the accurate determination of search directions for simple differentiable penalty functions*, IMA J. Numer. Anal. **6** (1986), 357–372.
10. Jean-Paul Gourret, Nadia Magnenat Thalmann, and Daniel Thalmann, *Simulation of object and human skin deformations in a grasping task*, SIGGRAPH'89, Comput. Graphics **23** (1989), no. 3, 21–30.
11. J. Hearle, P. Grosberg, and S. Bacher, *Structural mechanics of fibers, yarns, and fabrics*, Wiley-Interscience, vol. 1, New-York, 1969.
12. Shen Jianhua, Nadia Magnenat Thalmann, Pascal Volino, and Daniel Thalmann, *An evolving system for simulating clothes on virtual actors*, IEEE Comput. Graphics Appl. **16** (1996), no. 5, 42–51.
13. Hing N. Ng and Richard L. Grimsdale, *Computer graphics techniques for modeling cloth*, IEEE Comput. Graphics Appl. **16** (1996), no. 5, 28–41.
14. Hing N. Ng, Richard L. Grimsdale, and W. G. Allen, *A system for modelling and visualization of cloth material*, Comput. Graphics **19** (1995), no. 3, 423–430.
15. Nicolas Pfister, *Représentation des textiles par ordinateur: un modèle basé sur les splines*, Master's thesis, Univ. Sherbrooke, 1992.
16. John C. Platt and Alan H. Barr, *Constraint methods for flexible models*, SIGGRAPH'88, Comput. Graphics **22** (1988), no. 4, 279–288.
17. Xavier Provot, *Deformation constraints in a mass-spring model to describe rigid cloth behavior*, Graphics Interface'95, Quebec, Canada, 1995, pages 147–154.
18. Demetri Terzopoulos and Kurt Fleischer, *Modeling inelastic deformation, viscoelasticity, plasticity, fracture* SIGGRAPH'88, Comput. Graphics **22** (1988), no. 4, 269–278.
19. Demetri Terzopoulos, John Platt, Alan Barr, and Kurt Fleischer, *Elastically deformable models*, SIGGRAPH'87, Comput. Graphics **21** (1987), no. 4, 205–214.
20. Pascal Volino, Martin Courschesne, and Nadia Magnenat Thalmann, *Versatile and efficient techniques for simulating cloth and other deformable objects*, SIGGRAPH'95, Comput. Graphics (1995), 137–144.

21. Bernhard Eberhardt Andreas Weber and Wolfgang Strasser, *A fast, flexible, particle-system model for cloth draping*, IEEE Comput. Graphics Appl. **16** (1996), no. 5, 52–59.
22. Jerry Weil, *The synthesis of cloth objects*, SIGGRAPH'86, Comput. Graphics **20** (1986), no. 4, 49–54.

Département de mathématiques et informatique, Université de Sherbrooke, Sherbrooke, Québec, Canada J1K 2R1

*E-mail address*: jean-pierre.dussault@dmi.usherb.ca

Département de mathématiques et informatique, Université de Sherbrooke, Sherbrooke, Québec, Canada J1K 2R1

Centre de Recherches Mathématiques
CRM Proceedings and Lecture Notes
Volume **18**, 1999

# New Control Polygons for Polynomial Curves

J. C. Fiorot and P. Jeannin

ABSTRACT. The notion of mixed progressive sequence consisting of points and vectors is introduced. The elements of such a sequence are the natural arguments of a mixed polar form of a polynomial. Mixed progressive sequences allow us to define mixed control polygons of a polynomial curve or segment and a mixed blossoming algorithm. The latter leads to reversible schemes in Computer Aided Geometric Design. An isomorphism between polynomials and mixed polygons completes this study.

## 1. Introduction

Blossoming is an attractive and efficient new technique recently introduced in Computer Aided Geometric Design (CAGD) for analysing and unifying properties and algorithms related to polynomial curves and surfaces. Introduced by De Casteljau [**7, 8, 9**] and Ramshaw [**15, 16**], it is based on the well-known concept of polar form of a polynomial. The polar form or blossom is an $n$-affine symmetric function whose restriction to the diagonal gives the value of the polynomial.

Underlying the polar form, the mixed polar form (Seidel[**18**], Gormaz [**11**]) is a symmetric, vector-valued function, linear w.r.t. vector arguments, affine w.r.t. point arguments. It gives an elegant formula for calculating the derivatives of a polynomial. In a recent paper, Fiorot and Jeannin[**10**] using the mixed polar form define new control polygons consisting of points and vectors, blossoming algorithms and schemes which obtain a mixed polygon from another one, extending Ramshaw's algorithm (Ramshaw [**15, 16**], Barry and Goldman [**1**]) which derives a control polygon from another one.

In this paper we use the notion of mixed progressive sequence, in order to further exploit the capability of the mixed polar form. We know that a polynomial curve is determined by a control polygon of points (Bézier [**2**]—De Casteljau [**7**] representation) or traditionally by a Taylor polygon (the value and the successive derivatives at the origin). We show that between these two kinds of control polygons we may define mixed control polygons made up of points (at least one) and vectors by using the notion of mixed polar form instead of that of homogeneous

1991 *Mathematics Subject Classification.* 65D17, 41A10.
This is the final form of the paper.

polar form (Barry and Goldman [**1**]). A Mixed blossoming algorithm leads to reversible schemes giving different kinds of control polygons and allows to establish an isomorphism between polynomials and mixed polygons.

The paper is organized as follows. Section 2 gives a brief introduction to mixed polar forms. Section 3 introduces the notion of mixed progressive sequence of $\mathbb{R}$. This leads to define mixed control polygons in Section 4 and a mixed blossoming algorithm in Section 5. Section 6 is devoted to the isomorphism between polynomials and mixed polygons. The determination of a mixed polygon from another by changing mixed progressive sequences is obtained by reversible schemes in Section 7.

## 2. Mixed Polar Form

Let $\mathcal{P}$ and $\mathcal{E}$ be two real affine spaces of finite dimension, $\overrightarrow{\mathcal{P}}$ and $\overrightarrow{\mathcal{E}}$ designate their associated vector space.

### 2.1. Fundamental theorem.

THEOREM 2.1. *Let $F$ be a polynomial of degree $\leq n$, $F\colon \mathcal{P} \to \mathcal{E}$, then there exists a unique symmetric n-affine map $f\colon \mathcal{P}^n \to \mathcal{E}$ satisfying: $\forall x \in \mathcal{P}$, $F(x) = f(x, x, \ldots, x)$. $f$ is called the polar form of $F$ or the blossom of $F$.*

PROOF. De Casteljau [**7, 8, 9**], Ramshaw [**15, 16**], Lang[**13**]. $\square$

### 2.2. Definition of a mixed polar form.

*Mixed polar forms of order one associated with $f$.* Let $x_1, x_2, \ldots, x_n$, be a set of points belonging to $\mathcal{P}$ and let $\vec{h}$ be a vector of $\overrightarrow{\mathcal{P}}$. The difference $f(x_1 + \vec{h}, x_2, \ldots, x_n) - f(x_1, x_2, \ldots, x_n)$ is a vector of $\overrightarrow{\mathcal{E}}$ which does not depend on $x_1$ since $f$ is an affine map with respect to the first argument. We denote by:

$$f(\vec{h}, x_2, \ldots, x_n) = f(x_1 + \vec{h}, x_2, \ldots, x_n) - f(x_1, x_2, \ldots, x_n).$$

There is no ambiguity since the nature of the argument, point or vector, is clearly mentioned. The map: $\vec{h} \in \overrightarrow{\mathcal{P}} \mapsto f(\vec{h}, x_2, \ldots, x_n) \in \overrightarrow{\mathcal{E}}$ is the linear map associated with the affine map: $x_1 \in \mathcal{P} \mapsto f(x_1, x_2, \ldots, x_n) \in \mathcal{E}$.

The map: $(\vec{h}, x_2, \ldots, x_n) \in \overrightarrow{\mathcal{P}} \times \mathcal{P} \times \cdots \times \mathcal{P} \mapsto f(\vec{h}, x_2, \ldots, x_n) \in \overrightarrow{\mathcal{E}}$ is called a *mixed polar form of order one*. Similarly we define $f(x_1, \vec{h}, x_3, \ldots, x_n), f(x_1, x_2, \vec{h}, x_4, \ldots, x_n), \ldots, f(x_1, x_2, \ldots, x_{n-1}, \vec{h})$.

These mixed polar forms of order one are affine maps w.r.t. each point argument and linear maps w.r.t. the vector argument. They still have the symmetry property.

*Mixed polar forms of higher order.* The *mixed polar forms of order two* are defined from the mixed polar forms of order one by setting for example:

$$f(\overrightarrow{h}_1, \overrightarrow{h}_2, x_3, \ldots, x_n) = f(\overrightarrow{h}_1, x_2 + \overrightarrow{h}_2, x_3, \ldots, x_n) - f(\overrightarrow{h}_1, x_2, x_3, \ldots, x_n).$$

By induction we define the *mixed polar forms of order p* ($p \leq n$) from those of order $(p-1)$, for instance $f(\overrightarrow{h_1}, \overrightarrow{h_2}, \ldots, \overrightarrow{h_p}, x_{p+1}, \ldots, x_n)$.

All these mixed polar forms are affine maps w.r.t. each point argument and are linear maps w.r.t. each vector argument. They have the symmetry property and they are vector-valued functions.

**2.3. Example.**

$$\mathcal{P} = \mathbb{R}, \quad \mathcal{E} = \mathbb{R}, \quad F(x) = 1 + x + 2x^2 + x^3 \quad (n = 3)$$

$$f(x_1, x_2, x_3) = 1 + \frac{1}{3}(x_1 + x_2 + x_3) + \frac{2}{3}(x_1x_2 + x_1x_3 + x_2x_3) + x_1x_2x_3$$

$$f(\overrightarrow{h_1}, x_2, x_3) = f(0 + h_1, x_2, x_3) - f(0, x_2, x_3)$$

$$f(\overrightarrow{h_1}, x_2, x_3) = \frac{1}{3}h_1 + \frac{2}{3}h_1(x_2 + x_3) + h_1x_2x_3 \quad (= f(x_2, \overrightarrow{h}_1, x_3))$$

$f(\overrightarrow{h_1}, x_2, x_3)$ is linear w.r.t. $\overrightarrow{h_1}$ and affine w.r.t. $x_2$ (resp. $x_3$).

$$f(\overrightarrow{h_1}, x_2, \overrightarrow{h_3}) = f(\overrightarrow{h_1}, x_2, 0 + \overrightarrow{h_3}) - f(\overrightarrow{h_1}, x_2, 0)$$

$$f(\overrightarrow{h_1}, x_2, \overrightarrow{h_3}) = \frac{2}{3}h_1h_3 + h_1x_2h_3 \quad (= f(\overrightarrow{h}_1, \overrightarrow{h}_3, x_2)).$$

$f(\overrightarrow{h_1}, x_2, \overrightarrow{h_3})$ is linear w.r.t. $h_1$ (resp. $h_3$) and affine w.r.t. $x_2$.
The mixed polar form is well-suited for the calculation of a polynomial (or a spline) and its derivatives and, as we will see, leads to a mixed blossoming algorithm and reversible schemes.

PROPOSITION 2.1. *The derivative of order $k$ of $F$ at $x \in \mathcal{P}$ with respect to the vectors $\overrightarrow{h_1}, \ldots, \overrightarrow{h_k} \in \overrightarrow{\mathcal{P}}$ is given by:*

$$D^{(k)}_{\overrightarrow{h_1}, \ldots, \overrightarrow{h_k}} F(x) = \frac{n!}{(n-k)!} f(\overrightarrow{h_1}, \ldots, \overrightarrow{h_k}, \underbrace{x, \ldots, x}_{n-k}).$$

PROOF. Left to the reader or see Seidel [**18**]. □

NOTATION. If $\overrightarrow{h}^{(k)}$ (resp. $x^{(n-k)}$) is short for $k$-fold (resp. $(n-k)$-fold) repetition of $h$ (resp. $x$) we will write $f(\overrightarrow{h}^{(k)}, x^{(x-k)})$ for $f(\underbrace{\vec{h}, \ldots, \vec{h}}_{k}, \underbrace{x, \ldots, x}_{n-k})$.

PARTICULAR CASES.

$$D^{(k)}_{\overrightarrow{h}^{(k)}} F(x) = \frac{n!}{(n-k)!} f(\overrightarrow{h}^{(k)}, x^{(n-k)}).$$

For $\mathcal{P} = \mathbb{R}$ (there $\overrightarrow{\mathcal{P}} = \mathbb{R}$), $\vec{h} = \vec{1}$ (the standard unit vector), $x \in \mathbb{R}$, we have:

$$F^{(k)}(x) = \frac{n!}{(n-k)!} f(\overrightarrow{1}^{(k)}, x^{(n-k)}).$$

## 3. Mixed Progressive Sequence of $\mathbb{R}$

Now we consider the case $\mathcal{P} = \mathbb{R}$, $F\colon \mathbb{R} \to \mathcal{E}$ is a polynomial curve of degree $\leq n$ and $f\colon \mathbb{R}^n \to \mathcal{E}$ its polar form.

**3.1. Definition, progressive sequence of $\mathbb{R}^2$.** A sequence $\overrightarrow{\mathcal{E}_i}$, $i = 1, \ldots, 2n$ of vectors belonging to $\mathbb{R}^2$ (or $\mathbb{R}^n$) is a *progressive sequence* when $\overrightarrow{\mathcal{E}_j}$ is not collinear to $\overrightarrow{\mathcal{E}_{i+n}}$ for $1 \leq i \leq j \leq n$.
We write $\overrightarrow{\mathcal{E}_j} \nparallel \overrightarrow{\mathcal{E}_{i+n}}$ to mean that $\overrightarrow{\mathcal{E}_j}$ is not collinear to $\overrightarrow{\mathcal{E}_{i+n}}$. This definition comes from that given by Ramshaw [**15, 16**] and Barry and Goldman [**1**] for a scalar sequence $t_i$, $i = 1, 2, \ldots, 2n$: $t_j \neq t_{i+n}$, $1 \leq i \leq j \leq n$.

PROPOSITION 3.1. *If a sequence $\overrightarrow{\mathcal{E}_1}, \overrightarrow{\mathcal{E}_2}, \ldots, \overrightarrow{\mathcal{E}_{2n}}$ is progressive, the sequence $\overrightarrow{\mathcal{E}_{2n}}, \overrightarrow{\mathcal{E}_{2n-1}}, \ldots, \overrightarrow{\mathcal{E}_1}$ is also progressive.*

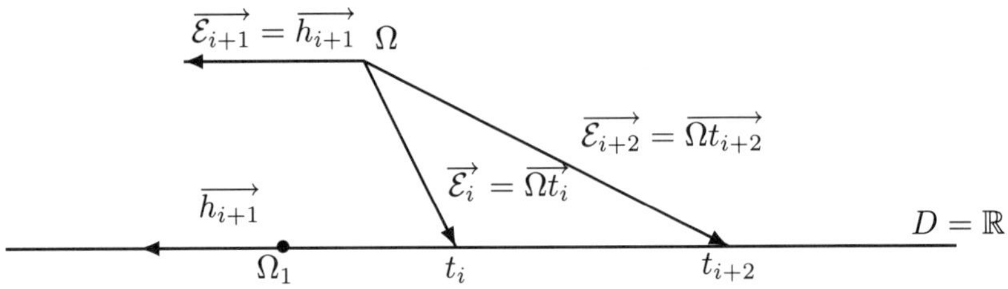


FIGURE 1. Three elements $t_i$, $\overrightarrow{h_{i+1}}$, $t_{i+2}$ of a mixed progressive sequence of $\mathbb{R}$ and the corresponding vectors $\overrightarrow{\mathcal{E}_i}$, $\overrightarrow{\mathcal{E}_{i+1}}$, $\overrightarrow{\mathcal{E}_{i+2}}$ of the progressive sequence of $\mathbb{R}^2$.

PROOF. By hypothesis we have:

$$\overrightarrow{\mathcal{E}_j} \nparallel \overrightarrow{\mathcal{E}_{1+n}}, \overrightarrow{\mathcal{E}_{2+n}}, \dots, \overrightarrow{\mathcal{E}_{j+n}}, \quad j = 1, 2, \dots, n$$

We conclude by writing these relations in the following way:

$$\overrightarrow{\mathcal{E}_{2n+1-j}} \nparallel \overrightarrow{\mathcal{E}}_n, \overrightarrow{\mathcal{E}_{n-1}}, \dots, \overrightarrow{\mathcal{E}_{n+1-j}}, \quad j = 1, 2, \dots, n. \quad \square$$

**3.2. Definition, mixed progressive sequence of $\mathbb{R}$.** Consider a sequence of points $t_i$ of $\mathbb{R}$, $i \in L$, and of vectors $\overrightarrow{h_i}$ of $\overrightarrow{\mathbb{R}}$, $i \in \bar{L}$, with $L \cup \bar{L} = \{1, 2, \dots, 2n\}$ and $L \cap \bar{L} = \emptyset$. Let $D$ be the line containing $\mathbb{R}$ (and consequently $t_i$, $\overrightarrow{h_i}$) and let $\Omega$ be a point of the plane not belonging to $D$. The sequence $t_i$, $i \in L$, $\overrightarrow{h_i}, i \in \bar{I}$ is said to be a *mixed progressive sequence* of $\mathbb{R}$ when the sequence of vectors $\overrightarrow{\mathcal{E}_i}$ of $\mathbb{R}^2$ defined by $\overrightarrow{\mathcal{E}_i} = \overrightarrow{\Omega t_i}$, $i \in L$, $\overrightarrow{\mathcal{E}_i} = \overrightarrow{h_i}$ , $i \in \bar{L}$, is a progressive sequence of $\mathbb{R}^2$ (Fig. 1).

REMARK 3.1. When $\bar{L} = \emptyset$ the mixed progressive sequence $t_i$, $i = 1, 2, \dots, n$ is a progressive sequence: $\overrightarrow{\mathcal{E}_j} \nparallel \overrightarrow{\mathcal{E}_{i+n}}$, $1 \leq i \leq j \leq n$ is equivalent to $t_j \neq t_{i+n}$, $1 \leq i \leq j \leq n$.

EXAMPLE 3.1. The sequence $\overrightarrow{h_1}, t_2, t_3, t_4, \overrightarrow{h_5}, t_6$ $(n = 3)$ is a mixed progressive sequence of $\mathbb{R}$ when $t_2 \neq t_4$, $t_3 \neq t_4$, $t_3 \neq t_6$: $\overrightarrow{\mathcal{E}_1} = \overrightarrow{h_1}$, $\overrightarrow{\mathcal{E}_2} = \overrightarrow{\Omega t_2}$, $\overrightarrow{\mathcal{E}_3} = \overrightarrow{\Omega t_3}$, $\overrightarrow{\mathcal{E}_4} = \overrightarrow{\Omega t_4}$, $\overrightarrow{\mathcal{E}_5} = \overrightarrow{h_5}$, $\overrightarrow{\mathcal{E}_6} = \overrightarrow{\Omega t_6}$ satisfy $\overrightarrow{\mathcal{E}_1} \nparallel \overrightarrow{\mathcal{E}_4}$; $\overrightarrow{\mathcal{E}_2} \nparallel \overrightarrow{\mathcal{E}_4}, \overrightarrow{\mathcal{E}_5}$; $\overrightarrow{\mathcal{E}_3} \nparallel \overrightarrow{\mathcal{E}_4}, \overrightarrow{\mathcal{E}_5}, \overrightarrow{\mathcal{E}_6}$.

PROPOSITION 3.2. *Let $v_i$, $i = 1, 2, \dots, 2n$ be a mixed progressive sequence of $\mathbb{R}$, then it contains at most $n$ vectors.*

PROOF. Suppose, contrary to our claim, that there are at least $(n+1)$ vectors. For $i$ varying from $(n+1)$ to $2n$, we denote by $i_0 + n$ $(i_0 \geq 1)$ the index of the first vector encountered (such an index exists). Consequently among the $v_i$, $i = i_0$, $i_0 + 1, \dots, n$, there exists a vector $\overrightarrow{h_{j_1}}$. Thus $\overrightarrow{h_{j_1}} \| \overrightarrow{h_{i_o+n}}$ with $1 \leq i_0 \leq j_1 \leq n$, which contradicts the fact that the mixed sequence is progressive. Our assumption is false and the proposition is proved. $\square$

## 4. Mixed Polygon

**4.1. Definition.** Let $F\colon \mathbb{R} \to \mathcal{E}$ be a polynomial of degree $\leq n$, let $f$ be its polar form and let $t_i$ $(i \in L)$, $\overrightarrow{h_i}(i \in \bar{L})$, $L \cup \bar{L} = \{1, 2, \dots, 2n\}$, $L \cap \bar{L} = \emptyset$, be a mixed progressive sequence of $\mathbb{R}$. Define $v_i = t_i$ $(i \in L)$, $v_i = \overrightarrow{h_i}$ $(i \in \bar{L})$, a *mixed* *polygon* of $F$ (or of $f$) relative to the mixed progressive sequence $v_i$, $i = 1$ to $2n$, is the $(n+1)$-tuple

$$\big(f(v_1, v_2, \dots, v_n), f(v_2, v_3, \dots, v_{n+1}), \dots, f(v_{n+1}, v_{n+2}, \dots, v_{2n})\big).$$

Let us recall that $f(v_i, v_{i+1}, \dots, v_{i+n-1})$ belongs to $\mathcal{E}$ if all the arguments are points of $\mathbb{R}$, otherwise $f(v_i, v_{i+1}, \dots, v_{i+n-1})$ belongs to $\overrightarrow{\mathcal{E}}$ if at least one argument is a vector of $\overrightarrow{\mathbb{R}}$.

**4.2. Proposition.** A mixed polygon contains at least a point of $\mathcal{E}$.

PROOF. Suppose the contrary: the polygon contains only vectors of $\overrightarrow{\mathcal{E}}$ i.e. in each set of $n$ consecutive $v_i$ there exists at least a vector $\overrightarrow{h_j}$ of $\overrightarrow{\mathbb{R}}$. As in the previous proof we denote by $i_{0+n}$ ($i_0 \geq 1$) the index of the first vector encountered when $i$ varies from $n+1$ to $2n$. Consequently there exists a vector denoted by $\overrightarrow{h_{j_1}}$ among the $n$ consecutive elements $v_{i_0}, v_{i_0+1}, \dots, v_{i_0+n-1}$ with $i_0 \leq j_1 \leq n$. So we have found two collinear vectors $\overrightarrow{h_{j_1}}$ and $\overrightarrow{h_{i_0+n}}$ with $1 \leq i_0 \leq j_1 \leq n$, which contradicts the fact that the mixed sequence is progressive. □

EXAMPLE 4.1. $F$ is a degree-three polynomial ($n = 3$), the mixed polygon

$$\big(f(t_1, \overrightarrow{h_2}, t_3), f(\overrightarrow{h_2}, t_3, t_4), f(t_3, t_4, t_5), f(t_4, t_5, \overrightarrow{h_6})\big)$$

is associated to the mixed progressive sequence

$$t_1, \overrightarrow{h_2}, t_3, t_4, t_5, \overrightarrow{h_6} \text{ with } t_1 \neq t_4, t_3 \neq t_4, t_3 \neq t_5.$$

EXAMPLE 4.2. The mixed polygon ($n = 3$)

$$\big(f(t_1, \overrightarrow{h_2}, \overrightarrow{h_3}), f(\overrightarrow{h_2}, \overrightarrow{h_3}, t_4), f(\overrightarrow{h_3}, t_4, t_5), f(t_4, t_5, t_6)\big)$$

is associated to the mixed progressive sequence $t_1$, $\overrightarrow{h_2}$, $\overrightarrow{h_3}$, $t_4$, $t_5$, $t_6$ with $t_1 \neq t_4$.

EXAMPLE 4.3. $f(t_1, t_2, \dots, t_n)$, $f(t_2, t_3, \dots, t_{n+1})$, $\dots$ , $f(t_{k+n-2}, \dots, t_k, \overrightarrow{h_{k+1}})$, $f(t_{k+n-1}, \dots, t_k, \overrightarrow{h_{k+1}}, \overrightarrow{h_{k+2}})$, $\dots$ , $f(t_{n+1}, \dots, t_k, \overrightarrow{h_{k+1}}, \dots, \overrightarrow{h_{2n}})$ is a mixed polygon of $F$ (or of $f$) associated to the mixed progressive sequence $t_1$, $\dots$ , $t_k$, $\overrightarrow{h_{k+1}}$, $\overrightarrow{h_{k+2}}$, $\dots$ , $\overrightarrow{h_{2n}}$ with $n \leq k \leq 2n-1$, $t_j \neq t_{i+n}$ for $1 \leq i \leq j \leq n$ and $i + n \leq k$. This example has been studied by Fiorot and Jeannin [**10**].

For $n = 3$ ($k = 3$), we have the mixed polygon

$$f(t_1, t_2, t_3), f(t_2, t_3, \overrightarrow{h_4}, f(t_3, \overrightarrow{h_4}), \overrightarrow{h_5}), f(\overrightarrow{h_4}, \overrightarrow{h_5}, \overrightarrow{h_6})$$

associated to the mixed progressive sequence $t_1, t_2, t_3, \overrightarrow{h_4}, \overrightarrow{h_5}, \overrightarrow{h_6}$.

For $n = 3$ ($k = 4$) we have $\big(f(t_1, t_2, t_3), f(t_2, t_3, t_4), f(t_3, t_4, \overrightarrow{h_5}), f(t_4, \overrightarrow{h_5}, \overrightarrow{h_6}\big)$ associated to $t_1, t_2, t_3, t_4, \overrightarrow{h_5}, \overrightarrow{h_6}$ with $t_1 \neq t_4$, $t_2 \neq t_4$, $t_3 \neq t_4$.

The de Boor polygon is defined relatively to the progressive sequence $t_1 < t_2 < \dots < t_{2n}$, it would correspond to $k = 2n$ (no vector $\overrightarrow{h_i}$).

**4.3. Definition: Taylor polygon.** In the Example 4.3 considering the case $k = n$ and $t_1 = t_2 = \dots = t_n = t$, $\overrightarrow{h_{n+1}} = \overrightarrow{h_{n+2}} = \dots = \overrightarrow{h_{2n}} = \overrightarrow{h}$ ($\neq 0$), the mixed polygon

$$\left(f(\underbrace{t, \dots, t}_{n}), f(\underbrace{t, \quad , t}_{n-1}, \overrightarrow{h}), \dots, f(t, \underbrace{\overrightarrow{h}, \dots, \overrightarrow{h}}_{n-1}), f(\underbrace{\overrightarrow{h}, \dots, \overrightarrow{h}}_{n})\right)$$

is called the *Taylor polygon* of $F$ at $t$ relative to $\overrightarrow{h}$.

It is equal to $\left(F(t), 1/nF'_{\overrightarrow{h}}(t), \ldots, 1/n!F^{(n-1)}_{\overrightarrow{h}(n-1)}(t), 1/n!F^{(n)}_{\overrightarrow{h}(n)}(t)\right)$ by Proposition 2.1.

## 5. Mixed Blossoming Algorithm

**Data:** $t_i (i \in L)$ are points of $\mathbb{R}$, $\overrightarrow{h_i} (i \in \bar{L})$ are vectors of $\overrightarrow{\mathbb{R}}$, $L \cup \bar{L} = \{1, 2, \ldots, 2n\}$, $L \cap \bar{L} = \emptyset$ a mixed progressive sequence of $\mathbb{R}$. Define $v_i$, $i = 1, 2, \ldots, 2n$ with $v_i = t_i (i \in L)$, $v_i = \overrightarrow{h_i} (i \in \bar{L})$, $F : \mathbb{R} \to \mathcal{E}$ a polynomial curve of degree $\leq n$, $f : \mathbb{R}^n \to \mathcal{E}$ its polar form, $(f(v_{j+1}, v_{j+2}, \ldots, v_{j+n}))$, $j = 0$ to $n$, the associated mixed polygon, $u_i$, $i = 1, 2, \ldots, n$ a sequence of points or vectors of $\mathbb{R}$. Define $P^i_j(u_1, \ldots, u_i) = f(u_1, \ldots, u_i, v_{i+j+1}, \ldots, v_{j+n})$. Then $P^i_j$ $(i = 0, 1, \ldots, n$, $j = 0, \ldots, n - i)$ are determined by the *mixed blossoming algorithm*:

**initialization:** for $j = 0$ to $n$ do $P^0_j = f(v_{j+1}, v_{j+2}, \ldots, v_{j+n})$. For $i = 1$ to $n$ for $j = 0$ to $n - i$ do

$$P^i_j = \lambda(u_i, v_{i+j}, v_{j+n+1}) P^{i-1}_j + \mu(u_i, v_{i+j}, v_{j+n+1}) P^{i-1}_{j+1} \tag{1}$$

$\lambda(u_i, v_{i+j}, v_{j+n+1})$ and $\mu(u_i, v_{i+j}, v_{j+n+1})$ are the coordinates of $u_i$ relatively to the frame of reference $(v_{i+j}, v_{j+n+1})$ of $\mathbb{R}$.

**Result:** $P^n_0 = f(u_1, u_2, \ldots, u_n)$ (and $P^i_j$, $i = 0$ to $n$, $j = 0$ to $n - i$).

We have $1 \leq j + 1 \leq i + j \leq n$ then $\overrightarrow{\mathcal{E}_{i+j}} \not\parallel \overrightarrow{\mathcal{E}_{j+1+n}}$, so we never encounter $v_{i+j} = \overrightarrow{h_{i+j}}$ and $v_{j+n+1} = \overrightarrow{h_{j+n+1}}$. Then to determine $\lambda$ and $\mu$ we need only consider six cases:

(1) $u_i \in \mathbb{R}$, $v_{i+j} = t_{i+j}$, $v_{j+n+1} = t_{j+n+1}$,

$$u_i = \underbrace{\frac{t_{j+n+1} - u_i}{t_{j+n+1} - t_{i+j}}}_{\lambda(u_i, t_{i+j}, t_{j+n+1})} t_{i+j} + \underbrace{\frac{u_i - t_{i+j}}{t_{j+n+1} - t_{i+j}}}_{\mu(u_i, t_{i+j}, t_{j+n+1})} t_{j+n+1} \quad (\lambda + \mu = 1).$$

(2) $u_i \in \mathbb{R}$, $v_{i+j} = t_{i+j}$, $v_{j+n+1} = \overrightarrow{h_{j+n+1}}$, $u_i = t_{i+j} + x(u_i, t_{i+j}, \overrightarrow{h_{j+n+1}}) \overrightarrow{h_{j+n+1}}$ $(\lambda = 1, \mu = x(u_i, t_{i+j}, \overrightarrow{h_{j+n+1}}))$ designates the abscissas of $u_i$, $t_{i+j}$ being the origin, $\overrightarrow{h_{j+n+1}}$ the reference vector of $\mathbb{R}$)

(3) $u_i \in \mathbb{R}$, $v_{i+j} = \overrightarrow{h_{i+j}}$, $v_{j+n+1} = t_{j+n+1}$,

$$u_i = t_{j+n+1} + x(u_i, t_{j+n+1}, \overrightarrow{h_{i+j}}) \overrightarrow{h_{i+j}} \quad (\lambda = x(u_i, t_{j+n+1}, \overrightarrow{h_{i+j}}), \mu = 1)$$

(4) $u_i \in \overrightarrow{\mathbb{R}}$, $v_{i+j} = t_{i+j}$, $v_{j+n+1} = t_{j+n+1}$,

$$u_i = \underbrace{\frac{-u_i}{t_{j+n+1} - t_{i+j}}}_{\lambda(u_i, t_{i+j}, t_{j+n+1})} t_{i+j} + \underbrace{\frac{u_i}{t_{j+n+1} - t_{i+j}}}_{\mu(u_i, t_{i+j}, t_{j+n+1})} t_{j+n+1} \quad (\lambda + \mu = 0).$$

(5) $u_i \in \overrightarrow{\mathbb{R}}$, $v_{i+j} = t_{i+j}$, $v_{j+n+1} = \overrightarrow{h_{j+n+1}}$, $u_i = \alpha_{j+n+1} \overrightarrow{h_{j+n+1}}$ $(\lambda = 0, \mu = \alpha_{j+n+1}$ is the coefficient of proportionality between vectors $u_i$ and $\overrightarrow{h_{j+n+1}})$,

(6) $u_i \in \overrightarrow{\mathbb{R}}$, $v_{i+j} = \overrightarrow{h_{i+j}}$, $v_{j+n+1} = t_{j+n+1}$,

$$u_i = \alpha_{i+j} \overrightarrow{h_{i+j}} \quad (\lambda = \alpha_{i+j}, \mu = 0).$$

REMARK 5.1. When we are always in situation 1 (all the $u_i$ and $v_i$ are points) we obtain the standard blossoming algorithm of Ramshaw [**15, 16**], see also Barry and Goldman [**1**], Seidel [**17, 19**], De Rose *et al.* [**5**]. Moreover if $u_i = u$, $i = 1$

to $n$, the algorithm is the well-known Cox-de Boor algorithm for splines (Cox [**6**], de Boor [**4**]). It gives $f(\underbrace{u,\dots,u}_{n}) = F(u)$.

REMARK 5.2. The order in which we introduce the $u_i$ in the algorithm does not matter, because we know that $f$ has the symmetry property.

*Some justifications.* Let us recall that the mixed polar form $f$ is affine w.r.t. point arguments, linear w.r.t. vector arguments and has the symmetry property.

For instance in case 2, we successively use the definition of the mixed polar form, the linear and the symmetry property:

$$
\begin{aligned}
&f(u_1,\dots,u_i,v_{i+j+1},\dots,v_{j+n})\\
&= f\big(u_1,\dots,u_{i-1},t_{i+j}+x(u_i,t_{i+j},\overrightarrow{h_{j+n+1}})\overrightarrow{h_{j+n+1}},v_{i+j+1},\dots,v_{j+n}\big)\\
&= f\big(u_1,\dots,u_{i-1},t_{i+j},v_{i+j+1},\dots,v_{j+n}\big)\\
&\qquad\qquad + f\big(u_1,\dots,u_{i-1},x\overrightarrow{h_{j+n+1}},v_{i+j+1},\dots,v_{j+n}\big)\\
&= f\big(u_1,\dots,u_{i-1},v_{i+j},\dots,v_{j+n}\big)\\
&\qquad\qquad + xf\big(u_1,\dots,u_{i-1},\overrightarrow{h_{j+n+1}},v_{i+j+1},\dots,v_{j+n}\big)\\
&= f\big(u_1,\dots,u_{i-1},v_{i+j},\dots,v_{j+n}\big) + xf\big(u_1,\dots,u_{i-1},v_{i+j+1},\dots,v_{j+n},\overrightarrow{h_{j+n+1}}\big)\\
&= f\big(u_1,\dots,u_{i-1},v_{i+j},\dots,v_{j+n}\big) + xf\big(u_1,\dots,u_{i-1},v_{i+j+1},\dots,v_{j+n+1}\big)
\end{aligned}
$$

Case 4 ($\lambda+\mu=0$) uses the linearity and the symmetry property:

$$
\begin{aligned}
&f(u_1,\dots,u_{i-1},u_i,v_{i+j+1},\dots,v_{j+n})\\
&= f(u_1,\dots,u_{i-1},\lambda t_{i+j}+\mu t_{j+n+1},v_{i+j+1},\dots,v_{j+n})\\
&= f(u_1,\dots,u_{i-1},\lambda\overrightarrow{\omega t_{i+j}}+\mu\overrightarrow{\omega t_{j+n+1}},v_{i+j+1},\dots,v_{j+n}) \qquad \forall\omega\\
&= \lambda f(u_1,\dots,u_{i-1},\overrightarrow{\omega t_{i+j}},v_{i+j+1},\dots,v_{j+n})\\
&\qquad\qquad + \mu f(u_1,\dots,u_{i-1},\overrightarrow{\omega t_{j+n+1}},v_{i+j+1},\dots,v_{j+n})\\
&= \lambda f(u_1,\dots,u_{i-1},\omega,v_{i+j+1},\dots,v_{j+n})f(u_1,\dots,u_{i-1},t_{i+j},v_{i+j+1},\dots,v_{j+n})\\
&\quad + \mu f(u_1,\dots,u_{i-1},\omega,v_{i+j+1},\dots,v_{j+n})f(u_1,\dots,u_{i-1},t_{j+n+1},v_{i+j+1},\dots,v_{j+n})\\
&= \lambda f(u_1,\dots,u_{i-1},t_{i+j},v_{i+j+1},\dots,v_{j+n})\\
&\qquad\qquad + \mu f(u_1,\dots,u_{i-1},t_{j+n+1},v_{i+j+1},\dots,v_{j+n})\\
&= \lambda f(u_1,\dots,u_{i-1},v_{i+j},\dots,v_{j+n}) + \mu f(u_1,\dots,u_{i-1},v_{i+j+1},\dots,v_{j+n+1}).
\end{aligned}
$$

Case 5 is the more simple, by linearity and symmetry we obtain:

$$
\begin{aligned}
&f(u_1,\dots,u_{i-1},u_i,v_{i+j+1},\dots,v_{j+n})\\
&\qquad = f(u_1,\dots,u_{i-1},\alpha_{j+n+1}\overrightarrow{h_{j+n+1}},v_{i+j+1},\dots,v_{j+n})\\
&\qquad = \alpha_{j+n+1}f(u_1,\dots,u_{i-1},\overrightarrow{h_{j+n+1}},v_{i+j+1},\dots,v_{j+n})\\
&\qquad = \alpha_{j+n+1}f(u_1,\dots,u_{i-1},v_{i+j+1},\dots,v_{j+n},v_{j+n+1}).
\end{aligned}
$$

In the different figures, an element which is not part of the data, is obtained from two neighbouring elements (sometimes one) where the arrows pointing into this element originate.

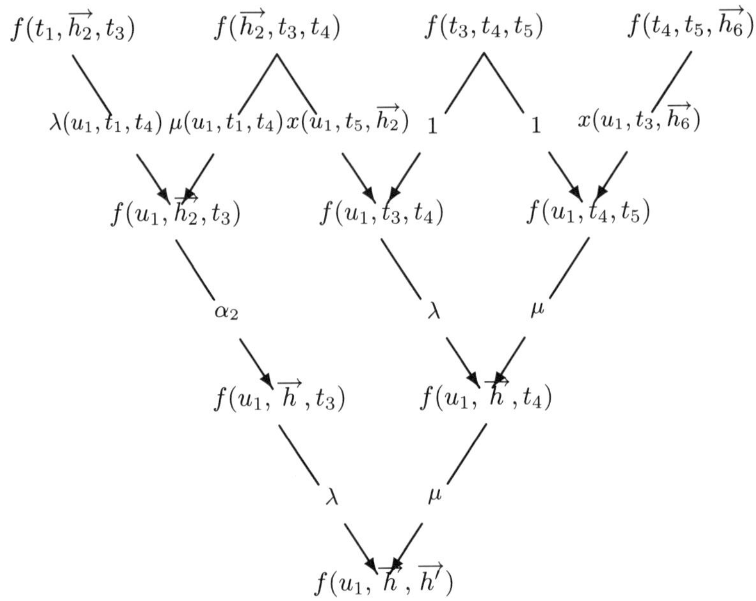


FIGURE 2. Mixed blossoming algorithm: recursive evaluation algorithm of a mixed polar form from a mixed polygon ($n = 3$)

EXAMPLE 5.1 ($n = 3$). We consider the mixed progressive sequence of Example 4.1. The knowledge of the mixed polygon (top line of Fig. 2) allows us to calculate $f(u_1, \overrightarrow{h}, \overrightarrow{h'})$, $u_1 \in \mathbb{R}$, by the mixed blossoming algorithm.

REMARK 5.3. In particular the blossoming algorithm gives

$$f(\underbrace{\overrightarrow{1}, \dots, \overrightarrow{1}}_{k}, \underbrace{x, \dots, x}_{n-k}) \text{ i.e. } \frac{(n-k)!}{n!} F^{(k)}(x) \quad \text{(Proposition 2.1)}$$

## 6. Isomorphism Between Polynomials and Mixed Polygons

NOTATION. We denote by $R_n[x]$ the polynomial vector space of degree $\leq n$, w.r.t. the variable $x$.

THEOREM 6.1. *Let $F\colon \mathbb{R} \to \mathbb{R}$ be a polynomial of degree $\leq n$, $f\colon \mathbb{R}^n \to \mathbb{R}$ be its polar form, $v_1, v_2, \dots, v_{2n}$, be a mixed progressive sequence of $\mathbb{R}$. Then the map*:

$$\begin{aligned} L\colon & R_n[x] \to \mathbb{R}^{n+1} \\ :\ & F \mapsto \big(f(v_1, \dots, v_n), f(v_2, \dots, v_{n+1}), \dots, f(v_{n+1}, \dots, v_{2n})\big) \end{aligned}$$

*is a linear isomorphism.*

PROOF. From Theorem 2.1, $\varphi\colon F \to f$ is a linear map. We prove by induction that the different maps $\varphi_i\colon F \to f(v_i, \dots, v_{i+n-1})$ are linear. If $v_j = t_j$ (points of $\mathbb{R}$) for $j = i$ to $i + n - 1$, $\psi_0\colon F \mapsto f(t_i, \dots, t_{i+n-1})$ is a linear map as a consequence of $\varphi$ being linear. If one $v_j$ is a vector $\overrightarrow{h_j}$, $\psi_1: F \to f(t_i, \dots, \overrightarrow{h_j}, \dots, t_{i+n-1})$, we write $\psi_1(F) = f(t_i, \dots, a_j + \overrightarrow{h_j}, \dots, t_{i+n-1}) - f(t_i, \dots, a_j, \dots, t_{i+n-1})$ for any

$a_j \in \mathbb{R}$, and $\psi_1$ being the difference of two linear maps of type $\psi_0$, is a linear map. Consider $\psi_2 \colon F \to f(t_i, \dots, \overrightarrow{h_k}, \dots, \overrightarrow{h_j}, \dots, t_{i+n-1})$ where two arguments are vectors, we write again $\psi_2(F) = f(t_i, \dots, a_k + \overrightarrow{h_k}, \dots, \overrightarrow{h_j}, \dots, t_{i+n-1}) - f(t_i, \dots, a_k, \dots, \overrightarrow{h_j}, \dots, t_{i+n-1})$ for any $a_k \in \mathbb{R}$. $\psi_2$ being the difference of two linear maps of type $\psi_1$, is a linear map. And so on, by induction, $\varphi_i$ is a linear map.

Therefore $L$ is a linear map. By the mixed blossoming algorithm (Section 5), the kernel of $L$ is zero and since $R_n[x]$ has same dimension as $\mathbb{R}^{n+1}$, it implies that $L$ is a linear isomorphism. □

THEOREM 6.2. *Let $v_1, v_2, \dots, v_n, \dots, v_{2n}$ be a mixed progressive sequence of $\mathbb{R}$, let $P_0, P_1, \dots, P_n$ be a sequence of $(n+1)$ elements of $\mathcal{E}$ or $\overrightarrow{\mathcal{E}}$ such that $P_i \in \mathcal{E}$ if $v_{i+1}, v_{i+2}, \dots, v_{i+n} \in \mathbb{R}$ and $P_i \in \overrightarrow{\mathcal{E}}$ if one at least of $v_{i+1}, v_{i+2}, \dots, v_{i+n}$ is a vector of $\overrightarrow{\mathbb{R}}$.*

*Then there exists a unique polynomial $F \colon \mathbb{R} \to \mathcal{E}$ of degree $\leq n$ such that $P_i = f(v_{i+1}, v_{i+2}, \dots, v_{i+n})$, $i = 0, 1, \dots, n$, where $f$ is the polar form of $F$. So $(P_0, P_1, \dots, P_n)$ is the mixed polygon of $F$ relative to the mixed progressive sequence $v_1, v_2, \dots, v_n, \dots, v_{2n}$.*

PROOF. We suppose $\dim \mathcal{E} = q$. Relatively to a frame of reference of $\mathcal{E}$, a point or a vector of $\mathcal{E}$ is defined by its $q$ coordinates $P_i = (P_i^1, P_i^2, \dots, P_i^q)$.

By Theorem 6.1 there exists a unique polynomial $F^j \colon \mathbb{R} \to \mathbb{R}$ of degree $\leq n$ such that:

$$f^j(v_{i+1}, v_{i+2}, \dots, v_{i+n}) = P_i^j, i = 0, 1, \dots, n,$$

$f^j$ being the polar form of $F^j$.

Consider

$$\begin{aligned} F \colon & \mathbb{R} \to \mathcal{E} \\ & \colon t \mapsto F(t) = \big(F^1(t), F^2(t), \dots, F^q(t)\big). \end{aligned}$$

The polar form of $F$ is $f = (f^1, f^2, \dots, f^q)$, then $f(v_{i+1}, v_{i+2}, \dots, v_{i+n}) = P_i$. The polynomials $F^j$, $j = 1, 2, \dots, q$, are unique, so is $F$ and $F$ is the solution of our problem. □

## 7. Changing a Mixed Progressive Sequence and a Mixed Polygon: A Reversible Scheme

**Data:** $F \colon \mathbb{R} \to \mathcal{E}$ a polynomial of degree $\leq n$, $f \colon \mathbb{R}^n \to \mathcal{E}$ its polar form.

Let $v_i$, $i = 1, 2, \dots, 2n$, and $w_i$, $i = 1, 2, \dots, 2n$ be two mixed progressive sequences of $\mathbb{R}$, let $\big(f(v_1, v_2, \dots, v_n), f(v_2, v_3, \dots, v_{n+1}), \dots, f(v_{n+1}, v_{n+2}, \dots, v_{2n})\big)$ be the mixed polygon of $F$ relative to $(v_i)$. We assume we have the compatibility condition: $w_{2n}, w_{2n-1}, \dots, w_{n+1}, v_{n+1}, \dots, v_{2n}$ is a mixed progressive sequence.

The algorithm illustrated in Fig. 3 for the case $n = 3$ determines by following the solid arrows the mixed polygon of $F$ relative to a given mixed progressive sequence $w_i$, $i = 1, 2, \dots, 2n$ (bottom line) from the mixed polygon of $F$ relative to $(v_i)$ (top line). Conversely starting from the mixed polygon of $F$ relative to $(w_i)$ (bottom line) we deduce by following the dashed arrows the mixed polygon of $F$ relative to $(v_i)$.

We operate as follows. Each element (see Fig. 3 or Fig. 4, for $n = 3$) is obtained with the formulae of the mixed blossoming algorithm (Section 5) from

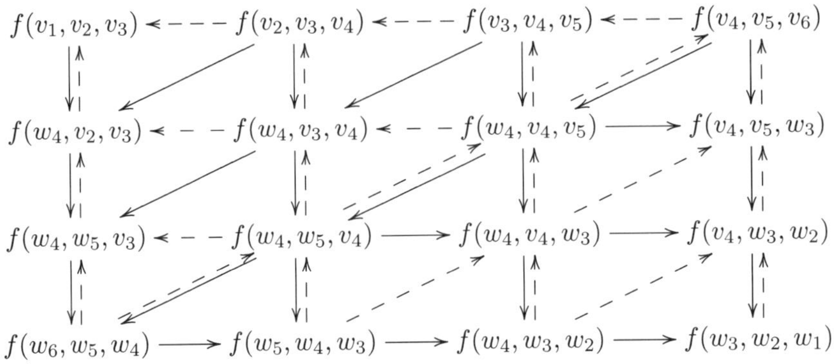


FIGURE 3. Case $n = 3$. By following the solid (resp. dashed) arrows the mixed polygon of $F$ at the bottom (resp. top) line relative to the mixed sequence $(w_i)$ (resp. $(v_i)$) is obtained from the mixed polygon of $F$ at the top (resp. bottom) line relative to the mixed sequence $(v_i)$ (resp. $w_i$). Elements $w_4$, $w_5$, $w_6$, $w_3$, $w_2$, $w_1$ (resp. $v_4$, $v_5$, $v_6$, $v_3$, $v_2$, $v_1$) are introduced in this order.

two neighbouring elements (sometimes one: cases 5 and 6 of Section 5) where the arrows of the same type (either solid arrows or dashed arrows) pointing into this element originate.

*Description of the scheme.* From the mixed polygon (top (resp. bottom) line in Fig. 3 for the case $n = 3$) relative to the mixed progressive sequence $v_i$ (resp. $w_i$), $i = 1$ to $2n$ we first compute the elements of an upper left (resp. lower right) triangle by introducing $w_{n+1}$, $w_{n+2}, \dots, w_{2n}$ (resp. $v_{n+1}$, $v_{n+2}, \dots, v_{2n}$) in this order to obtain on the ascending diagonal the mixed polygon $f(w_{2n}, \dots, w_{n+1})$, $f(w_{2n-1}, \dots, w_{n+1}, v_{n+1})$, $\dots$, $f(w_{n+1}, v_{n+1}, \dots, v_{2n-1})$, $f(v_{n+1}, v_{n+1}, \dots, v_{2n})$ relative to the sequence $w_{2n}$, $w_{2n-1}$, $\dots$, $w_{n+1}$, $v_{n+1}, \dots, v_{2n}$.

As we have assumed that this mixed sequence is progressive (the compatibility condition), we can then compute from this diagonal the elements of a lower right (resp. upper left) triangle by introducing $w_n$, $w_{n-1}$, $\dots$, $w_1$ (resp. $v_n$, $v_{n-1}$, $\dots$, $v_1$) in this order. Finally we get on the bottom (resp. top) line the mixed polygon of $F$ relative to $(w_i)$ (resp. $(v_i)$).

*A variant of the previous scheme.* We assume (compatibility condition) that $w_1$, $w_2$, $\dots$, $w_n$, $v_{n+1}$, $v_{n+2}$, $\dots$, $v_{2n}$ is a mixed progressive sequence.

Starting from the mixed polygon (top (resp. bottom) line in Fig. 4 for the case $n = 3$) relative to $(v_i)$ (resp. $(w_i)$) we first introduce $w_n$, $w_{n-1}$, $\dots$, $w_1$ (resp. $v_{n+1}$, $v_{n+2}$, $\dots$, $v_{2n}$) in this order to calculate the elements of an upper left (resp. lower right) triangle. On the ascending diagonal we find the mixed polygon $f(w_1, w_2, \dots, w_n)$, $f(w_2, w_3, \dots, w_n, v_n)$, $\dots$, $f(w_n, v_{n+1}, \dots, v_{2n-1})$, and $f(v_{n+1}, v_{n+2}, \dots, v_{2n})$. Secondly, the compatibility condition allows us, starting from this diagonal, to compute the elements of a lower right (resp. upper left) triangle by introducing $w_{n+1}$, $w_{n+2}$, $\dots$, $w_{2n}$ (resp. $v_n$, $v_{n-1}$, $\dots$, $v_1$) in this order. Finally we get on the bottom (resp. top) line the mixed polygon relative to $(w_i)$ (resp. $(v_i)$).

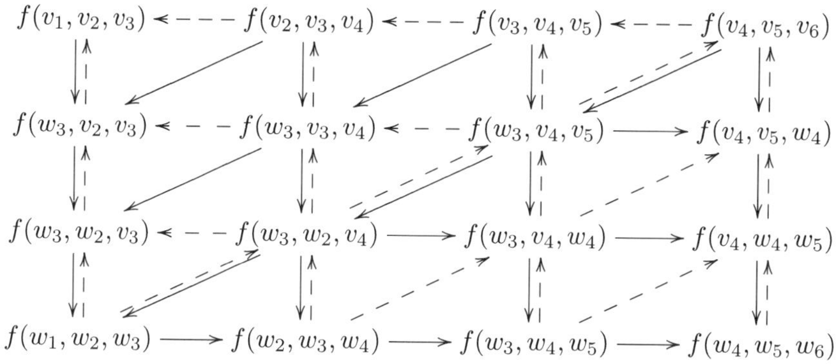


FIGURE 4. Case $n = 3$. By following the solid (resp. dashed) arrows the mixed polygon of $F$ at the bottom (resp. top) line relative to the mixed sequence $(w_i)$ (resp. $(v_i)$) is obtained from the mixed polygon of $F$ at the top (resp. bottom) line relative to the mixed sequence $(v_i)$ (resp. $w_i$). Elements $w_3$, $w_2$, $w_1$, $w_4$, $w_5$, $w_6$ (resp. $v_4$, $v_5$, $v_6$, $v_3$, $v_2$, $v_1$) are introduced in this order.

*An application.* The previous schemes can be used to calculate the derivatives of a polynomial or a spline at $u \in \mathbb{R}$.

(i) Taking $w_{n+1} = w_{n+2} = \cdots = w_{2n} = \vec{1}$, $w_1 = w_2 = \cdots = w_n = u$ (resp. $w_{n+1} = w_{n+2} = \cdots = w_{2n} = u$, $w_1 = w_2 = \cdots = w_n = \vec{1}$) the first scheme gives the Taylor polygon of $F$ at $u$ (bottom line in Fig. 3, case $n = 3$) from the mixed polygon relative to the mixed sequence $(v_1, v_2, \ldots, v_{2n})$ (top line in fig. 3, case $n = 3$): $f(\vec{1}^{(n)})$, $f(\vec{1}^{(n-1)}, u), \ldots, f(\vec{1}, u^{(n-1)})$, $f(u^{(n)})$, i.e. respectively $1/n!F^{(n)}(u)$, $1/n!F^{(n-1)}(u), \ldots, 1/nF'(u)$, $F(u)$.

The compatibility condition is: $v_{n+1}, v_{n+2}, \ldots, v_{2n}$ are points (resp. $u \neq v_{n+1}, v_{n+2}, \ldots, v_{2n}$).

(ii) Taking $v_i = t_i$, $i = 1$ to $2n$, $t_i$ being a point of $\mathbb{R}$ and as previously $w_{n+1} = w_{n+2} = \cdots = w_{2n} = \vec{1}$, $w_1 = w_2 = \cdots = w_n = u$, the first scheme gives the Taylor polygon of $F$ at $u$ (bottom line in Fig. 3, case $n = 3$) from the control polygon of $F$ relative to $(t_i)$ (top line in Fig. 3, $n = 3$). Similarly, points (i) and (ii) are obtained by the variant of the scheme.

(iii) Let $s(u) = \sum_i N_i^{n+1}(u)P_i$ be a $B$-spline curve of degree $\leq n$, $(n+1)$ is the order, defined by its control points $P_i$ (called de Boor polygon) and let $(t_i)$ be the underlying subdivision of $\mathbb{R}$. Consider $t_n < t_{n+1}$, we call $F$ the polynomial segment which is the restriction of $s$ to the interval $[t_n, t_{n+1}[$ and $f$ its polar form. Following Ramshaw [**15, 16**] see also Seidel [**17, 19**], Lee [**14**] we have:

$$P_j = f(t_{j+1}, \ldots, t_{j+n}) \text{ for } \quad j = 0, 1, \ldots, n.$$

The above application (ii) gives the values of the derivatives of $F = s$ at any point $u \in [t_n, t_{n+1}[$: $s^{(k)}(u) = n!/(n-k)!f(\vec{1}^{(k)}, u^{(n-k)})$, $k = 0, 1, \ldots, n$. So we have illustrated in a natural way Boehm's algorithm (Boehm [**3**]) by using the mixed polar form. A similar illustration has been given by Kirmann [**12**], Barry and Goldman [**1**] with the multilinear or the homogeneous form of a polar form.

## References


1. P. J. Barry and R. N. Goldman, *Algorithms for progressive curves: extending B-spline and blossoming techniques to the monomial, power, and Newton dual bases*, Knot Insertion and Deletion Algorithms for $B$-spline Curves and Surfaces (R. N. Goldman and T. Lyche, eds.) , Geom. Des. Publ., SIAM, Philadelphia, PA, 1993, pp. 11–63.
2. P. Bézier, *Définition numérique des courbes et surfaces.* I, Automatisme **11** (1966), 625–632.
3. W. Boehm, *Efficient evaluation of splines*, Computing **33** (1984), 171–177.
4. C. de Boor, *On calculating with B-splines*, J. Approx. Theory **6** (1972), 50–62.
5. T. De Rose, R. Goldman, and M. Lounsbery, *A tutorial introduction to blossoming*, Geometric Modeling Methods and Applications (H. Hagen and D. Roller, eds.) (Böblingen, 1990), Comput. Graph. Systems Appl., Springer, Berlin, 1991, pp. 267–286.
6. M. Cox, *The numerical evaluation of B-splines*, J. Inst. Math. Appl. **10** (1972), 134–149.
7. P. De Casteljau, *Outillage, méthode de calcul*, Tech. report A. Citroën, Paris, 1959.
8. ———, *Courbes et surfaces à pôles*, Tech. report A. Citroën Paris, 1963.
9. ———, *Formes à pôles*, Hermès, Paris, 1985; English transl. *Shape mathematics in* CAD, Kogan Page, London, 1986.
10. J. C. Fiorot and P. Jeannin, *Mixed polar form and mixed polygon*, Les cahiers du Laboratoire de Mathématiques Appliquées, LMA N2, Univ. Littoral, Calais, France, 1994.
11. R. Gormaz, *Floraisons polynomiales: Applications à l'étude des B-splines à plusieurs variables*, Thesis, Univ. de Grenoble I, France, 1993.
12. G. Kirmann, *A short proof of an algorithm by Boehm*, Comput. Aided Geom. Design **9** (1992), 69–71.
13. S. Lang, *Algebra*, Addison-Wesley, Reading, Mass, 1965.
14. E. T. Y. Lee, *A note on blossoming*, Comput. Aided Geom. Design **6** (1989), 359–362.
15. L. Ramshaw, *Blossoming: a connect-the-dots approach to splines*, Digital Systems Research Center, Palo Alto, 1987.
16. ———, *Blossoms are polar forms*, Comput. Aided Geom. Design **6** (1989), 323–358.
17. H. P. Seidel, *A new multiaffine approach to B-splines*, Comput. Aided Geom. Design **6** (1989), 23–52.
18. ———, *An introduction to polar forms*, IEEE. Comput. Graph. Appl. **13** (1993), no. 1, 38–46.
19. ———, *Computing B-spline control points using polar forms*, Comput. Aided Design **23** (1991), 634–640



Université de Valenciennes et du Hainaut-Cambrésis, ENSIMEV, Laboratoire LIMAV, BP 311, 59304 Valenciennes Cedex, France

*E-mail address*: fiorot@univ_valenciennes.fr

Université du Littoral, Centre Universitaire de la Mi-Voix, Bâtiment H. Poincaré, 50 Rue F. Buisson, 62100, Calais, France

Centre de Recherches Mathématiques
CRM Proceedings and Lecture Notes
Volume **18**, 1999

# Splines in Approximation and Differential Operators: $(m, \ell, s)$ Interpolating–Spline

A. Le Méhauté and A. Bouhamidi

Abstract. The theory of interpolating $(m, \ell, s)$-spline, is an extension of Duchon's theory to a more general setting for multivariate splines. More precisely, we consider the problem of interpolation with respect to a finite number of linear and continuous functionals defined on a semi-Hilbert space $D_\ell^{-m}(\widetilde{H}^s)$ subject to the minimization of the semi-norm. We prove existence and uniqueness for the interpolating $(m, \ell, s)$-spline and we investigate some of its properties. The $(m, \ell, s)$-spline is explicitly given in terms of translates of a fundamental solution of a (pseudo) differential operators. We emphasis on some practical cases, including thin plates, Hardy multiquadrics, thin plates and curves under tension.

## 1. Introduction

This paper deals with interpolation problems using multivariate splines. Following the abstract spline theory of reproducing kernels in (semi-) Hilbert spaces [**6**], we prove that it is possible to define a spline, i.e. an interpolating function with a minimal norm, which solve an interpolation problem on non regular data; those data does not need to be pointwise evaluations of functions or derivatives.

The norm in the Hilbert space is defined from a differential elliptic operator. The justification behind this approach is that there exist many cases where the usual norms are not relevant. Spaces with those norms arise easily in different areas of applications: geometric invariance of surfaces, mechanical properties of surfaces, such as surfaces under tension, etc., leads to minimizing problems subject to interpolation constraints of Lagrange or Hermite type (or others) and the norm which is involved is related to a differential operator [**1**].

In the first section, we define the spaces where we'll search the splines, then there reproducing kernels. We define the interpolating splines and investigate some of their properties in the next two sections, and finish with a short review of examples, examples of fundamental solutions for differential operators and practical examples of $(m, \ell, s)$-splines.

---

1991 *Mathematics Subject Classification.* 05D05, 65D07, 65D10, 41A15.
This is the final form of the paper.

## 2. The Functional Spaces for $(m, \ell, s)$-Spline

Let $s$ be a real number such that $s < d/2$ and let $\widetilde{H}^s(\mathbb{R}^d)$ be the space of tempered distributions $u$ whose Fourier transforms $\hat{u}$ are measurable locally integrable valued complex functions on $\mathbb{R}^d$ and satisfying

$$\int_{\mathbb{R}^d} |2\pi\xi|^2 \left|\hat{u}(\xi)\right|^2 d\xi < +\infty.$$

On this space, we consider the following scalar product

$$(u,v)_{0,s} = \int_{\mathbb{R}^d} |2\pi\xi|^{2s} \hat{u}(\xi)\overline{\hat{v}(\xi)}\, d\xi \tag{2.1}$$

and the associated norm $|u|_{0,s} = (u,u)_{0,s}^{1/2}$. We have the following result (see Duchon [**3**])

Proposition 2.1. *The space $\widetilde{H}^s(\mathbb{R}^d)$ endowed with the scalar product* (2.1) *and its associated norm is a hilbertian subspace of $\mathcal{S}'$.*

For any integers $m \in \mathbb{N}^*$ and $\ell \in \mathbb{N}$, we introduce the Beppo–Levi type space denoted $D_\ell^{-m}(\widetilde{H}^s)$ or $X^{m,\ell,s}$ [**1**] of distributions on $\mathbb{R}^d$ with all derivatives of order $k$, with $m \le k \le m+\ell$, are in $\widetilde{H}^s(\mathbb{R}^d)$, namely

$$D_\ell^{-m}(\widetilde{H}^s) = \left\{v \in \mathcal{D}' \colon \forall \alpha \in \mathbb{N}^d, m \le |\alpha| \le m+\ell,\ D^\alpha v \in \widetilde{H}^s(\mathbb{R}^d)\right\}. \tag{2.2}$$

In the space $D_\ell^{-m}(\widetilde{H}^s)$ a semi-inner product is defined for any $u \in D_\ell^{-m}(\widetilde{H}^s)$ and $v \in D_\ell^{-m}(\widetilde{H}^s)$ by:

$$(u,v)_{m,\ell,s} = \sum_{m \le |\alpha| \le m+\ell} C_\alpha \int_{\mathbb{R}^d} |2\pi\xi|^{2s} \widehat{D^\alpha u}(\xi)\overline{\widehat{D^\alpha v}(\xi)}\, d\xi \tag{2.3}$$

where $C_\alpha$ are real positive constants. The associated semi-norm is

$$|u|_{m,\ell,s} = (u,u)_{m,\ell,s}^{1/2}. \tag{2.4}$$

Let us remark that for $\ell = 0$ and $C_\alpha = (|\alpha|!)/\alpha!$ the space $D_0^{-m}(\widetilde{H}^s)$ is the space $X^{m,s}$ in [**4**], endowed with the semi-scalar product:

$$(u,v)_{m,0,s} = \sum_{|\alpha|=m} \frac{m!}{\alpha!} \int_{\mathbb{R}^d} |2\pi\xi|^{2s} \widehat{D^\alpha u}(\xi)\overline{\widehat{D^\alpha v}(\xi)}\, d\xi \tag{2.5}$$

and the associated semi-norm

$$|u|_{m,0,s} = (u,u)_{m,\ell,s}^{1/2}. \tag{2.6}$$

We can prove the following:

Proposition 2.2. *The space $D_\ell^{-m}(\widetilde{H}^s)$ endowed with the topology associated to the semi-norm* (2.4) *is continuously imbedded in $X^{m,s}$ with the semi-norm* (2.6).

The next results may be considered as a consequence of the previous one, but it is also possible to have a direct proof.

Corollary 2.3. *If $-m - d/2 < s < d/2$, the elements of $D_\ell^{-m}(\widetilde{H}^s)$ are tempered distributions.*

PROPOSITION 2.4. *If $-m - d/2 < s < d/2$, the space $\mathcal{S}$ (functions in $C^\infty(\mathbb{R}^d)$ which are rapidly decreasing) is algebraically enclosed in $D_\ell^{-m}(\widetilde{H}^s)$, and the space $D_\ell^{-m}(\widetilde{H}^s)$ separates points in $\mathbb{R}^d$.*

Let us assume from now that there exists an integer $r$, with $0 \le r < m$ such that

$$-m + r + \frac{d}{2} < s < \frac{d}{2}. \tag{2.7}$$

Note that the condition (2.7) obviously implies the condition of the previous propositions.

PROPOSITION 2.5. *The space $D_\ell^{-m}(\widetilde{H}^s)$ with the semi-scalar product (2.3) and the associated semi-norm (2.4) is a semi-Hilbert space which is a subspace of $C^r(\mathbb{R}^d)$.*

The question now is to associate a (pseudo) differential operator to the semi-scalar product

It is well known that for any distribution $T$ on $\mathbb{R}^d$ and any positive integer $k$, the Fourier transform of the $k$-th iterate of the usual Laplace operator in $\mathbb{R}^d$ is given by

$$\widehat{\Delta^k T} = (-1)^k |2\pi\xi|^{2k} \widehat{T}. \tag{2.8}$$

When $k$ is a real number and $T$ a tempered distribution on $\mathbb{R}^d$, we can also define the $\Delta^k T$ [**9**] as the tempered distribution whose Fourier transform is given by

$$\begin{aligned} \widehat{\Delta^k T} &= e^{i\pi k} |2\pi\xi\,|^{2k} \widehat{T}, \quad \text{i.e.} \\ \Delta^k T &= \mathcal{F}^{-1}\big[e^{i\pi k}|2\pi\xi|^{2k}\widehat{T}\big]. \end{aligned} \tag{2.9}$$

The formula (2.9) is a natural extension of (2.8).

Let us now introduce a linear differential operator $\Delta_{m,\ell,s}$, defined from $\mathcal{S}'$ onto itself and such that, for any tempered distribution $T$ on $\mathbb{R}^d$,

$$\Delta_{m,\ell,s} T = e^{-i\pi s} \Delta^s \Bigg[ \sum_{m \le |\alpha| \le m+\ell} (-1)^{|\alpha|} C_\alpha D^{2\alpha} T \Bigg] \tag{2.10}$$

where the constants $C_\alpha$ are those involved in (2.3). It is obvious that, for any tempered distribution $T$ on $\mathbb{R}^d$ we have

$$\mathcal{F}(\Delta_{m,\ell,s} T) = g_{m,\ell,s}(\xi) \widehat{T} \tag{2.11}$$

where $g_{m,\ell,s}$ is the function given by $g_{m,\ell,s}(\xi) = |2\pi\xi|^{2s}\big[\sum_{m \le |\alpha| \le m+\ell} C_\alpha (2\pi\xi)^{2\alpha}\big]$. We obtain, from the relation $\sum_{|\alpha|=k} k!/\alpha!\xi^{2\alpha} = |\xi|^{2k}$, that the function $g_{m,\ell,s}$ satisfies the condition

$$\forall k = m, \dots, m+\ell, \ u \cup \exists A_k > 0 \colon g_{m,\ell,s}(\xi) \ge A_k |\xi|^{2k+2s} \quad \forall \xi \in \mathbb{R}^d. \tag{2.12}$$

(We can choose $A_k = \min_{|\alpha|=k}\big((C_\alpha . \alpha!)/k!)\big)$.

From (2.12), it is obvious that $g_{m,\ell,s}(\xi) = 0$ if and only if $\xi = 0$.

The next proposition gives us the connection between the semi-scalar product (2.3) and the operator $\Delta_{m,\ell,s}$.

PROPOSITION 2.6. *For any element $u$ of $D_\ell^{-m}(\widetilde{H}^s)$ and for any function $\varphi$ of $\mathcal{S}$, we have the following equalities*:

$$(u,\varphi)_{m,\ell,s} = \langle \Delta_{m,\ell,s}u, \overline{\varphi}\rangle = \langle u, \overline{\Delta_{m,\ell,s}\varphi}\rangle.$$

Let $\mathcal{P}_{m-1}$ be the space of polynomial of total degree $\leq m-1$, in $d$ real variables. The spline will be obtained through a projection in the Hilbert space , so we need to know what is the kernel of the projection in the corresponding semi-Hilbert space.

PROPOSITION 2.7. *The null space $\mathcal{N}$ of $\big(D_\ell^{-m}(\widetilde{H}^s), |\,.\,|_{m,\ell,s}\big)$ is $\mathcal{P}_{m-1}$. and the space $\mathcal{S}$ is dense in $D_\ell^{-m}(\widetilde{H}^s)$.*

## 3. Fundamental Solutions and Reproducing Kernels

A fundamental solution for the differential operator $\Delta_{m,\ell,s}$ on $\mathbb{R}^d$ is a distribution $K$ on $\mathbb{R}^d$ such that

$$\Delta_{m,\ell,s}K = \delta. \tag{3.1}$$

In practice, a fundamental solution is not always easy to exhibit. However, in our case, in order to compute a fundamental solution, we use the Fourier transform of (3.1) to get

$$g_{m,\ell,s}(\xi)\widehat{K(\xi)} = 1. \tag{3.2}$$

Then we search for a fundamental solution which is a tempered distribution. As $g_{m,\ell,s}$ vanishes only at the origin, and following Schwartz [**10**, pp. 281–284] there exists a fundamental solution $K$ which is a tempered distribution and is the limit, in $\mathcal{S}'$ of a finite linear combination of exponentials and polynomials which are also fundamental solutions.

Let us notice that a fundamental solution is not unique.

Thereafter, we assume the existence of a fundamental solution $K_{m,\ell,s}$, which is associated (up to a continuous extension) to a function of class $\mathcal{C}^{2r}$ on $\mathbb{R}^d$ and still denoted by $K_{m,\ell,s}$. We assume, furthermore that the Fourier transform $\widehat{K}_{m,\ell,s}$ of $K_{m,\ell,s}$ is a locally integrable function on $\mathbb{R}^d$. Hereafter, $r \geq 0$ denotes such an integer and is assumed to be such that

$$-2m + 2r + d < 2s < d. \tag{3.3}$$

PROPOSITION 3.1. *For any compactly supported distribution $T$ of order $\leq r$ orthogonal to $\mathcal{P}_{m-1}$ in the distributional sense , the convolution product $T * K_{m,\ell,s}$ element of $\mathcal{C}^r$ is an element of $D_\ell^{-m}(\widetilde{H}^s)$.*

In the next section we will need to apply the operator $\Delta_{m,\ell,s}$ to the convolution product $S * T$ where $S$ is a compactly supported distribution and $T$ is a tempered distribution, thus we prove the next proposition:

PROPOSITION 3.2. *If $S$ is a rapidly decreasing distribution (in particular if $S$ is compactly supported distribution) and $T$ is a tempered distribution, then*

$$\Delta_{m,\ell,s}(S * T) = S * \Delta_{m,\ell,s}T.$$

PROOF. The result is a simple consequence of $\Delta^s(S*T) = S*\Delta^s T$, which will be proved. If $k$ is a positive integer, and $S$, $T$ are two distributions such that the convolution product $S*T$ exists, it is usual that $\Delta^k(S*T) = \Delta^k S*T = S*\Delta^k T$. But when $k$ is a real number, this is not always true.

Now we suppose $S$ is a rapidly decreasing distribution and $T$ is a tempered distribution, then $S*T$ is a tempered distribution and we have

$$\mathcal{F}\big[\Delta^s(S*T)\big] = e^{i\pi s}|2\pi\xi|^{2s}\mathcal{F}(S*T)$$

and because $S$ is a rapidly decreasing distribution and $T$ is a tempered distribution we have $\mathcal{F}(S*T) = \widehat{S}\widehat{T}$ , then

$$\mathcal{F}\big[\Delta^s(S*T)\big] = \widehat{S}\widehat{\Delta^s T}.$$

But $\widehat{S}$ is a slowly increasing $\mathcal{C}^\infty$ function and $\widehat{\Delta^s T}$ is a tempered distribution then $\widehat{S}\widehat{\Delta^s T} = \mathcal{F}(S*\Delta^s T)$. Since $\mathcal{F}$ and $\overline{\mathcal{F}}$ are isomorphics in $\mathcal{S}'(\mathbb{R}^d)$ we obtain the aim result.

## 4. Interpolating $(m, \ell, s)$-Splines

Let $d(m)$ be the dimension of the space $\mathcal{P}_{m-1}$.Let us consider a set $L = (\ell_1, \cdots, \ell_p)$ of $p \geq d(m)$ compactly supported distributions of order $\leq r$, i.e. a set of $p$ continuous linear functionals defined on $\mathcal{C}^r$. We assume the following hypotheses:

**(H):** The set $L$ is linearly independent in $\mathcal{D}'$ and contains a subset $\widetilde{L}$ which is $\mathcal{P}_{m-1}$-unisolvent. For convenience, we assume that $\widetilde{L}$ is the set of the first $d(m)$-th distributions in $L$.

Namely, the hypothesis means that for any $(\lambda_1, \dots, \lambda_p) \in \mathbb{C}^p$ we have

$$\left(\left\langle \sum_{i=1}^{p} \lambda_i \ell_i, \varphi \right\rangle = 0 \ \forall \varphi \in \mathcal{D}\right) \quad \Longrightarrow (\lambda_1 = \cdots = \lambda_p = 0) \text{ and}$$

$$\left(\left\langle \sum_{i=1}^{d(m)} \lambda_i \ell_i, p \right\rangle = 0 \ \forall p \in \mathcal{P}_{m-1}\right) \Longrightarrow (\lambda_1 = \cdots = \lambda_{d(m)} = 0).$$

We obtain the following proposition :

PROPOSITION 4.1. *The hypotheses* **(H)** *is equivalent to the two hypothesis:*

**(H1):** *There exists a family $(\varphi_j)_{1\leq j\leq p}$ of functions in $\mathcal{D}$ such that:*

$$\langle \ell_i, \varphi_j \rangle = \delta_{ij}, \quad i, j = 1, \dots, p.$$

**(H2):** *There exists a basis $(\nu_j)_{1\leq j\leq d(m)}$ in $\mathcal{P}_{m-1}$ such that:*

$$\langle \ell_i, \nu_j \rangle = \delta_{ij}, \quad i, j = 1, \dots, d(m)$$

PROOF. It is obvious that the hypotheses **(H1)** and **(H2)** together imply **(H)**.

Suppose **(H)** holds, then **(H2)** also holds [**7**]. Let $i \in \{1, \dots, p\}$ and $F_i$ be the closure of the linear space $\operatorname{span}\{\ell_1, \dots, \ell_{i-1}, \ell_{i+1}, \dots, \ell_p\}$, $F_i$ is a closed linear subspace of the Hausdorff locally convex space $\mathcal{D}'$ and $\ell_i$ is not in $F_i$. By the Hahn-Banach theorem and its consequences (see [**13**, pp. 186–187]), there is a function $\varphi_i$ in $\mathcal{D}$ such that $\langle \ell_i, \varphi_i \rangle = 1$ and $\langle \ell_j, \varphi_i \rangle = 0$ for $j = 1, \dots, p$ and $j \neq i$.

PROPOSITION 4.2. *In particular, the assumption* (**H1**) *is fulfilled whenever the support of the* $(\ell_i)$ *are mutually disjoints.*

PROOF. Let $(S_i)_{1\le i\le p}$ the supports of the $\ell_i$'s and let us assume that we have $S_i \cap S_j = \emptyset$ for $i \neq j$. Let us fixed $j, j = 1, \dots, p$ and let $R_j = \bigcup_{k\neq j} S_k$. We have obviously $R_j \cap S_j = \emptyset$. But $R_j$ is a closed subset of $\mathbb{R}^d$ as a union of finite closed subsets and $S_j$ is a compact of $\mathbb{R}^d$, thus by Urysohn's lemma, there exists a function $\theta_j$ in $\mathcal{D}$ such that $0 \le \theta(x) \le 1$ for all $x \in \mathbb{R}^d$, $\theta \equiv 1$ on a neighborhood of $S_j$ and $\theta \equiv 0$ on a neighborhood of $R_j$. Let $\psi_j$ a function of $\mathcal{D}$ such that $\langle \ell_j, \psi_j\rangle \neq 0$ and let $\varphi_j = \theta_j / \big(\langle \ell_j, \psi_j\rangle\big)\psi_j$. For $i = 1, \dots, p$, we have $\langle \ell_i, \theta_j\psi_j\rangle = \delta_{i,j}\langle \ell_j, \psi_j\rangle$, which prove the result.

EXAMPLE. Given any set of distinct points $a_1, \dots, a_p$ in $\mathbb{R}^d$ $\big(p > d(m)\big)$ and

$$\ell_i = \begin{cases} \delta_{a_i} & \text{for } i = 1, \dots, d(m) \\ \sum_{|\alpha|\le k_i} C_{\alpha,i} D^\alpha \delta_{a_i} & \text{for } i = d(m)+1, \dots, p \end{cases}$$

where $k_i \in \mathbb{N}$, $C_{\alpha,i} \in \mathbb{R}$ and $D^\alpha\delta_{a_i}$ is the derivative of order $\alpha \in \mathbb{N}^d$ of the Dirac measure at the point $a_i$. The support of $\ell_i$ is the singleton $\{a_i\}$, then it is obvious the family $L = (\ell_1, \dots, \ell_p)$ satisfies (**H**). In the particular case where $k_i = 0$ and $C_{0,i} = 1$, the family $L$ is an interpolation scheme of Lagrange type.

For $z = (z_1, \dots, z_p) \in \mathbb{C}^p$, let $G(z)$ be the subset of $D_\ell^{-m}(\widetilde{H}^s)$ such that

$$G(z) = \big\{u \in D_\ell^{-m}(\widetilde{H}^s)\langle \ell_i, u\rangle = z_i;\ i = 1, \dots, p\big\}. \tag{4.1}$$

Let us remark that for a given $z = (z_1, \dots, z_p) \in \mathbb{C}^p$ the subset $G(z)$ is non empty: in any case $\sum_{i=1}^p z_i\varphi_i$ is in $G(z)$, where the $(\varphi_j)_{1\le j\le p}$ are defined in Proposition 4.1 (**H1**)

DEFINITION 4.3. *For any $z \in \mathbb{C}^p$, an interpolating $(m,\ell,s)$-spline is an element $\sigma$ of $D_\ell^{-m}(\widetilde{H}^s)$ which is solution of the problem:*

$$\begin{cases} \sigma \in G(z) \text{ and} \\ |\sigma|_{m,\ell,s} = \min_{u\in G(z)} |u|_{m,\ell,s}. \end{cases} \tag{4.2}$$

Using the abstract spline theory [**6**] we obtain a characterization of $(m,\ell,s)$-splines [**2**]

THEOREM 4.4. *Given $z = (z_1, \dots, z_p) \in \mathbb{C}^p$, there is a unique interpolating $(m,\ell,s)$-spline function $\sigma$ in $D_\ell^{-m}(\widetilde{H}^s)$ solution of the problem (4.2). Moreover, $\sigma$ is the unique element of minimal semi-norm in $D_\ell^{-m}(\widetilde{H}^s)$ which interpolates $z$ with respect the $\ell_i$'s and $\sigma$ is explicitely given by*

$$\sigma = \sum_{i=1}^p \lambda_i \ell_i * K_{m,\ell,s} + q, \ \textit{with} \ \sum_{i=1}^p \lambda_i\langle \ell_i, \xi^\alpha\rangle = 0, \quad \forall\alpha,\ |\alpha| \le m-1$$

*and $q$ is a polynomial in $\mathcal{P}_{m-1}$.*

Let $(q_j)_{j=1,\dots,d(m)}$ be any basis of $\mathcal{P}_{m-1}$, the $(m,\ell,s)$-spline $\sigma$ may be written as

$$\sigma = \sum_{i=1}^p \lambda_i \ell_i * K_{m,\ell,s} + \sum_{i=1}^{d(m)} \alpha_i q_i. \tag{4.3}$$

The coefficients $\lambda_i$, $i = 1, \ldots, p$ and $\alpha_j$, $j = 1, \ldots, d(m)$ satisfying the interpolating conditions and the orthogonality conditions, are solutions of the system of linear equations

$$(4.4)\quad \begin{cases} \langle \ell_k, \sigma \rangle = \sum_{i=1}^{p} \lambda_i \langle \ell_k, \ell_i * K_{m,\ell,s} \rangle + \sum_{j=1}^{d(m)} \alpha_j \langle \ell_k, q_j \rangle = z_k & \text{for } k = 1, \ldots, p \\ \sum_{i=1}^{p} \lambda_i \langle \ell_i, q_j \rangle = 0 & \text{for } j = 1, \ldots, d(m). \end{cases}$$

It is possible write this system as

$$\begin{pmatrix} K & Q \\ Q^T & 0 \end{pmatrix} \begin{pmatrix} \Lambda \\ \alpha \end{pmatrix} = \begin{pmatrix} Z \\ 0 \end{pmatrix},$$

where $K = \big(\langle \ell_i, \ell_j * K_{m,\ell,s} \rangle\big)_{1 \le i,j \le p}$ is the $p \times p$ symmetric matrix, $Q$ is the $p \times d(m)$ matrix $Q = \big(\langle \ell_i, q_j \rangle\big)_{\substack{1 \le i \le p \\ 1 \le j \le d(m)}}$, $\Lambda = (\lambda_1, \ldots, \lambda_p)^T$, $\alpha = (\alpha_1, \ldots, \alpha_{d(m)})^T$ and $Z = (z_1, \ldots, z_p)^T$ is the interpolating vector and it follows from Proposition 2.5, and the Hypotheses **(H1)** and **(H2)** that the matrix

$$A = \begin{pmatrix} K & Q \\ Q^T & 0 \end{pmatrix}$$

is nonsingular.

## 5. Examples

In this section we give some examples, some plots of this examples are presented in [**1**].

5.1 Lagrange interpolation with thin plate spline. Let us first consider the $(m, s)$-spline of Duchon [**4**]. Here we choose $\ell = 0$ and $C_\alpha = m!/\alpha!$ for $|\alpha| = m$. Given a set $\{A_i, i = 1, \ldots, N\}$ of distinct points, $\big(N > d(m)\big)$, we define the $\ell_i$ by $\langle \ell_i, f \rangle = f(A_i)$ for $i = 1, \ldots, N$. The usual thin plate splines are obtained with the kernel $\Delta_{2,0,0}$ and more generally the $(m, s)$-spline of Duchon with $\Delta_{m,0,s}$. Another example, the pseudo-cubic splines in $\mathbb{R}^2$, are related to $m = 2$ and $s = 1/2$.

5.2 Hermite interpolation. Here the data set is splitted in subsets of distinct points, each related to a different degree of Hermite conditions. It becomes rapidly tedious to write it down, but in the simplest case of hermite interpolation of degree one, we consider the data points

$$\{A_i, i = 1, \ldots, N_0\} \cup \{B_j, j = 1, \ldots, N_1\}$$

and the interpolation conditions $\{\ell_i, \ell_j^1, \ell_j^2\}$ such that

$$\langle \ell_i, f \rangle = f(A_i), \quad i = 1, \ldots, N_0,$$
$$\langle \ell_j^1, f \rangle = \frac{\partial}{\partial x} f(B_j), \ \langle \ell_j^2, f \rangle = \frac{\partial}{\partial y} f(B_j), \quad j = 1 \ldots, N_1.$$

The thin plates involved need to be $\mathcal{C}^1$, and we consider $\Delta_{2,0,1/2}$, together with $r = 1$; thus $K_{2,0,1/2}(X) = |X|^3$ and

$$\sigma(M) = \sum_{i=1}^{N_0} \lambda_i \ell_i * K_{2,0,1/2}(M) + \sum_{j=1}^{N_1} \lambda_j^1 \ell_j^1 * K_{2,0,1/2}(M) + \lambda_j^2 \ell_j^2 * K_{2,0,1/2}(M) + q(M)$$

where $q$ is a polynomial of degree 2.

5.3 Local averaging splines. Another case related to these splines, whatever the differential operator could be (and its related kernel whenever it is available) involves data conditions which are not pointwise evaluations.

Let us consider a set $\{\Omega_i, i = 1; \ldots, N\}$ of disjoint domains in $\mathbb{R}^d$, we still use Urysohn's lemma to prove that there exists an interpolating spline subject to the conditions

$$\langle \ell_i, f \rangle = \frac{1}{\operatorname{meas} \Omega_i} \int_{\Omega_i} f(t)\, dt.$$

and the splines are obtained the same manners for Lagrange interpolation, solving a linear system.

## 6. Examples of Operators $\Delta_{m,l,s}$ and Fundamental Solutions $K_{m,l,s}$.

The main drawback for using $(m, \ell, s)$-splines is that our approach relies on the explicit knowledge of fundamental solutions associated to pseudo differential operators. However, there are many practical cases where such solutions are available and we try to collect some results in this section.

From Garnir [**5**], we call meta-harmonic the operator $E_k = \Delta - k^2 I$, where $I$ is the identity. The harmonic operator $E_0 = \Delta$ is singular and is not obtain from $E_k$ as $k \to 0$. The $4^{th}$-order operators are $E_k E_\ell = (\Delta - k^2 I)(\Delta - \ell^2 I)$, including $E_k E_0$ and the singular bi-harmonic operator $E_O E_0 = \Delta^2$.

Proposition 6.1. *Let $K(X)$ be a fundamental solution for the operator$E_{k^*}$ where $k^* = \sqrt{k^2 + 1/4 \sum_{i=1}^d a_i^2}$. Then a fundamental solution for the operator $\left(E_k + \sum_{i=1}^d a_i \partial/(\partial x_i)\right)$ is given by $e^{(-1/2 \sum_{i=1}^d a_i . x_i)} K(X)$.*

Proposition 6.2. *Let $K_k(.)$ (resp. $K_\ell(.)$) be a fundamental solution for $E_k$ (resp. $E_\ell$). Whenever $k$ and $\ell$ are not both equal to 0, a fundamental solution for the operator $E_k E_\ell$ is given by $1/(k^2 - \ell^2)\left(K_k(.) - K_\ell(.)\right)$.*

Let us consider now some practical cases of $(m, \ell, s)$-splines.

In 1976, Duchon introduces the $(m, 0, s)$-splines, where m is a positive integer and s a real number such that, in $\mathbb{R}^d$, $2s - d < 0$ and $2m + 2s - d > 0$. They are related to the operator

$$\Delta_{m,0,s} = e^{-i\pi(m+s)} \Delta^s[\Delta^m], \quad m \text{ positive integer, } s \text{ real.}$$

The fundamental solution is well known [**10**] and the hilbertian kernel $K_{m,0,s}$ in $\mathbb{R}^d$ depends upon $2m + 2s - d$:

(i) Whenever $2m + 2s - d$ is an even integer:

$$K_{m,0,s}(\xi) = c_{m,s} |\xi|^{2m+2s-d} \ln |\xi| \quad \text{where}$$

$$c_{m,s} = \frac{(-1)^{1/2+m+s+1}}{2^{2m+2s-1} \pi^{d/2} \Gamma(m+s)(m+s-d/2)!} \quad \text{is a constant.}$$

Here is a list of some fundamental solutions and operators in $\mathbb{R}^d$, $d = 1, 2, 3$. (the notation $r$ stand for $r = \|x\|$, the euclidean norm of $x$ in $\mathbb{R}^d$.)

| Operator | Dimension | Solution |
|---|---|---|
| $E_k = \Delta - k^2 I$ | $d = 1$ | $-\frac{1}{2}\frac{1}{k}e^{-k\|x\|}$ |
| | $d = 2$ | $-\frac{1}{2}\mathcal{K}_0(kr)$ |
| | $d = 3$ | $-\frac{1}{4\pi}\frac{1}{r}e^{-kr}$ |
| $E_0 = \Delta$ | $d = 1$ | $-\frac{1}{2}\|x\|$ |
| | $d = 2$ | $-\frac{1}{2\pi}\ln r$ |
| | $d = 3$ | $-\frac{1}{4\pi}\frac{1}{r}$ |
| $E_0 E_k = \Delta(\Delta - k^2 I)$ | $d = 1$ | $-\frac{1}{2}\frac{1}{k^3}\{e^{-k\|x\|} + k\|x\|\}$ |
| | $d = 2$ | $-\frac{1}{2\pi}\{\mathcal{K}_0(kr) + \ln r\}$ |
| | $d = 3$ | $-\frac{1}{4\pi}\frac{1}{k^2 r}\{1 - e^{-kr}\}$ |
| $E_{u(k)} E_{\bar{u}(k)} = \Delta^2 + k^2 I$ | $d = 1$ | $-\frac{\sqrt{k}}{2k^2}\{e^{(-\sqrt{k/2}\|x\|)}\sin(\sqrt{k/2}\|x\| + \pi/4)\}$ |
| | $d = 2$ | $-\frac{1}{2\pi k}\{\mathrm{kei}\sqrt{k}r\}$ |
| $u(k) = (1+i)\sqrt{\frac{k}{2}}$ | $d = 3$ | $-\frac{1}{4\pi k}\frac{1}{r}\{e^{-\sqrt{k/2}r}\sin(\sqrt{k/2}r)\}$ |
| $E_0 E_0 = \Delta^2$ | $d = 1$ | $-\frac{1}{12}\|x\|^3$ |
| | $d = 2$ | $-\frac{1}{8\pi}r^2 \ln r$ |
| | $d = 3$ | $-\frac{1}{8\pi}r$ |

(ii) When $2m + 2s - d$ is not an even integer (and is not necessarily an integer):

$$K_{m,0,s}(\xi) = d_{m,s}|\xi|^{2m+2s-d} \quad \text{where}$$
$$d_{m,s} = \frac{\Gamma(d/2 - m - s)}{2^{2m+2s}\pi^{d/2}\Gamma(m+s)}.$$

The usual practical cases in $\mathbb{R}^2$ are the following:
(1) $m = 2$, $d = 2$, $s = 0$, which corresponds to the usual thin plate splines:

$$\Delta_{2,0,0} = \Delta^2, \quad K_{2,0,0}(x) = \frac{1}{2\pi}r^2 \ln r.$$

(2) $m = 2$, $d = 2$, $s = 1/2$, which gives the pseudo-cubic splines:

$$\Delta_{2,0,1/2} = e^{-i5/2\pi}\Delta^{1/2}[\Delta^2], \quad K_{2,0,1/2}(x) = r^3.$$

(3) $m = 3$, $d = 2$, $s = 1/2$, which gives the pseudo-quintic splines:

$$\Delta_{3,0,1/2} = e^{-i7/2\pi}\Delta^{1/2}[\Delta^3], \quad K_{3,0,1/2}(x) = r^5.$$

(4) The so-called direct multiquadrics are also a particular case of $(m,\ell,s)$-splines. They are using in $\mathbb{R}^d$ the kernel $K(r) = \sqrt{r^2 = c^2}$ where $c$ is a given constant. In order to write them as a $(m,\ell,s)$-splines, one has to go to $\mathbb{R}^{d+1}$ and associate to the point $M \in \mathbb{R}^d$ the new point $\widetilde{M} = (M,c) \in \mathbb{R}^{d+1}$. Then $\|\widetilde{M}\| = \sqrt{r^2+c^2}$ and the multiquadric is nothing but the restriction to $\mathbb{R}^d$ of the $(m,\ell,s)$-splines defined in $\mathbb{R}^{d+1}$ with $m = 1, \ell = 0, s = (d-1)/2$ and the operator and kernel:

$$\Delta_{1,0,(d-1)/2} = e^{-i(d-1)/2\pi}\Delta^{(d-1)/2}[\Delta],$$
$$K_{1,0,(d-1)/2}(\xi) = \frac{(-1)^{(d+1)/2}}{2^{d+1}\pi^{(d+1)/2}\Gamma\big((d+1)/2\big)}.$$

Let us remark that this presentation explain why one has to add a constant to the linear combination of translates of the kernel when using multiquadrics (as for thin plate, one adds a polynomial of degree one). It is because all the points $\widetilde{M} = (M,c)$ lie on the same hyperplane in $\mathbb{R}^{d+1}$ and thus we are only allowed to use $m = 1$.

(5) Thin plates under tension (and for the particular case of $d = 1$, curves under tension) provide another example of $(m,\ell,s)$-splines, this time with $\ell$ different from 0. Let us consider:

$$\Delta_{2,1,0} = \Delta^2 - \tau^2\Delta = \Delta(\Delta - \tau^2 I) = E_0 E_\tau$$

where $k$ is a given constant. Explicitely,

$$\Delta_{2,1,0}(u) = \sum_{|\alpha|=2} \frac{2!}{\alpha!}\int_{\mathbb{R}^d} \big|D^\alpha u(\xi)\big|^2 \, d\xi + k^2 \sum_{|\alpha|=1}\int_{\mathbb{R}^d} \big|D^\alpha u(\xi)\big|^2 \, d\xi.$$

There are different ways to exhibit a fundamental solution. For instance:

(1) $d = 1$; $K_k(\xi) = ae^{-k|\xi|} + be^{+k|\xi|}$ where $a$ and $b$ are real constants. This particular case of curves under tension provides us a theoretical construction for the exponential splines [**12**] which are usually introduced as piecewise splines with each piece defined as $\sigma_k^{(i)}(x) = a_i + b_i x + c_i \operatorname{sh}(kx) + d_i \operatorname{ch}(kx)$ for $x \in [x_i, x_{i+1}]$.

(2) $d = 2$ ; $K_k(x) = a\mathcal{K}_0(kr) + b\mathcal{I}_0(kr)$, here $a$ and $b$ are real constants and $\mathcal{K}_0$ and $\mathcal{I}_0$ are the Bessel functions of first (resp. second) kind.

(3) $d = 3$ ; $K_k(x) = 1/r(ae^{-kr} + be^{+kr})$ where $a$ and $b$ are real constants.

As we have already noticed, those fundamental solutions are not unique and it is interesting to consider the following:

(4) $d = 1$; $K_k(\xi) = -1/(2k^3)\big(e^{-k|\xi|} + k|\xi|\big)$.

(5) $d = 2$; $K_k(x) = -1/(2\pi k^2)\big(\mathcal{K}_0(kr) + \ln r\big)$, and even more interesting:

$$K_k(x) = -\frac{1}{8\pi}r^2\ln r - \frac{1}{8\pi}r^2\Big(\gamma + \ln\frac{r}{2}\Big) + \big(\mathcal{K}_0(kr) + \ln r\big) + \frac{1}{2\pi k^2}\mathcal{K}_0(kr),$$

where the thin plate under tension appear as a perturbation of the usual thin plate.

(6) $d = 3$; $K_k(x) = -1/(4\pi k^2 r)(e^{-kr} - 1)$.

## Conclusion

As a conclusion, we want to emphasis on the fact that this approach of multivariate splines give us the opportunity to enlarge the list of differential problems which can be consider for the minimizing properties of the splines, but also that the explicit solutions rely on the ability to exhibit a practical fundamental solution for the underlying differential operator.

## References


1. A. Bouhamidi, *Approximation et interpolation par des fonctions splines radiales à plusieurs variables*, Dissertation Thesis, Univ. Nantes (1992).
2. A. Bouhamidi and A. Le Méhauté, *Spline curves and surfaces with tension*, Wavelets, Images, and Surface Fitting (Chamonix-Mont-Blanc, 1993), A K Peters, Wellesley, MA, 1994, pp. 51–58.
3. J. Duchon, *Fonctions-spline à énergie invariante par rotation*, rapport de recherche no. 27, Grenoble (1976).
4. ———, *Interpolation des fonctions de deux variables suivant le principe de la flexion des plaques minces*, Rev. Franaise Automat. Informat. Recherche Oprationnelle Sr. Rouge Anal. Numer. **10** (1976), no. R-3, 5–12. (French)
5. H. G. Garnir, *Sur les distributions résolvantes des opérateurs de la physique mathématique*, Bull. Soc. Roy. Sci. Liège (1951), 174–195 and 271–296.
6. P. J. Laurent, *Approximation et optimisation*, Hermann, Paris, 1972.
7. A. Le Méhauté, *Interpolation et approximation par des fonctions polynomiales par morceaux dans* $\mathbb{R}^n$, Thèse d'État, Univ. Rennes (1984).
8. A. Le Méhauté and A. Bouhamidi, $L^{m,l,s}$*-splines in* $\mathbb{R}^d$, Internat. Ser. Numer. Math. **105** (1992), Birkhauser, Basel, 135–154.
9. C. Rabut, *B-splines polyharmoniques cardinales: Interpolation, quasi-interpolation, filtrage*, Thèse d'État, Toulouse (1990).
10. L. Schwartz, *Théorie des distributions*, Hermann, Paris, 1966.
11. ———, *Sous-espaces hilbertiens d'espaces vectoriels topologiques et noyaux associés (noyaux reproduisants)*, J. Anal. Math. **13** (1964), 115–256. (French)
12. D. G. Schweikert, *An interpolation curve using a spline in tension*, J. Math. Phys. **45** (1966), 312–317.
13. F. Trèves, *Topological vector spaces*, Distributions and Kernels, Academic Press, New York, 1967.



Département de Mathématiques, Faculté des Sciences, Université de Nantes, BP 92208, 44322 Nantes Cedex 3, France.

*E-mail address*: alm@math.univ-nantes.fr

L.M.A. Université du Littoral, BP 699, 62228 Calais Cedex , France

*E-mail address*: bouhamid@lma.univ-littoral.fr

Centre de Recherches Mathématiques
CRM Proceedings and Lecture Notes
Volume **18**, 1999

# New Families of B-Splines on Uniform Meshes of the Plane

Paul Sablonnière

ABSTRACT. This talk is a survey of some results obtained recently by the author and coworkers on the existence and the properties of some families of piecewise polynomial functions with bounded supports (B-splines) on the two classical uniform triangulations of the plane. Given an integer $k \geq 0$ and a simple polygon $P$ consisting of a small number of triangles (e.g. a triangle, a square or an hexagon), we first study the existence of $C^k$ splines with support $P$ and of minimal degree (simple B-splines) whose integer translates form (or do not form) a partition of unity. Then, we define composed B-splines by repeated convolution of simple B-splines with the piecewise affine pyramid, for the first triangulation, or with characteristic functions of small squares or lozenges, for the second triangulation. Various properties of these new B-splines are discussed, in particular the local and global linear independence of the families of integer translates, the approximation order in the space generated by this family and the solution of some Lagrange interpolation problems on rectangular domains.

## 1. Introduction

Let $\tau_1$ denote the uniform triangulation of $\mathbb{R}^2$having $\mathbb{Z}^2$ as set of vertices and whose edges are parallel to the three directions $e_1 = (1,0)$, $e_2 = (0,1)$, $e_3 = (1,1)$. Similarly, let $\tau_2$ denote the uniform triangulation whose set of vertices is $\mathbb{Z}^2 \cup (\mathbb{Z} + \frac{1}{2})^2$ and whose edges are parallel to the four directions $e_1$, $e_2$, $e_3$ and $e_4 = (1,-1)$. For $\tau = \tau_1$ or $\tau_2$, $\mathbb{P}_d^r(\tau)$ denotes the space of piecewise polynomial functions of degree $d$ and class $C^r$ in two variables.

We are interested in the construction of *splines with bounded support* (called B-splines in the following) obtained by repeated convolution of *simple* B-*splines* with the piecewise affine pyramid $\pi$ on $\tau_1$ and the characteristic function $\sigma$ (resp. $\lambda$) of the unit square (resp. the unit lozenge) on $\tau_2$. Classical *box-splines* (see [**2**]) are important examples of such B-splines . Simple B-splines have *simple supports* $P$ with various shapes. On $\tau_1$, we have restricted our study to the cases $P = \Delta_1$ (one triangle), $P = \Sigma_1$ (one unit square), and $P = H_1$ (the hexagon formed by the six triangles surrounding a given vertex). The corresponding simple B-splines are called respectively $\Delta_1$-splines, $\Sigma_1$-splines and $H_1$-splines. On $\tau_2$, we consider simple

1991 *Mathematics Subject Classification.* Primary: 41A15; Secondary:41A63, 41A05, 41A35, 65D07.

This is the final form of the paper.

supports $P = \Sigma_1$ (one unit square centered at some point in $(\mathbb{Z}+\frac{1}{2})^2$) and $P = \Lambda_1$ (one unit lozenge centered at some point in $\mathbb{Z}^2$). The corresponding B-splines are called respectively $\Sigma_1$ and $\Lambda_1$-splines. Some simple B-splines have been already described in [**5**] under the name of *vertex-splines.*

Once the support $P$ and the smoothness $r \geq 0$ are given, our first task is to determine the *minimal degree* $d = d(r)$ for which the construction of a simple B-spline of class $C^r$ is possible. For sake of simplicity, we assume the invariance of the B-spline with respect to the isometries of its support. Otherwise it would be much more difficult to obtain satisfying results of existence and uniqueness. In general, the *family of integer translates* $B(\phi) = \{\phi(. - \alpha), \alpha \in \mathbb{Z}^2\}$ of a simple B-spline $\phi$ does not form a partition of unity. When it is the case, we say that $\phi$ is a *complete* simple B-spline, otherwise, we use the term of *reduced* simple B-spline.

Then, we construct *composed* B-*splines* $\phi$ by repeated convolution of one of the previous simple B-splines with $\pi$, $\sigma$ or $\lambda$. It is not hard to compute their smoothness and their degree, to describe their support, or to determine the space $\mathbb{P}(\phi) = \mathbb{P}_n$ of polynomials of maximal total degree $n$ included in the space $S(\phi)$ generated by the set $B(\phi)$. Given a smooth function, we deduce immediately its approximation order $n+1$ in the space $S(\phi)$. These properties and the construction of quasi-interpolants are direct consequences of the *Strang-Fix theory* [**17**] applied to Fourier transforms of B-splines (see also [**2,3**] and [**4**] for the case of classical box-splines). For B-splines which are not box-splines, we have not in general a refinement equation, therefore there is no subdivision algorithm. However, once the BB (Bernstein-Bézier form) of simple B-splines is known, the one of composed B-splines can be computed by standard convolution algorithms (similar to those proposed e.g. by Lai in [**9**]). The approximation order is in general proportional to the size of the support. Thus B-splines with small supports are easy to manipulate, but have a low approximation order while B-splines with larger supports are more difficult to construct, but have a higher approximation order. Moreover, the degrees are also decreasing when the support increases, for a given smoothness. So, we have at our disposal a great choice of piecewise polynomial functions with bounded supports of various shapes which can be used for the solution of approximation and interpolation problems and eventually for multiresolution analysis. The construction of such B-splines has been given in references [**12,13,14,15**] and [**16**]. For the sake of completeness, we recall their main results in Sections 2 and 3. We give some additional results on B-splines with octagonal supports on $\tau_2$. In Section 4, we study the global linear independence of B-splines : new results are obtained on $\tau_2$. In Section 5, we study the local linear independence of certain families of B-splines with small supports. In Section 6, we give some properties of two quasi-interpolants which are used in the next section. Finally, Section 7 is devoted to the study of two specific Lagrange interpolation problems on rectangular domains. Existence and uniqueness results are given together with error bounds. Similar topics are also studied in [**8**], with complementary results.

## 2. Simple B-Splines

**2.1 Triangulation $\tau_1$.** For a given smoothness $r \geq 0$, we denote respectively by $\beta_r$, $\chi_r$ and $\tilde{\pi}_r$ the $\Delta_1$,$\Sigma_1$ and *reduced* $H_1$-splines of minimal degree, when they exist. We denote by $\pi_r$ the *complete* $H_1$-spline for which $B(\pi_r)$ is a partition of unity. They are characterized by the following theorem (see [**14,15,16**]).

THEOREM 1.
(i) *For all* $r \geq 0$, *there exists a unique* $\Delta_1$*-spline* $\beta_r$ *of degree* $3r+3$ *(up to a multiplicative constant) and this degree is minimal.*
(ii) *For all* $r \geq 0$, *there exist* $\Sigma_1$*-splines* $\chi_r$ *of minimal degree* $3r+2$ *for* $r$ *even and* $3r+3$ *for* $r$ *odd.*
(iii) *For all* $r \geq 0$, *there exists a unique complete* $H_1$*-spline* $\pi_r$ *of minimal degree* $3r+1$ *for* $r$ *even and* $3r+2$ *for* $r$ *odd. In particular,* $\pi = \pi_0$.
(iv) *There is no reduced* $H_1$*-spline* $\tilde{\pi}_r$ *for* $(r,d) = (1,3)$ *and* $(2,4)$, *and for pairs* $(2r-1, 4r)$ *and* $(2r, 4r+1)$ *whenever* $r \geq 2$. *There exist reduced* $H_1$*-splines for* $(r,d) = (1,4), (2,5), (3,9), (4,10), (5,13), (6,14)$ *and the degrees obtained are minimal, but there is not yet a general result in this case.*

**2.2 Triangulation $\tau_2$.** For $r \geq 0$, we denote by $\tilde{\sigma}_r$ and $\tilde{\lambda}_r$ (resp. $\sigma_r$ and $\lambda_r$) the *reduced* (resp. *complete*) $\Sigma_1$- and $\Lambda_1$-splines of class $C^r$ which are characterized by the following theorem (see [**12**]). For $r=-1$, $\tilde{\sigma}_{-1} = \sigma$ and $\tilde{\lambda}_{-1} = \lambda$, are the characteristic functions of $\Sigma_1$ and $\Lambda_1$.

THEOREM 2.
(i) *Up to a multiplicative constant, there exists a unique* $\Sigma_1$*-spline* $\tilde{\sigma}_r$ *(resp.* $\Lambda_1$*-spline* $\tilde{\lambda}_r$*) of class* $C^r$, *of degree* $d = 2r+1$ *for* $r$ *even and* $d = 2r+2$ *for* $r$ *odd. These degrees are minimal. Moreover the integer translates of the two functions do not form a partition of unity.*
(ii) *There exist* $\Sigma_1$*-splines* $\sigma_r$ *and* $\Lambda_1$*-splines* $\lambda_r$ *of degree* $d = 3r+1$ *for* $r$ *even and* $d = 3r+2$ *for* $r$ *odd, whose integer translates form a partition of unity. For* $r = 2s-1$ *or* $2s$, $s \geq 1$, *their construction depends on* $s(s+1)/2$ *arbitrary parameters.*

Now, by convolution of these simple B-splines with convolution powers of $\pi$, $\sigma$ and $\lambda$, we can construct different families of *composed* B-splines on the two triangulations.

## 3. Some Families of Composed B-splines

**3.1 Triangulation $\tau_1$.**
Let $\pi = \pi_0$ be the $H_1$-spline in $\mathbb{P}_1^0(\tau_1)$, i.e. the classical piecewise affine pyramid (Courant's function). Then $\pi^k = \pi * \cdots * \pi$ (k times) is the box-spline in $\mathbb{P}_{3k+1}^{2k}(\tau_1)$ (see e.g. [**2**], Chapter 1, or [**4**], Chapter 2). For $k=0$, we define $\beta_r^0 = \beta_r$, $\chi_r^0 = \chi_r$, $\pi_r^0 = \pi_r$ and for $k \geq 1$, $\beta_r^k = \beta_r * \pi^k$, $\chi_r^k = \chi_r * \pi^k$, $\tilde{\pi}_r^k = \tilde{\pi}_r * \pi^k$. Note that the powers are convolution powers.

THEOREM 3. *The three families of composed* B*-splines defined above have the following properties*:
(i) $\beta_r^k$ *is of degree* $3(r+k+1)$, $\chi_r^k$ *is of degree* $3(r+k)+2$ *for* $r$ *even and of degree* $3(r+k+1)$ *for* $r$ *odd. Moreover* $\mathbb{P}(\beta_r^k) = \mathbb{P}(\chi_r^k) = \mathbb{P}_{2k-1}$ *for* $k \geq 1, r \geq 0$.
(ii) $\tilde{\pi}_r^k$ *is of degree* $3(r+k)+1$ *for* $r$ *even, of degree* $3(r+k)+2$ *for* $r$ *odd, and* $\mathbb{P}(\tilde{\pi}_r^k) = \mathbb{P}_{2k}$ *for* $k > 1$ *and* $r > 1$.

They have hexagonal supports of various shapes which are described e.g. in [**15**].

As an example, we now give the list of 7 different B-splines of class $C^3$. Let $\beta$ (resp. $\chi$) be the characteristic function of $\Delta_1$ (resp. $\Sigma_1$).

(1) The box-spline $\phi_1 = \chi * \pi^2$ and the B-spline $\phi_2 = \beta * \pi^2$ are of degree 6 and $\mathbb{P}(\phi_1) = \mathbb{P}_4$, but $\mathbb{P}(\phi_2) = \mathbb{P}_3$.

(2) $\phi_3 = \pi_1^1 = \pi_1 * \pi$ and $\phi_4 = \chi_1^1 = \chi_1 * \pi$ are B-splines of degree 8 and $\mathbb{P}(\phi_3) = \mathbb{P}_2$, but $\mathbb{P}(\phi_4) = \mathbb{P}_1$.

(3) $\phi_5 = \beta_1^1 = \beta_1 * \pi$ and the *reduced* $H_1$-spline $\phi_6 = \tilde{\pi}_3$ are of degree 9 and $\mathbb{P}(\phi_5) = \mathbb{P}_1$, but $S(\phi_6)$ does not contain $\mathbb{P}_0$).

(4) $\phi_7 = \pi_3$ is the *complete* $H_1$-spline of degree 11 such that $\mathbb{P}(\phi_7) = \mathbb{P}_0$.

### 3.2 Triangulation $\tau_2$.

DEFINITIONS 1.

(i) *Let $\sigma^k$ (resp. $\lambda^l$) be the $k-th$ (resp. $l-th$) convolution power of $\sigma$ (resp. $\lambda$), and define the box-splines $\omega^{k,0} = \sigma^k$, $\omega^{0,l} = \lambda^l$ and $\omega^{k,l} = \sigma^k * \lambda^l$ for $k,l \geq 1$ ($\omega = \omega^{1,1} = \sigma * \lambda$ is the classical $C^1$ quadratic box-spline).*

(ii) *The composed* B-*splines that we consider are defined as follows: $\tilde{\sigma}_r^{0,0} = \tilde{\sigma}_r$, $\tilde{\lambda}_r^{0,0} = \tilde{\lambda}_r$ and for $(k,l) \neq (0,0)$, $\tilde{\sigma}_r^{k,l} = \tilde{\sigma}_r * \omega^{k,l}$, $\tilde{\lambda}_r^{k,l} = \tilde{\lambda}_r * \omega^{k,l}$.*

(iii) *We denote $\Omega_{k,l}$ the octagon whose sides parallel to $e_1$ and $e_2$ (resp. $e_3$ and $e_4$) are composed of $k$ (resp. $l$) edges of $\tau_2$. For example $\Omega = \Omega_{11}$ is the octagonal support of the quadratic box-spline $\omega$. $\Omega_{k,0} = \Sigma_k$ denotes the square whose sides parallel to the axes are of length $k$. Similarly, $\Omega_{0,l} = \Lambda_l$ denotes the lozenge whose sides parallel to the diagonals are of length $l$.*

We observe that, for $r = -1$ and $r = 0$, composed B-splines are box-splines. Let $d_r$ be the common degree of $\tilde{\sigma}_r$ and $\tilde{\lambda}_r$ as given in Theorem 2(i) of Section 2. Using known results on box-splines (see e.g. [**4**], Chapters 2 and 3), we obtain for $\phi = \omega^{k,l}$: $d^{k,l} = \deg(\phi) = 2(k+l-1)$, $\text{supp}(\phi) = \Omega_{k,l}$, $s^{k,l}$=smoothness order of $\phi$=$2k+l-2$ for $k \geq l$ and $k+2l-2$ for $l \geq k$ and $\mathbb{P}(\phi) = \mathbb{P}_{s^{k,l}+1}$. Using now classical results on the convolution product of piecewise polynomial functions and the Strang-Fix theory (see e.g. [**4**], Chapters 3 and 8, and [**17**]), we are able to prove the following

THEOREM 4.

(i) *The support of $\tilde{\sigma}_r^{k,l}$ (resp. $\tilde{\lambda}_r^{k,l}$) is the octagon $\Omega_{k+1,l}$ (resp. $\Omega_{k,l+1}$).*

(ii) *$\tilde{\sigma}_r^{k,l}$ and $\tilde{\lambda}_r^{k,l}$ are positive* B-*splines of degree $d_r + 2(k+l) = 2(k+l+r)+1$ for $r$ even and $2(k+l+r)+2$ for $r$ odd. They are of class $s^{k,l}+r+2 = 2k+l+r$ for $l \geq k$ and $k+2l+r$ for $k \geq l$.*

(iii) *For $r = -1$, $\mathbb{P}(\sigma * \omega^{k,l}) = \mathbb{P}(\lambda * \omega^{k,l}) = \mathbb{P}_{s^{k,l}+2}$. For $r \geq 0$, $(\mathbb{P}\tilde{\sigma}_r^{k,l}) = \mathbb{P}(\tilde{\lambda}_r^{k,l}) = \mathbb{P}(\omega^{k,l}) = \mathbb{P}_{s^{k,l}+1}$.*

## 4. Global Linear Dependence or Independence of $B(\phi)$

DEFINITIONS 2. *$B(\phi)$ is globally linearly independent* (GLI) *iff $\sum_\alpha c(\alpha)\phi(\cdot - \alpha) = 0$ implies $c(\alpha) = 0$ for all $\alpha \in \mathbb{Z}^2$. Else, $B(\phi)$ is globally linearly dependent* (GLD) *and there exists at least one sequence $\{c(\alpha), \alpha \in \mathbb{Z}^2\}$ such that $\sum_\alpha c(\alpha)\phi(\cdot - \alpha) = 0$.*

For the triangulation $\tau_1$, Sbibih [**16**] proved the following result, by using techniques similar to those in Chui and Ron [**6**] and Ron [**10**].

THEOREM 5. *For each composed* B-*spline $\phi$ on $\tau_1$ described in Theorem 3, the family $B(\phi)$ is* GLI.

For the triangulation $\tau_2$, things are more complicated (this fact is well known for box-splines, see e.g. [**4**], Chapter 3 or [**7**]), but we can prove the following result:

THEOREM 6.
(i) *For* $\phi = \tilde{\sigma}_r^{k,0}$ *or* $\phi = \tilde{\lambda}_r^{k,0}$, $r \geq 0$, $k \geq 0$, *then* $B(\phi)$ *is* GLI.
(ii) *For* $\phi = \tilde{\sigma}_r^{k,l}$ *or* $\phi = \tilde{\lambda}_r^{k,l}$, *with* $r, k \geq 0$ *and* $l \geq 1$, *then* $B(\phi)$ *is* GLD *and there holds the following dependence relation:*

(DR)
$$\sum_{\alpha} (-1)^{|\alpha|} \phi(. - \alpha) = 0.$$

PROOF.
(i) According to Corollary 1.1 of Ron [**10**], p. 298, it suffices to prove that $B(\phi)$ is GLI for $\phi = \tilde{\sigma}_r * \sigma$ and $\phi = \tilde{\lambda}_r * \sigma$. In the first case $\phi = \tilde{\sigma}_r * \sigma$, we assume that $f(x) = \sum c(\alpha)\phi(x - \alpha) = 0$ for all $x \in \mathbb{R}^2$and that $\Sigma_2 = \text{supp}(\phi)$ is the square $[-1, 1]^2$ centered at $x = 0$. Since $f(0) = c(0) = 0$, the result is obvious. Now, in the case $\phi = \tilde{\lambda}_r * \sigma$, we assume that $g(x) = \sum c(\beta)\phi(x - \beta) = 0$ for all $x \in \mathbb{R}^2$. We also assume that $\text{supp}(\phi) = \Omega$ is centered at the origin. We set $a = \phi(0)$ and $b = \phi(e_1) = \phi(-e_1) = \phi(e_2) = \phi(-e_2)$. The equation $g(\alpha) = 0$ gives
$$ac(\alpha) + b\big[c(\alpha - e_1) + c(\alpha + e_1) + c(\alpha - e_2) + c(\alpha + e_2)\big] = 0$$
for all $\alpha \in \mathbb{Z}^2$. The unique solution of this infinite system of equations is the null sequence if $a > 4b$: according to [**2**] (Chapter 4, pp. 81–82), this is due to the fact that the symbol of the difference equation $g(\alpha) = 0$, which is equal to $a + 2b\big(\cos(x_1) + \cos(x_2)\big)$, never vanishes in $\mathbb{R}^2$.
(ii) According to Ron [**10**], we first establish that $B(\phi)$ is GLD and we give the corresponding dependence relation for $\phi = \tilde{\sigma}_r * \lambda, \phi = \tilde{\sigma}_r * \omega$, $\phi = \tilde{\lambda}_r * \lambda$ and $\phi = \tilde{\lambda}_r * \omega$, where $\omega = \sigma * \lambda$ is the classical quadratic box-spline.

For $\phi = \tilde{\sigma}_r * \lambda$, let us prove that $B(\phi)$ is GLD and (DR) holds. Assume that $\Sigma_1 = [-\frac{1}{2}, +\frac{1}{2}]^2$ and denote $\phi_1, \ldots, \phi_9$ the nine translates of $\phi$ whose octagonal supports intersect $\Sigma_1$. Proving (DR) is equivalent to prove that the two functions $F = \phi_1 + \phi_3 + \phi_5 + \phi_7 + \phi_9$ and $G = \phi_2 + \phi_4 + \phi_6 + \phi_8$ coincide on $\Sigma_1$. Now, let $f = s_1 + s_3 + s_5 + s_7 + s_9$ and $g = s_2 + s_4 + s_6 + s_8$, where $s_i$ is the translate of $\tilde{\sigma}_r$ whose support is centered at the point number $i$. Thus, $\text{supp}(f)$ and $\text{supp}(g)$ are included in the square $\Sigma_3 = [-3/2, 3/2]^2$. Moreover, $F = f * \lambda$ and $G = g * \lambda$, i.e. for $x \in \Sigma_1$, we have:
$$F(x) = \int_{\mathbb{R}^2} f(u)\lambda(x - u)\, du = \int_{x+\Lambda_1} f(u)\, du$$
and similarly
$$G(x) = \int_{\mathbb{R}^2} g(u)\lambda(x - u)\, du = \int_{x+\Lambda_1} g(u)\, du.$$
The figure shows that the unions of intersections of $x + \Lambda_1$ with $\text{supp}(f)$ and $\text{supp}(g)$ are both equal to $\Sigma_1$, therefore
$$F(x) = G(x) = \int_{\Sigma_1} \tilde{\sigma}_r(u)\, du$$
therefore the first result holds. In a similar way, we can also prove that for $\phi = \tilde{\lambda}_r * \lambda$, $\tilde{\sigma}_r * \omega$ or $\tilde{\lambda}_r * \omega$, $B(\phi)$ is GLD and (DR) holds.

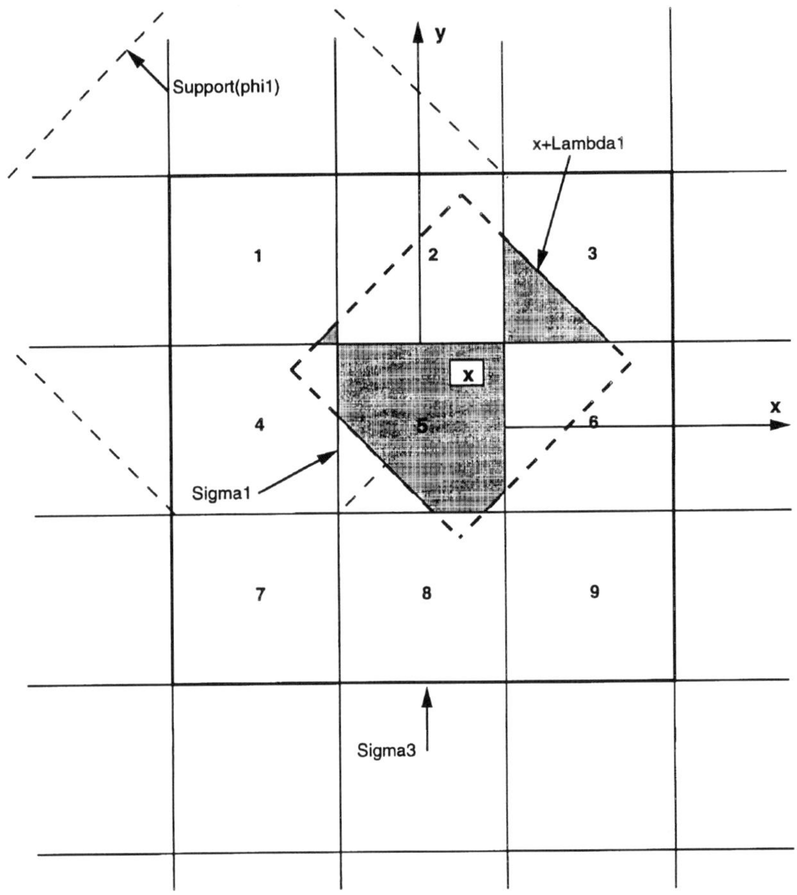


Let us give the proof for $\phi = \tilde{\sigma}_r * \omega$: we assume that $\text{supp}(\phi) = \Omega_{2,1}$ is centered at the origin. Let $I = \{1, 2, \ldots, 16\}$ and let $\{\phi_i, i \in I\}$ be the set of the sixteen translates of $\phi$ whose supports intersect the square $\Sigma_1$ centered at the origin. Let $I_1 = \{1, 3, 6, 8, 9, 11, 14, 16\}, I_2 = \{2, 4, 5, 7, 10, 12, 13, 15\}$, and define $F = \sum_{i \in I_1} \phi_i$ and $G = \sum_{i \in I_2} \phi_i$. Now, let $\{\omega_i, i \in I\}$ be the sixteen translates of $\omega$ whose octagonal support is centered at the point number $i$. It is well known that these translates are GLD and satisfy (DR) (see e.g. [**4**], Chapter 3), therefore we have $f = \sum_{i \in I_1} \omega_i = g = \sum_{i \in I_2} \omega_i$ on $\Sigma_1$ and, by convolution with $\tilde{\sigma}_r$, we finally obtain $F = f * \tilde{\sigma}_r = g * \tilde{\sigma}_r = G$, which proves our result and concludes the proof of Theorem 6.

## 5. Local Linear Independence of Translates

DEFINITIONS 3.

(i) $B(\phi)$ *is locally linearly independent* (LLI) *if and only if, for all domain* $K \subset \mathbb{R}^2$, $\sum c(\alpha)\phi(. - \alpha) = 0$ *on* $K$ *implies* $c(\alpha) = 0$ *for all* $\alpha \in \mathbb{Z}^2$ *such that* $\text{mes}\big(\text{supp}\big(\phi(. - \alpha)\big) \cap K\big) > 0$.

(ii) *Let* $\tau = \tau_1$ *or* $\tau_2$. $B(\phi)$ *is T-linearly independent* (TLI) *iff for all* $T \in \tau$, $\sum c(\alpha)\phi(. - \alpha) = 0$ *on* $T$ *implies* $c(\alpha) = 0$ *for all* $\alpha \in I$, *where* $I = \{\alpha \in \mathbb{Z}^2, \mathrm{mes}(\mathrm{supp}(\phi(. - \alpha)) \cap T) > 0\}$.

It is clear that $B(\phi)$ LLI implies $B(\phi)$ TLI and that the latter implies $B(\phi)$ GLI. Note that for box-splines on $\tau_1$, LLI is equivalent to GLI (see e.g. [**2**], Chap. 2).

In this section, we restrict our study to the T-linear independence of $B(\phi)$ for three families of B-splines on $\tau_1$ and for five families of B-splines on $\tau_2$.

THEOREM 7 (Triangulation $\tau_1$). *Let* $r \geq -1$ *and let* $\phi$ *be* $\beta_r^1 = \beta_r * \pi$, $\chi_r^1 = \chi_r * \pi$ *or* $\pi_r^1 = \pi_r * \pi$. *Then* $B(\phi)$ *is* TLI.

PROOF.

(i) Case of splines $\phi = \beta_r * \pi$. Let $a > 0$ be the common value of $\phi$ at the three interior knots of $\Delta_2 = \mathrm{supp}(\phi)$. We assume that these points are 0, $e_1$, $e_3 = e_1 + e_2$ and that they are also the vertices of the triangle $T$. Let

$$d = \frac{\partial \phi}{\partial x}(0) = \frac{\partial \phi}{\partial y}(e_1) = -\frac{\partial \phi}{\partial x}(e_1) = -\frac{\partial \phi}{\partial y}(e_3) > 0.$$

The seven $\Delta_2$-splines whose supports satisfy $\mathrm{mes}(\mathrm{supp}(\phi(. - \alpha)) \cap T) > 0$ correspond to $\alpha = e_2, e_3, -e_1, 0, e_1, -e_3, -e_2$ respectively. For sake of simplicity, we denote them $\phi_1, \phi_2, \ldots, \phi_7$. We now prove that $f = \sum \lambda_i \phi_i = 0$ on $T$ implies $\lambda_i = 0$ for $1 \leq i \leq 7$. By computing successively:

$$\begin{aligned}
f(0) &= a(\lambda_3 + \lambda_4 + \lambda_6) = 0 \\
f(e_1) &= a(\lambda_4 + \lambda_5 + \lambda_7) = 0 \\
f(e_3) &= a(\lambda_1 + \lambda_2 + \lambda_4) = 0 \\
\partial^{10} f(0) &= d(-\lambda_3 + \lambda_4) = 0 \\
\partial^{10} f(e_1) &= d(\lambda_5 - \lambda_4) = 0 \\
\partial^{01} f(e_1) &= d(\lambda_4 - \lambda_6) = 0 \\
\partial^{01} f(e_3) &= d(\lambda_1 - \lambda_4) = 0
\end{aligned}$$

we readily deduce the desired result. A similar proof holds for the triangle $T$ with vertices $\{0, e_2, e_3\}$.

(ii) Case of splines $\phi = \chi_r * \pi$. We assume that the four interior knots of $\Sigma_2 = \mathrm{supp}(\phi)$ are the points $\{0, e_1, e_2, e_3\}$ and that $T$ is the triangle with vertices $\{0, e_1, e_3\}$. We denote $\{\phi_i, i = 1, \ldots, 8\}$ the eight translates of $\phi$ whose supports contains $T$ and we set $a = \phi(0) = \phi(e_3), b = \phi(e_1) = \phi(e_2), d = \partial^{10}\phi(0) = \partial^{01}\phi(0) = \partial^{01}\phi(e_1) = \partial^{10}\phi(e_2) = -\partial^{10}\phi(e_1) = -\partial^{01}\phi(e_2) = -\partial^{10}\phi(e_3) = -\partial^{01}\phi(e_3)$. It can be proved that $a > b$ and that $d = \frac{1}{2}$ if we choose the normalization $\int \tilde{\sigma}_r = 1$. Now assume that $f = \sum \lambda_i \phi_i = 0$ on $T$. From the following equations:

$$\begin{aligned}
f(0) &= a(\lambda_4 + \lambda_6) + b(\lambda_3 + \lambda_7) = 0 \\
\partial^{10} f(0) &= d(-\lambda_3 + \lambda_4 - \lambda_6 + \lambda_7) = 0 \\
\partial^{01} f(0) &= d(\lambda_3 + \lambda_4 - \lambda_6 - \lambda_7) = 0 \\
f(e_1) &= a(\lambda_5 + \lambda_7) + b(\lambda_4 + \lambda_8) = 0 \\
\partial^{10} f(e_1) &= d(-\lambda_4 + \lambda_5 - \lambda_7 + \lambda_8) = 0
\end{aligned}$$

$$\partial^{01} f(e_1) = d(\lambda_4 + \lambda_5 - \lambda_7 - \lambda_8) = 0$$
$$f(e_3) = a(\lambda_2 + \lambda_4) + b(\lambda_1 + \lambda_5) = 0$$
$$\partial^{10} f(e_3) = d(-\lambda_1 + \lambda_2 - \lambda_4 + \lambda_5) = 0$$
$$\partial^{01} f(e_3) = d(-\lambda_1 + \lambda_2 - \lambda_4 - \lambda_5) = 0$$

we deduce $a\lambda_1 + b\lambda_2 = a\lambda_2 + b\lambda_1 = 0$, and the assumption $a > b > 0$ implies $\lambda_i = 0$ for $1 \le i \le 8$. The same result holds for the triangle $T$ with vertices $\{0, e_2, e_3\}$.

(iii) Case of splines $\phi = \pi_r * \pi$. We assume that $H_2 = \operatorname{supp}(\phi)$ is centered at the origin and we set $a = \phi(0)$, $b = \phi(e_1)$, $b_1 = \partial^{10}\phi(-e_3) = \partial^{01}\phi(-e_3)$, $b_2 = \partial^{10}\phi(-e_1)$. It will be proved elsewhere that $a > \frac{1}{2}$ (or, since $a + 6b = 1$, that $b < 1/12$), $b_2 = 2b_1 = 1/3$, $\partial^{11}\phi(e_1) = \partial^{11}\phi(e_2) = 0$, and $\partial^{11}\phi(e_3) = -\frac{1}{2}\partial^{11}\phi(0) > 0$ when $\phi$ is twice continuously differentiable. Now, let $h = \sum \lambda_i \phi_i = 0$ on the triangle $T$ with vertices $\{0, e_1, e_3\}$. By computing the values of $h, \partial^{10}h, \partial^{01}h$ and $\partial^{11}h$ at the vertices of $T$, we obtain a linear system of 12 equations for the 12 coefficients of $h$. Taking into account the symmetries of $\operatorname{supp}(\phi) = H_2$, this system is reduced to the following one with 7 equations and unknowns (where $c = a/b > 0$):

$$\lambda_1 + \lambda_2 + \lambda_3 + (1 + c)\lambda_4 + \lambda_5 + \lambda_8 = 0$$
$$-\lambda_1 + \lambda_2 - 2\lambda_3 - \lambda_4 + 2\lambda_5 + \lambda_8 = 0$$
$$2\lambda_1 + \lambda_2 + \lambda_3 - \lambda_4 - \lambda_5 - 2\lambda_8 = 0$$
$$\lambda_2 - \lambda_4 = 0$$
$$2\lambda_4 + 2\lambda_5 + c\lambda_8 + 2\lambda_9 = 0$$
$$-\lambda_4 + \lambda_9 = 0$$
$$\lambda_5 - \lambda_8 = 0.$$

Reducing this new system leads to $\lambda_i = 0$ for $1 \le i \le 12$ provided $c \ne 2$, i.e. $a \ne 1/4$, but this cannot occur since $a > \frac{1}{2}$. The proof being similar for the triangle T with vertices $\{0, e_2, e_3\}$, we obtain the T-linear independence of $H_2$-splines.

THEOREM 8 (Triangulation $\tau_2$).

(i) *For $\Sigma_2$-splines $\phi = \tilde{\sigma}_r * \sigma$ or $\Omega$-splines $\phi = \tilde{\lambda}_r * \sigma$, $B(\phi)$ is* TLI.

(ii) *For $\Omega$-splines $\phi = \tilde{\sigma}_r * \lambda$, for $\Lambda_2$-splines $\phi = \tilde{\lambda}_r * \lambda$, for $\Omega_{2,1}$-splines $\phi = \tilde{\sigma}_r * \omega$ and for $\Omega_{1,2}$-splines $\tilde{\lambda}_r * \omega$, $B(\phi)$ is* TLD, *i.e. linearly dependent on each triangle $T$ of $\tau_2$. The dependence relation is* (DR) *given in Theorem* 6.

PROOF.

(i) For $\Sigma_2$-splines, the proof is easy and we only give it for $\Omega$-splines $\phi = \tilde{\lambda}_r * \sigma$. Assuming that $\Omega$ is centered at the origin, we set $a = \phi(0)$, $b = \phi(e_1) = \phi(e_2) = \phi(-e_1) = \phi(-e_2)$. Since $B(\phi)$ is a partition of unity, we have $a + 4b = 1$, and it can be proved that $a > \frac{1}{2}$ or equivalently that $b < 1/8$. Let $c = \partial^{10}\phi(-e_1) = -\partial^{10}\phi(e_1)$ and $d = \partial^{10}\phi(-e_3/2) = \partial^{01}\phi(-e_3/2) = -\partial^{10}\phi(-e_4/2) = -\partial^{01}\phi(-e_4/2)$ $(c, d \ge 0)$. Since $\phi(-e_3/2) = \phi(-e_4/2) = \phi(e_3/2) = \phi(e_4/2) = 1/4$ and $\operatorname{grad}\phi(0) = 0$, we obtain the following equations by taking into account the symmetries of $\phi$ on its support:

$$f(0) = a\lambda_3 + b(\lambda_1 + \lambda_2 + \lambda_4 + \lambda_6) = 0$$
$$\partial^{01} f(0) = c(\lambda_1 - \lambda_6) = 0$$
$$f(-e_3/2) = (\lambda_2 + \lambda_3 + \lambda_5 + \lambda_6)/4 = 0$$
$$\partial^{10} f(-e_3/2) = d(-\lambda_2 + \lambda_3 - \lambda_5 + \lambda_6) = 0$$
$$\partial^{01} f(-e - 3/2) = d(\lambda_2 + \lambda_3 - \lambda_5 - \lambda_6) = 0$$
$$\partial^{10} f(-e_4/2) = d(-\lambda_3 + \lambda_4 - \lambda_6 + \lambda_7) = 0$$
$$\partial^{01} f(-e_4/2) = d(\lambda_3 + \lambda_4 - \lambda_6 - \lambda_7) = 0.$$

From the six last equations, we deduce: $\lambda_1 = \lambda_2 = \lambda_4 = \lambda_6 = -\lambda_3 = -\lambda_5 = -\lambda_7$ and the first one gives $(4b - a)\lambda_1 = 0$. Since $4b < \frac{1}{2} < a$, we get $\lambda_1 = 0$, which implies that the other coefficients are also equal to zero.

(ii) Since the translates of these functions are GLD, they are automatically LLD (and TLD) and the same dependence relation holds.

## 6. Quasi-Interpolants

Let $\phi$ be either some $H_2$-spline on $\tau_1$ or some $\Omega$-spline on $\tau_2$, the supports being centered at the origin. For $H_2$-splines, let $a = \phi(0)$ and $b = \phi(e_1)$, with $a + 6b = 1$, $a > \frac{1}{2}$. In this section, we define and study the properties of two particular quasi-interpolants which will be useful in Section 7 for the estimation of interpolation errors on rectangular domains. The first one is the well-known Schoenberg operator (see e.g. [**2**], Chapter 3). The second one is a particular case of a family of discrete quasi-interpolants already introduced in [**11**]. We now use grids with mesh size $h > 0$.

DEFINITIONS 4.

(i) *The discrete operator $\delta$ is defined on the triangulation $\tau_1$ by:*

$$\delta f(h\alpha) = -6f(h\alpha) + f\big(h(\alpha+e_1)\big) + f\big(h(\alpha-e_1)\big) + f\big(h(\alpha+e_2)\big) + f\big(h(\alpha-e_2)\big) + f\big(h(\alpha+e_3)\big) + f\big(h(\alpha-e_3)\big),$$

*for all $\alpha \in \mathbb{Z}^2$.*

(ii) *The quasi-interpolants $S_1$ and $S_2$ are respectively defined by:*

$$S_1 f = \sum_\alpha f(h\alpha)\phi(./h - \alpha),$$
$$S_2 f = \sum_\alpha \big[f(h\alpha) - b\delta f(h\alpha)\big]\phi(./h - \alpha).$$

THEOREM 9 (Approximation order of quasi-interpolants).

(i) *On $\tau_1$ (resp. $\tau_2$) $S_1$ is exact on $\mathbb{P}_1$ (resp. $\mathbb{P}_0$). Hence for $f$ smooth enough $S_1 f - f = O(h^2)$ (resp. $O(h)$).*

(ii) *On $\tau_1$, the operator $S_2$ is exact on $\mathbb{P}_2$, therefore, for $f$ smooth enough, $S_2 f - f = O(h^3)$.*

PROOF. We only give a sketch of the proof.

(i) From Section 3, we know that $\mathbb{P}(\phi) = \mathbb{P}_2$ for $H_2$-splines and $\mathbb{P}(\phi) = \mathbb{P}_0$ for $\Omega$-splines. From [**4**], Chapter 8, these properties imply the results on $S_1$ because $H_2$ is symmetrical w.r.t. the origin.

(ii) Since $\mathbb{P}(\phi) = \mathbb{P}_2$ for $H_2$-splines, the result is a particular case of Theorem 3.1 in [**11**]. With the notations of this paper, the difference operator coinciding with $S_1$ on $\mathbb{P}_2$ is $D = I + b\delta$. Its formal inverse in $\mathbb{P}_2$ is $D^{-1} = I - b\delta$ and the associated discrete quasi-interpolant is $Qf = S_1 D^{-1} f = \sum D^{-1} f(h\alpha)\phi(. - \alpha)$, i.e. $Q = S_2$.

The results on the approximation orders are straightforward.

## 7. Lagrange Interpolation on Rectangular Domains

In this section, we study two particular Lagrange interpolation problems: the first one by translates of $H_2$-splines on the triangulation $\tau_1$, the second one by translates of $\Omega$-splines on the triangulation $\tau_2$ of a rectangular domain $R = [1, m] \times [1, n]$. This study is, in some sense, an extension of [**1**] to these new families of B-splines .

Let $\phi$ denote one of these B-splines, let $A = R \cap \mathbb{Z}^2$ (hence $\mathrm{Card}(A) = mn$) and let $B = \{\beta \in \mathbb{Z}^2 : \mathrm{mes}(\mathrm{supp}(\phi(. - \beta)) \cap R) > 0\}$. For given values $\{f(\alpha), \alpha \in A\}$, we want to find a spline

$$S = \sum_{\beta \in B} c(\beta)\phi(. - \beta)$$

in $S(\phi)$ satisfying $S(\alpha) = f(\alpha)$ for $\alpha \in A$.

When $\phi$ is a $H_2$-spline (i.e. $\phi = \pi_r^1$ for some $r \geq 0$), then $B = [0, m+1] \times [0, n+1] \backslash \{(m+1, 0), (0, n+1)\}$ and $\mathrm{Card}(B) = mn + 2(m+n) + 2$. When $\phi$ is an $\Omega$-spline (i.e. $\phi = \tilde{\sigma}_r * \lambda$ or $\phi = \tilde{\lambda}_r * \sigma$ for some $r \geq 0$), then $B = [0, m+1] \times [0, n+1] \backslash \{(0,0), (m+1, 0), (0, n+1), (m+1, n+1)\}$ and $\mathrm{Card}(B) = mn + 2(m+n)$.

Since $f$ is not defined on $B \backslash A$, we have to approximate its values on this set: this will be specified below. The problem is to solve the linear system:

$$\sum_{\beta \in B} c(\beta)\phi(\alpha - \beta) = f(\alpha)$$

for $\alpha \in A$, $c(\beta)$ being given for $\beta \in B \backslash A$.

For $H_2$-splines, the equations have the form:

$$ac(\alpha) + b\big(c(\alpha + e_1) + c(\alpha - e_1) + c(\alpha + e_2) + c(\alpha - e_2) + c(\alpha + e_3) + c(\alpha - e_3)\big) = f(\alpha),$$

where $a + 6b = 1$ and $a > \frac{1}{2}$.

For $\Omega$-splines, they have the form:

$$ac(\alpha) + b\big(c(\alpha + e_1) + c(\alpha - e_1) + c(\alpha + e_2) + c(\alpha - e_2)\big) = f(\alpha),$$

where $a + 4b = 1$, $a > \frac{1}{2}$ for $\phi = \tilde{\lambda}_r * \sigma$ and $a = \frac{1}{2}$ and $b = 1/8$ for $\phi = \tilde{\sigma}_r * \lambda$. The matrix $M$ of the linear system is block-tridiagonal: the matrices of the main diagonal are tridiagonal and those of the upper and lower diagonals are bidiagonal or diagonal.

In the cases of $H_2$-splines and of $\Omega$-splines $\tilde{\lambda}_r * \sigma$, $M$ is strictly diagonally dominant, hence invertible. Moreover, by Gershgorin's theorem, the norm of its inverse is uniformly bounded by $(2a - 1)^{-1}$. For $\Omega$-splines $\tilde{\sigma}_r * \lambda$, $a = \frac{1}{2}$ and $b = 1/8$, $M$ is the collocation matrix associated with quadratic box-splines and is invertible (see e.g. [**1**]), but the the norm of its inverse depends on $m$ and $n$.

THEOREM 10. *For $H_2$-splines and $\Omega$-splines, the Lagrange interpolation problem has a unique solution, provided the values $\{c(\beta), \beta \in B\backslash A\}$ are fixed. We denote $Lf \in S(\phi)$ the interpolant of $f$ on $A$.*

For grids with mesh size $h > 0$, we use the same notations with the modifications given in the previous section. In particular, we denote by $\mu(\alpha)$ the generic coefficient in $S_2$, i.e. the bracket $[I - b\delta]f(h\alpha)$. We obtain the following theorems.

THEOREM 11 (Error estimates for $H_2$-splines).
(i) *If $c(\beta)$ is such that $c(\beta) - f(h\beta) = O(h^2)$ for all $\beta \in B\backslash A$, then $Lf - f = O(h^2)$ for $f \in C^2(R)$.*
(ii) *If $c(\beta)$ is such that $c(\beta) - \mu(\beta) = O(h^3)$ for all $\beta \in B\backslash A$, then $Lf - f = O(h^3)$ for $f \in C^4(R)$.*

PROOF. The idea is to introduce quasi-interpolants as intermediate operators between $f$ and $Lf$, i.e. to estimate the infinite norms of $Lf - Sf$ and $Sf - f$, where $S = S_1$ or $S_2$, and then to use Theorem 9. We only sketch the proof.
(i) With the hypothesis on $\{c(\beta), \beta \in B\backslash A\}$, we first prove that $Lf - S_1 f = O(h^2)$ for $f \in C^2(R)$. Let $e(\alpha) = c(\alpha) - f(h\alpha)$ for $\alpha \in B$. For $\alpha \in A$, these quantities are solutions of the linear system:

$$ae(\alpha) + b\sum_{i=1}^{3}\big(e(\alpha + e_i) + e(\alpha - e_i)\big) = -b\delta f(h\alpha).$$

Since $f$ is $C^2$, then, by Taylor's expansions, it is easy to see that the second member of this equation is $O(h^2)$. Moreover, by assumption, we have also $e(\beta) = O(h^2)$ for $\beta \in B\backslash A$. Therefore, the uniform boundedness of the norm of the inverse of $M$ implies that the unique solution of the above linear system is also $O(h^2)$. Using the properties of the translates of B-splines (positivity and partition of unity), we deduce that $Lf - S_1 f = O(h^2)$. By Theorem 9, we know that $S_1 f - f = O(h^2)$ and finally we conclude that $Lf - f = O(h^2)$.
(ii) With the hypothesis on $\{c(\beta), \beta \in B\backslash A\}$, we first prove that $Lf - S_2 f = O(h^3)$ for $f \in C^4(R)$. Let $\mu(\alpha)$ denote the coefficient of $\phi(./h - \alpha)$ in the expression of $S_2$ and let $e(\alpha) = c(\alpha) - \mu(\alpha)$ for all $\alpha \in B$, i.e. $Lf - S_2 f = \sum e(\alpha)\phi(.-h\alpha)$. After some algebraic computations, we obtain the following system of equations for $\alpha \in A$:

$$ae(\alpha) + b\sum_{i=1}^{3}\big[e(\alpha + e_i) + e(\alpha - e_i)\big] = b^2\delta^2 f(h\alpha).$$

Since $f$ is $C^4$, then, by using Taylor's expansions, we easily prove that the second member of this equation is $O(h^4)$. Moreover, by assumption, we have $e(\beta) = O(h^3)$ for $\beta \in B\backslash A$, therefore the uniform boundedness of the norm of the inverse of $M$ implies that the unique solution of the above linear system is also $O(h^3)$. From this result, we deduce that $Lf - S_2 f = O(h^3)$. By Theorem 9, we know that $S_2 f - f = O(h^3)$ and finally, we conclude that $Lf - f = O(h^3)$.

In a similar way, we can prove the following theorem (see also **[8]** for related results).

THEOREM 12 (Error estimates for $\Omega$-splines).

(i) *For $\Omega$-splines $\tilde{\lambda}_r * \sigma$, if $c(\beta)$ is chosen so that $c(\beta) - f(h\beta) = O(h)$ for all $\beta \in B\backslash A$, then $Lf - f = O(h^2)$ for $f \in C^2(R)$.*

(ii) *For $\Omega$-splines $\tilde{\sigma}_r * \lambda$, we have not yet results on the approximation order since the condition number of the collocation matrix is unbounded.*

## References

1. E. Arge, M. Daehlen, and A. Tveito, *Box-spline interpolation: a computational study*, J. Comput. Appl. Math. **44** (1992), 303–329.
2. C. de Boor, K. Höllig, and S. Riemenschneider, *Box-splines*, Applied Mathematical Sciences 98, Springer Verlag, New York, 1993.
3. B. D. Bojanov, H. A. Hakopian, and A. A. Sahakian, *Spline functions and multivariate interpolation*, Kluwer, Dordrecht, 1993.
4. C. K. Chui, *Multivariate splines. With an appendix by Harvey Diamond*, CBMS-NSF Regional Conference Series in Applied Mathematics, vol. 54, SIAM, Philadelphia, PA, 1988.
5. ______, *Vertex splines and their applications to interpolation of discrete data*, Computation of Curves and Surfaces (W. Dahmen *et al.*, ed.), Kluwer, Dordrecht, 1990, pp. 137–181.
6. C. K. Chui and A. Ron, *On the convolution of a box-spline with a compactly supported distribution: linear independence for the integer translates*, Canad. J. Math. **43** (1991), 19–33.
7. W. Dahmen and C. A. Micchelli, *On the optimal approximation rates for criss-cross finite element spaces*, J. Comput. Appl. Math. **10** (1984), 255–273.
8. X. Guillet, *Interpolation by new families of* B*-splines on uniform meshes of the plane*, Surface Fitting and Multiresolution Methods (A. Le Méhauté, Ch. Rabut, and L. L. Schumaker, eds.), Vanderbilt University Press, Nashville, 1997, pp. 183–190.
9. M. J. Lai, *Fortran subroutines for* B*-nets of box-splines on three and four directional meshes*, Numer. Algorithms **2** (1992), 33–38.
10. A. Ron, *A necessary and sufficient condition for the linear independence of integer translates of a compactly supported distribution*, Constr. Approx. **5** (1989), 297–308.
11. P. Sablonnière, *Quasi-interpolants associated with H-splines on a three-direction mesh*, J. Comput. Appl. Math. **66** (1996), 433–442.
12. ______, B*-splines on uniform meshes of the plane*, Advanced Topics in Multivariate approximation (F. Fontanella, K. Jetter, and L. L. Schumaker, eds.), World Scientific Publishing Co., 1996, pp. 323–338.
13. P. Sablonnière and D. Sbibih, *Existence and unicity of $H_1$-splines on an equilateral triangulation of the plane*, Publication LANS 43, INSA de Rennes, 1992.
14. ______, B*-splines à supports hexagonaux sur un réseau tridirectionnel régulier du plan*, C. R. Acad. Sci. Paris I Math. **319** (1994), 277–282.
15. ______, *Some families of* B*-splines with hexagonal support on a three-direction mesh*, Mathematical Methods for Curves and Surfaces (M. Daehlen, T. Lyche, and L. L. Schumaker, eds.), Vanderbilt University Press, 1995, pp. 467–475.
16. D. Sbibih, B*-splines et quasi-interpolants sur un réseau tridirectionnel du plan*, Thèse d'état, Univ. d' Oujda, Maroc, 1995.
17. G. Strang and G. Fix, *A Fourier analysis of the finite element variational method*, Constructive Aspects of Functional Analysis (G. Geymonat *et al.*, eds.), CIME II Ciclo, 1971, pp. 793–840.

I.N.S.A., LABORATOIRE L.A.N.S., LOGICIELS, ANALYSE NUMÉRIQUE ET STATISTIQUES, 20, AVENUE DES BUTTES DE COËSMES, CS 14315, 35043 RENNES CEDEX, FRANCE

*E-mail address*: `sablonni@perceval.univ-rennes1.fr`

# Part 2

# Theory of Wavelets

## Introduction and Summary

Shortly after their introduction by Morlet and Grossmann at the beginning of the 80's, wavelet decompositions were soon related with techniques which had developed previously in different fields. Though the wavelet framework provided a common language and created many interactions between these fields, different origins can still be traced back in the present research and provide different motivations which are well represented in this selection of papers. One of these starting points is the subdivision schemes introduced by Deslauriers and Dubuc. These schemes turned out to be a cornerstone for understanding the properties of orthonormal wavelets when the scaling function generating a multiresolution analysis was shown to satisfy such an equation.

Recent developments in subdivision schemes go in two different directions which are both present here: either prove uniform regularity of the solutions (see the paper of Dyn and Levin) or study precisely their irregularity, which often allows to uncover the multifractal nature of these solutions (see the paper of Lau and Ma). It is interesting to notice that wavelets are at the chore of multifractal analysis from two opposite points of view: as an analysing tool (see the paper by Arneodo *et al.* in the Physics section) or as the analyzed object, as investigated in the paper of Lau and Ma. Multidimensional constructions are much more difficult to deal with, and research towards 'truly' multidimensional wavelets bases (i.e. bases which are not deduced from 1-D wavelets by the standard tensor product technique) is still flourishing; it is the purpose of the general construction proposed by Karoui and Vaillancourt. Such constructions are of course an important issue in image signal processing, in cases where one doesn't wish the coordinates axes to play a too specific role.

Another trend which has its roots simultaneously in theoretical physics and signal analysis is the determination of bases which have a given time frequency localization. Standard wavelets give an answer for hyperbolic type pavings of the time-frequency plane; Malvar wavelets give another answer for rectangular pavings. Two generalizations are proposed here; first, the paper of Kovacevic and Bernardini, motivated by signal analysis, yields a construction adapted to arbitrary pavings. The standard wavelet paving and the Malvar-type paving can be interpreted as a regular sampling, either of the $ax + b$ group, in the first case, or of the Weyl-Heisenberg group, in the second case. Generalizations to more general groups have been proposed since the beginning of wavelet analysis, since one of its founders (A.Grossmann) was a theoretical physicist. Such constructions are the purpose of the paper by Schulz and Taylor

It has often been argued that a naive approach consists in believing that a simple wavelet decomposition will solve any problem in technology. Indeed, wavelet techniques are now often just an ingredient in a long technological chain. One example is supplied by the paper of S.Maes which shows how continuous wavelet decompositions improve previous techniques of speech recognition. On a more theoretical level, the paper of Holschneider shows why directional wavelet techniques might be a help to study directional regularity of microlocal type.

Stéphane Jaffard

Centre de Recherches Mathématiques
CRM Proceedings and Lecture Notes
Volume **18**, 1999

# Analysis of Hermite-Interpolatory Subdivision Schemes

Nira Dyn and David Levin

ABSTRACT. The theory of matrix subdivision schemes provides tools for the analysis of general uniform stationary matrix schemes. The special case of Hermite-interpolatory subdivision schemes deals with refinement algorithms for the function and the derivatives' values, with matrix masks depending upon the refinement level, i.e., non-stationary matrix masks. Here we first show that a Hermite-interpolatory subdivision scheme can be transformed into a stationary process. Then, using special schemes for generating some Hermite-type divided differences, we give the theory and the tools for analyzing the convergence and smoothness of Hermite-interpolatory schemes.

## 1. Stationary Hermite-Interpolatory Schemes

Hermite-type subdivision schemes of order 2 were already considered in [**14**] and in [**12**]. In the present paper we are discussing the basic properties and the proper analysis tools for higher order Hermite-type subdivision schemes. The analysis presented here is an adaptation of the methods in [**2,3,10,15**] and especially [**12**], exploiting the structure and the special significance of Hermite-type data. Examples and numerical implementation of the analysis tools are presented in [**6**].

The Hermite-interpolatory scheme of order $m$ is of the form

$$f_n^{k+1} = \sum_{j\in\mathbb{Z}} A_{n-2j}^{(k)} f_j^k, \quad n \in \mathbb{Z}, \tag{1.1}$$

where $\{f_n^k\}$ are vectors in $\mathbb{R}^m$, $f_n^k$ is the vector attached to the diadic point $t = 2^{-k}n$ at level $k$ of the subdivision. The $m \times m$ matrices $\{A_n^{(k)}, n \in \mathbb{Z}\}$, which form the "mask" of the scheme at level $k$, are non-zero only for the finite set of indices $I$. In particular

$$A_{2n}^{(k)} = I_{m\times m}\delta_{n,0},$$

due to the interpolatory nature of the scheme. The scheme converges to a $C^s$ limit function with $s \geq m-1$ if there exists $f \in C^s$ such that

$$f_n^k = \left( f^{(0)}(2^{-k}n), f^{(1)}(2^{-k}n), \frac{f^{(2)}(2^{-k}n)}{2!}, \dots, \frac{f^{(m-1)}(2^{-k}n)}{(m-1)!} \right)^t, \quad n \in \mathbb{Z},\ k \in \mathbb{Z}_+.$$

1991 *Mathematics Subject Classification.* Primary: 41A05; Secondary: 65D10.
This is the final form of the paper.

The normalization with the factorials is for later convenience.

In the following we consider schemes which reproduce the space $\Pi_\ell$ of polynomials of degree $\ell$ or less, for $\ell \geq m-1$, namely schemes with the property that if the vectors at level $k$ are of the form

$$f_i^k = \left(p(2^{-k}i), p'(2^{-k}i), \ldots, \frac{p^{(m-1)}(2^{-k}i)}{(m-1)!}\right)^t, \quad i \in \mathbb{Z},$$

for $p \in \Pi_\ell$, then application of (1.1) results in

$$f_i^{k+1} = \left(p(2^{-k-1}i), p'(2^{-k-1}i), \ldots, \frac{p^{(m-1)}(2^{-k-1}i)}{(m-1)!}\right)^t, \quad i \in \mathbb{Z}.$$

Later we show that this property is necessary for such a scheme to converge to $C^\ell$ limit functions. In fact we consider a subset of such schemes which we term stationary. Let us assume that the degree $\ell$ of the polynomial space is chosen so that a mask $\{A_n^{(0)}, n \in I\}$ exists with the property of reproduction of polynomials in $\Pi_\ell$, namely

$$P(2^{-1}n) = \sum_{j\in\mathbb{Z}} A_{n-2j}^{(0)} P(j), \quad n \in \mathbb{Z}, \tag{1.2}$$

where $P(x)$ is a matrix of order $m \times \ell$ with polynomial elements of the form

$$\big(P(x)\big)_{i,j} = \frac{1}{i!}\frac{d^i}{dx^i}x^j, \quad i = 0, \ldots, m-1, \; j = 0, \ldots, \ell-1. \tag{1.3}$$

We would like to establish the connection between the mask at level $k$ to the mask at level zero. The system of equations that the mask at level $k > 0$ must satisfy can be written as

$$P(2^{-k-1}n) = \sum_{j\in\mathbb{Z}} A_{n-2j}^{(k)} P(2^{-k}j), \quad n \in \mathbb{Z}. \tag{1.4}$$

By the form (1.3) of $P$ it follows that

$$P(hx) = M_m(h)^{-1}P(x)M_\ell(h), \tag{1.5}$$

where $M_j(h) = \operatorname{diag}\{1, h, \ldots, h^{-j+1}\}$. Multiplying (1.2) by $M_m(2^{-k})^{-1}$ from the left and by $M_\ell(2^{-k})$ from the right we get,

$$\begin{aligned} &M_m(2^{-k})^{-1}P(2^{-1}n)M_\ell(2^{-k}) \\ &\qquad = \sum_{j\in\mathbb{Z}} M_m(2^{-k})^{-1}A_{n-2j}^{(0)}M_m(2^{-k})M_m(2^{-k})^{-1}P(j)M_\ell(2^{-k}). \end{aligned}$$

Now, using relation (1.5) with $h = 2^{-k}$, it follows that (1.4) is satisfied for $k > 0$ with

$$A_n^{(k)} = M_m(2^{-k})^{-1}A_n^{(0)}M_m(2^{-k}), \quad n \in I. \tag{1.6}$$

Thus given the mask at level $k = 0$ we can choose the mask at level $k > 0$ according to (1.6). We define such schemes as "stationary Hermite schemes". Although the scheme as given is not stationary since the mask depends on the level $k$, yet by a simple scaling the scheme becomes stationary. Defining the scaled vectors by $M_m(2^{-k})f_n^k$ these vectors are generated by a stationary scheme with the level independent mask

$$A_n = M_m(2^{-1})A_n^{(0)}, \quad n \in I. \tag{1.7}$$

## 2. Divided-Difference Operators for Hermite-Interpolatory Schemes

Our analysis of the convergence of stationary Hermite-interpolatory schemes which reproduce $\Pi_\ell$, $\ell \geq m-1$ to $C^s$ limit functions with $s \geq m-1$, is based on the existence of stationary subdivision schemes for the divided differences of the data vectors $f_n^k$, $n \in \mathbb{Z}$, of order $j$ for $m-1 \leq j \leq \ell+1$. These divided differences are defined regarding every diadic point $t = n2^{-k}$ as having multiplicity $m$ and interpreting the data there as

$$f_n^k = \left(f^{(0)}(t), f^{(1)}(t), \frac{f^{(2)}(t)}{2!}, \dots, \frac{f^{(m-1)}(t)}{(m-1)!}\right)^t, \quad n \in \mathbb{Z},\ k \in \mathbb{Z}_+,$$

for some function $f \in C^{m-1}$. Related to the data vectors $f_n^k$, $n \in \mathbb{Z}$, we consider the $(m-1)$th order divided differences defined as

$$u_k^{[m-1]}(mn+j) = [\tau_{j+1}, \tau_{j+2}, \dots, \tau_{j+m}]f, \quad j = 0, \dots, m-1, \tag{2.1}$$

where

$$\tau_0 = \tau_1 = \dots = \tau_{m-1} = (n-1)2^{-k},$$

and

$$\tau_m = \tau_{m+1} = \dots = \tau_{2m-1} = n2^{-k}.$$

Thus, if a scalar stationary subdivision of the form

$$u_{k+1}^{[m-1]} = S_B u_k^{[m-1]}, \quad k \in \mathbb{Z},$$

exists and converges uniformly to $C^0$ limit functions in the sense of convergence of scalar subdivision schemes, then the original Hermite-interpolatory scheme converges to $C^{m-1}$ limit functions.

The vector $u_k^{[m-1]}$ can be obtained recursively from the original data vectors in the following way. Let

$$v_k^{[0]}(n) = f_n^k, \quad n \in \mathbb{Z},\ k \in \mathbb{Z}_+, \tag{2.2}$$

and let

$$\left(v_k^{[r]}(n)\right)_0 = 2^k\left(v_k^{[r-1]}(n)\right)_0 - \left(\left(v_k^{[r-1]}(n-1)\right)_{r-1}\right), \tag{2.3}$$

$$\left(v_k^{[r]}(n)\right)_j = 2^k\left(\left(v_k^{[r-1]}(n)\right)\right)_j - \left(\left(v_k^{[r-1]}(n)\right)_{j-1}\right), \quad j = 1, \dots, r-1, \tag{2.4}$$

$$\left(v_k^{[r]}(n)\right)_j = \left(v_k^{[r-1]}(n)\right)_j, \quad j = r, \dots, m-1, \tag{2.5}$$

for $r = 1, 2, \dots, m-1$. We note that

$$v_k^{[r]}(n) = \big([\tau_0, \dots, \tau_{r-1}, \tau_m]f, [\tau_0, \dots, \tau_{r-2}, \tau_m, \tau_{m+1}]f, \dots,$$
$$[\tau_0, \tau_m, \dots, \tau_{m+r-1}]f, f^{(r)}(n), \dots, f^{(m-1)}(n)\big)^t,$$

and $u_k^{[m-1]}$ as defined in (2.1) is related to $v_k^{[m-1]}$ by

$$u_k^{[m-1]}(mn+j) = \left(v_k^{[m-1]}(n)\right)_j, \quad j = 0, \dots, m-1,\ n \in \mathbb{Z}.$$

The vectors $v_k^{[r]}(n)$, $n \in \mathbb{Z}$, can be obtained recursively by using the formalism of vector generating functions and of matrix Laurent polynomials. Let us introduce the vector generating functions

$$F_k^{[r]}(z) = \sum_{n\in\mathbb{Z}} v_k^{[r]}(n)z^n, \quad r = 0,\dots,m-1,\ k \in \mathbb{Z}_+,$$

and the matrix Laurent polynomials

$$\Gamma_k^{m,r}(z) = \left(2^k\Gamma^r(z)\right) \oplus I_{(m-r-1)\times(m-r-1)},$$

where $\Gamma^r(z)$ is a matrix of order $(r+1)\times(r+1)$, which for $r = 1,\dots,m-1$ is given by its non-zero elements:

$$\Gamma_{1,1}^r = 1,\ \Gamma_{1,r+1}^r = -z,\ \Gamma_{i,i}^r = 1,\ \Gamma_{i,i-1}^r = -1, \quad i = 2,\dots,r+1,$$

while for $r = 0$ its only element is $1 - z$. For this formalism we also define the symbol of a mask $\{A_n^{(k)}\ n \in I\}$ as the matrix Laurent polynomial

$$D_k^{[0]}(z) \equiv A_k(z) = \sum_{n\in I} A_n^{(k)} z^n.$$

With these notations we can rewrite relations (2.2)–(2.5) as relations between the generating functions,

$$F_k^{[r+1]}(z) = \Gamma_k^{m,r}(z)F_k^{[r]}(z), \quad r = 0,\dots,m-2. \tag{2.6}$$

Formally, let us assume that the symbol of the subdivision scheme for the vectors $v_k^{[r]}$ is $D_k^{[r]}(z)$. That is we have

$$F_{k+1}^{[r]}(z) = D_k^{[r]}(z)F_k^{[r]}(z^2).$$

Then, using (2.6), the symbol of the subdivision scheme for $v_k^{[r+1]}$ should be given by

$$D_k^{[r+1]}(z) = \Gamma_{k+1}^{m,r}(z)D_k^{[r]}(z)\left(\Gamma_k^{m,r}(z^2)\right)^{-1}. \tag{2.7}$$

We are now ready to state and prove the existence of subdivision schemes for the vectors $v_k^{[r]}$, $r = 0,\dots,m-1$.

THEOREM 1. *Consider a Hermite-interpolatory subdivision scheme of order $m$, of the form* (1.1). *If the scheme reproduces $\Pi_\ell$, $\ell \le m-2$, then there exist subdivision schemes, with finitely supported masks, defined by their symbol $D_k^{[r]}(z)$ for the refinement of the vectors $v_k^{[r]}$, for $r = 1,\dots,\ell+1$.*

The proof is by recursion on $r$. First we observe that since the original scheme reproduces the constant polynomials, the data of constant vector $f_n^k = (1,0,\dots,0)^t$, $n \in \mathbb{Z}$, is an eigenvector of the subdivision scheme, namely if $f_n^0 = (1,0,\dots,0)^t$, $n \in \mathbb{Z}$, then $f_n^k = (1,0,\dots,0)^t$, $n \in \mathbb{Z}$, for all $k \in \mathbb{Z}_+$. This is equivalent to the following property of the symbols of the scheme,

$$A_k(1)(1,0,\dots,0)^t = (2,0,\dots,0)^t, \quad A_k(-1)(1,0,\dots,0)^t = (0,0,\dots,0)^t. \tag{2.8}$$

As shown below, property (2.8) guarantees that

$$D_k^{[1]}(z) = \Gamma_{k+1}^{m,1}(z)A_k(z)\left(\Gamma_k^{m,1}(z^2)\right)^{-1}, \tag{2.9}$$

is a finite matrix Laurent polynomial. Hence, by (2.6), there is a finite mask consisting of the matrix coefficients of $D_k^{[1]}(z)$ which maps $v_k^{[1]}$ on $v_{k+1}^{[1]}$.

To see that indeed $D_k^{[1]}(z)$ in (2.9) is indeed a finite matrix Laurent polynomial, we recall that $\Gamma_k^{m,1}(z) = \operatorname{diag}\{2^k(1-z), 1, \dots, 1\}$. By condition (2.8) it follows that all the elements in the first column of $A_k(-1)$ vanish and that all but the first element of $A_k(1)$ vanish. Therefore, all the elements of the first column of $\Gamma_{k+1}^{m,1}(z)A_k(z)$ are divisible by $1-z^2$, and hence $D_k^{[1]}(z)$ is a Laurent polynomial.

To do the recursive step we assume that for a given $r$, $0 < r < m-1$, there exists a matrix Laurent polynomial $D_k^{[r]}(z)$ defining the subdivision scheme generating the vectors $v_k^{[r]}$. We observe that the first $r+1$ components of $v_k^{[r]}(n)$ are the divided difference of order $r$ of the original data $f^k$. Hence if $f_n^0$ is taken from $f(x) = x^r$ then by the polynomial reproduction property of the original scheme the corresponding vectors $v_k^{[r]}$ are all of the same form, namely

$$v_k^{[r]}(n) = e^{(r)}, \quad n \in \mathbb{Z},\ k \in \mathbb{Z}_+,$$

where $e^{(r)} = (1, 1, \dots, 1, 0, 0, \dots, 0)^t$, with 1 repeated $r+1$ times and 0 repeated $m-r-1$ times. This is equivalent to

$$D_k^{[r]}(1)e^{(r)} = 2e^{(r)}, \quad D_k^{[r]}(-1)e^{(r)} = 0. \tag{2.10}$$

We further observe that

$$\Gamma_k^{m,r}(z)^{-1} = \left(\frac{1}{1-z}2^{-k}E^{(r)}(z)\right) \oplus I_{(m-r-1)\times(m-r-1)},$$

where $E^{(r)}(z)$ is a matrix of order $(r+1)\times(r+1)$ with $E^{(r)}(z)_{i,j} = 1$ for $i \geq j$ and $E^{(r)}(z)_{i,j} = z$ for $i < j$. It is enough to show that the first $r+1$ columns of

$$\Gamma_{k+1}^{m,r}(z)D_k^{[r]}(z)\big(E^{(r)}(z^2)\big) \oplus I_{(m-r-1)\times(m-r-1)}$$

vanish at $z = \pm 1$. This follows by observing that the first $r+1$ columns of $E^{(r+1)}(z^2)$ equal $e^{(r)}$. Hence, by (2.10), the first $r+1$ columns of $D_k^{[r]}(z)\big(E^{(r+1)}(z^2)\big) \oplus I_{(m-r-1)\times(m-r-1)}$ are zero at $z=-1$ and equal $e^{(r)}$ at $z = 1$. The result now follows by noting that left operation with $\Gamma_{k+1}^{m,r}(1)$ gives differences of pairs of the first $r+1$ rows.

## 3. Conditions for Smoothness of Hermite-Interpolatory Schemes

In this section we show that a $C^s$ continuous stationary Hermite-interpolatory subdivision scheme must reproduce $\Pi_s$. Then we give necessary and sufficient conditions for $C^s$ continuity, which can be put in as an algorithm for checking smoothness of Hermite-interpolatory schemes.

To analyze the smoothness of the scheme we have to introduce higher order divided differences of the vector $u_k^{[m-1]}$, denoted by $u_k^{[r]}$, representing the divided differences of $f$ of order $m \leq r \leq s+1$, taking into account that each point $2^{-k}n$, $n \in \mathbb{Z}$, has multiplicity $m$. $u_k^{[r]}$ are defined recursively by

$$u_k^{[r+1]}(n) = 2^k(\Lambda^{m,r})^{-1}\big(u_k^{[r]}(n) - u_k^{[r]}(n-1)\big), \quad n \in \mathbb{Z}, \tag{3.1}$$

where $\Lambda^{m,r}$ is a diagonal matrix of order $m \times m$ with elements

$$\Lambda^{m,r}_{j,j} = \left\lceil \frac{m+r-j+1}{m} \right\rceil, \quad j = 1,2,\dots,m. \tag{3.2}$$

THEOREM 2. *Consider a stationary Hermite-interpolatory subdivision scheme of order $m$, of the form (1.1), with $A_i^{(k)}$ satisfying (1.6). If the scheme is $C^s$ then it reproduces polynomials of degree $s$, i.e., is satisfies (1.4) with $\ell = s+1$. Furthermore, there exist subdivision schemes, with finitely supported masks, for the refinement of the divided differences of orders $\leq s+1$.*

PROOF. The proof is iterative. First we observe that the original scheme must reproduce the constant polynomial, i.e., $(1,0,\dots,0)^t$ is an eigenvector of the subdivision scheme. To see this we transform the process (1.1), using (1.6), into the stationary process

$$\begin{aligned} M(2^{-k})f_i^{k+1} &= \sum_{j\in\mathbb{Z}} M(2^{-k})A^{(k)}_{i-2j}M(2^{-k})^{-1}M(2^{-k})f_j^k \\ &= \sum_{j\in\mathbb{Z}} A^{(0)}_{i-2j}M(2^{-k})f_j^k, \quad i \in \mathbb{Z}. \end{aligned} \tag{3.3}$$

Since $M(2^{-k}) = \mathrm{diag}\{1, 2^{-k}, 2^{-2k}, \dots, 2^{-(m-1)k}\}$ and the vectors $\{f_j^k\}$ are values on the continuous limit function $(f, f', \dots, f^{(m-1)})^t$, the values $g_j^k = M(2^{-k})f_j^k$ are on the continuous function $(f, 0, \dots, 0)^t$. Therefore, as in the scalar case, it follows that the constant vector data $(1,0,\dots,0)^t$ is a fixed point of the process. It thus follows that the scheme reproduces constant polynomials. The original scheme generates the vector values $f_n^k = \left(f_k^{(0)}(n), f_k^{(1)}(n), \dots, f_k^{(m-1)}(n)/(m-1)!\right)^t$ forming the limit vector function $f(t) = \left(f^{(0)}(t), f^{(1)}(t), \dots, f^{(m-1)}(t)/(m-1)!\right)^t$. The vectors $v_k^{[1]}(n) = \left(2^k\left(f_k^{(0)}(n) - f_k^{(0)}(n-1)\right), f_k^{(1)}(n), \dots, f_k^{(m-1)}(n)/(m-1)!\right)^t$ form the limit vector function $v^{[1]}(t) = \left(f^{(1)}(t), f^{(1)}(t), \dots, f^{(m-1)}(t)/(m-1)!\right)^t$. By Theorem 1, since $\Pi_0$ is reproduced, the vectors $\left\{v_k^{[1]}(n)\right\}$ are generated by a subdivision process defined by the finite Laurent polynomial matrix

$$D_k^{[1]}(z) = \Gamma^{m,1}_{k+1}(z)A_k(z)\left(\Gamma^{m,1}_k(z^2)\right)^{-1}.$$

Using this form and the form of $\Gamma^{m,1}_k(z)$ it can be shown here that the scheme $D_k^{[1]}$ is also stationary, but here the normalization matrix is

$$M^{(1)}(2^{-k}) = \mathrm{diag}\{1, 1, 2^{-k}, 2^{-2k}, \dots, 2^{-(m-2)k}\},$$

i.e., $D_k^{[1]} = M^{(1)}(2^{-k})^{-1}D_0^{[1]}M^{(1)}(2^{-k})$. Normalizing the process generating the vectors $\left\{v_k^{[1]}(n)\right\}$, and using the assumption that the limit vector function $v^{[1]}(t)$ is continuous, we now obtain that the constant vector data $(1,1,0,\dots,0)^t$ is reproduced by the subdivision defined by $D_k^{[1]}$. Since this vector data is the data $\left\{v_k^{[1]}(n)\right\}$ corresponding to the linear data $\left\{f_n^k = (n2^{-k}, 1, \dots, 0)^t\right\}$, it thus follows that the original scheme reproduces $\Pi_1$. We have thus completed the first step of the iterative proof and we are ready to prove the general step:

For $1 < r < m$ we assume that we have a finite polynomial $D_k^{[r]}(z)$ defining the subdivision scheme generating the data $\left\{v_k^{[r]}(n)\right\}$. Normalizing the process $\{D_k^{[r]}\}$

with

$$M^{(r)}(2^{-k}) = \operatorname{diag}\{1,\dots,1,2^{-k},2^{-2k},\dots,2^{-(m-r-1)k}\},$$

with the 1 repeated $r+1$, and using the continuity of $f^{(r)}$, it follows that

$$v_k^{[r]} \equiv (1,1,\dots,1,0,0,\dots,0)^t,$$

with the 1 repeated $r+1$ times, is a fixed point of the process. Therefore, it can be concluded that the original process reproduces $\Pi_r$, and Theorem 1 can be used again to establish the existence of $D_k^{[r+1]}(z)$.

The symbols of the subdivision schemes generating divided differences $u_k^{[r]}$ of order $m \le r \le s+1$, are independent of the level $k$, and are recursively given by

$$D^{[r+1]}(z) = 2(\Lambda^{m,r+1})^{-1}\Gamma^m(z)D^{[r]}(z)\big(\Gamma^m(z^2)\big)^{-1}\Lambda^{m,r}. \tag{3.4}$$

These schemes exist and are of finite support by the same arguments used in Theorem 1, and the above inductive process can thus be continued. Finally, the continuity of $f^{(s)}$ implies the existence of $D_k^{[s+1]}$.

THEOREM 3. *Consider a stationary Hermite-interpolatory subdivision scheme of order $m$, of the form (1.1), with $A_i^{(k)}$ satisfying (1.7). The scheme is $C^s$ if and only if the subdivision scheme for the refinement of the divided differences of orders $s$ is converging, or equivalently, the scheme for the differences of the divided differences of orders $s$ is contractive.*

To prove the above Theorem we need the following lemma on divided differences:

LEMMA 4. *Consider the set of distinct points $X = \{x_i\}_{i\ge 0} \subset \mathbb{R}$, and another set of points $T = \{t_j\}$ such that*

$$t_j = x_i, \quad m_i \le j < m_{i+1},$$

*where $m_0 < m_1 < \cdots < m_i < m_{i+1} < \dots$. Then the divided difference operators of order $k$ on the set $X$ can be expressed as a finite combination of the divided difference operators of order $k$ on the set $T$. In particular,*

$$[x_0,x_1,\dots,x_k]f = \sum_j w_j\ [t_j,t_{j+1},\dots,t_{j+k}]f, \quad \sum_j w_j = 1. \tag{3.5}$$

We note that in the case of equidistant points $X$ and uniform multiplicity the combination is even convex, $w_j > 0$.

PROOF. The proof is by induction on $k$. For $k=1$ the result is obvious. Let us assume that (3.5) holds, together with

$$[x_1,x_2,\dots,x_{k+1}]f = \sum_j v_j[t_j,t_{j+1},\dots,t_{j+k}]f, \quad \sum_j v_j = 1. \tag{3.6}$$

Hence, it follows that

$$\begin{aligned}[x_1,x_2,\dots,x_{k+1}]f &- [x_0,x_1,\dots,x_k]f \\ &= \sum_j c_j\big([t_{j+1},t_{j+2},\dots,t_{j+k+1}]f - [t_j,t_{j+1},\dots,t_{j+k}]f\big),\end{aligned} \tag{3.7}$$

with $c_j = \sum_{i=i_0}^{j}(v_i - w_i)$. Since both $\{w_j\}$ and $\{v_j\}$ sum up to 1 and are of finite support, then $\{c_j\}$ are also of finite support. The terms in the right-hand side of (2.22) with $t_{j+k+1} = t_j$ are zero and can be removed from the sum. Now each term in the right-hand side can be divided by $t_{j+k+1} - t_j$ and the left-hand side by $x_{k+1} - x_0$ to give a representation relating the finite differences of order $(k+1)$

$$[x_0, x_1, \ldots, x_{k+1}]f = \sum_j d_j \, [t_j, t_{j+1}, \ldots, t_{j+k+1}]f. \tag{3.8}$$

That $\sum_j d_j = 1$ follows from the application of (3.8) to $f(x) = x^{k+1}$.

PROOF OF THEOREM 3. The proof here follows the same lines of the proof of a similar result for the scalar case in [**1**,**8**] and [**11**]. The only difference is in the step of proving that $f$ is $C^s$ if the differences of the divided differences of orders $s$ tend uniformly to zero. Here the divided differences involve also derivatives' values, being based on a set of points of multiplicity $m$. Lemma 4 provides us the missing link since it says that any simple divided difference can be expressed as a finite linear combination of divided differences of the same order, based on points with multiplicity. The same holds for differences of divided differences, and thus we can conclude that if the scheme for the differences of the divided differences of orders $s$ is contractive then the original scheme is $C^s$. The existence of that scheme follows by Theorem 2, and its matrix Laurent polynomial is given by

$$\widehat{D}^{[s]}(z) = \Gamma^m(z) D^{[s]}(z) \left(\Gamma^m(z^2)\right)^{-1}. \tag{3.9}$$

Theorem 3, together with the recursive relations (2.7) and (3.4) and the definition (3.9) of $\widehat{D}^{[s]}$, form the basis for a smoothness checking algorithm for stationary Hermite-interpolator schemes.

## References


1. A. S. Cavaretta, W. Dahmen, and C. A. Micchelli, *Stationary subdivision*, Mem. Amer. Math. Soc, vol. 93, 1991.
2. A. Cohen, I. Daubechies, and G. Plonka, *Regularity of refinable function vectors*, J. Fourier Anal. Appl. **3** (1997), no. 3, 295–324.
3. A. Cohen, N. Dyn, and D. Levin, *Stability and inter-dependence of matrix subdivision schemes*, Advanced Topics in Multivariate Approximation (F. Fontanella, K. Jetter, and P. J. Laurent, eds.), World Scientific, 1996, pp. 33–45.
4. I. Daubechies and C. Lagarias, *Two-scale difference equations.* II. *Local regularity, infinite products of matrices and fractals*, SIAM J. Math. Anal. **23** (1992), no. 4, 1031–1079.
5. G. Deslauriers and S. Dubuc, *Interpolation dyadique*, Fractals, dimensions non entières et applications (G. Cherbit, ed.), Masson, Paris, 1989, pp. 44–55.
6. D. L. Donoho, N. Dyn, D. Levin, and T. Yu, *Smooth multiwavelet duals of Alpert bases by moment-interpolating refinement and applications in recursive partitioning*, preprint (1996).
7. S. Dubuc, *Interpolation through an iterative scheme*, J. Math. Anal. Appl. **114** (1986), no. 1, 185–204.
8. N. Dyn, *Subdivision schemes in computer-aided geometric design*, Advances in Numerical Analysis. II. Wavelets, Subdivision Algorithms, and Radial Basis Functions (W. A. Light, ed.), Clarendon Press, Oxford, 1992, pp. 36–104.
9. N. Dyn, J. A. Gregory, and D. Levin, *A 4-point interpolatory subdivision scheme for curve design*, Comput. Aided Geom. Design **4** (1987), no. 4, 257–268.
10. ———, *Analysis of uniform binary subdivision schemes for curve design*, Constr. Approx. **7** (1991), no. 2, 127–147.

11. N. Dyn and D. Levin, *Interpolating subdivision schemes for the generation of curves and surfaces*, Multivariate Interpolation and Approximation (Duisburg, 1989) (W. Haussmann and K. Jetter, eds.), Internat. Ser. Numer. Math., vol. 94, Birkhäuser, Basel, 1990, pp. 91–106.
12. ———, *Analysis of Hermite-type subdivision schemes*, Approximation Theory. VIII, vol. 2 (College Station, TX, 1995) (Charles K. Chui and Larry L. Schumaker, eds.), Ser. Approx. Decompos., vol. 6, World Sci. Publishing, River Edge, NJ, 1995, pp. 117–124.
13. T. N. T. Goodman and S. L. Lee, *Wavelets of multiplicity $r$*, Trans. Amer. Math. Soc. **342** (1994), 307–324.
14. J. L. Merrien, *A family of Hermite interpolants by bisection algorithms*, Numerical Algorithms **2** (1992), 187–200.
15. G. Plonka, *Approximation order provided by refinable function vectors*, Universität Rostock (1995) (preprint).
16. Z. Shen, *Regularity and stability of refinable vectors*, Singapore University (1995) (preprint).
17. G. Strang and V. Strela, *Orthogonal multiwavelets with vanishing moments*, Optical Engineering (Orlando 1994), SPIE, Bellingham, WA.

School of Mathematical Sciences, Tel-Aviv University, Ramat-Aviv, Tel Aviv 69978, Israel
*E-mail address*: niradyn@math.tau.ac.il

School of Mathematical Sciences, Tel-Aviv University, Ramat-Aviv, Tel Aviv 69978, Israel
*E-mail address*: levin@math.tau.ac.il

Centre de Recherches Mathématiques
CRM Proceedings and Lecture Notes
Volume **18**, 1999

# Some Directional Microlocal Classes Defined Using Wavelet Transforms

Matthias Holschneider

Abstract. In this paper we discuss how the position—scale half-space of wavelet analysis may be cut into different regions. We discuss conditions under which they are independent in the sense that the Toeplitz operators associated with their characteristic functions commute modulo smoothing operators. This shall be used to define microlocal classes of distributions having a well defined behavior along lines in wavelet space. This allows us the description of singular and regular directions in distributions. As an application we discuss elliptic regularity for these microlocal classes for domains with cusp-like singularities.

## 1. Introduction

The classical definition of local singular directions of a distribution $\eta$ is given by the wavefront set (e.g. [**Hör82**]). At a given point $x$ it is roughly speaking the cone of all directions in which the Fourier transform of the localized distribution $\phi \cdot \eta$ does not decay rapidly where $\phi$ is any smooth function that is supported by some neighborhood of $x$. More precisely for fixed $\phi \in C_0^\infty(\mathbb{R}^n)$, $x \in \operatorname{supp}\phi$, a direction $\xi \in \mathbb{R}^n - \{0\}$ is regular if in some conic neighborhood $\gamma \ni \xi$ we have

$$k \in \gamma \implies \left|(\phi \cdot \eta)\widehat{\ }(k)\right| \le c_N\bigl(1+|k|\bigr)^{-N}, \quad N = 0, 1, 2, \dots, .$$

The complement of the regular directions is denoted by $\Sigma_{x,\phi}$. The singular directions at $x$ are then defined as

$$\Sigma_x = \bigcap_\phi \Sigma_{x,\phi},$$

where the the intersection is over all $\phi \in C_0^\infty(\mathbb{R}^n)$ with $x \in \operatorname{supp}\phi$. However this concept of singular and regular directions does not always fit with what one would intuitively call a regular direction in a distribution. What we mean is best illustrated by an example in $\mathbb{R}^2$: consider the set $K = \{(x,y) \in \mathbb{R}^2 : x \ge 0, |y| \le x^2\}$ and let $\chi$ be a function that is of very low regularity in the complement of the cusp $K$, whereas inside the cusp it is smooth. It is plain to see that there is no direction in which the Fourier transform of the localized function $\phi\chi$ decays rapidly and so all directions are singular. This however is in contradiction with our intuition in the sense that if we approach the singularity along any path contained in the set $K$

1991 *Mathematics Subject Classification*. 35B40, 35B65, 35S99.
This is the final form of the paper.

no irregularity is to be noticed and one would like to call the direction of the cusp regular.

This is clearly only a very vague statement. To give a more precise definition we have to introduce the wavelet transform. We shall be very brief and we refer to the literature for a more detailed discussion (e.g. **[GM84]**, **[Dau92]**, **[Mey90]**, **[CM90]**, **[Tor95]**, **[Hol95]**).

Let $g \in \mathcal{S}(\mathbb{R}^n)$, the class of Schwartz of rapidly decaying functions. In addition suppose that $g$ has all moments vanishing

$$\int dx\, x^m g(x) = 0$$

for all multi indices $m$. Then the wavelet transform of $s \in L^p(\mathbb{R}^n)$ with respect to $g$ is defined as the following convolutions

$$\mathcal{W}_g s(b,a) = \mathcal{W}[g,s](b,a) = (\tilde{g}_a * s)(b) = \int dx \frac{1}{a^n} \bar{g}\left(\frac{x-b}{a}\right) s(x) \tag{1.1}$$

with $a > 0$ and $b \in \mathbb{R}^n$.

Here we have introduced the following notations, that we shall use in the sequel

$$\tilde{g}(t) = \bar{g}(-t), \quad g_a = g(\cdot/a)/a^n, \quad g_{b,a} = g_a(\cdot - b).$$

The wavelet transform thus maps functions over the real line to functions over the open half-space $\mathbb{H}^n = \{(b,a) \colon b \in \mathbb{R}^n, a > 0\}$.

From the definition it is clear that the wavelet transform is a sort of mathematical microscope whose position is fixed by $b$ and whose enlargement is given by $1/a$ or to put it differently $\mathcal{W}[g,s](b,a)$ is obtained by "looking at $s$ at position $b$ and at scale $a$". As a general statement one can say that local regularity of $s$ is mirrored in a certain speed of decay of $\mathcal{W}_g s$. So is for instance a uniform (in $b$) decay of $O(a^\infty)$ as $a \to 0$, of the wavelet coefficients equivalent to $C^\infty$ regularity of $s$. More quantitative information is available. So a uniform decay of $O(a^\alpha)$ with $\alpha \in (0,1)$ is equivalent to $s \in \Lambda^\alpha$, the space of Hölder continuous functions of exponent $\alpha$.

A temptative definition of regular direction at $x$ is therefore any direction $\xi$ for which the wavelet transform decays faster than any power of $a$ if the microscope approaches the singularity along a path that is tangent to $\xi$ in $x$ in such a way, that it looks at a scale that is small compared to the distance to $x$. That is we say, vaguely speaking, a direction is regular if along a parabolic line we have rapid decay of the wavelet coefficients

$$\mathcal{W}_g s(\lambda\xi, \lambda^\gamma) \leq O(\lambda^\infty), \quad (\lambda \to 0).$$

This idea will be made more precise in Section 5. In particular the definition will we modified in such a way that it becomes independent of the choice of the wavelet $g$.

## 2. The Basic Formulas of Continuous Wavelet Analysis

For the convenience of the reader we shall list here the basic formulas of wavelet analysis. We limit ourselves to formal expressions. They actually have a precise meaning when we consider the wavelet analysis in $\mathcal{S}_0(\mathbb{R}^n)$ or $\mathcal{S}_0'(\mathbb{R}^n)$ (see below).

Let $f$ be a complex valued function over $\mathbb{R}^n$. Let $g$ be another such function. The wavelet transform of $f$ with respect to the analyzing wavelet $g$ is defined through (we write $dx$ for $n$-dimensional Lebesgue measure)

$$\mathcal{W}_g f(b,a) = \int_{\mathbb{R}^n} dx\, \frac{1}{a^n} \overline{g}\left(\frac{x-b}{a}\right) f(x), \quad b \in \mathbb{R}^n,\ a > 0.$$

We also write $\mathcal{W}[g,f](b,a)$ instead of $\mathcal{W}_g f(b,a)$. Here $b \in \mathbb{R}^n$ is a position parameter and $a \in \mathbb{R}_+$ is a scale parameter. The wavelet transform of a function over $\mathbb{R}^n$ is thus a function over the position scale half-space $\mathbb{H}^n = \mathbb{R}^n \times \mathbb{R}_+$.

If we introduce the dilation $(D_a)$ and translation operators $(T_b)$

$$T_b s(x) = s(x-b), \quad D_a s(x) = s(x/a)/a^n,$$

then we may also write the wavelet transform as a family of scalar products

$$\mathcal{W}_g f(b,a) = \langle g_{b,a}, f \rangle, \quad g_{b,a} = T_b D_a g,$$

or as a family of convolutions indexed by a scale parameter

$$\mathcal{W}_g f(b,a) = (\tilde{g}_a * f)(b), \quad \tilde{g}_a = D_a \tilde{g}, \quad \tilde{g}(x) = \overline{g}(-x).$$

The convolution product is defined as usual

$$(s*r)(x) = \int_{\mathbb{R}^n} dy\, s(x-y) r(y) = (r*s)(x).$$

If we introduce the Fourier transform

$$\hat{s}(k) = \int_{\mathbb{R}^n} dx e^{-ikx} s(x), \quad s(x) = \frac{1}{(2\pi)^n} \int_{\mathbb{R}^n} dk, e^{ikx} \hat{s}(k)$$

then the wavelet transform may also be written as

$$\mathcal{W}_g f(b,a) = \frac{1}{(2\pi)^n} \int_{\mathbb{R}^n} dk\, \overline{\hat{g}(ak)}\, e^{ibk}\, \hat{f}(k).$$

The wavelet synthesis $\mathcal{M}$ maps functions over the position—scale half-space to functions over $\mathbb{R}^n$. Let $r = r(b,a)$ be a complex valued function over $\mathbb{H}^n$ and $h$ a function over $\mathbb{R}^n$. Then the wavelet synthesis of $r$ with respect to the synthesizing wavelet $h$ is defined as

$$\mathcal{M}_h r(x) = \int_{\mathbb{H}^n} \frac{db\, da}{a} r(b,a) \frac{1}{a^n} h\left(\frac{x-b}{a}\right).$$

**2.1. Relation between $\mathcal{W}$ and $\mathcal{M}$.** We now list some relations between $\mathcal{W}$ and $\mathcal{M}$. The wavelet synthesis is the adjoint of the wavelet transform—both with respect to the same wavelet—$\mathcal{W}_g^* = \mathcal{M}_g$

$$\int_{\mathbb{H}^n} \frac{db\, da}{a} \overline{\mathcal{W}_g s(b,a)} r(b,a) = \int_{\mathbb{R}^n} dx\, \overline{s(x)}\, \mathcal{M}_g r(x).$$

The combination $\mathcal{M}_h \mathcal{W}_g$ reads in Fourier space

$$\mathcal{M}_h \mathcal{W}_g \colon \hat{s}(k) \mapsto m_{g,h}(k) \hat{s}(k), \quad m_{g,h}(k) = \int_0^\infty \frac{da}{a} \overline{\hat{g}(ak)}\, \hat{h}(ak).$$

Note that the Fourier multiplier $m_{g,h}$ depends only on the direction of $k$, $m_{g,h} = m_{g,h}(k/|k|)$. This is because the measure $da/a$ is scaling invariant. In case that $q$ and $h$ are such that

$$m_{g,h}(k) = \int_0^\infty \frac{da}{a} \overline{\hat{g}(ak)}\, \hat{h}(ak) = c_{g,h}$$

with $0 < |c_{g,h}| < \infty$, we say that $g$, $h$ are an analysis reconstruction pair, or that $h$ is a reconstruction wavelet for $g$. In this case we have

$$\mathcal{M}_h \mathcal{W}_g = c_{g,h} \mathbb{1}.$$

We say that $g$ is strictly admissible if $g$ is its own reconstruction wavelet, or (what is the same) if

$$\forall k \in \mathbb{R}^n \backslash \{0\} \colon \int_0^\infty \frac{da}{a} \left|\hat{g}(ak)\right|^2 = c_g.$$

A sufficient condition for $g$ to have a reconstruction wavelet $r$ is that for some $c > 1$ we have

$$c^{-1} \leq \int_0^\infty \frac{da}{a} \left|\hat{g}(ak)\right|^2 \leq c.$$

If this condition holds, we call $g$ admissible. In this case the following function $r$ will be a reconstruction wavelet for $g$

$$\hat{r}(k) = \hat{g}(k) \Big/ \sqrt{\int_0^\infty \frac{da}{a} \left|\hat{g}(ak)\right|^2}. \tag{2.1}$$

If $g$ and $h$ are an analysis reconstruction pair, then the following formula holds

$$\int_{\mathbb{H}^n} \frac{db\, da}{a} \overline{\mathcal{W}_g s(b,a)} \mathcal{W}_h r(b,a) = c_{g,h} \int_{\mathbb{R}^n} dx\, \overline{s}(x)\, r(x).$$

In particular if $g$ is strictly admissible, then we have conservation of energy ($c_g = c_{g,g}$)

$$\int_{\mathbb{H}^n} \frac{db\, da}{a} \left|\mathcal{W}_g s(b,a)\right|^2 = c_g \int_{\mathbb{R}^n} dx \left|s(x)\right|^2.$$

Let $\Pi$ and $r$ be functions over the half-space $\mathbb{H}^n$. We define a (non-commutative) convolution for functions over the half-space via

$$(\Pi * r)(b,a) = \int_{\mathbb{H}^n} \frac{db' da'}{a'} \frac{1}{a'^n} \Pi\left(\frac{b-b'}{a'}, \frac{a}{a'}\right) r(b', a'). \tag{2.2}$$

Formally we can write

$$(\mathcal{W}_g \mathcal{M}_h s)(b,a) = \Pi_{g,h} * s(b,a), \quad \Pi_{g,h} = \mathcal{W}_g h.$$

In the case where $g$, and $h$ are an analysis reconstruction pair with $c_{g,h} = 1$ we clearly have $(\mathcal{W}_g \mathcal{M}_h)^2 = \mathcal{W}_g \mathcal{M}_h$ and hence

$$\Pi_{g,h} * \Pi_{g,h} = \Pi_{g,h}.$$

The mapping $r \mapsto \Pi_{g,h} * r$ is a projector into the range of the wavelet transform $\mathcal{W}_g$. In case that $g$ is strictly admissible with $c_{g,g} = 1$, we have that $\Pi_{g,g}$ is an orthogonal projector.

The most important formula for our work is the following. Consider a function $s$ over $\mathbb{R}^n$. Suppose that $g$ is admissible. It thus has a reconstruction wavelet $r$. An explicit formula has been given in 2.1. Then, since $\mathcal{W}_h = (\mathcal{W}_h \mathcal{M}_r)\mathcal{W}_g$, the wavelet transform of $s$ with respect to $g$ and the one with respect to $h$ are related via the so called cross kernel equation.

$$\mathcal{W}_h s = \Pi_{g \to h} * \mathcal{W}_g s, \quad \Pi_{g \to h} = \mathcal{W}_h r. \tag{2.3}$$

## 3. Some Function Spaces

All formulas that we have given so far have a well defined meaning if the wavelets are taken in some subset of the class of Schwartz as we will recall now.

**3.1. The analysis of $\mathcal{S}_0(\mathbb{R}^n)$.** Let $\mathcal{S}(\mathbb{R}^n)$ denote the class of Schwartz consisting of those functions that together with their derivatives decay faster than any polynomial such that the following norms are all finite for all multi-indices $\alpha$ and $\beta$

$$\|s\|_{\alpha,\beta} = \sup_{t\in\mathbb{R}^n} \left|t^\alpha \partial^\beta s(t)\right| < \infty.$$

They generate a locally convex topology which makes $\mathcal{S}(\mathbb{R}^n)$ a Fréchet space. We denoted by $\mathcal{S}_0(\mathbb{R}^n)$ the closed set of functions in $\mathcal{S}(\mathbb{R}^n)$ for which all moments vanish

$$\forall\alpha \in \mathbb{N}^n, \qquad \int dx\, s(x)x^\alpha = 0 \iff \forall m > 0,\ \hat{s}(k) = o(k^m)\ \big(|k| \to 0\big).$$

The Schwartz space of functions over the half-plane $\mathbb{H}^n$ shall be denoted by $\mathcal{S}(\mathbb{H}^n)$. It consists of those functions $r$ for which the following norms are all finite

$$\|r\|_{k,l,m,n} = \sup_{(b,a)\in\mathbb{H}^n} \left|(a + 1/a)^k \big(1 + |b|\big)^l \partial_b^m \partial_a^n\, r(b,a)\right| < \infty.$$

Note that this means that $r$ together with all its derivatives decays rapidly for large $b$ and for large or small $a$. It can be shown that

$$\mathcal{W}\colon \mathcal{S}_0(\mathbb{R}^n) \times \mathcal{S}_0(\mathbb{R}^n) \to \mathcal{S}(\mathbb{H}^n), \quad (g,s) \mapsto \mathcal{W}_g s$$

is continuous. The same holds for the wavelet synthesis defined through

$$\mathcal{M}_h r(x) = \int_{\mathbb{H}^n} \frac{db\,da}{a} r(b,a) \frac{1}{a^n} h\left(\frac{x-b}{a}\right),$$

and we have that

$$\mathcal{M}\colon \mathcal{S}_0(\mathbb{R}^n) \times \mathcal{S}(\mathbb{H}^n) \to \mathcal{S}_0(\mathbb{R}), \quad (h,r) \mapsto \mathcal{M}_h r$$

is continuous too. However that in this paper we will not discuss topologies on the microlocal classes we define. This can be done in an obvious way and we want to streamline the discussion.

We note here the following important fact. For admissible $g \in \mathcal{S}_0(\mathbb{R}^n)$, and arbitrary $h \in \mathcal{S}_0(\mathbb{R}^n)$ the crosskernel 2.3 is a function in $\mathcal{S}(\mathbb{H}^n)$. It thus is very well localized.

**3.2. Wavelet analysis of $\mathcal{S}_0'(\mathbb{R}^n)$.** We denote the space of linear continuous functionals $\eta\colon \mathcal{S}_0(\mathbb{R}^n) \to \mathbb{C}$ by $\mathcal{S}_0'(\mathbb{R}^n)$. We consider it together with its natural weak-$*$ topology. The space $\mathcal{S}_0'(\mathbb{R}^n)$ can canonically be identified with $\mathcal{S}'(\mathbb{R})/P(\mathbb{R}^n)$, where $P(\mathbb{R}^n)$ is the space of polynomials in $n$ variables. The wavelet transform of $\eta \in \mathcal{S}_0'(\mathbb{R}^n)$ can now be defined pointwise as

$$\mathcal{W}_g\eta(b,a) = \eta(\overline{g}_{b,a}).$$

This is a smooth function (e.g. [**Hol95**]) that satisfies at

$$\left|\mathcal{W}_g\eta(b,a)\right| \le c\big(1 + |b|\big)^m (a + a^{-1})^m \tag{3.1}$$

for some $m > 0$. By duality we have that the mapping (this time for fixed wavelet $g \in \mathcal{S}_0(\mathbb{R}^n)$)

$$\mathcal{W}_g\colon \mathcal{S}_0'(\mathbb{R}^n) \to \mathcal{S}'(\mathbb{H}^n), \quad \eta \mapsto \mathcal{W}_g\eta$$

is continuous. Here $\mathcal{S}'(\mathbb{H}^n)$ is the dual of $\mathcal{S}(\mathbb{H}^n)$ together with the weak-$*$ topology Vice versa any let $r \in \mathcal{S}'(\mathbb{H}^n)$. Then we may set for $s \in \mathcal{S}_0(\mathbb{R}^n)$

$$(\mathcal{M}_h r)(s) = r(\mathcal{W}_{\bar{h}} s).$$

Clearly again $\big(h \in \mathcal{S}_0(\mathbb{R}^n)\big)$

$$\mathcal{M}_h \colon \mathcal{S}'(\mathbb{H}^n) \to \mathcal{S}_0'(\mathbb{R}^n), \quad r \mapsto \mathcal{M}_h r$$

is continuous. In the case of a locally integrable function $r$ of at most polynomial growth we have

$$\mathcal{M}_h r(s) = \int_{\mathbb{H}^n} \frac{db\,da}{a} r(b,a) \mathcal{W}_{\bar{h}} s(b,a).$$

**3.3. More general spaces.** Many function spaces can be characterized in terms of wavelet coefficients. As a rule, the faster the wavelet coefficients decay, the more the analyzed function is regular.

We come to the details. For this consider a vector space $B(\mathbb{H}^n)$ of locally integrable functions

$$\mathcal{S}(\mathbb{H}^n) \subset B(\mathbb{H}^n) \subset \mathcal{S}'(\mathbb{H}^n).$$

Suppose in addition that $B(\mathbb{H}^n)$ is invariant under convolutions with highly localized kernels

$$r \in \mathcal{S}(\mathbb{H}^n),\ s \in B(\mathbb{H}^n) \implies r * s,\ s * r \in B(\mathbb{H}^n).$$

The convolution of two functions over $\mathbb{H}^n$ is defined through 2.2. It then makes sense to pull back $B(\mathbb{H}^n)$ to a vector space of distributions over $\mathbb{R}^n$. We shall denote this space of distributions by $B(\mathbb{R}^n)$. It is defined through the following theorem

THEOREM 3.1. *For a distribution $\eta \in \mathcal{S}_0'(\mathbb{R}^n)$ the following is equivalent.*

- *There is a wavelet $g \in \mathcal{S}_0(\mathbb{R}^n)$ which is admissible in that it satisfies*

$$\int_0^\infty \frac{da}{a} |\hat{g}(ak)|^2 \sim 1$$

*for which we have*

$$\mathcal{W}_g \eta \in B(\mathbb{H}^n).$$

- *For all $h \in \mathcal{S}_0(\mathbb{R}^n)$ we have*

$$\mathcal{W}_h \eta \in B(\mathbb{H}^n).$$

PROOF. The passage from $\mathcal{W}_g\eta$ to $\mathcal{W}_h\eta$ is given by the highly localized cross kernel 2.3. By definition, this operation leaves invariant $B(\mathbb{H}^n)$. □

Therefore it makes sense to speak of the space $B(\mathbb{R}^n)$ associated with $B(\mathbb{H}^n)$. It is precisely the space of distributions for which $\mathcal{W}_g\eta \in B(\mathbb{H}^n)$ where $g$ is as given in the theorem.

We still shall need an additional technical assumption on the spaces $B(\mathbb{H}^n)$: their multiplier algebra should contain the bounded functions

$$m \in L^\infty(\mathbb{H}^n),\ s \in B(\mathbb{H}^n) \implies m \cdot s \in B(\mathbb{H}^n).$$

This allows us to define the Toeplitz operators

$$\mathcal{M}_h m \mathcal{W}_g \colon B(\mathbb{R}^n) \to B(\mathbb{R}^n).$$

For the rest of the paper we refer to these spaces $B(\mathbb{H}^n)$ satisfying all stated properties and their pulled back counter part over $\mathbb{R}^n$, $B(\mathbb{R}^n)$, as admissible local regularity spaces.

We end this section with a remark concerning topology. Suppose that in addition $B(\mathbb{H}^n)$ is a Banach space. Suppose in addition that it is a Banach lattice

$$\big\||s|\big\|_{B(\mathbb{H}^n)} = \|s\|_{B(\mathbb{H}^n)}.$$

Suppose further that for fixed $\Pi \in \mathcal{S}(\mathbb{H}^n)$ we have

$$r \mapsto \Pi * r$$

is continuous. Then we can define a norm on $B(\mathbb{R}^n)$ which makes it a Banach space by setting

$$\|s\|_{B(\mathbb{R}^n)} = \|\mathcal{W}_g s\|_{B(\mathbb{H}^n)}.$$

This is well defined, since for different wavelets satisfying the hypothesis of Theorem 3.1 we obtain equivalent norms. There is an easy to verify sufficient condition for $B(\mathbb{H}^n)$ to be stable under convolution with localized kernels in case that $B(\mathbb{H}^n)$ is a Banach space. It is enough to find $K$ and $c$ such that for all $s \in B(\mathbb{H}^n)$ we can estimate

$$\left\|s(\alpha \cdot +\beta, \alpha \cdot)\right\|_{B(\mathbb{H}^n)} \leq \|s\|_{B(\mathbb{H}^n)}\, c(\alpha + 1/\alpha)^K \big(1 + |b|\big)^K.$$

Indeed, by a simple change of variables, we may write

$$\|\Pi * s\|_{B(\mathbb{H}^n)} \leq \|s\|_{B(\mathbb{H}^n)}\, c \int_{\mathbb{H}^n} \frac{d\beta\, d\alpha}{\alpha} \Pi(-\beta, 1/\alpha)(\alpha + 1/\alpha)^K \big(1 + |\beta|\big)^K.$$

The last integral is a finite constant. For the sake of simplicity, we shall not give a detailed discussion of possible topologies on the microlocal classe we are going to define in the next chapter.

**3.4. Some examples of local regularity spaces.** Many functions spaces of day to day functional analysis can be characterized with this easy concept. Most of them are contained in the following two scales of spaces. Let $\phi\colon \mathbb{R}_+ \to \mathbb{R}_+$ and $\kappa\colon \mathbb{R}^n \to \mathbb{R}_+$ be two tempered weight functions over $\mathbb{R}_+$ and $\mathbb{R}^n$ respectively. By this we understand that they satisfy at

$$\phi(aa') \leq O(1)(a + 1/a)^n\, \phi(a'), \quad \kappa(b + b') \leq O(1)\, \big(1 + |b|\big)^m\, \kappa(b'),$$

for some $m > 0$. Then the following expressions define norms for functions over the half-space $1 \leq p, q \leq \infty$

$$\|r\| = \left\{ \int_0^\infty \frac{da}{a} \phi(a) \Big( \int_{-\infty}^{+\infty} db\, \kappa(b)\, |r(b,a)|^q \Big) \right\}^{p/q}$$

or

$$\|r\| = \left\{ \int db\, \kappa(b) \Big( \int \frac{db\, da'}{a'} \phi(a') \frac{1}{a'^n} \Big( 1 + \frac{|b - b'|}{a'} \Big)^{-m-n-1} |r(b,a)|^q \Big)^{p/q} \right\}.$$

The associated Banach spaces are stable under convolution with highly regular kernels and thus they may be pulled back to $\mathbb{R}^n$ giving rise to two scales of spaces. The first scale of spaces contains the Besov spaces, whereas the second scale contains the $L^p$-spaces and Sobolev spaces (see e.g. [**Mey90**], [**CM90**], [**Hol95**], [**Tri84**] for more details).

## 4. More General Microlocal Classes

In this section we shall be concerned with the problem of constructing new local regularity spaces out of old ones. The idea is easily explained. Consider an arbitrary set $\Omega \subset \mathbb{H}^n$. Eventually we want to consider lines $\Omega = \big\{(b = \lambda\xi, a = \lambda^\gamma)\big\}$ in order to define regularity in the direction $\xi \in \mathbb{R}^n$. But for the moment we stay general since it renders the discussion more easy to understand. Now fix in addition to $\Omega$, an admissible, local regularity space $B(\mathbb{R}^n)$. It is characterized by the fact that the wavelet transforms of its members are in some vector space $B(\mathbb{H}^n)$. The regularity classes we want to construct are roughly speaking distributions whose wavelet coefficients have on $\Omega$ a growth behavior governed by the class $B(\mathbb{H}^n)$.

A slightly more general concept is obtained if we take two local regularity spaces $B_1(\mathbb{R}^n)$ and $B_2(\mathbb{R}^n)$, both of the type considered in the previous section. We now want to cut the half-space $\mathbb{H}^n$ into two parts, say $\Omega$ and its complement $\Omega^c$. In some sense, to be made precise below, we consider classes of distributions whose wavelet coefficients behave inside $\Omega$ like the wavelet coefficients of functions in $B_1(\mathbb{R}^n)$, whereas in $\Omega^c$, these distributions have regularity governed by $B_2(\mathbb{R}^n)$.

A naive approach might be to require that the restriction of $\mathcal{W}_g\eta$ to $\Omega$ satisfies $\mathcal{W}_g\eta \in B_1(\mathbb{H}^n)$, whereas its restriction to the complement of $\Omega$ should correspond (via $\mathcal{M}_h$) to a function in $B_2(\mathbb{R}^n)$. However this definition might depend on the wavelets we use and thus it is not useful.

To get around this difficulty we first construct a suitable family of neighborhoods for $\Omega$. With these neighborhoods it turns out that we can define vector spaces that are independent of the wavelets $g$ and $h$.

**4.1. A non-Euclidian distance.** The first step of the construction is to introduce a non-euclidian distance function adapted to the geometry of the half-space. A suitable choice is given by dist as defined via

$$\operatorname{dist}\big((b,a),(b',a')\big) = \max\{a/a', a'/a\} + \big|(b-b')/a'\big| + \big|(b'-b)/a\big|.$$

This clearly is not a distance in the usual sense, since $\operatorname{dist}\big((b,a),(b,a)\big) = 1$. However, a kind of multiplicative triangular inequality holds (see Lemma 4.1 below), so that $\log\operatorname{dist}\big((b,a),(b',a')\big)$ actually is a distance. Note However, that the distance function is symmetric.

$$\operatorname{dist}\big((b,a),(b',a')\big) = \operatorname{dist}\big((b',a'),(b,a)\big).$$

Clearly the upper half-space carries a natural group structure. It is given by the following composition law

$$(b,a)\circ(b',a') = \delta_a\tau_b(b',a') = (ab'+b, aa'),$$

where $\tau_b$ and $\delta_a$ stand for the translation and dilation as left actions of $\mathbb{R}_+$ and $\mathbb{R}^n$ on $\mathbb{H}^n$ via

$$\delta_\alpha\colon (b,a)\mapsto(\alpha b, \alpha a), \qquad \tau_\beta\colon (b,a)\mapsto(b+\beta, a'). \tag{4.1}$$

The inverse element reads

$$(b,a)^{-1} = (-b/a, 1/a),$$

and the neutral element clearly is $(0,1)$.

REMARK 4.1. We could have taken the logarithm of the hyperbolic distance instead of the distance above. However in order to simplify certain proofs, we have opted for the distance function above.

If we denote by $\Delta\big((b,a)\big)$ the distance of a point $(b,a)$ from the point $(0,1)$

$$\begin{aligned}\Delta\big((b,a)\big) &= \operatorname{dist}\big((b,a),(0,1)\big) = \exp\big(|\log a|\big) + |b|(1+1/a)\\ &= \max\{a, 1/a\} + |b|(1+1/a),\end{aligned}$$

then we have the following relation

$$\operatorname{dist}\big((b,a),(b',a')\big) = \Delta\big((b,a)^{-1}(b',a')\big).$$

Note also the following identity

$$\Delta\big((b,a)^{-1}\big) = \Delta\big((b,a)\big).$$

Moreover, the function $\Delta$ is a tempered weight function

$$\Delta\big((b+a\beta,\alpha a)\big) \le (\alpha + 1/\alpha)\big(1+|\beta|\big)\Delta\big((b,a)\big).$$

The next lemma shows that a kind of triangular inequality holds

LEMMA 4.1. *We have the following triangular inequalities*

$$\max\big\{\Delta\big((b,a)\big)/\Delta\big((b',a')\big), \Delta\big((b',a')\big)/\Delta\big((b,a)\big)\big\} \\ \le \Delta\big((b,a)(b',a')\big) \le \Delta\big((b,a)\big)\Delta\big((b',a')\big).$$

PROOF. To prove this inequality note that an elementary direct computation shows that

$$\begin{aligned} aa'\Delta\big((b,a)^{-1}(b',a')\big) &= aa'\Delta\big(((b'-b)/a, a'/a)\big) \\ &= \max\{a^2, a'^2\} + a|b-b'| + a'|b-b'| \\ &\le \max\{a^2, a'^2\} + a|b| + a|b'| + a'|b| + a'|b'|. \end{aligned}$$

On the other hand

$$aa'\Delta\big((b,a)\big)\Delta\big((b',a')\big) = \big(\max\{1,a^2\} + |b| + a|b|\big)\big(\max\{1,a'^2\} + |b'| + a'|b'|\big).$$

Upon neglecting positive terms, we have

$$aa'\Delta\big((b,a)\big)\Delta\big((b',a')\big) \ge \max\{1,a^2\}\max\{1,a'^2\} + a|b| \\ + \max\{1,a'^2\}|b| + a'|b'| + \max\{1,a^2\}||b'|.$$

Therefore the difference between the last and the previous expression is majorized by

$$\max\{1,a^2\}\max\{1,a'^2\} - \max\{a^2,a'^2\} \\ + |b|(\max\{1,a'^2\} - a') + |b'|\big(\max\{1,a^2\} - a\big) \ge 0.$$

The proof of the right most inequality follows now from the identity

$$\Delta\big((b,a)^{-1}\big) = \Delta\big((b,a)\big).$$

The remaining inequality follows as usual from the previous one, namely

$$\Delta\big((b,a)\big) \le \Delta\big((b,a)(b',a')\big)\Delta\big((b',a')^{-1}\big) = \Delta\big((b,a)(b',a')\big)\Delta\big((b',a')\big).$$

□

This immediately implies the following relation for the distance function

$$\operatorname{dist}\big((b,a),(b',a')\big) = \Delta\big((b,a)^{-1}(b',a')\big) \le \Delta\big((b,a)\big)\Delta\big((b',a')\big),$$

and

$$\operatorname{dist}\big((b,a),(b',a')\big) \ge \Delta\big((b,a)\big)/\Delta\big((b',a')\big).$$

Thus the following triangular inequality holds

$$\operatorname{dist}\big((b,a),(b',a'')\big) < \operatorname{dist}\big((b,a),(b',a')\big)\operatorname{dist}\big((b',a'),(b'',a'')\big),$$

and

$$\operatorname{dist}\big((b,a),(b'',a'')\big) \ge \operatorname{dist}\big((b,a),(b',a')\big)\operatorname{dist}\big((b',a'),(b'',a'')\big).$$

**4.2. A family of neighborhoods.** Let us introduce the closed non-euclidian balls

$$U\big((b,a),r\big) = \big\{(b',a') \in \mathbb{H}^n \colon \operatorname{dist}\big((b,a),(b',a')\big) \le r\big\}.$$

Note that they all are obtained by dilations and translations of the balls around the point $(0,1)$. More precisely, since the distance function satisfies

$$\operatorname{dist}\big((\gamma b+\beta,\gamma a),(\gamma b'+\beta,\gamma a')\big) = \operatorname{dist}\big((b,a),(b',a')\big), \quad \gamma>0, \beta\in\mathbb{R}^n,$$

we have

$$U\big((b,a),r\big) = \tau_b\delta_a U\big((0,1),r\big).$$

An equivalent system of neighborhoods $U'$ is obtained by translating and dilating the family of balls defined via the following inequality

$$\big(a - [1+r+1/(1+r)]/2\big)^2 + |b|^2 \le \big(1+r+1/(1+r)\big)^2/4.$$

They are euclidian balls with the "south-pole" at the point $\big(b=0, a=1/(1+r)\big)$ and the "north-pole" at the point $(b=0, a=1+r)$. The equivalence being expressed by the fact that for some constant $c>1$ we have

$$U'\big((0,1),r/c\big) \subset U\big((0,1),r\big) \subset U'\big((0,1),ccr\big).$$

We may leave the elementary calculations to the reader.

We now are interested in when it makes sense to speak of a certain regularity in one set and an other regularity in an other set of the half-space. Consider therefore two arbitrary subsets $\Omega$, $\Sigma \subset \mathbb{H}^n$. We now say that $\Omega$ and $\Sigma$ are well separated if the following holds. For $(b,a)\in\Omega$ consider a non-euclidian ball $U\big((b,a),r\big)$ with center $(b,a)$ and radius $r$. Choose $r$ small enough so that $U$ does not meet $\Sigma$. Well separated means for us that for some $\varepsilon>0$ we may choose $r$ such that the following estimate holds true for small $a$

$$r > a^{-\varepsilon}.$$

In other terms we define more formally and slightly more general

DEFINITION 4.1. We say that two sets $\Omega$ and $\Sigma$ are well separated if for some $\varepsilon>0$ we have

$$(b,a)\in\Omega \implies \operatorname{dist}\big((b,a),\Sigma\big) > \Delta\big((b,a)\big)^{\varepsilon}.$$

Here the distance between a point and a set $\Omega\subset\mathbb{H}^n$ is defined as usual

$$\operatorname{dist}\big((b,a),\Omega\big) = \inf_{(b',a')\in\Omega} \operatorname{dist}\big((b,a),(b',a')\big).$$

Note that although the non-euclidian distance diverges at small scale, the euclidian distance might tend to 0 as $a\to 0$. As an example, which is somehow typical, consider in $\mathbb{H}^2$ the sets

$$\Omega = \big\{a < |b|^{\alpha}\big\} \cap \{a<1/2\}, \quad \Sigma = \big\{a > |b|^{\beta}\big\} \cap \{a<1/2\}.$$

They are well separated iff $\alpha>\beta$. However, their euclidian distance tends always to 0 as $a\to 0$.

We have still an other useful characterization of well separateness of two sets $\Omega$ and $\Sigma$. For this consider the sets

$$\delta_{1/a}\tau_{-b}\Omega, \quad (b,a)\in\Sigma.$$

Now both sets are well separated iff each of these sets is contained in the complement of a non-Euclidian ball $U\big((0,1),r(b,a)\big)$, with $r(b,a) > \Delta\big((b,a)\big)^{\varepsilon}$.

$$\delta_{1/a}\tau_{-b}\Omega \subset U\big((0,1),\Delta(b,a)\big)^{\varepsilon}\big)^{c}. \tag{4.2}$$

**4.3. More about well separated sets.** Since the distance function is continuous in the euclidian topology, it is clear that the distance of a point and a set and its euclidian closure are the same. Therefore a set is well separated from another if and only if its closures are.

The notion of well separated is inherited by subsets. If $\Omega \subset \Sigma$ and $\Sigma$ is well separated from $\Xi$, then $\Omega$ is well separated from $\Xi$, too.

The notion of well separated is symmetric and we have

LEMMA 4.2. *If $\Omega$ is well separated from $\Sigma$ then $\Sigma$ is well separated from $\Omega$.*

PROOF. By hypothesis we have that $(b, a) \in \Omega$ and $(b', a') \in \Sigma$ implies that

$$\operatorname{dist}\big((b,a),(b',a')\big) > \Delta\big((b,a)\big)^{\varepsilon}. \tag{4.3}$$

We claim that this implies that

$$\operatorname{dist}\big((b,a),(b',a')\big) > \Delta\big((b',a')\big)^{\varepsilon'},$$

for $\varepsilon' = \varepsilon/(1+\varepsilon)$. For suppose that on the contrary for some points we have

$$\operatorname{dist}\big((b,a),(b',a')\big) = \Delta\big((b,a)^{-1}(b',a')\big) \le \Delta\big((b',a')\big)^{\varepsilon'}.$$

This implies, via the second triangular inequality, in particular, that

$$\Delta\big((b',a')\big)/\Delta\big((b,a)\big) \le \Delta\big((b',a')\big)^{\varepsilon'},$$

and therefore by the choice of $\varepsilon'$ that $\Delta\big((b',a')\big)^{\varepsilon'} \le \Delta\big((b,a)\big)^{\varepsilon}$. This implies

$$\operatorname{dist}\big((b,a),(b',a')\big) \le \Delta\big((b,a)\big)^{\varepsilon},$$

which is in contradiction with (4.3). □

For every $\varepsilon > 0$ let us introduce the following non-Euclidian neighborhoods of a set $\Omega \subset \mathbb{H}^n$

$$\Gamma_\varepsilon(\Omega) = \bigcup_{(b,a)\in\Omega} U\big((b,a),\Delta\big((b,a)\big)^{\varepsilon}\big).$$

The system of such neighborhoods constitutes a fundamental family of neighborhoods in the following sense. We have that $\Gamma_\varepsilon(\Omega)^c$ is well separated from $\Omega$. In addition, if $\Sigma$ is well separated from $\Omega$, then for some $\varepsilon > 0$ we have that

$$\Sigma \cap \Gamma_\varepsilon(\Omega) = \phi.$$

Thanks to the triangular inequalities we have the following associativity for the $\varepsilon$ neighborhoods.

LEMMA 4.3. *For any set $\Omega \subset \mathbb{H}^n$ the following holds true. For $\varepsilon_1 > 0$, $\varepsilon_2 > 0$ and $\varepsilon_3 \ge \varepsilon_1 + \varepsilon_2(1+\varepsilon_1)$ we have*

$$\Gamma_{\varepsilon_2}\big(\Gamma_{\varepsilon_1}(\Omega)\big) \subset \Gamma_{\varepsilon_3}(\Omega).$$

*On the other hand for $\varepsilon_1 > 0$, $0 < \varepsilon_2 \le \varepsilon_1/(1+\varepsilon_1)$ and $\varepsilon_3 > 0$ small enough such that*

$$\varepsilon_3 - \varepsilon_1 + \varepsilon_2 \frac{1+\varepsilon_3}{1 \quad \varepsilon_2} < 0,$$

*we have*

$$\Gamma_{\varepsilon_2}(\Gamma_{\varepsilon_1}(\Omega)^c)^c \supset \Gamma_{\varepsilon_3}(\Omega).$$

PROOF. We show the first part. If $(b'', a'') \in \Gamma_{\varepsilon_2}(\Gamma_{\varepsilon_1}(\Omega))$ then for some points $(b', a') \in \mathbb{H}^n$ and $(b, a) \in \Omega$ we have

$$\operatorname{dist}\big((b'', a''), (b', a')\big) \leq \Delta\big((b', a')\big)^{\varepsilon_2}, \quad \operatorname{dist}\big((b', a'), (b, a)\big) \leq \Delta\big((b, a)\big)^{\varepsilon_1}.$$

Therefore by the triangular relation we have

$$\operatorname{dist}\big((b'', a''), (b, a)\big) \leq \Delta\big((b', a')\big)^{\varepsilon_2} \Delta\big((b, a)\big)^{\varepsilon_1}.$$

Now as before by the reverse triangular inequality we have

$$\Delta\big((b', a')\big)/\Delta\big((b, a)\big) \leq \Delta\big((b, a)\big)^{\varepsilon_1},$$

and therefore finally as claimed

$$\operatorname{dist}\big((b'', a''), (b, a)\big) \leq \Delta\big((b, a)\big)^{\varepsilon_1 + \varepsilon_2(1+\varepsilon_1)}.$$

To show the second statement we have to prove that

$$\Gamma_{\varepsilon_3}(\Omega) \cap \Gamma_{\varepsilon_2}(\Gamma_{\varepsilon_1}(\Omega)^c) = \emptyset.$$

Suppose on the contrary, there where $(b, a)$ such that

$$\exists (b', a') \in \Omega\colon\ \operatorname{dist}\big((b, a), (b', a')\big) \leq \Delta\big((b', a')\big)^{\varepsilon_3}, \tag{4.4}$$

and simultaneously we have some $(b'', a'') \in \Gamma_{\varepsilon_2}(\Omega)^c$ satisfying therefore

$$\forall (\beta, \alpha) \in \Omega\colon\ \operatorname{dist}\big((\beta, \alpha), (b'', a'')\big) > \Delta\big((\beta, \alpha)\big)^{\varepsilon_1},$$

and for which we have

$$\operatorname{dist}\big((b, a), (b'', a'')\big) \leq \Delta\big((b'', a'')\big)^{\varepsilon_2}. \tag{4.5}$$

In particular we have

$$\operatorname{dist}\big((b', a'), (b'', a'')\big) > \Delta\big((b', a')\big)^{\varepsilon_1}.$$

Therefore we have by the triangular inequality

$$\begin{aligned}\Delta\big((b', a')\big)^{\varepsilon_1} &< \operatorname{dist}\big((b', a'), (b'', a'')\big) \\ &\leq \operatorname{dist}\big((b', a'), (b, a)\big) \operatorname{dist}\big((b, a), (b'', a'')\big) \\ &\leq \Delta\big((b', a')\big)^{\varepsilon_3} \Delta b\big((b'', a'')\big)^{\varepsilon_2}.\end{aligned}$$

Now, relation 4.4 implies

$$\Delta\big((b, a)\big) \leq \Delta\big((b', a')\big)^{1+\varepsilon_3}.$$

Moreover, (4.5) implies

$$\Delta\big((b'', a'')\big)^{1-\varepsilon_2} \leq \Delta\big((b, a)\big).$$

It follows that

$$\Delta\big((b'', a'')\big) < \Delta\big((b', a')\big)^{(1+\varepsilon_3)/(1-\varepsilon_2)}.$$

Therefore

$$1 < \Delta\big((b', a')\big)^{\varepsilon_3 - \varepsilon_1 + \varepsilon_2(1+\varepsilon_3)/(1-\varepsilon_2)},$$

which is impossible by the choice of $\varepsilon_3$ since $\Delta\big((b', a')\big) \geq 1$. □

This immediately implies the following corollary that we shall use in the next section.

LEMMA 4.4. *Let $\Sigma \supset \Omega$ be such that $\Sigma^c$ is well separated from $\Omega$. Then there is a set $\Xi$, $\Sigma \supset \Xi \supset \Omega$ such that $\Xi$ is well separated from $\Sigma^c$ and $\Omega$ is well separated from $\Xi^c$.*

PROOF. Some $\Gamma_\varepsilon(\Omega)$ with $\varepsilon$ small enough will do. □

**4.4. Cutting the half-space.** Let us come back to our original goal of dividing the half-space into two sets of different regularity. As we said already, it is not possible to speak of regularity $B(\mathbb{H}^n)$ inside a given set $\Omega \subset \mathbb{H}^n$, since this notion is not independent under highly regular Calderón Zygmund operators, or to put it simpler, it might depend on the given wavelet we use for the definition.

However, if we require regularity $B(\mathbb{H}^n)$ in a region that that is slightly larger than $\Omega$, it then follows that the same regularity holds true in $\Omega$ for any wavelet.

We denote by abuse of notation $\Sigma$, (respectively $\Omega$) the operator that restricts functions over $\mathbb{H}^n$ to the set $\Sigma$ (respectively $\Omega$). That is we have

$$\Sigma\colon r \mapsto \chi_\Sigma r,$$

where $\chi_\Sigma$ is the characteristic function of $\Sigma$.

THEOREM 4.1. *Consider two sets $\Sigma$ and $\Omega$ and suppose that $\Sigma \supset \Omega$ in such a way that $\Sigma^c$ and $\Omega$ are well separated. Let $g$, $g'$, $h$, $h' \in \mathcal{S}_0(\mathbb{R}^n)$ satisfy for some $c > 1$ ($s = g, g', h, h'$),*

$$c^{-1} < \int_0^\infty \frac{da}{a} \left|\hat{s}(ak)\right|^2 < c.$$

*Suppose that $\eta \in \mathcal{S}'(\mathbb{R}^n)$ satisfies*

$$\mathcal{M}_h \Sigma \mathcal{W}_g \eta \in B(\mathbb{R}^n).$$

*Then*

$$\mathcal{M}_{h'} \Omega \mathcal{W}'_g \eta \in B(\mathbb{R}^n).$$

Therefore it makes sense to separate $\mathbb{H}^n$ into regions of different regularity provided, the regions are well separated.

The proof is based on the following lemma. It estimates the influence under convolution operators over the half-space of a nasty function inside some region $\Xi$ on $\Xi'$ when both are well separated.

LEMMA 4.5. *Suppose $\Xi'$ and $\Xi$ are well separated. Let $r$ be a locally integrable function over $\mathbb{H}^n$ that is equal to 0 except on $\Xi'$, where it satisfies for some $M > 0$ and some $c > 0$*

$$(b,a) \in \Xi' \implies \left|r(b,a)\right| \le c\,\Delta\big((b,a)\big)^M.$$

*Then, for $P \in \mathcal{S}(\mathbb{H}^n)$ we have that $\tilde{r} = P * r$ satisfies*

$$(b,a) \in \Xi \implies \left|\tilde{r}(b,a)\right| \le c_k \Delta\big((b,a)\big)^{-k},$$

*for all $k \in \mathbb{N}$.*

PROOF. We have to estimate the localization of $P * r(b,a)$ for $(b,a) \in \Xi$. By definition $P * r(b,a)$ equals

$$\int_{\Xi'} \frac{db'da'}{a'} \frac{1}{a'^n} P\left(\frac{b-b'}{a'}, \frac{a}{a'}\right) r(b',a').$$

By dilation and translation we also may write using the action (4.1) of dilation and translation on $\mathbb{H}^n$

$$\int_{\delta_{1/a}\tau_{-b}\Xi'} \frac{db'da'}{a'} P(-b'/a', 1/a')\, r(ab'+b, aa').$$

Now by hypothesis on $r$ we may write with some $K$ and some $c > 0$

$$\left|r(ab' + b, aa')\right| \leq c\,\Delta\big((b,a)\big)^K \Delta\big((b',a')\big)^K.$$

Plugging this estimate into the previous expression we have to estimate for $(b, a) \in \Xi$

$$\Delta\big((b,a)\big)^K \int_{\delta_{1/a}\tau_{-b}\Xi'} \frac{db'da'}{a'} \left|P'(b',a')\right|, \tag{4.6}$$

with $P'(b', a') = \Delta\big((b', a')\big)^K P(-b'/a', 1/a')$. Together with $P$ we have that $P'$ is highly localized.

For $\lambda \geq 0$ let us look at the following integral running over the complement of a non-euclidian ball centered at $(0, 1)$

$$F(\lambda) = \int_{\Delta((b',a'))>\lambda} \frac{db'da'}{a'} \left|P'(b',a')\right|.$$

Thanks to the high localization of $P'$, this function is faster decreasing than any power of $1/\lambda$ as $\lambda \to \infty$. Now, since $\Xi$ and $\Xi'$ are well separated we may use characterization (4.2) to conclude that the integral in (4.6) is estimated by $F\big(\Delta((b,a))^\varepsilon\big)$ for some $\varepsilon > 0$. But this function is again rapidly decreasing as $\Delta\big((b,a)\big) \to \infty$ and the proof is finished. □

Note that the lemma we went to prove may be rephrased as follows: for all $P \in \mathcal{S}(\mathbb{H}^n)$ we have that

$$s \mapsto \Xi'\big(P * (\Xi s)\big)$$

is infinitely smoothing in the sense that it maps functions of polynomial growth into rapidly decreasing functions over the half-space.

PROOF THEOREM 4.6. These previous considerations imply the following: if we have that $\Sigma\mathcal{W}_g\eta \in B(\mathbb{H}^n)$, with $g$ admissible, then for all $g' \in \mathcal{S}_0(\mathbb{R}^n)$ we have that $\Omega\mathcal{W}'_g\eta \in B(\mathbb{H}^n)$. To show this note the transition from $\mathcal{W}_g\eta$ to $\mathcal{W}'_g\eta$ is done by convoluting with a highly localized kernel $P$. Now we may write,

$$\Omega\mathcal{W}'_g\eta = \Omega\big(P * (\Sigma\mathcal{W}_g\eta)\big) + \Omega\big(P * (\Sigma^c\mathcal{W}_g\eta)\big).$$

Since by hypothesis $B(\mathbb{H}^n)$ is invariant under multiplications with bounded functions and convolutions with $P$ the first term is again in $B(\mathbb{H}^n)$, whereas the second term is arbitrary smooth.

A slightly more complicated situation occurs in our theorem, since we can not conclude from $\mathcal{M}_h r \in B(\mathbb{R}^n)$ that $r \in B(\mathbb{H}^n)$, since the wavelet synthesis is not injective.

Now we can find a set $\Xi$ between $\Sigma$ and $\Omega$

$$\Omega \subset \Xi \subset \Sigma,$$

such that $\Xi$ is well separated from the complement of $\Sigma$ and $\Omega$ is well separated from the complement of $\Xi$. This follows from Lemma 4.5. We may conclude that

$$\Xi\mathcal{W}_g\mathcal{M}_h\Sigma\mathcal{W}_g\eta = \Xi\big(P_1 * (\Sigma\mathcal{W}_g\eta)\big) \in B(\mathbb{H}^n).$$

Where $P_1 = \mathcal{W}_f h$ for any admissible $f \in \mathcal{S}_0(\mathbb{R}^n)$. In particular, we may choose $f$ to be a reconstruction wavelet for $g$ and thus it follows that $P_1 * \mathcal{W}_g\eta = \mathcal{W}_g\eta$. Now writing (as characteristic functions!) $\Sigma = 1 - \Sigma^c$ the last expression equals

$$\Xi(P_1 * \mathcal{W}_g\eta) - \Xi\big(P_1 * (\Sigma^c\mathcal{W}_g\eta)\big).$$

The set $\Xi$ is well separated from $\Sigma^c$ and thus the second term has rapid decay as $\Delta\big((b,a)\big)$ gets large. Let us call this function $u$. Then, since $u$ is well localized we have

$$r * u \in \mathcal{S}(\mathbb{H}^n),$$

for all $r \in \mathcal{S}(\mathbb{H}^n)$. We therefore obtain, up to a function of rapid decay

$$\Xi \mathcal{W}_g \eta \in B(\mathbb{H}^n).$$

Now $\mathcal{W}_g' \eta = P * \mathcal{W}_g \eta$ for some $P \in \mathcal{S}(\mathbb{H}^n)$. Therefore, since $\Omega \subset \Xi$ is well separated from the complement of $\Xi$ we have as at the beginning of the proof that $\Omega \mathcal{W}_g' \eta \in B(\mathbb{H}^n)$ up to the well localized function $\Omega u$. But then clearly $\mathcal{M}_{h'} \Omega \mathcal{W}_g' \eta \in B(\mathbb{R}^n)$. □

Let $\Sigma \supset \Omega$ be open and let again $\Omega$ be well separated from the complement of $\Sigma$ as before. Consider the two Toeplitz operators

$$T_\Sigma = \mathcal{M}_h \Sigma \mathcal{W}_g, \quad T_\Omega = \mathcal{M}_h \Omega \mathcal{W}_g.$$

We then have proved the following

COROLLARY 4.1. *We have that*

$$[T_\Sigma, T_\Omega] = T_\Sigma T_\Omega - T_\Omega T_\Sigma,$$

*is infinitely smoothing in the sense that it maps the tempered distributions in $\mathcal{S}_0'(\mathbb{R}^n)$ into smooth function in $\mathcal{S}_0(\mathbb{R}^n)$.*

**4.5. Microlocal classes.** The theorems of the previous section may be used to define some very general microlocal classes. Suppose we are given two regularity spaces $A(\mathbb{R}^n)$ and $B(\mathbb{R}^n)$ and Suppose in addition that $B(\mathbb{R}^n) \subset A(\mathbb{R}^n)$. Consider a set $\Omega \subset \mathbb{H}^n$. Since we are only interested in local properties we may suppose that $\Omega$ is bounded in the euclidian norm. In order to avoid technicalities we suppose $\Omega$ is closed.

The first type of local regularity classes corresponds to the idea the globally, a distribution has a regularity described by $A(\mathbb{H}^n)$ whereas locally, in $\Omega$ we have some higher regularity of type $B(\mathbb{H}^n)$.

A dual idea would be to have the wavelet coefficients concentrated on the subset $\Omega$. That is that outside of $\Omega$, the wavelet coefficients are small, hence correspond to the higher regularity $B(\mathbb{H}^n)$, whereas inside $\Omega$, the coefficients are in $A(\mathbb{H}^n)$.

We now want to make these statements more precise. Consider first the case of higher local regularity. Suppose that there is a sequence of closed sets $\{\Omega_k\}$, $k = 1, 2, \dots$ with

$$\Omega_1 \subset \cdots \Omega_k \subset \Omega_{k+1} \cdots \subseteq \Omega.$$

We suppose that $\Omega_k$ converges to $\Omega$ in the sense that

$$\Omega = \bigcup_k \Omega_k.$$

Suppose that $\Omega_k$ and $\Omega_{k+1}^c$ are well separated for each $k$. Then clearly $\Omega_k$ and $\Omega_l^c$ are well separated for $k < l$. We then say that $\eta \in \mathcal{S}_0'(\mathbb{R}^n)$ belongs to the microlocal class $\Lambda(\Omega; A, B)$ iff for some admissible $g$ and all $k$ we have

$$\eta \subset A(\mathbb{R}^n), \text{ and } \mathcal{M}_g \Omega_k \mathcal{W}_g \eta \in B(\mathbb{R}^n).$$

By the results of the previous theorem it is clear that the definition does not depend on the specific wavelets nor on the family of approximating sets $\Omega_k$. Indeed, by

Lemma 4.4 we may take the family $\Gamma_{1/k}(\Omega)$ as universal family of approximating sets.

Note however, that for arbitrary $\Omega$, the previous class might coincide with $A(\mathbb{R}^n)$. Indeed, in order to have an approximating sequence from the interior, of mutually well separated sets the "smoother" region can not be arbitrary thin. It must contain at least some non-euclidian neighborhood of some set.

Frequently one takes $A = \mathcal{S}_0'(\mathbb{R}^n)$ in which case one is only interested in the behavior of the wavelet coefficients around $\Omega$. In the next section we shall use this kind of classes to define directional regularity in distributions.

Consider now the dual approach, where we want to formalize the idea of wavelet coefficients concentrated on $\Omega$. Suppose now that a sequence of open sets $\{\Omega^k\}$, $k = 1, 2, \ldots$ with

$$\Omega^1 \supset \ldots \Omega^k \supset \Omega^{k+1} \cdots \supset \Omega$$

converges to $\Omega$ in the sense that

$$\Omega = \bigcap_k \Omega^k.$$

Again we require that $(\Omega^k)^c$ and $\Omega^{k+1}$ are well separated for each $k$. We then say that $\eta \in \mathcal{S}_0'(\mathbb{R}^n)$ belongs to the microlocal class $\Sigma(\Omega; A, B)$ iff for some admissible $g$ and all $k$ we have

$$\eta \in A(\mathbb{R}^n), \text{ and } \mathcal{M}_g(\Omega^k)^c \mathcal{W}_g \eta \in B(\mathbb{R}^n).$$

Note that in the case where $B(\mathbb{R}^n) = \mathcal{S}_0(\mathbb{R}^n)$, this corresponds to the idea having the wavelet coefficients concentrated on the set $\Omega$, where they satisfy the less restrictive regularity estimate given by $A(\mathbb{H}^n)$.

## 5. Some Directional Microlocal Classes

We now propose to look at more specific examples of regularity classes. In particular to those we mentioned in the beginning of the paper, that is to classes related to the notion of singular or regular directions in distributions.

Particular useful examples arise when we consider parabolic regions or lines in wavelet space. As measure of regularity it is useful to consider the Hölder-Zygmund scale $\Lambda^\alpha$ of spaces defined in wavelet space via

$$\|s\|_\alpha = \sup_{(b,a)\in\mathbb{H}^n} \left|a^{-\alpha} s(b,a)\right|.$$

Now fix a vector $\xi \in \mathbb{R}^n$, $|\xi| > 0$ and consider the set

$$\Xi = \Xi(\xi, \gamma) = \left\{(b = \lambda\xi, a = \lambda^\gamma) \colon 1/2 > \lambda > 0\right\}$$

for some $\gamma > 1$. We now say that $\eta \in \Lambda^\alpha(\mathbb{R}^n)$ is locally of type $(\alpha, \xi, \gamma)$ if it belongs to the microlocal class $\Omega^{A,B}$ with $\Omega = \Gamma_\varepsilon\big(\Xi(\xi,\gamma)\big)$ for some $\varepsilon > 0$, $A = \mathcal{S}(\mathbb{R}^n)$ and $B = \Lambda^\alpha(\mathbb{R}^n)$. Explicitly, this means—let us recall it once more—that the wavelet transform of $s$ satisfies for some $\varepsilon > 0$ at

$$(b,a) \in \Gamma_\varepsilon\big(\Xi(\xi,\gamma)\big) \implies \left|\mathcal{W}_g\eta(b,a)\right| \le ca^\alpha$$

and

$$(b,a) \notin \Gamma_\varepsilon\big(\Xi(\xi,\gamma)\big) \implies \left|\mathcal{W}_g\eta(b,a)\right| \le c(a+1/a)^k\big(1+|b|\big)^k.$$

This corresponds to looking at behavior of the wavelet coefficients under the following non-homogeneous dilations

$$\mathcal{W}_g\eta(\lambda\xi, \lambda^\gamma), \quad \lambda > 0.$$

Here $c$ depends on $\varepsilon$ only. Actually we may choose $\xi$ such that $|\xi = 1|$. Indeed suppose $\xi' = \beta\xi$ with $\beta > 0$ and denote by $\Xi'$ the corresponding line. Then if $(b, a) \in \Xi$ it follows that $(b, \beta^{-\alpha}a) \in \Xi'$. Therefore the non-euclidian distance between the two points is uniformly bounded by $\beta^{\alpha} + \beta^{-\alpha}$. Therefore, they define the same micro-local classes.

Let us explain in which sense these classes are linked to singular and regular directions. For this replace for the moment the wavelet at position $b$ and scale $a$ by the characteristic function of the euclidian ball centered at $b$ and of radius $a$. As $(b, a)$ tends to $(0, 0)$, while always $(b, a) \in \big(\Omega(\xi, \gamma)\big)$, the support of the wavelets is contained in a cusp-like region, around the line in direction $\xi$. This shows, that the micro-local class $(\alpha, \xi, \gamma)$ quantifies the regularity of $s$ in direction $\xi$.

**5.1. Some elliptic regularity.** We now want to apply the classes introduced above to a problem of elliptic regularity. For the sake of simplicity, we only discuss the Laplace equation and leave more general elliptic operators for subsequent papers. We say an open domain $\Omega \subset \mathbb{R}^n$ satisfies the cusp condition of degree $\delta > 0$ at $x \in \partial\Omega$ in direction $\xi$, $\xi \in \mathbb{R}^n - \{0\}$ if there is some $c > 0$ such that

$$\{y\colon |y - x| \leq c\big|(y - x \mid \xi)\big|^{\delta}, |y - x| \leq c\} \subset \Omega.$$

THEOREM 5.1. *Let $\Omega$ satisfy a cusp condition at 0 of type $\delta$ in direction $\xi$. Suppose that $\eta$ is a tempered distribution that satisfies inside $\Omega$ at*

$$\Delta\eta = f$$

*for some distribution $f$ supported by $\overline{\Omega}$. Now if $f$ is of type $(\xi, \gamma, \alpha)$, with $\gamma > \delta$, then it follows that $\eta$ is of type $(\xi, \gamma, \alpha + 2)$.*

This theorem is a special case of a more general theorem. Let $B(\mathbb{R}^n)$ be a local regularity space of the type we have considered before with $B(\mathbb{H}^n)$ the associated space of functions over $\mathbb{H}^n$. It is plain to see that together with $B(\mathbb{H}^n)$, the space of functions which consists the functions $a^{\gamma}r(b, a)$ with $r \in B(\mathbb{H}^n)$ is again an admissible regularity space. It shall be denoted by $(a^{\gamma}B)(\mathbb{H}^n)$ respectively $(a^{\gamma}B)(\mathbb{R}^n)$,

For a set $\Omega \subset \mathbb{R}^n$ we consider the set

$$\bigcup_{b\in\Omega} K(b),$$

where $K(b) \subset \mathbb{H}^n$ is the cone of opening angle 1 with top in $b$

$$K(b) = \big\{(\beta, \alpha)\colon |\beta - b| \leq \alpha\big\}.$$

We call this set the influence region of $\Omega$ in the upper half-space.

The general theorem can now be stated as follows.

THEOREM 5.2. *Let $\Omega \subset \mathbb{R}^n$ be open. Suppose that $\Xi \subset \mathbb{H}^n$ is well separated from the influence region of $\Omega$. Suppose that $\Xi' \subset \Xi$ such that $\Xi'$ and $\Xi^c$ are well separated. Suppose that $\eta$ is a tempered distribution that satisfies inside $\Omega$ at*

$$\Delta\eta = f$$

*for some distribution $f \in \mathcal{S}(\mathbb{R}^n)$ supported by $\overline{\Omega}$. If now $f$ satisfies at*

$$\mathcal{M}_g\Xi\mathcal{W}_g f \in B(\mathbb{R}^n),$$

*with some admissible wavelet $g \in \mathcal{S}_0(\mathbb{R}^n)$, then it follows that*

$$\mathcal{M}_g\Xi'\mathcal{W}_g\eta \in (a^2B)(\mathbb{R}^n).$$

PROOF. We may suppose that $g \in \mathcal{S}_0(\mathbb{R}^n)$ be spherically symmetric, $g = g(|x|)$. Then with $h = \Delta g$ we may write

$$\mathcal{W}_g \Delta \eta = -a^{-2} \mathcal{W}_h \eta.$$

Now $g$ and $h$ are both admissible in the sense that

$$\int_0^\infty \frac{da}{a} |\hat{g}(ak)|^2 \sim \int_0^\infty \frac{da}{a} |\hat{h}(ak)|^2 \sim 1.$$

From this it follows immediately that $s \in \mathcal{S}'(\mathbb{R}^n)$ and $\Delta s$ satisfying at

$$\mathcal{M}_g \Xi \mathcal{W}_g \Delta s \in B(\mathbb{R}^n),$$

implies that $s$ satisfies at

$$\mathcal{M}_g \Xi' \mathcal{W}_g s \in (a^2 B)(\mathbb{R}^n).$$

The theorem is therefore proved if we can show that any distribution $f'$ that coincides with $f$ inside $\Omega$ satisfies again at

$$\mathcal{M}_g \Xi \mathcal{W}_g \Delta f \in B(\mathbb{R}^n).$$

But this is follows from the next lemma, which justifies the name influence region for $\Omega$. □

LEMMA 5.1. *Let $\Xi \subset \mathbb{H}^n$ be well separated from the influence region of $\Omega$. Then for all $\rho \in \mathcal{S}'(\mathbb{R}^n)$ with support in $\Omega$ we have that*

$$\mathcal{M}_g \Xi \mathcal{W}_g \rho \in \mathcal{S}_0(\mathbb{R}^n).$$

PROOF. By hypothesis there is an $\varepsilon > 0$ such that for $(b,a) \in \Gamma_\varepsilon(\Xi)$ there is an $n$ dimensional Euclidian ball of radius $> \Delta\big((b,a)\big)^\varepsilon$ around $b$ that is contained in the complement of the influence region of $\Omega$. Denote by $\phi$ a $C^\infty$ function which is identically 1 on the complement of the unit ball of radius 1 and which is supported on the complement of a slightly smaller ball. Then denote by $\phi_{b,a}$ the family of translates and dilates of $\phi$. Therefore if $\rho$ is supported by $\Omega$, it follows that

$$(b,a) \in \Xi \implies \mathcal{W}_h \rho(b,a) = \mathcal{W}_g\big(\phi_{b,\Delta((b,a))^\varepsilon} \rho\big)(b,a) = \big\langle g_{b,a} \phi_{b,\Delta((b,a))^\varepsilon} \rho \big\rangle.$$

But now for every $\varepsilon > 0$ we have that $g_{b,a}\phi_{b,\Delta((b,a))^\varepsilon}$ tends to 0 as $\Delta\big((b,a)\big) \to \infty$ in $\mathcal{S}(\mathbb{R}^n)$ in such a way that for all semi-norms in $\mathcal{S}_0(\mathbb{R}^n)$ and all $M > 0$ we have

$$\big\| g_{b,a} \phi_{b,\Delta((b,a))^\varepsilon} \big\|_{l,m} \le c_{l,m,M} \Delta\big((b,a)\big)^{-M}.$$

This proofs the lemma □

The theorem is proved.

## References


[AAGM95] S. T. Ali, J. P. Antoine, J. P. Gazeau, and U. A. Mueller, *Coherent states and their generalizations: A mathematical overview*, Rev. Math. Phys. **7** (1975), no. 7, 1013–1104.

[CZ57] A. P. Calderón and A. Zygmund, *Singular integral operators and differential equations*, Amer. J. Math. **79** (1957), 901–921.

[Chu92] C. K. Chui, *An introduction to wavelets*, Academic Press, 1992.

[CM90] R. Coifman and Y. Meyer, *Ondelettes et opérateurs.* III, Hermann, 1990.

[CGT89] J. M. Combes, A. Grossmann, and Ph. Tchamitchiam. *Wavelets, time-frequency methods and phase space*, Springer-Verlag, 1989.

[Dau92] I. Daubechies, *Ten lectures on wavelets*, CBMS-NSF Regional Conference Series in Applied Mathematics, vol. 61, SIAM, Philadelphia, PA, 1992.

[FG89a] H. G. Feichtinger and K. Gröchenig, *Banach spaces related to integrable group representations and their atomic decompositions.* I, J. Funct. Anal. **86** (1989), no. 2, 307–340.

[FG89b] ———, *Banach spaces related to integrable group representations and their atomic decompositions.* II, Monatsh. Math. **108** (1989), no. 2-3, 129–148.

[GM84] A. Grossmann and J. Morlet, *Decomposition of Hardy functions into square integrable wavelets of constant shape*, SIAM J. Math. Anal. **15** (1984), no. 4, 723–736.

[Hol93b] M. Holschneider, *Localization properties of wavelet transforms*, J. Math. Phys. **34** (1993), no. 7, 3227–3244.

[Hol95] ———, *Wavelets: An analysis tool*, Oxford University Press, Oxford, 1995.

[HT91] M. Holschneider and Ph. Tchamitchian, *Pointwise regularity of Riemann "nowhere differentiable" function*, Invent. Math. **105** (1991), 157–175.

[Hör82] L. Hörmander, *The analysis of partial differential operators.* I. *Distribution theory and Fourier analysis*, Fundamental Principles of Mathematical Sciences, vol. 256, Springer Verlag, Berlin-New York, 1983.

[Jaf89] S. Jaffard, *Exposants d'Hölder en des points donnés et coefficients d'ondelettes*, C. R. Acad. Sci., Paris Sér. Math. I **308** (1989), 79–81.

[JK82] D. Jerison and C. Kenig, *Boundary behavior of harmonic functions in nontangentially accessible domains*, Adv. in Math **46** (1982), no. 1, 80–147.

[Mey90] Y. Meyer, *Ondelettes et opérateurs.* I, II, Hermann, Paris, 1990.

[Mey95] ———, *Chirps n-dimensionnels et analyse 2-microlocale*, Séminaire sur les équations aux derivées partielles, École polytechnique, 1994.

[Sar94] G. Saracco, *Propagation of transient waves through a stratified fluid medium: Wavelet analysis of a nonasymptotic decomposition of the propagator*, J. Acoust. Soc. Amer. **95** (1994), no. 3, 1191–1205.

[Tor95] B. Torrésani, *Analyse continue par ondelettes*, CNRS Editions, 1995.

[Tri84] H. Triebel, *Theory of function spaces.* II, Monographs in Mathematics, vol. 84. Birkhäuser Verlag, Basel, 1992.



Centre de Physique Théorique, CNRS, Luminy, Case 907, F–13288 Marseilles Cedex 9, France

*E-mail address*: hols@cpt.univ-mrs.fr

Centre de Recherches Mathématiques
CRM Proceedings and Lecture Notes
Volume **18**, 1999

# Nonseparable Biorthogonal Wavelet Bases of $L^2(\mathbb{R}^n)$

Abderrazek Karoui and Rémi Vaillancourt

ABSTRACT. A procedure is presented for the construction of nonseparable biorthogonal wavelet bases of $L^2(\mathbb{R}^n)$ for $n > 1$ with arbitrary dilation matrices $\boldsymbol{D} \in \mathbb{Z}^{n\times n}$. Gröchenig's nonconstructive characterization of $n$-dimensional orthonormal wavelets with $\boldsymbol{D} = 2\boldsymbol{I}_n$ is generalized to biorthogonal wavelets with an arbitrary dilation matrix and used as a construction guide. McClellan's transformation is used to construct nonseparable $n$-dimensional compactly supported and symmetric wavelet filters from a one-dimensional symmetric filter of finite length. It is shown by examples that the regularity of wavelets depends on the dilation matrix, and in particular, on the norm, $\|\boldsymbol{D}^{-1}\|_2$, of its inverse. An example of wavelets in $L^2(\mathbb{R}^2)$, where the dilation matrix $\boldsymbol{D}$ satisfies $|\det \boldsymbol{D}| = 4$, is constructed.

## 1. Introduction

Wavelets are functions generated by translations and dilations of one basic function. Separable wavelets in $L^2(\mathbb{R}^2)$, which are constructed by tensor products of wavelets in $L^2(\mathbb{R})$, have good horizontal and vertical resolution properties but they do not perform so well in other directions. Thus, there is a growing interest [**3, 9, 12, 13**] in nonseparable multidimensional wavelets for the applications.

The design of nonseparable wavelet bases of $L^2(\mathbb{R}^n)$ for $n > 1$ is complicated because many techniques and results used in the 1-D (one-dimensional) case do not generalize to higher dimensions. Moreover, this design critically depends on the choice of the dilation matrix $\boldsymbol{D} \in \mathbb{Z}^{n\times n}$ which defines the $n$-D ($n$-dimensional) sampling lattice [**18**]. If $\sigma_1 \geq \sigma_2 \geq \cdots \geq \sigma_n$ denote the singular values of $\boldsymbol{D}$, then both the condition $\sigma_n > 1$, which implies that $\|\boldsymbol{D}^{-1}\| = 1/\sigma_n < 1$, and the absolute value of the determinant $d = |\det \boldsymbol{D}| = \sigma_1\sigma_2\cdots\sigma_n$ play an important role in the design process and regularity estimates of the wavelets and in the convergence of the $n$-D version of the cascade algorithm. It will appear, in an example, that the condition $|\lambda_k| > 1$ on the eigenvalues of $\boldsymbol{D}$ is not sufficient to insure regularity of the wavelets.

Unlike in the 1-D case, where it is relatively easy to design lowpass wavelet filters which generate the scaling functions and, consequently, the wavelets [**4, 5, 11, 17**], in the $n$-D case, with $n \geq 2$, the construction of the equivalent general

---

1991 *Mathematics Subject Classification.* Primary: 42C15; Secondary: 04A12.
This is the final form of the paper.

lowpass filter is hard to achieve without using special techniques. In certain cases, McClellan's transformation [**15**], which is well known in the filter bank theory, can be efficiently used to generate $n$-D compactly supported symmetric wavelet filters from a 1-D symmetric filter of finite length. But, because of the zero phase restriction, only compactly supported biorthogonal wavelets will be constructed.

In Section 2, we briefly review multiresolution analysis and wavelets of $L^2(\mathbb{R}^n)$ only as needed for the sequel. In Section 3, we characterize wavelet filters for general dilation matrices $\boldsymbol{D}$. In Section 4, we briefly describe the McClellan transformation and construct numerical examples in 2-D.

We let ${}^t\boldsymbol{D}$ denote the transpose of the matrix $\boldsymbol{D}$ with $l^2$-norm $\|\boldsymbol{D}\|_2$ and $\boldsymbol{\xi} = {}^t(\xi_1, \xi_2)$ denote a column vector with $l^2$-norm $|\boldsymbol{\xi}|$. The Fourier and inverse Fourier transforms are, respectively,

$$\hat{f}(\boldsymbol{\xi}) = \frac{1}{(2\pi)^{n/2}} \int_{\mathbb{R}^n} e^{-i\boldsymbol{x}\cdot\boldsymbol{\xi}} f(\boldsymbol{x})\, d\boldsymbol{x}, \quad f(\boldsymbol{x}) = \frac{1}{(2\pi)^{n/2}} \int_{\mathbb{R}^n} e^{i\boldsymbol{x}\cdot\boldsymbol{\xi}} \hat{f}(\boldsymbol{\xi})\, d\boldsymbol{\xi}.$$

## 2. Multiresolution Analysis and Wavelets of $L^2(\mathbb{R}^n)$

**2.1. Biorthogonal basis of a Hilbert space.** We say that two sequences $\{f_n\}_{n\in\mathbb{N}}$, $\{g_n\}_{n\in\mathbb{N}}$ of elements of a Hilbert space $\mathcal{H}$ form a biorthogonal system if

$$\langle f_n, g_m\rangle = \delta_{n-m}. \tag{2.1}$$

This system is said to be complete if each of the systems $\{f_n\}$, $\{g_n\}$ is complete in $\mathcal{H}$; in that case, we have the $L^2$-convergent biorthogonal expansions:

$$f = \sum_{n=1}^{\infty} \langle f, g_n\rangle f_n, \quad f = \sum_{n=1}^{\infty} \langle f, f_n\rangle g_n, \quad f \in L^2(\mathbb{R}^n). \tag{2.2}$$

**2.2. Biorthogonal multiresolution analysis.** One way of designing $n$-D wavelets is to generalize the 1-D multiresolution analysis (MRA) [**14**] to the $n$-D case, where the choice of the dilation matrix $\boldsymbol{D}$ defined below is critical.

DEFINITION 1. A matrix $\boldsymbol{D} \in \mathbb{Z}^{n\times n}$ is said to be a dilation matrix if all its singular values are larger than 1: $\sigma_i > 1$, $i = 1, 2, \ldots, n$.

The condition on the singular values ensures that every direction in $\mathbb{R}^n$ is dilated, which is a very desirable property of the multiresolution concept. Moreover, $\|\boldsymbol{D}\|_2 = \sigma_1 > 1$, $\|\boldsymbol{D}^{-1}\|_2 = 1/\sigma_n < 1$ and $d = |\det \boldsymbol{D}| = \sigma_1\sigma_2\cdots\sigma_n$ is an integer. The condition $\|\boldsymbol{D}^{-1}\|_2 = 1/\sigma_n < 1$ is used in Theorems 1 and 2. In Example 2 of Section 4, the eigenvalues have modulus larger than 1 but $\sigma_2 < 1$.

Biorthogonal $n$-D MRA, defined below, generate $n$-D biorthogonal wavelet bases.

DEFINITION 2. For a given dilation matrix $\boldsymbol{D}$, an $n$-D biorthogonal MRA is a pair of decreasing families $\{V_j\}$, $\{\widetilde{V}_j\}$, $j \in \mathbb{Z}$, of closed linear subspaces of $L^2(\mathbb{R}^n)$, each satisfying

$$V_{j+1} \subset V_j, \quad \bigcap_{-\infty}^{j=\infty} V_j = \{0\}, \quad \overline{\bigcup_{-\infty}^{j=\infty} V_j} = L^2(\mathbb{R}^n), \tag{2.3}$$

$$f(\boldsymbol{x}) \in V_0 \iff f(\boldsymbol{x} - \boldsymbol{k}) \in V_0, \quad f(\boldsymbol{x}) \in V_{j+1} \iff f(\boldsymbol{D}\boldsymbol{x}) \in V_j. \tag{2.4}$$

and similar relations with $\widetilde{V}_j$. Moreover there exist a pair of functions $\phi \in V_0$ and $\widetilde{\phi} \in \widetilde{V}_0$ such that $\{\phi_{0,\boldsymbol{n}}\}_{\boldsymbol{n}\in\mathbb{Z}^n}$ and $\{\widetilde{\phi}_{0,\boldsymbol{n}}\}_{\boldsymbol{n}\in\mathbb{Z}^n}$ form a biorthogonal basis of $V_0$, where

$$\phi_{0,\boldsymbol{n}}(\boldsymbol{x}) = \phi(\boldsymbol{x}-\boldsymbol{n}), \quad \widetilde{\phi}_{0,\boldsymbol{n}}(\boldsymbol{x}) = \widetilde{\phi}(\boldsymbol{x}-\boldsymbol{n}).$$

Finally, the two families of subspaces are related by the orthogonality conditions:

$$\widetilde{W}_j \perp V_j, \quad W_j \perp \widetilde{V}_j, \tag{2.5}$$

where $W_j$ and $\widetilde{W}_j$ are the complements of $V_j$ and $\widetilde{V}_j$ in $V_{j-1}$ and $\widetilde{V}_{j-1}$, respectively.

**2.3. Biorthogonal scaling functions and wavelets.** Given an $n$-D biorthogonal MRA, the first step in designing biorthogonal wavelets of $L^2(\mathbb{R}^n)$ is described as follows. The inclusions

$$\phi \in V_0 \subset V_{-1}, \quad \widetilde{\phi} \in \widetilde{V}_0 \subset \widetilde{V}_{-1},$$

imply the existence of two sequences of real numbers $\{\alpha^0_{\boldsymbol{n}}\}_{\boldsymbol{n}\in\mathbb{Z}^n}$, $\{\widetilde{\alpha}^0_{\boldsymbol{n}}\}_{\boldsymbol{n}\in\mathbb{Z}^n}$, such that

$$\phi(\boldsymbol{x}) = \sum_{\boldsymbol{n}\in\mathbb{Z}^n} \alpha^0_{\boldsymbol{n}} \phi(\boldsymbol{D}\boldsymbol{x}-\boldsymbol{n}), \quad \widetilde{\phi}(\boldsymbol{x}) = \sum_{\boldsymbol{n}\in\mathbb{Z}^n} \widetilde{\alpha}^0_{\boldsymbol{n}} \widetilde{\phi}(\boldsymbol{D}\boldsymbol{x}-\boldsymbol{n}). \tag{2.6}$$

Taking Fourier transforms, we have

$$\widehat{\phi}(\boldsymbol{\xi}) = \frac{1}{|\det \boldsymbol{D}|} \sum_{\boldsymbol{n}\in\mathbb{Z}^n} \alpha^0_{\boldsymbol{n}} e^{-i\boldsymbol{D}^{-1}\boldsymbol{n}\cdot\boldsymbol{\xi}} \widehat{\phi}({}^t\boldsymbol{D}^{-1}\boldsymbol{\xi}) =: m_0({}^t\boldsymbol{D}^{-1}\boldsymbol{\xi})\widehat{\phi}({}^t\boldsymbol{D}^{-1}\boldsymbol{\xi}), \tag{2.7}$$

$$\widehat{\widetilde{\phi}}(\boldsymbol{\xi}) = \frac{1}{|\det \boldsymbol{D}|} \sum_{\boldsymbol{n}\in\mathbb{Z}^n} \widetilde{\alpha}^0_{\boldsymbol{n}} e^{-i\boldsymbol{D}^{-1}\boldsymbol{n}\cdot\boldsymbol{\xi}} \widehat{\widetilde{\phi}}({}^t\boldsymbol{D}^{-1}\boldsymbol{\xi}) =: \widetilde{m}_0({}^t\boldsymbol{D}^{-1}\boldsymbol{\xi})\widehat{\widetilde{\phi}}({}^t\boldsymbol{D}^{-1}\boldsymbol{\xi}), \tag{2.8}$$

where $m_0(\boldsymbol{\xi})$ and $\widetilde{m}_0(\boldsymbol{\xi})$ are $2\pi\mathbb{Z}^n$-periodic functions, which reduce to trigonometric polynomials if the scaling functions have compact support.

If $\widehat{\phi}(0) = 1$ and $\widehat{\widetilde{\phi}}(0) = 1$, by iterating (2.7) and (2.8) infinitely often, one obtains dual scaling functions from the dual low pass filters $m_0$ and $\widetilde{m}_0$,

$$\widehat{\phi}(\boldsymbol{\xi}) = \prod_{j=1}^{\infty} m_0({}^t\boldsymbol{D}^{-j}\boldsymbol{\xi}), \quad \widehat{\widetilde{\phi}}(\boldsymbol{\xi}) = \prod_{j=1}^{\infty} \widetilde{m}_0({}^t\boldsymbol{D}^{-j}\boldsymbol{\xi}). \tag{2.9}$$

Note that these infinite products converge in $L^2(\mathbb{R}^n)$ whenever the conditions of Theorem 2 below are satisfied. For more detail, see [**2**].

Following [**9**], since $|\det \boldsymbol{D}| = d$ is the number of cosets in $\mathbb{Z}^n/\boldsymbol{D}\mathbb{Z}^n$, then two families, each containing $d-1$ elementary wavelets $\{\psi^i\}_{i=1}^{d-1}$, $\{\widetilde{\psi}^i\}_{i=1}^{d-1}$, are needed for the following pair of $d$-families:

$$\{\phi_{0,\boldsymbol{k}}, \psi^i_{0,\boldsymbol{k}}\}_{\boldsymbol{k}\in\mathbb{Z}^n}, \quad \{\widetilde{\phi}_{0,\boldsymbol{k}}, \widetilde{\psi}^i_{0,\boldsymbol{k}}\}_{\boldsymbol{k}\in\mathbb{Z}^n}, \quad i = 1, \ldots, d-1, \tag{2.10}$$

to form a biorthogonal basis of $V_{-1}$. For convenience, we shall denote the scaling functions $\phi$ and $\widetilde{\phi}$ by $\psi^0$ and $\widetilde{\psi}^0$, respectively.

By definition, biorthogonal wavelet bases of $L^2(\mathbb{R}^n)$ are biorthogonal Riesz bases, that is,

$$\langle \psi^i_{p,\boldsymbol{k}}, \widetilde{\psi}^j_{q,\boldsymbol{l}} \rangle = \delta_{i-j}\delta_{p-q}\delta_{\boldsymbol{k}-\boldsymbol{l}}, \quad p,q \in \mathbb{Z}, \quad \boldsymbol{k},\boldsymbol{l} \in \mathbb{Z}^n, \quad i,j = 0,\dots,d-1, \tag{2.11}$$

$$f = \sum_{i=1}^{d-1} \sum_{j\in\mathbb{Z}} \sum_{\boldsymbol{k}\in\mathbb{Z}^n} \langle f, \widetilde{\psi}^i_{j,\boldsymbol{k}} \rangle \psi^i_{j,\boldsymbol{k}} = \sum_{i=1}^{d-1} \sum_{j\in\mathbb{Z}} \sum_{\boldsymbol{k}\in\mathbb{Z}^n} \langle f, \psi^i_{j,\boldsymbol{k}} \rangle \widetilde{\psi}^i_{j,\boldsymbol{k}}, \quad \forall f \in L^2(\mathbb{R}^n), \tag{2.12}$$

$$\sum_{j\in\mathbb{Z}} \sum_{\boldsymbol{k}\in\mathbb{Z}^n} \left|\langle f, \psi^i_{j,\boldsymbol{k}} \rangle\right|^2 \le C\|f\|_2^2, \quad \sum_{j\in\mathbb{Z}} \sum_{\boldsymbol{k}\in\mathbb{Z}^n} \left|\langle f, \widetilde{\psi}^i_{j,\boldsymbol{k}} \rangle\right|^2 \le C'\|f\|_2^2, \tag{2.13}$$

for some constants $C$, $C'$, $i = 1,\dots,d-1$, and any $f \in L^2(\mathbb{R}^n)$. Note that condition (2.13) ensures that the system $\{\psi^i_{j,\boldsymbol{k}}, \widetilde{\psi}^i_{j,\boldsymbol{k}}\}_{i=1}^{d-1}$ gives rise to a stable biorthogonal wavelet basis of $L^2(\mathbb{R}^n)$ (see [**1**] and [**2**]).

## 3. Biorthogonal Wavelets with a General Dilation Matrix

We shall generalize Gröchenig's nonconstructive algorithm [**8**] for characterizing $n$-D orthonormal wavelet bases with dilation matrix $\boldsymbol{D} = 2\boldsymbol{I}_n$ to $n$-D dual biorthogonal wavelet bases with an arbitrary dilation matrix $\boldsymbol{D} \in \mathbb{Z}^{n\times n}$.

**3.1. Necessary existence condition.** Given a dilation matrix $\boldsymbol{D}$, we assume that we have an $n$-D biorthogonal wavelet basis and show that the different coefficients have to satisfy a matrix identity.

Since $V_0$, $W_0 \subset V_{-1}$ and $\widetilde{V}_0$, $\widetilde{W}_0 \subset \widetilde{V}_{-1}$, then, for each $i$, $i = 0,\dots,d-1$, there are two sequences, $\{\alpha^i_{\boldsymbol{m}}\}_{\boldsymbol{m}\in\mathbb{Z}^n}$ and $\{\widetilde{\alpha}^i_{\boldsymbol{m}}\}_{\boldsymbol{m}\in\mathbb{Z}^n}$, of real numbers such that

$$\psi^i(\boldsymbol{x}) = \sum_{\boldsymbol{m}\in\mathbb{Z}^n} \alpha^i_{\boldsymbol{m}} \psi^0(\boldsymbol{D}\boldsymbol{x} - \boldsymbol{m}), \quad \widetilde{\psi}^i(\boldsymbol{x}) = \sum_{\boldsymbol{m}\in\mathbb{Z}^n} \widetilde{\alpha}^i_{\boldsymbol{m}} \widetilde{\psi}^0(\boldsymbol{D}\boldsymbol{x} - \boldsymbol{m}). \tag{3.1}$$

Taking Fourier transforms, we have

$$\widehat{\psi}^i({}^t\boldsymbol{D}\boldsymbol{\xi}) = m_i(\boldsymbol{\xi})\widehat{\psi}^0(\boldsymbol{\xi}), \quad \widehat{\widetilde{\psi}}^i({}^t\boldsymbol{D}\boldsymbol{\xi}) = \widetilde{m}_i(\boldsymbol{\xi})\widehat{\widetilde{\psi}}^0(\boldsymbol{\xi}).$$

where $m_i(\boldsymbol{\xi})$ and $\widetilde{m}_i(\boldsymbol{\xi})$ are $2\pi\mathbb{Z}^n$-periodic functions.

By standard technique, the wavelet filters have to satisfy the matrix equation

$$\begin{bmatrix} \overline{\widetilde{m}_0(\boldsymbol{\xi})} & \overline{\widetilde{m}_0(\boldsymbol{\xi}+\boldsymbol{\eta}_1)} & \cdots & \overline{\widetilde{m}_0(\boldsymbol{\xi}+\boldsymbol{\eta}_{d-1})} \\ \overline{\widetilde{m}_1(\boldsymbol{\xi})} & \overline{\widetilde{m}_1(\boldsymbol{\xi}+\boldsymbol{\eta}_1)} & \cdots & \overline{\widetilde{m}_1(\boldsymbol{\xi}+\boldsymbol{\eta}_{d-1})} \\ \vdots & \vdots & & \vdots \\ \overline{\widetilde{m}_{d-1}(\boldsymbol{\xi})} & \overline{\widetilde{m}_{d-1}(\boldsymbol{\xi}+\boldsymbol{\eta}_1)} & \cdots & \overline{\widetilde{m}_{d-1}(\boldsymbol{\xi}+\boldsymbol{\eta}_{d-1})} \end{bmatrix}$$

$$\times \begin{bmatrix} m_0(\boldsymbol{\xi}) & m_1(\boldsymbol{\xi}) & \cdots & m_{d-1}(\boldsymbol{\xi}) \\ m_0(\boldsymbol{\xi}+\boldsymbol{\eta}_1) & m_1(\boldsymbol{\xi}+\boldsymbol{\eta}_1) & \cdots & m_{d-1}(\boldsymbol{\xi}+\boldsymbol{\eta}_1) \\ \vdots & \vdots & & \vdots \\ m_0(\boldsymbol{\xi}+\boldsymbol{\eta}_{d-1}) & m_1(\boldsymbol{\xi}+\boldsymbol{\eta}_{d-1}) & \cdots & m_{d-1}(\boldsymbol{\xi}+\boldsymbol{\eta}_{d-1}) \end{bmatrix} = \boldsymbol{I}_d,$$

where $\boldsymbol{\eta}_j$ denote the different points of the set $\{2\pi\, {}^t\boldsymbol{D}^{-1}\boldsymbol{\eta};\ \boldsymbol{\eta} \in \mathbb{Z}^n/{}^t\boldsymbol{D}\mathbb{Z}^n\}$.

**3.2. Sufficient existence condition.** The hypotheses in the following generalization of Theorem 3.2 [4] to the $n$-D case are sufficient conditions for the construction of a stable biorthogonal wavelet basis of $L^2(\mathbb{R}^n)$.

THEOREM 1. *Let $\boldsymbol{D}$ be a dilation matrix and set $d = |\det \boldsymbol{D}|$. Assume that for some positive numbers, $\varepsilon$ and $\widetilde{\varepsilon} > 0$, and $0 \le i \le d-1$, we have*

$$|\widehat{\psi^i}(\boldsymbol{\xi})| < c\big(1 + |\boldsymbol{\xi}|^2\big)^{-\varepsilon - n/4}, \quad |\widehat{\widetilde{\psi}^i}(\boldsymbol{\xi})| < c\big(1 + |\boldsymbol{\xi}|^2\big)^{-\widetilde{\varepsilon} - n/4}, \tag{3.2}$$

*for some constant $c$. For $j \in \mathbb{Z}$ and $\boldsymbol{k} \in \mathbb{Z}^n$, define*

$$\psi^i_{j,\boldsymbol{k}}(\boldsymbol{x}) = |\det \boldsymbol{D}|^{-j/2} \psi^i(\boldsymbol{D}^{-j}\boldsymbol{x} - \boldsymbol{k}), \quad \widetilde{\psi}^i_{j,\boldsymbol{k}}(\boldsymbol{x}) = |\det \boldsymbol{D}|^{-j/2} \widetilde{\psi}^i(\boldsymbol{D}^{-j}\boldsymbol{x} - \boldsymbol{k}),$$

*where the compactly supported wavelets $\psi^i$, $\widetilde{\psi}^i$ are as defined in* (3.1), *and assume that*

$$\sum_{i=0}^{d-1} \sum_{\boldsymbol{l} \in \mathbb{Z}^n} \alpha^i_{\boldsymbol{l}-\boldsymbol{D}\boldsymbol{j}} \widetilde{\alpha}^i_{\boldsymbol{l}-\boldsymbol{D}\boldsymbol{k}} = \delta_{\boldsymbol{j}-\boldsymbol{k}}. \tag{3.3}$$

*Then for $f \in L^2(\mathbb{R}^n)$, the following series converge in the weak sense,*

$$f = \sum_{i=1}^{d-1} \sum_{j \in \mathbb{Z}} \sum_{\boldsymbol{k} \in \mathbb{Z}^n} \langle f, \widetilde{\psi}^i_{j,\boldsymbol{k}} \rangle \psi^i_{j,\boldsymbol{k}} = \sum_{i=1}^{d-1} \sum_{j \in \mathbb{Z}} \sum_{\boldsymbol{k} \in \mathbb{Z}^n} \langle f, \psi^i_{j,\boldsymbol{k}} \rangle \widetilde{\psi}^i_{j,\boldsymbol{k}}. \tag{3.4}$$

PROOF. The proof is carried out in three parts denoted by (a), (b) and (c).
**(a)** We first prove that, for all $f_1$, $f_2 \in L^2(\mathbb{R}^n)$,

$$\begin{aligned}\sum_{\boldsymbol{k} \in \mathbb{Z}^n} &\langle f_1, \psi_{-1,\boldsymbol{k}} \rangle \langle \widetilde{\psi}_{-1,\boldsymbol{k}}, f_2 \rangle \\ &= \sum_{\boldsymbol{l} \in \mathbb{Z}^n} \Big[ \langle f_1, \psi_{0,\boldsymbol{l}} \rangle \langle \widetilde{\psi}_{0,\boldsymbol{l}}, f_2 \rangle + \sum_{i=1}^{d-1} \langle f_1, \psi^i_{0,\boldsymbol{l}} \rangle \langle \widetilde{\psi}^i_{0,\boldsymbol{l}}, f_2 \rangle \Big].\end{aligned} \tag{3.5}$$

To do this, we remark that

$$\begin{aligned}\langle f_1, \psi^i_{0,\boldsymbol{l}} \rangle &= \int f_1(\boldsymbol{x}) \overline{\psi^i(\boldsymbol{x} - \boldsymbol{l})}\, d\boldsymbol{x} \\ &= \int f_1(\boldsymbol{x}) \overline{\sum_{\boldsymbol{n} \in \mathbb{Z}^n} \alpha^i_{\boldsymbol{l}-\boldsymbol{D}\boldsymbol{n}} \psi^0(\boldsymbol{D}\boldsymbol{x} - \boldsymbol{n})}\, d\boldsymbol{x}, \quad i = 0, \dots, d-1,\end{aligned}$$

and, similarly,

$$\begin{aligned}\langle \widetilde{\psi}^i_{0,\boldsymbol{l}}, f_2 \rangle &= \int \widetilde{\psi}^i(\boldsymbol{x} - \boldsymbol{l}) \overline{f_2(\boldsymbol{x})}\, d\boldsymbol{x} \\ &= \int \sum_{\boldsymbol{n} \in \mathbb{Z}^n} \big[\widetilde{\alpha}^i_{\boldsymbol{l}-\boldsymbol{D}\boldsymbol{n}} \widetilde{\psi}^0(\boldsymbol{D}\boldsymbol{x} - \boldsymbol{n})\big] \overline{f_2(\boldsymbol{x})}\, d\boldsymbol{x}, \quad i = 0, \dots, d-1.\end{aligned}$$

Hence,

$$\begin{aligned}\sum_{\boldsymbol{l} \in \mathbb{Z}^n} &\sum_{i=0}^{d-1} \langle f_1, \psi^i_{0,\boldsymbol{l}} \rangle \langle \widetilde{\psi}^i_{0,\boldsymbol{l}}, f_2 \rangle \\ &= \sum_{\boldsymbol{l} \in \mathbb{Z}^n} \sum_{i=0}^{d-1} \Big[ \sum_{\boldsymbol{n} \in \mathbb{Z}^n} \alpha^i_{\boldsymbol{l}-\boldsymbol{D}\boldsymbol{n}} \langle f_1, \psi^0_{-1,\boldsymbol{n}} \rangle \Big] \Big[ \sum_{\boldsymbol{n} \in \mathbb{Z}^n} \widetilde{\alpha}^i_{\boldsymbol{l}-\boldsymbol{D}\boldsymbol{n}} \langle \widetilde{\psi}^0_{-1,\boldsymbol{n}}, f_2 \rangle \Big] \\ &= \sum_{\boldsymbol{l} \in \mathbb{Z}^n} \sum_{i=0}^{d-1} \sum_{\boldsymbol{j},\boldsymbol{k} \in \mathbb{Z}^n} \alpha^i_{\boldsymbol{l}-\boldsymbol{D}\boldsymbol{j}} \widetilde{\alpha}^i_{\boldsymbol{l}-\boldsymbol{D}\boldsymbol{k}} \langle f_1, \psi^0_{-1,\boldsymbol{j}} \rangle \langle \widetilde{\psi}^0_{-1,\boldsymbol{k}}, f_2 \rangle.\end{aligned}$$

Finally, equality (3.5) follows from (3.3).

**(b)** We prove that, for all $f_1$, $f_2 \in L^2(\mathbb{R}^n)$, the following equality holds:

$$\sum_{\boldsymbol{k}\in\mathbb{Z}^n}\sum_{j\in\mathbb{Z}}\sum_{i=1}^{d-1}\langle f_1, \psi^i_{j,\boldsymbol{k}}\rangle\langle\widetilde{\psi}^i_{j,\boldsymbol{k}}, f_2\rangle = \langle f_1, f_2\rangle.$$

Using the techniques of part (a), one shows that

$$\sum_{\boldsymbol{k}\in\mathbb{Z}^n}\langle f_1, \psi_{j-1,\boldsymbol{k}}\rangle\langle\widetilde{\psi}_{j-1,\boldsymbol{k}}, f_2\rangle = \sum_{\boldsymbol{l}\in\mathbb{Z}^n}\sum_{i=0}^{d-1}\langle f_1, \psi^i_{j,\boldsymbol{l}}\rangle\langle\widetilde{\psi}^i_{j,\boldsymbol{l}}, f_2\rangle \tag{3.6}$$

which implies that

$$\begin{aligned}\sum_{\boldsymbol{k}\in\mathbb{Z}^n}\langle f_1, \psi_{-J-1,\boldsymbol{k}}\rangle\langle\widetilde{\psi}_{-J-1,\boldsymbol{k}}, f_2\rangle &= \sum_{j=-J}^{J}\sum_{\boldsymbol{l}\in\mathbb{Z}^n}\sum_{i=1}^{d-1}\langle f_1, \psi^i_{j,\boldsymbol{l}}\rangle\langle\widetilde{\psi}^i_{j,\boldsymbol{l}}, f_2\rangle \\ &\quad + \sum_{\boldsymbol{l}\in\mathbb{Z}^n}\langle f_1, \psi^0_{J,\boldsymbol{l}}\rangle\langle\widetilde{\psi}^0_{J,\boldsymbol{l}}, f_2\rangle.\end{aligned} \tag{3.7}$$

To prove the $d$ inequalities

$$\lim_{J\to+\infty}\sum_{j=-J}^{J}\sum_{\boldsymbol{l}\in\mathbb{Z}^n}\left|\langle f_1, \psi^i_{j,\boldsymbol{l}}\rangle\langle\widetilde{\psi}^i_{j,\boldsymbol{l}}, f_2\rangle\right| < \infty, \qquad i = 0, \dots, d-1,$$

it suffices to consider the case $i = 0$. By the $n$-D version of Parseval's equality, we have

$$\sum_{j=-J}^{J}\sum_{\boldsymbol{l}\in\mathbb{Z}^n}\left|\langle f_1, \psi^0_{j,\boldsymbol{l}}\rangle\langle\widetilde{\psi}^0_{j,\boldsymbol{l}}, f_2\rangle\right| = \sum_{j=-J}^{J}\sum_{\boldsymbol{l}\in\mathbb{Z}^n}\left|\langle \hat{f}_1, \widehat{\psi}^0_{j,\boldsymbol{l}}\rangle\langle\widehat{\widetilde{\psi}}^0_{j,\boldsymbol{l}}, \hat{f}_2\rangle\right|.$$

Considering the set $\mathcal{C}_j = \left\{2\pi\, {}^t\boldsymbol{D}^{-j}\boldsymbol{x};\ \boldsymbol{x} \in [0,1]^n\right\}$, and using elementary operations we obtain

$$\begin{aligned}&\sum_{j=-J}^{J}\sum_{\boldsymbol{l}\in\mathbb{Z}^n}\left|\langle \hat{f}_1, \widehat{\psi}^0_{j,\boldsymbol{l}}\rangle\langle\widehat{\widetilde{\psi}}^0_{j,\boldsymbol{l}}, \hat{f}_2\rangle\right| \\ &\quad = \sum_{-J}^{J}\Bigg|\int_{\mathcal{C}_j}\sum_{\boldsymbol{l},\boldsymbol{k}\in\mathbb{Z}^n}\Big[\overline{\hat{f}_2(\boldsymbol{\xi}+2\pi\,{}^t\boldsymbol{D}^{-j}\boldsymbol{k})}\hat{f}_1(\boldsymbol{\xi}+2\pi\,{}^t\boldsymbol{D}^{-j}\boldsymbol{l}) \\ &\qquad\qquad \overline{\widehat{\psi}^0({}^t\boldsymbol{D}^j\boldsymbol{\xi}+2\pi\boldsymbol{l})}\widehat{\widetilde{\psi}}^0({}^t\boldsymbol{D}^j\boldsymbol{\xi}+2\pi\boldsymbol{k})\,d\boldsymbol{\xi}\Big]\Bigg| \\ &\quad \le \sum_{-J}^{J}\int_{\mathcal{C}_j}\sum_{\boldsymbol{l},\boldsymbol{k}\in\mathbb{Z}^n}\left|\hat{f}_1(\boldsymbol{\xi}+2\pi\,{}^t\boldsymbol{D}^{-j}\boldsymbol{l})\overline{\hat{f}_2(\boldsymbol{\xi}+2\pi\,{}^t\boldsymbol{D}^{-j}\boldsymbol{k})}\right| \\ &\qquad\qquad \left|\overline{\widehat{\psi}^0({}^t\boldsymbol{D}^j\boldsymbol{\xi}+2\pi\boldsymbol{l})}\widehat{\widetilde{\psi}}^0({}^t\boldsymbol{D}^j\boldsymbol{\xi}+2\pi\boldsymbol{k})\right|^{\delta} \\ &\qquad\qquad \times\sum_{\boldsymbol{l},\boldsymbol{k}\in\mathbb{Z}^n}\left|\overline{\widehat{\psi}^0({}^t\boldsymbol{D}^j\boldsymbol{\xi}+2\pi\boldsymbol{l})}\widehat{\widetilde{\psi}}^0({}^t\boldsymbol{D}^j\boldsymbol{\xi}+2\pi\boldsymbol{k})\right|^{1-\delta}d\boldsymbol{\xi},\end{aligned} \tag{3.8}$$

where $\delta \in (0,1)$ is to be fixed later. By (3.8), the series

$$\sum_{\boldsymbol{l},\boldsymbol{k}\in\mathbb{Z}^n}\left|\overline{\widehat{\psi}^0({}^t\boldsymbol{D}^j\boldsymbol{\xi}+2\pi\boldsymbol{l})}\widehat{\widetilde{\psi}}^0({}^t\boldsymbol{D}^j\boldsymbol{\xi}+2\pi\boldsymbol{k})d\boldsymbol{\xi}\right|^{1-\delta}$$

is uniformly bounded in $\boldsymbol{\xi}$ for all $\delta < 2\varepsilon(n+2\varepsilon)^{-1}$. Hence the left-most term of (3.8) is bounded by

$$\|f_1\|_2\|f_2\|_2\left\{\sup_{\boldsymbol{\xi}\in[1,2]^n}\sum_{j\in\mathbb{Z}}\left|\widehat{\psi}^0({}^t\boldsymbol{D}^j\boldsymbol{\xi})\right|\right\}^\delta\left\{\sup_{\boldsymbol{\xi}\in[1,2]^n}\sum_{j\in\mathbb{Z}}\left|\widehat{\widetilde{\psi}}^0({}^t\boldsymbol{D}^j\boldsymbol{\xi})\right|\right\}^\delta.$$

Since $\psi$ has compact support, then $\psi \in L^1(\mathbb{R}^n)$. By the $n$-D version of the Paley–Wiener theorem, $\widehat{\psi}(\boldsymbol{\xi})$ can be extended to a holomorphic function satisfying

$$\left|\widehat{\psi}(\boldsymbol{\xi}+i\boldsymbol{\eta})\right| \le c\big(1+|\boldsymbol{\xi}|+|\boldsymbol{\eta}|\big)\,e^{|\boldsymbol{\eta}|}, \quad \boldsymbol{\xi},\boldsymbol{\eta}\in\mathbb{R}^n.$$

Since $\widehat{\psi}(0)=0$, there exists a constant $A$ such that

$$\left|\widehat{\psi}^0(\boldsymbol{\xi})\right| \le A|\boldsymbol{\xi}|, \quad \left|\widehat{\widetilde{\psi}}^0(\boldsymbol{\xi})\right| \le A|\boldsymbol{\xi}|.$$

Therefore, for $\boldsymbol{\xi}\in[1,2]^n$, we have

$$\sum_{j=-\infty}^{0}\left|\widehat{\psi}^0({}^t\boldsymbol{D}^j\boldsymbol{\xi})\right|^\delta \le CA^\delta\sum_{j=-\infty}^{0}|{}^t\boldsymbol{D}^j\boldsymbol{\xi}|^\delta \le CA^\delta\sum_{j=-\infty}^{0}\|\boldsymbol{D}\|_2^{\delta j}\le C',$$

because $\|\boldsymbol{D}\|_2>1$. Similarly

$$\sum_{j=-\infty}^{0}\left|\widehat{\widetilde{\psi}}^0({}^t\boldsymbol{D}^j\boldsymbol{\xi})\right|^\delta \le C''.$$

On the other hand, since $\|\boldsymbol{D}\|_2>\eta>1$, then for $\boldsymbol{\xi}\in[1,2]^n$, we have

$$\sum_{j=1}^{\infty}\left|\widehat{\psi}^0({}^t\boldsymbol{D}^j\boldsymbol{\xi})\right|^\delta \le C^\delta\sum_{j=1}^{\infty}(1+\eta^{2j})^{-n\delta/4}<\infty \quad\text{and}\quad \sum_{j=1}^{\infty}\left|\widehat{\widetilde{\psi}}^0({}^t\boldsymbol{D}^j\boldsymbol{\xi})\right|^\delta<\infty.$$

Combining these results we have that for all integer $J\ge1$

$$\sum_{j=-J}^{J}\sum_{\boldsymbol{k}\in\mathbb{Z}^n}\left|\langle f_1,\psi^0_{j,\boldsymbol{k}}\rangle\langle\widetilde{\psi}^0_{j,\boldsymbol{k}},f_2\rangle\right|<\infty.$$

Finally, by the same techniques one can prove that

$$\sum_{j=-J}^{J}\sum_{\boldsymbol{k}\in\mathbb{Z}^n}\left|\langle f_1,\psi^i_{j,\boldsymbol{k}}\rangle\langle\widetilde{\psi}^i_{j,\boldsymbol{k}},f_2\rangle\right|<\infty, \quad i=1,\dots,d-1.$$

**(c)** To complete the proof of the theorem, we first show that

$$\lim_{J\to+\infty}\sum_{\boldsymbol{k}\in\mathbb{Z}^n}\langle f_1,\psi^0_{J,\boldsymbol{k}}\rangle\langle\widetilde{\psi}^0_{J,\boldsymbol{k}},f_2\rangle=0.$$

Using the Cauchy–Schwarz inequality, we have

$$\sum_{\boldsymbol{k}\in\mathbb{Z}^n}\langle f_1,\psi^0_{J,\boldsymbol{k}}\rangle\langle\widetilde{\psi}^0_{J,\boldsymbol{k}},f_2\rangle \le \left[\sum_{\boldsymbol{k}\in\mathbb{Z}^n}\left|\langle f_1,\psi^0_{J,\boldsymbol{k}}\rangle\right|^2\right]^{1/2}\left[\sum_{\boldsymbol{k}\in\mathbb{Z}^n}\left|\langle\widetilde{\psi}^0_{J,\boldsymbol{k}},f_2\rangle\right|^2\right]^{1/2}.$$

Hence, following the proof in (b), we have

$$\sum_{\boldsymbol{k}\in\mathbb{Z}^n}\langle f_1,\psi^0_{J,\boldsymbol{k}}\rangle\langle\widetilde{\psi}^0_{J,\boldsymbol{k}},f_2\rangle \le K\left[\int|\hat{f}_1(\boldsymbol{\xi})|\left(1+|{}^t\boldsymbol{D}^J\boldsymbol{\xi}|^2\right)^{-n\delta/2}d\boldsymbol{\xi}\right]^{1/2} \times\left[\int|\hat{f}_2(\boldsymbol{\xi})|\left(1+|{}^t\boldsymbol{D}^J\boldsymbol{\xi}|^2\right)^{-n\delta/2}d\boldsymbol{\xi}\right]^{1/2}\to 0,\quad \text{as}\quad J\to+\infty, \tag{3.9}$$

where the limit to zero follows by the Lebesgue dominated convergence theorem and the fact $\|\boldsymbol{D}\|_2>1$. It remains to be proved that

$$\lim_{J\to+\infty}\sum_{\boldsymbol{k}\in\mathbb{Z}^n}\langle f_1,\psi^0_{-J-1,\boldsymbol{k}}\rangle\langle\widetilde{\psi}^0_{-J-1,\boldsymbol{k}},f_2\rangle=\langle f_1,f_2\rangle.$$

Again, using the techniques of part (b), one can easily show that

$$\begin{aligned}\sum_{\boldsymbol{k}\in\mathbb{Z}^n}&\langle f_1,\psi^0_{-J-1,\boldsymbol{k}}\rangle\langle\widetilde{\psi}^0_{-J-1,\boldsymbol{k}},f_2\rangle\\ &=\int\hat{f}_1(\boldsymbol{\xi})\overline{\hat{f}_2(\boldsymbol{\xi})\widehat{\psi}^0({}^t\boldsymbol{D}^{-J-1}\boldsymbol{\xi})}\widehat{\widetilde{\psi}}^0({}^t\boldsymbol{D}^{-J-1}\boldsymbol{\xi})\,d\boldsymbol{\xi}\\ &\quad+\sum_{0\neq\boldsymbol{k}\in\mathbb{Z}^n}\int\overline{\hat{f}_2(\boldsymbol{\xi}+2\pi{}^t\boldsymbol{D}^{J+1}\boldsymbol{k})}\hat{f}_1(\boldsymbol{\xi})\overline{\widehat{\psi}^0({}^t\boldsymbol{D}^{-J-1}\boldsymbol{\xi})}\widehat{\widetilde{\psi}}^0({}^t\boldsymbol{D}^{-J-1}\boldsymbol{\xi}+2\pi\boldsymbol{k})\,d\boldsymbol{\xi}.\end{aligned}\tag{3.10}$$

Since $\widehat{\psi}^0(0)=1$ and $\widehat{\widetilde{\psi}}(0)=1$, and $\|\boldsymbol{D}^{-1}\|_2|<1$, then the limit of the first term on the right-hand side of (3.10), as $J$ goes to infinity, is equal to

$$\int\hat{f}_1(\boldsymbol{\xi})\overline{\hat{f}_2(\boldsymbol{\xi})}\,d\boldsymbol{\xi}=\langle f_1,f_2\rangle.$$

From the results of part (b) one concludes that the absolute value of the second term (that is, the summation) on the right-hand side of (3.10) is bounded by a constant.

Hence by approximating $\hat{f}_1(\boldsymbol{\xi})$ and $\hat{f}_2(\boldsymbol{\xi})$ by compactly supported functions in $L^2(\mathbb{R}^n)$ and remembering that $\|\boldsymbol{D}\|_2>1$, we may assume that for $\boldsymbol{k}\in\mathbb{Z}^n$, with $\boldsymbol{k}\neq 0$,

$$\operatorname{supp}\hat{f}_1(\cdot)\cap\operatorname{supp}\hat{f}_2(\cdot+2\pi\,{}^t\boldsymbol{D}^J\boldsymbol{k})=\emptyset,\quad \text{as } J\to+\infty;$$

thus, the second term of the sum in (3.10) goes to 0 as $J$ goes to infinity.

Combining the above results, we conclude that

$$\lim_{J\to+\infty}\sum_{\boldsymbol{k}\in\mathbb{Z}^n}\langle f_1,\psi^0_{-J-1,\boldsymbol{k}}\rangle\langle\widetilde{\psi}^0_{-J-1,\boldsymbol{k}},f_2\rangle=\langle f_1,f_2\rangle,$$

and, consequently, (3.4) follows. □

COROLLARY 1. *Under the conditions of Theorem* 1, *the system* $\{\psi^i_{j,\boldsymbol{k}},\widetilde{\psi}^i_{j,\boldsymbol{k}}\}_{i=1}^{d-1}$ *is a Riesz basis of* $L^2(\mathbb{R}^n)$.

PROOF. It suffices to check that $\psi^0_{j,\boldsymbol{k}}$ and $\widetilde{\psi}^0_{j,\boldsymbol{k}}$ satisfy condition (2.13). Taking $f_2=f_1=f$ and substituting $\widetilde{\psi}^0$ with $\psi^0$ in part (b) of the proof of Theorem 1, one can easily see that the first inequality in (2.13) is satisfied. The second inequality follows in a similar way. □

REMARK 1. As explained in [**2**], the Fourier based technique used to deduce the stability of the wavelets is not sharp. Nevertheless, it is practical for long filters and in multidimensional situations. A sharper technique is based on the spectral radius of a transition operator (see, for example, [**2**]–[**5**]). Slightly different conditions on $\widehat{\psi}^0(\xi)$ and $\widehat{\widetilde{\psi}}^0(\xi)$ have been used in [**2**] and [**1**] to prove the stability of wavelets in the univariate and multivariate cases, respectively. These conditions are

$$\sup_{\xi\in\mathbb{R}}\sum_{k\in\mathbb{Z}}\left|\widehat{\psi}^0(\xi+2k\pi)\right|^{2-\sigma}<\infty,\quad \sup_{\xi\in\mathbb{R}}(1+|\xi|)^{\sigma}\left|\widehat{\psi}^0(\xi)\right|<\infty$$

for some $\sigma > 0$. Note that for wavelets with compact supports, conditions (3.2) implies the above two conditions.

**3.3. Regular $L^2(\mathbb{R}^n)$ wavelets.** We give an $n$-D generalization of the decay estimate on $|\widehat{\phi}(\boldsymbol{\xi})|$ given in [**3**].

THEOREM 2. *Let $\boldsymbol{D}$ be a dilation matrix and assume that the frequency response, $H(\boldsymbol{\xi})$, of an $n$-D FIR wavelet filter factors in the form*

$$H(\boldsymbol{\xi}) = f^{2N}(\boldsymbol{\xi})\mathcal{F}(\boldsymbol{\xi}), \tag{3.11}$$

*where*

$$\prod_{j=1}^{\infty} f^2({}^t\boldsymbol{D}^{-j}\boldsymbol{\xi}) \leq \frac{c}{1+|\boldsymbol{\xi}|^2}, \tag{3.12}$$

*and for some $k \geq 1$ we have the bound*

$$B_k := \max_{\boldsymbol{\xi}\in\mathbb{R}^n}\left|\mathcal{F}(\boldsymbol{\xi})\mathcal{F}({}^t\boldsymbol{D}\boldsymbol{\xi})\cdots\mathcal{F}({}^t\boldsymbol{D}^{k-1}\boldsymbol{\xi})\right|^{1/k} < \|\boldsymbol{D}^{-1}\|_2^{-2N+n/2}. \tag{3.13}$$

*Then, the Fourier transform*

$$\widehat{\phi}(\boldsymbol{\xi}) = \prod_{j=1}^{\infty} H({}^t\boldsymbol{D}^{-j}\boldsymbol{\xi}),$$

*of the scaling function $\phi(\boldsymbol{\xi})$ satisfies the inequality*

$$\left|\widehat{\phi}(\boldsymbol{\xi})\right| \leq C(1+|\boldsymbol{\xi}|^2)^{-\varepsilon-n/2},\quad \varepsilon < N - \frac{n}{2} + \frac{\log(B_k)}{2\log(\|\boldsymbol{D}^{-1}\|_2)}. \tag{3.14}$$

PROOF. See [**11**]. □

REMARK 2. Under the hypotheses of Theorem 2, it is easily seen that the scaling function and, consequently, the wavelets are $r$-Hölder continuous with

$$r < 2N - n + \frac{\log(B_k)}{\log(\|\boldsymbol{D}^{-1}\|_2)}.$$

REMARK 3. The above brute force method of estimating the regularity of wavelets is not optimal. However, unlike other methods [**7, 16**], it can be used in the $n$-D case, where $n \geq 1$, and for an arbitrarily long filter $H(\boldsymbol{\xi})$.

## 4. Numerical Construction in Particular Cases

In this section, we give some tricks for the construction of $n$-D lowpass wavelet filters that generate nonseparable $n$-D biorthogonal wavelet bases of $L^2(\mathbb{R}^n)$, and, consequently, the corresponding wavelets by using Gröchenig's algorithm as a guide.

**4.1. McClellan transformation and the construction of nonseparable multidimensional wavelets.** We briefly describe the McClellan transformation as an efficient tool for designing nonseparable multidimensional biorthogonal wavelets.

DEFINITION 3. Write the frequency response of a zero-phase FIR filter in the form

$$h(\xi)=\sum_{n=0}^{N}\alpha_n\cos(n\xi)=\sum_{n=0}^{N}\alpha_n T_n\big[\cos(\xi)\big],$$

where $T_n$ is the $n^{th}$-degree Chebyshev polynomial. If $F(\xi_1,\xi_2)$ is the frequency response of a 2-D zero-phase FIR filter, then

$$H(\xi_1,\xi_2)=\sum_{n=0}^{N}\alpha_n T_n\big[F(\xi_1,\xi_2)\big]$$

is McClellan's transformation associated to $F(\xi_1,\xi_2)$ and applied to $h(\xi)$.

The design technique is illustrated by the following example.

EXAMPLE 1. Consider the quincunx dilation matrix

$$\boldsymbol{D}_1=\begin{bmatrix}1 & 1\\ 1 & -1\end{bmatrix} \tag{4.1}$$

with $|\det \boldsymbol{D}_1|=2$ and $\|\boldsymbol{D}_1^{-1}\|_2\approx 0.7071<1$. We construct a pair of dual lowpass filters $m_0^2(\boldsymbol{\xi})$ and $\widetilde{m}_0^2(\boldsymbol{\xi})$ such that

$$m_0^2(\boldsymbol{\xi})\overline{\widetilde{m}_0^2(\boldsymbol{\xi})}+m_0^2(\boldsymbol{\xi}+{}^t(\pi,\pi))\overline{\widetilde{m}_0^2(\boldsymbol{\xi}+{}^t(\pi,\pi))}=1. \tag{4.2}$$

According to McClellan's transformation, we need to choose an appropriate transformation function $F(\xi_1,\xi_2)$ and one pair of 1-D dual filters $m_0(\xi)=\sum_0^N\alpha_n\cos(n\xi)$, $\widetilde{m}_0(\xi)=\sum_0^{\tilde{N}}\widetilde{\alpha}_n\cos(n\xi)$ such that

$$m_0^2(\boldsymbol{\xi})=\sum_{n=0}^{N}\alpha_n T_n\big[F(\boldsymbol{\xi})\big],\quad \widetilde{m}_0^2(\boldsymbol{\xi})=\sum_{n=0}^{\tilde{N}}\widetilde{\alpha}_n T_n\big[F(\boldsymbol{\xi})\big].$$

One possible choice, as done in [**12**], is

$$F(\xi_1,\xi_2)=\frac{1}{2}(\cos\xi_1+\cos\xi_2). \tag{4.3}$$

The 1-D dual wavelet filters, $m_0(\xi)$, $\widetilde{m}_0(\xi)$, which are constructed by the technique described in [**11**] with the coefficients listed in Table 1, factor in the form

$$m_0(\xi)=\cos^8\left(\frac{\xi}{2}\right)\mathcal{L}(\xi),\quad \widetilde{m}_0(\xi)=\cos^8\left(\frac{\xi}{2}\right)\widetilde{\mathcal{L}}(\xi),$$

where $\mathcal{L}$ and $\widetilde{\mathcal{L}}$ are polynomials in $\cos(\xi/2)$.

If we take for the function $f^2$ of Theorem 2 the function

$$f^2(\xi_1,\xi_2)=1-\frac{1}{2}\left[\sin^2\left(\frac{\xi_1}{2}\right)-\sin^2\left(\frac{\xi_2}{2}\right)\right],$$

then the pair of dual lowpass filters are

$$m_0^2(\boldsymbol{\xi})=f^8(\boldsymbol{\xi})\mathcal{F}(\boldsymbol{\xi}),\quad \widetilde{m}_0^2(\boldsymbol{\xi})=f^8(\boldsymbol{\xi})\widetilde{\mathcal{F}}(\boldsymbol{\xi}).$$

TABLE 1. The coefficients $\alpha_n$, $-6 \le n \le 6$, of $m_0(\boldsymbol{\xi})$ and $\widetilde{\alpha}_n$, $-15 \le n \le 15$, of $\widetilde{m}_0(\boldsymbol{\xi})$.

| $n$ | $\alpha_n$ | $\widetilde{\alpha}_n$ | $n$ | $\widetilde{\alpha}_n$ |
|---|---|---|---|---|
| 0 | 0.546 875 000 | 0.522 229 947 4 | ±8 | 0.034 058 024 9 |
| ±1 | 0.332 031 250 | 0.298 668 044 0 | ±9 | 0.009 710 727 0 |
| ±2 | −0.003 906 250 | −0.057 538 581 2 | ±10 | −0.016 147 369 2 |
| ±3 | −0.091 796 875 | −0.098 144 662 4 | ±11 | 0.000 294 776 7 |
| ±4 | −0.023 437 500 | 0.075 332 835 0 | ±12 | 0.004 466 500 4 |
| ±5 | 0.009 765 625 | 0.076 441 686 0 | ±13 | −0.001 039 368 3 |
| ±6 | 0.003 906 250 | −0.050 787 826 9 | ±14 | −0.000 526 358 6 |
| ±7 | | −0.036 145 207 6 | ±15 | 0.000 214 169 9 |

It is proved in [**3**] that

$$\prod_{j=1}^{\infty} f^2({}^t\boldsymbol{D}_1^{-j}\boldsymbol{\xi}) \le \frac{c}{1+|\boldsymbol{\xi}|^2}.$$

Guided by the results of Section 3.1, we choose the following dual highpass filters:

$$m_1^2(\xi_1,\xi_2) = e^{i\xi_1}\widetilde{m}_0^2(\xi_1+\pi,\xi_2+\pi), \quad \widetilde{m}_1^2(\xi_1,\xi_2) = e^{i\xi_1}m_0^2(\xi_1+\pi,\xi_2+\pi).$$

Six iterations of the 2-D cascade algorithm (see [**11**]) are used for the numerical approximation of the dual scaling functions, $\phi_1(\boldsymbol{x})$, $\widetilde{\phi}_1(\boldsymbol{x})$, and wavelets, $\psi_1(\boldsymbol{x})$, $\widetilde{\psi}_1(\boldsymbol{x})$, shown in Fig. 1.

Does McClellan transformation preserve the stability of 1-D wavelets? A partial answer to this question is given in the following proposition.

PROPOSITION 1. *Let*

$$h(\xi) = \sum_{n=0}^{N} \alpha_n \cos(n\xi) = \sum_{n=0}^{N} \alpha_n T_n\big[\cos(\xi)\big] = \left(\cos^2\frac{\xi}{2}\right)^{\rho} P(\cos\xi)$$

*be a lowpass filter, where $\rho$ is a positive integer and $T_n$ is the $n^{th}$-degree Chebyshev polynomial. Assume that the polynomial $P$ satisfies the following two conditions:*

$$\max_{-1\le x\le 1}\left|P(x)P\big(T_2(x)\big)\cdots P\big(T_{2^{k-1}}(x)\big)\right|^{1/k} \le 2^{2\rho-1/2} \tag{4.4}$$

*and*

$$\max_{-1\le x,y\le 1}\left|P\left(\frac{x+y}{2}\right)P\left(\frac{T_2(x)+T_2(y)}{2}\right)\cdots P\left(\frac{T_{2^{k-1}}(x)+T_{2^{k-1}}(y)}{2}\right)\right|^{1/k} \le 2^{2\rho-1} \tag{4.5}$$

*for some integer $k \ge 1$. Then*

$$H(\xi_1,\xi_2) = \sum_{n=0}^{N} \alpha_n T_n\left[\frac{\cos(\xi_1)+\cos(\xi_2)}{2}\right]$$

*generates a stable wavelet basis of $L^2(\mathbb{R}^2)$ with dilation matrix $\boldsymbol{D}_1$.*

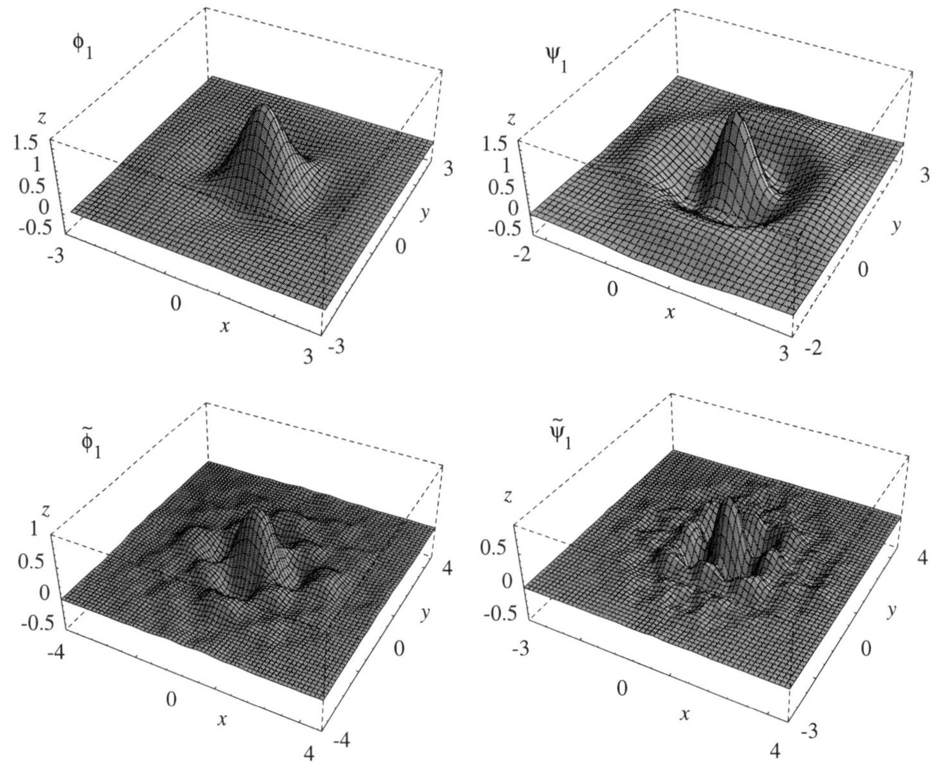


FIGURE 1. Graphs of regular dual scaling functions $\phi_1$, $\widetilde{\phi}_1$ and wavelets $\psi_1$, $\widetilde{\psi}_1$.

PROOF. From (4.4) we have

$$\max_{\xi\in\mathbb{R}}\left|P(\cos\xi)P(\cos 2\xi)\cdots P(\cos 2^{k-1}\xi)\right|^{1/k}\leq 2^{2\rho-1/2}.$$

Using Theorem 8 in [**2**], we see that if the dual, $\tilde{h}(\xi)$, of $h(\xi)$ satisfies the conditions of the proposition, then $h$ and $\tilde{h}$ generate biorthogonal Riesz bases of $L^2(\mathbb{R})$. By [**12**], the 2-D lowpass filters can be written in the form

$$H(\boldsymbol{\xi})=\left\{1-\frac{1}{2}\left[\sin^2\frac{\xi_1}{2}+\sin^2\frac{\xi_2}{2}\right]\right\}^{\rho}P\left(\frac{\cos\xi_1+\cos\xi_2}{2}\right)=\left[f^2(\boldsymbol{\xi})\right]^{\rho}P\big(F(\boldsymbol{\xi})\big).$$

A similar formula holds for the dual $\widetilde{H}(\boldsymbol{\xi})$. By Theorem 2, to prove the decay of

$$\widehat{\phi}(\boldsymbol{\xi})=\prod_{j-1}^{\infty}H({}^t\boldsymbol{D}_1^{-j}\boldsymbol{\xi})$$

it suffices that the following inequality holds,

$$\prod_{j=1}^{\infty}f^2({}^t\boldsymbol{D}_1^{-j}\boldsymbol{\xi})\leq\frac{C}{1+|\boldsymbol{\xi}|^2},\tag{4.6}$$

and there exists a positive integer $k$ such that

$$B_k = \max_{\boldsymbol{\xi}\in\mathbb{R}^2} \left| P\big(F(\boldsymbol{\xi})\big) P\big(F({}^t\boldsymbol{D}_1\boldsymbol{\xi})\big) \cdots P\big(F({}^t\boldsymbol{D}_1^{k-1}\boldsymbol{\xi})\big) \right|^{1/k} < 2^{2\rho-1}.$$

Since $\boldsymbol{D}_1^2 = 2\boldsymbol{I}_2$, then

$$\left|\prod_{l=0}^{2k-1} P[F({}^t\boldsymbol{D}_1^l\boldsymbol{\xi})]\right|^{1/(2k)} = \left\{\left|\prod_{l=0}^{k-1} P[F(2^l\boldsymbol{\xi})]\right| \left|\prod_{l=0}^{k-1} P[F({}^t\boldsymbol{D}_1^l(2^l\boldsymbol{\xi}))]\right|\right\}^{1/(2k)}$$
$$= \left\{\left|\prod_{l=0}^{k-1} P\left(\frac{\cos 2^l\xi_1 + \cos 2^l\xi_2}{2}\right)\right| \left|\prod_{l=0}^{k-1} P\left(\frac{\cos 2^l(\xi_1+\xi_2) + \cos 2^l(\xi_1-\xi_2}{2}\right)\right|\right\}^{1/(2k)}$$
$$= \left|\prod_{l=0}^{k-1} P\left(\frac{T_{2^l}(\cos\xi_1) + T_{2^l}(\cos\xi_2)}{2}\right)\right|^{1/(2k)}$$
$$\times \left|\prod_{l=0}^{k-1} P\left(\frac{T_{2^l}(\cos(\xi_1+\xi_2)) + T_{2^l}(\cos(\xi_1-\xi_2))}{2}\right)\right|^{1/(2k)}.$$

Consequently, using (4.5), we conclude that

$$B_k \le \left[\max_{-1\le x,y\le 1} \left|\prod_{l=0}^{k-1} P\left(\frac{T_{2^l}(x) + T_{2^l}(y)}{2}\right)\right|^{1/(2k)}\right]^2 < 2^{2\rho-1}.$$

Moreover, Proposition 5.1 in [**3**] gives (4.6), By Theorem 2 and Corollary 1, we conclude that

$$|\widehat{\psi}^0(\boldsymbol{\xi})| = \left|\prod_{j=1}^{\infty} H({}^t\boldsymbol{D}_1^{-j}\boldsymbol{\xi})\right| \le C\big(1+|\boldsymbol{\xi}|^2\big)^{-1-\varepsilon}$$

for some $\varepsilon > 0$. A similar formula holds for $\widehat{\widetilde{\psi}}^0(\boldsymbol{\xi})$. Consequently, $\psi^0$ and $\widetilde{\psi}^0$ generate biorthogonal Riesz bases of $L^2(\mathbb{R}^2)$ with dilation matrix $\boldsymbol{D}_1$. □

In [**10**], McClellan's transformation is used to solve the dilation equation (2.6) in dimension $n \ge 2$ with dilation matrix $\boldsymbol{D} = 2\boldsymbol{I}_n$. In this case, the $n$-D lowpass filter $\mathcal{H}_0(\boldsymbol{\xi})$ is given by

$$\mathcal{H}_0(\boldsymbol{\xi}) = \prod_{k=0}^{n-1} H_0(\boldsymbol{D}_k\boldsymbol{D}_{k-1}\cdots\boldsymbol{D}_1\boldsymbol{D}_0\boldsymbol{\xi}),$$

where $\{\boldsymbol{D}_i\}_{i=0}^{n-1}$ is a set of well chosen square matrices,

$$H_0(\boldsymbol{\xi}) = \sum_{m=0}^{N} \alpha_m T_m\left[\frac{1}{n}\sum_{i=1}^{n}\cos\xi_i\right],$$

and $\{\alpha_m\}_m$ is a 1-D lowpass filter. Similar formulae hold for the dual filter $\widetilde{\mathcal{H}}_0(\boldsymbol{\xi})$. Furthermore, under the condition that $\prod_{j=1}^{\infty}|H_0(2^{-j}\boldsymbol{\xi})|$ and $\prod_{j=1}^{\infty}|\widetilde{H}_0(2^{-j}\boldsymbol{\xi})|$ do not increase for large $|\boldsymbol{\xi}|$, the resulting scaling functions generate Riesz bases of $L^2(\mathbb{R}^n)$. The highpass filters are also easily constructed in this case.

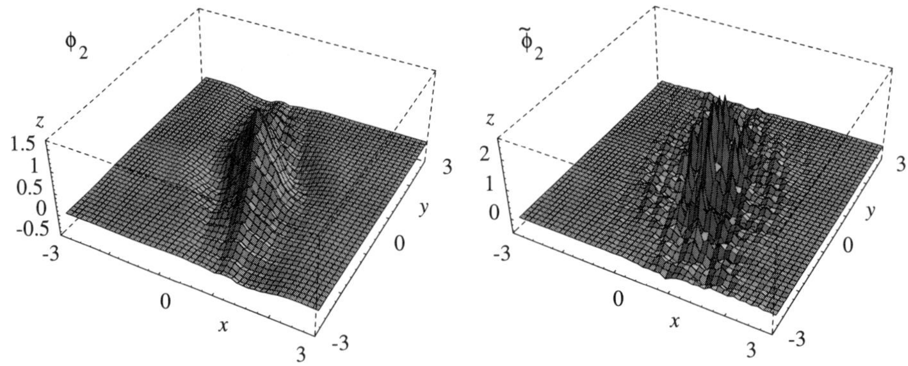


FIGURE 2. Graphs of nonregular dual scaling functions $\phi_2$, $\widetilde{\phi}_2$.

### 4.2. Numerical examples.

EXAMPLE 2. The dilation matrix

$$\boldsymbol{D}_2 = \begin{bmatrix} 1 & 3 \\ -1 & -1 \end{bmatrix} \tag{4.7}$$

gives rise to the same sampling lattice as $\boldsymbol{D}_1$. Since $\|\boldsymbol{D}_2^{-1}\|_2 \approx 1.7071 > 1$, $\boldsymbol{D}$ is not a dilation matrix according to Definition 1. Since this condition of Theorem 2 is not satisfied, the wavelets may not be regular. McClellan's transformation is used with the same 1-D dual filters and transformation function $F(\xi_1, \xi_2)$ as in Example 1. Six iterations of the 2-D cascade algorithm have been used to approximate the dual scaling functions $\phi_2(\boldsymbol{x})$, $\widetilde{\phi}_2(\boldsymbol{x})$, whose graphs, shown in Fig. 2, indicate that they are not as regular as those of Example 1, in agreement with the fact mentioned in [**3**] and [**13**] that the regularity of multidimensional wavelets depends on the choice of dilation matrices.

EXAMPLE 3. We consider the dilation matrix

$$\boldsymbol{D}_3 = \begin{bmatrix} 1 & 1 \\ 3 & -1 \end{bmatrix}. \tag{4.8}$$

Here $|\det \boldsymbol{D}| = 4$ and $\|{\boldsymbol{D}_3}^{-1}\|_2 \approx 0.8090 < 1$. By the results of Section 3.1, we find that

$$\boldsymbol{\eta}_0 = {}^t(0,0), \quad \boldsymbol{\eta}_1 = {}^t\left(\frac{\pi}{2}, \frac{\pi}{2}\right), \quad \boldsymbol{\eta}_2 = {}^t(\pi, \pi), \quad \boldsymbol{\eta}_3 = {}^t\left(\frac{3\pi}{2}, \frac{3\pi}{2}\right).$$

With $\boldsymbol{\xi} = {}^t(\xi_1, \xi_2) \in [0, 2\pi]^2$, the 2-D lowpass filters $m_0^2(\boldsymbol{\xi})$, $\widetilde{m}_0^2(\boldsymbol{\xi})$ have to satisfy the condition

$$\begin{aligned} m_0^2(\boldsymbol{\xi} + \boldsymbol{\eta}_0)\overline{\widetilde{m}_0^2(\boldsymbol{\xi} + \boldsymbol{\eta}_0)} &+ m_0^2(\boldsymbol{\xi} + \boldsymbol{\eta}_1)\overline{\widetilde{m}_0^2(\boldsymbol{\xi} + \boldsymbol{\eta}_1)} \\ &+ m_0^2(\boldsymbol{\xi} + \boldsymbol{\eta}_2)\overline{\widetilde{m}_0^2(\boldsymbol{\xi} + \boldsymbol{\eta}_2)} + m_0^2(\boldsymbol{\xi} + \boldsymbol{\eta}_3)\overline{\widetilde{m}_0^2(\boldsymbol{\xi} + \boldsymbol{\eta}_3)} = 1. \end{aligned} \tag{4.9}$$

Once these filters have been designed, the highpass filters $m_1^2(\boldsymbol{\xi})$, $m_2^2(\boldsymbol{\xi})$ and $m_3^2(\boldsymbol{\xi})$, together with their duals are to be constructed. The approach taken to solve this problem is first to design symmetric 1-D biorthogonal wavelet filters $m_0(\xi)$ and

$\widetilde{m}_0(\xi)$ that satisfy the identity

$$m_0(\xi)\overline{\widetilde{m}_0(\xi)} + m_0\left(\xi + \frac{\pi}{2}\right)\overline{\widetilde{m}_0\left(\xi + \frac{\pi}{2}\right)} + m_0(\xi+\pi)\overline{\widetilde{m}_0(\xi+\pi)} + m_0\left(\xi + \frac{3\pi}{2}\right)\overline{\widetilde{m}_0\left(\xi + \frac{3\pi}{2}\right)} = 1.$$

Then applying the McClellan transformation with the transformation function (4.3), one generates nonseparable 2-D biorthogonal lowpass wavelet filters $m_0^2(\boldsymbol{\xi})$ and $\widetilde{m}_0^2(\boldsymbol{\xi})$ satisfying (4.9). The remaining difficult task is to choose the right highpass wavelet filters $m_i^2(\boldsymbol{\xi})$ and their duals $\widetilde{m}_i^2(\boldsymbol{\xi})$, $i = 1, 2, 3$. This problem can be considerably simplified if we assume that the 1-D biorthogonal wavelet filters $m_0(\xi)$, $\widetilde{m}_0(\xi)$ factor in the form

$$m_0(\xi) = m(\xi)m(2\xi), \quad \widetilde{m}_0(\xi) = \widetilde{m}(\xi)\widetilde{m}(2\xi), \tag{4.10}$$

where $m(\xi)$ and $\widetilde{m}(\xi)$ are biorthogonal wavelet filters.

In this case, one possible choice for the highpass wavelet filters is

$$\begin{aligned} m_1(\xi_1, \xi_2) &= e^{i\xi_1}\widetilde{m}(\xi_1 + \pi, \xi_2 + \pi)m(2\xi_1, 2\xi_2), \\ m_2(\xi_1, \xi_2) &= e^{2i\xi_1}m(\xi_1, \xi_2)\widetilde{m}(2\xi_1 + \pi, 2\xi_2 + \pi), \\ m_3(\xi_1, \xi_2) &= e^{3i\xi_1}\widetilde{m}(\xi_1 + \pi, \xi_2 + \pi)\widetilde{m}(2\xi_1 + \pi, 2\xi_2 + \pi), \end{aligned}$$

where $m^2(\boldsymbol{\xi})$ and $\widetilde{m}^2(\boldsymbol{\xi})$ are the McClellan transformations of $m(\xi)$ and $\widetilde{m}(\xi)$, respectively.

Similar formulae are used to compute the corresponding highpass wavelet filters $\widetilde{m}_i^2(\boldsymbol{\xi})$, $i = 1, 2, 3$.

The above technique has been used first to compute 1-D biorthogonal wavelet filters $m_0(\xi)$ and $\widetilde{m}_0(\xi)$, whose coefficients are given in Table 2.

TABLE 2. The coefficients $\alpha_n$, $-9 \le n \le 9$, of $m_0(\xi)$ and $\beta_n$, $-12 \le n \le 12$, of $\widetilde{m}_0(\xi)$.

| $n$ | $\alpha_n$ | $\beta_n$ |
|---|---|---|
| 0 | 0.323 974 609 375 000 0 | 0.251 374 173 755 488 1 |
| ±1 | 0.259 643 554 687 500 0 | 0.224 497 771 460 162 7 |
| ±2 | 0.155 273 437 500 000 0 | 0.134 238 377 090 327 9 |
| ±3 | 0.049 072 265 625 000 0 | 0.047 392 411 665 482 9 |
| ±4 | −0.039 550 781 250 000 0 | −0.022 533 101 483 810 0 |
| ±5 | −0.047 607 421 875 000 0 | −0.036 047 628 103 208 9 |
| ±6 | −0.030 273 437 500 000 0 | −0.005 776 712 717 103 5 |
| ±7 | −0.014 099 121 093 750 0 | 0.003 434 299 437 467 7 |
| ±8 | 0.002 563 476 562 500 0 | 0.020 361 466 841 264 2 |
| ±9 | 0.002 990 722 656 250 0 | 0.011 995 615 052 782 7 |
| ±10 | | −0.003 461 664 373 224 4 |
| ±11 | | −0.001 272 469 512 687 2 |
| ±12 | | 0.001 484 547 764 801 7 |

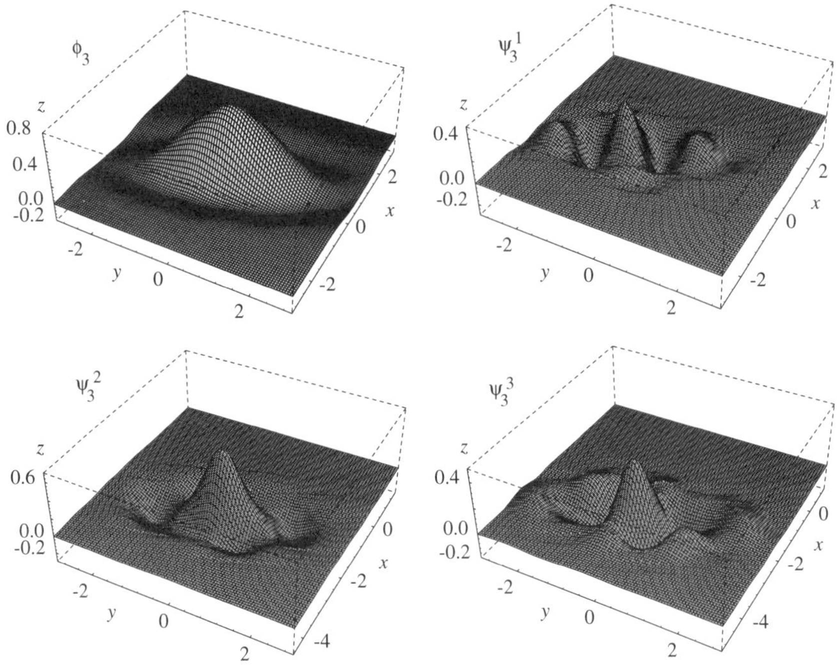


FIGURE 3. Graphs of regular dual scaling functions $\phi_3$ and wavelets $\psi_3^i$, $i = 1, 2, 3$.

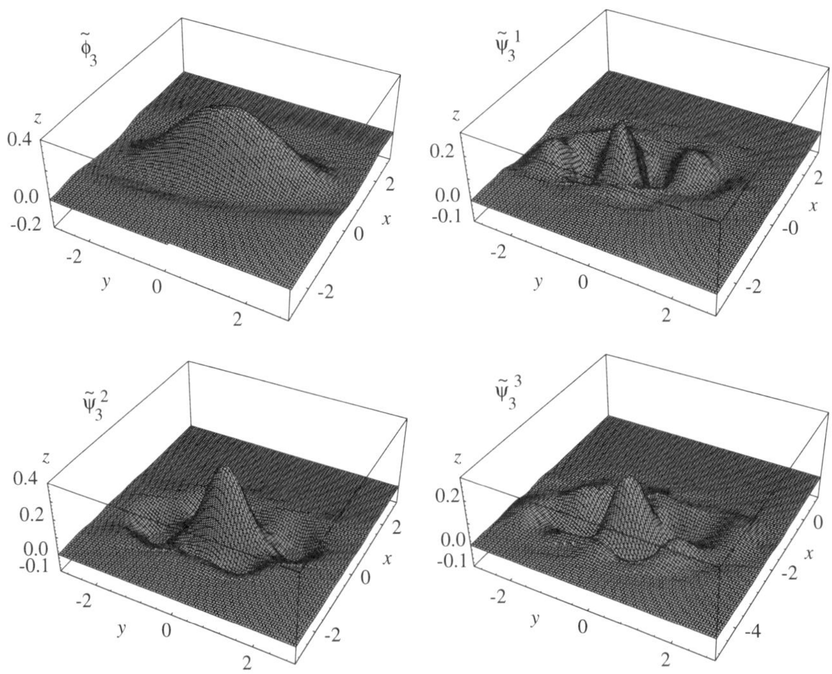


FIGURE 4. Graphs of regular dual scaling functions $\widetilde{\phi}_3$ and wavelets $\widetilde{\psi}_3^i$, $i = 1, 2, 3$.

The lowpass and highpass wavelet filters have been computed via the McClellan transformation. The 2-D cascade algorithm has been used for the numerical approximation of the analysing and synthesizing scaling functions and wavelets $\phi_3(\boldsymbol{x})$, $\psi_3^i(\boldsymbol{x})$, $i = 1, 2, 3$, and $\widetilde{\phi}_3(\boldsymbol{x})$, $\widetilde{\psi}_3^i(\boldsymbol{x})$, $i = 1, 2, 3$, shown in Figs. 3 and 4, respectively.

## References


1. C. K. Chui and C. Li, *A general framework of multiwavelets with duals*, Appl. Comput. Harmon. Anal. **1** (1994), 368–390.
2. A. Cohen, *Biorthogonal wavelets*, Wavelets: A Tutorial in Theory and Applications (C. K. Chui, ed.), Academic Press, San Diego, CA, 1992, pp. 123–152.
3. A. Cohen and I. Daubechies, *Non-separable bidimensional wavelet bases*, Rev. Mat. Iberoamericana **9** (1993), no. 151–137.
4. A. Cohen, I. Daubechies, and J.-C. Feauveau, *Biorthogonal bases of compactly supported wavelets*, Comm. Pure Appl. Math. **45** (1992), no. 5, 485–560.
5. I. Daubechies, *Orthonormal bases of compactly supported wavelets*, Comm. Pure Appl. Math. **41** (1988), no. 7, 909–996.
6. ———, *Ten lectures on wavelets*, SIAM, Philadelphia, PA, 1992.
7. I. Daubechies and J. C. Lagarias, *Two scale difference equations. I. Existence and global regularity of solutions*, SIAM. J. Math. Anal. **22** (1991), no. 5, 1388–1410.
8. K. Gröchenig, *Analyse multi-échelle et bases d'ondelettes*, C. R. Acad. Sci. Paris Sér. I Math. **305** (1987), no. 1, 13–15.
9. G. Gröchenig and W. Madych, *Multiresolution analysis, Haar bases, and self similar tilings of* $\mathbb{R}^n$, IEEE Trans. Inform. Theory **38** (1992), no. 2, part 2, 556–568.
10. A. Karoui, *A technique for the construction of smooth and compactly supported biorthogonal wavelets of* $L^2(\mathbb{R}^n)$, $n \geq 2$, 1996 (preprint).
11. A. Karoui and R. Vaillancourt, *Nonseparable biorthogonal wavelet bases of* $L^2(\mathbb{R}^n)$, Tech. report no. CRM–2220, Univ. de Montréal, 1994.
12. A. Karoui and R. Vaillancourt, *McClellan transformation and the construction of biorthogonal wavelet bases of* $L^2(\mathbb{R}^2)$, Comput. Math. Appl. **29** (1995), no. 12, 13–26.
13. J. Kovačević and M. Vetterli, *Nonseparable multidimensional perfect reconstruction filter banks and wavelet bases for* $R^n$, IEEE Trans. Inform. Theory **38** (1992), no. 2, 533–555.
14. S. Mallat, *Multiresolution approximations and wavelet orthonormal bases of* $L^2(\mathbb{R})$, Trans. Amer. Math. Soc. **315** (1989), 69–88.
15. J. H. McClellan and D. S. K. Chan, *A 2-D* FIR *filter structure derived from the Chebyshev recursion*, IEEE Trans. Circuits Systems I Fund. Theory Appl. **CAS-24** (1977), no. 7, 372–378.
16. O. Rioul, *Simple regularity criteria for subdivision schemes*, SIAM. J. Math. Anal. **23** (1992), no. 6, 1544–1576.
17. M. Vetterli and C. Herley, *Wavelets and filter banks: Theory and design*, IEEE Trans. Signal Process. **40** (1992), no. 9, 2207–2232.
18. E. Viscito and J. P. Allebach, *The analysis and design of multidimensional* FIR *perfect reconstruction filter banks for arbitrary sampling lattices*, IEEE Trans. Circuits Systems I Fund. Theory Appl. **38** (1991), no. 1, 29–42.



Département de mathématiques, Faculté des sciences de Bizerte, Jarzouna, 7021, Tunisie

*Current address*: Institut préparatoire aux études d'ingénieur de Mateur, Mateur, 7030, Tunisie

*E-mail address*: `hechmi.said@fsb.rnu.tn`

Département de mathématiques et de statistique, Université d'Ottawa, 585, av. King Edward, C.P. 450, Succ. A, Ottawa, Ontario, K1N 6N5, Canada

*E-mail address*: `remiv@mathstat.uottawa.ca`

Centre de Recherches Mathématiques
CRM Proceedings and Lecture Notes
Volume **18**, 1999

# Local Bases: Theory and Applications

Jelena Kovačević and Riccardo Bernardini

ABSTRACT. We present an overview of the work on local orthogonal bases in the past couple of years and their applications in signal processing. Local orthogonal bases can be seen as a generalization of local cosine bases both in continuous and discrete time. Their basis functions are obtained from a single prototype window and there exists a fast algorithm for implementation. We show how to construct such bases including an account on window design.

To demonstrate their usefulness we give two different applications: In the first, we show how to obtain arbitrary tilings of the time-frequency plane using local orthogonal bases. We give an example of design, that of a critical-band system for use in audio coding. The second application is concerned with finding a measure for evaluating image quality where we need a system consisting of a bank of filters obtained by modulations and rotations of a single prototype filter. We develop the theoretical machinery necessary to obtain such bases and we show a specific design example.

## 1. Introduction

Discrete-time local cosine bases, or, cosine modulated filter banks[1], have been in use for some time [**18, 20**]. Due to a few of their properties, they have become quite popular; For example, all filters (basis functions) of a filter bank are obtained by appropriate modulation of a single prototype filter. Then, fast algorithms exist, making them very attractive for implementation. Finally, they have been used recently to achieve time-varying splittings of the time-frequency plane [**12**].

Their continuous counterpart, termed "Malvar's wavelets", has found use in decomposing a signal into a linear combination of time-frequency atoms [**19, 7**].

Local cosine bases have been used extensively in audio coding [**24**]. They have also found use in image coding, due to the reduction of blocking effects [**17**] when compared to the DCT. Some video works contain local cosine bases as well [**13**]. However, in all image and video applications, one-dimensional local cosine bases are used separately.

In multiple dimensions one has more freedom than in one dimension, and using only separable cosine bases is very restrictive. In this paper we give an overview of

1991 *Mathematics Subject Classification.* Primary: 94A12; Secondary: 94A29.

This is the final form of the paper.

[1]These constructions are known under various names: MDCT, LOT, modulated lapped transforms, time—domain aliasing cancellation filter banks, Princen and Bradley filter banks, Malvar wavelets.

local orthogonal bases (not necessarily cosine; some cosine constructions in multiple dimensions can be found in [**28, 27**]). We first show how to build local orthogonal bases in a very general way by using a formalism that includes the one-dimensional case as well the multidimensional one, the continuous as well the discrete, the separable as well the nonseparable and, moreover, can handle nonrectangular decompositions of the multidimensional domain [**5**]. We also explain the design of the prototype window both in continuous and discrete time [**6**].

We then show the first application of these bases in trying to build arbitrary tilings of the time-frequency plane. The solutions we offer in [**2**] are in between two known classes of time-frequency atoms—wavelet packets and local cosine bases. We discuss how, given a particular tiling, we can find a local orthogonal basis that will produce it. As an example, we show a design of a critical-band local orthogonal basis to be used in audio coding. Note that this particular basis achieves nonuniform frequency decomposition, a hard task in itself.

As the second example, we discuss how one can build local orthogonal sets and bases on fairly general groups [**3, 4**]. This work was triggered by the need to find good measures of image quality by passing a signal first through a set of orthonormal filters which are all obtained as modulations and rotations of single prototype filter.

## 2. Local Orthogonal Bases

### 2.1. Review of local cosine bases.

2.1.1. *Discrete time.* The usual one-dimensional local cosine bases are a class of perfect reconstruction filter banks which uses a single prototype filter—window—$w[n]$ of length $2N$ (where $N$ is the number of channels and is even) to construct all of the filters $h_0, \dots, h_{N-1}$ as follows:

$$h_k[n] = \frac{w[n]}{\sqrt{N}} \cdot \cos\left[\frac{2k+1}{4N}(2n-N+1)\pi\right],$$

with $k = 0, \dots, N-1$, $n = 0, \dots, 2N-1$, and where the prototype lowpass filter $w[n]$ is symmetric

$$w[n] = w[2N-1-n], \quad n = N, \dots, 2N-1$$

and satisfies the power-complementarity condition [**20**]

$$w^2[n] + w^2[N-1-n] = 2, \tag{1}$$

for $n = 0, \dots, N-1$. This last condition, imposed on the window, ensures that the resulting modulated lapped transform is orthogonal. The two symmetric halves of the window are called "tails".

2.1.2. *Continuous time.* In [**1**] the problem of the cosine bases in the continuous-time case is approached in an interesting way. The authors divide the vector space of $\mathcal{V}_{\mathbb{R}} = L_2(\mathbb{R})$ into the orthogonal sum of "local" vector spaces. These vector spaces have functions of finite support, living on an interval $I = [\alpha - \varepsilon_1, \beta + \varepsilon_2]$. In [**1**] it is shown that the vector space relative to interval $I$ is obtained via projections of the form

$$(2)\quad P_\alpha^\beta f(x) = w_I^2(x)f(x) + w_I(x)w_I(2\alpha - x)f(2\alpha - x) + w_I(x)w_I(2\beta - x)f(2\beta - x).$$

The well-known constraints of symmetry and power-complementarity on the windows relative to adjacent intervals follow from the conditions of self-adjointness and idempotency of (2) and the condition of orthogonality between two spaces relative to adjacent intervals.

It can be shown [**1**] that these "local" vector spaces are the sets of functions with support $I$ that can be written as $f(x) = w_I(x)S(x)$, where $w_I$ is the window with support on the interval $I$ and $S(x)$ is a function, with support $I$, that enjoys the following symmetries:

$$(3)\quad \begin{aligned} S(x)\chi_{[\alpha-\varepsilon_1,\alpha+\varepsilon_1]} &= \pm S(2\alpha - x)\chi_{[\alpha-\varepsilon_1,\alpha+\varepsilon_1]}, \\ S(x)\chi_{[\beta-\varepsilon_2,\beta+\varepsilon_2]} &= \pm S(2\beta - x)\chi_{[\beta-\varepsilon_2,\beta+\varepsilon_2]}, \end{aligned}$$

where $\chi_{[\alpha,\beta]}$ is the characteristic function of the interval $[\alpha, \beta]$.[2] The approach in [**1**] can be generalized in a more abstract fashion that fits many cases that are of interest for applications.

**2.2. Construction of local orthogonal bases.** In this section we want to mimic the approach in [**1**] in a more general fashion that fits almost every case of interest. Consider a set $X$ and a vector space $\mathcal{V}_X$ of functions on $X$ and let $\langle \cdot, \cdot \rangle$ be an inner product for $\mathcal{V}_X$. For example, in the one-dimensional continuous-time case $X = \mathbb{R}$ and $\mathcal{V}_X = L_2(\mathbb{R})$; on the other hand, in discrete time $X = \mathbb{Z}$ and $\mathcal{V}_X = \ell_2(\mathbb{Z})$.

Our goal is to find a local orthonormal basis for $\mathcal{V}_X$, that is, a set of functions $g_{i,k}(x)$ such that

$$(4)\quad f = \sum_{i,k\in\mathbb{Z}} \langle g_{i,k}, f\rangle g_{i,k}, \quad \text{for all } f \in \mathcal{V}_X$$

and each $g_{i,k}$ has finite support $B_i$. For example, in the one-dimensional discrete-time case,

$$(5)\quad g_{i,k}[n] = h_k[n - Ni]$$

and $B_i = Ni + \{0, \ldots, 2N - 1\}$.

Call $\mathcal{V}_{B_i}$ the vector space spanned by $g_{i,k}, k \in \mathbb{Z}$. Vector space $\mathcal{V}_{B_i}$, by construction, has functions with finite support $B_i$ and $g_{i,k}$, $k \in \mathbb{Z}$ as an orthonormal basis. It is worth noting that $\mathcal{V}_{B_i}$ is a subspace of $\chi_{B_i}\mathcal{V}_X$, the subspace of $\mathcal{V}_X$ of functions with support $B_i$. In the one-dimensional continuous-time case $\mathcal{V}_{B_i}$ is the vector space of the functions $w_i(x)S(x)$, with $S(x)$ satisfying (3). With such an assumption (4) is equivalent to

$$(6)\quad \mathcal{V}_X = \oplus_i \mathcal{V}_{B_i}.$$

It is clear that (6) implies $X = \cup_i B_i$. Therefore, our goal can be achieved as follows:

1. Choose a decomposition $X = \cup_i B_i$.

[2] Note that the above equations mean that the function $S$ is symmetric around $\alpha$ (or $\beta$) on interval $[\alpha - \varepsilon_1, \alpha + \varepsilon_1]$ (or $[\beta - \varepsilon_2, \beta + \varepsilon_2]$).

2. Choose vector spaces $\mathcal{V}_{B_i}$ such that (6) is met.
3. Find an orthonormal basis for each $\mathcal{V}_{B_i}$.

The first step is very simple and we will not elaborate on it. In the one-dimensional case one can choose each $B_i$ as an interval, in two dimensions $B_i$ can be, for example, a rectangle. Note that, in general, two sets $B_i$ and $B_j$ will have a nonempty intersection. For example, in the one-dimensional case $B_i \cap B_{i+1} = [\alpha_{i+1} - \varepsilon_{i+1}, a_{i+1} + \varepsilon_{i+1}]$.

The key point is the second step. To accomplish it we need to investigate the following two points:

1. When are two spaces $\mathcal{V}_{B_i}$ and $\mathcal{V}_{B_j}$ orthogonal?
2. If they are orthogonal, what is the structure of $\mathcal{V}_{B_i} \oplus \mathcal{V}_{B_j}$?

The first point is necessary since we want the sum in (6) to be direct and this means that each $\mathcal{V}_{B_i}$ is orthogonal to every other $\mathcal{V}_{B_j}$. The second point is motivated by the fact that we want the sum in (6) to be complete or, in other words, we want to be able to write each function of $\mathcal{V}_X$ as a sum of functions $g_i \in \mathcal{V}B_i$, $i \in \mathbb{Z}$.

2.2.1. *Orthogonality of two vector spaces.* It can be shown that[3] if $B_i \cap B_j = \emptyset$ then $\mathcal{V}_{B_i} \perp \mathcal{V}_{B_j}$, so we need to study only the case when $B_i \cap B_j = C \neq \emptyset$. In the one-dimensional continuous-time case, we have to study only the case relative to two adjacent intervals $[\alpha_0 - \varepsilon_0, \alpha_1 + \varepsilon_1]$ and $[\alpha_{-1} - \varepsilon_{-1}, a_{-1+1} + \varepsilon_{-1+1}]$, where the set $C$ is $[\alpha_0 - \varepsilon_0, \alpha_0 + \varepsilon_0]$.

Suppose that $\chi_C \mathcal{V}_{B_i}$ is a subspace of $\mathcal{V}_{B_i}$ (this requirement will be satisfied by construction) and define $D_i = B_i \setminus C$. Then, $\chi_{D_i} \mathcal{V}_{B_i}$ and $\chi_C \mathcal{V}_{B_i}$ are orthogonal. Indeed, if $f \in \chi_C \mathcal{V}_{B_i}$ and $g \in \chi_{D_i} \mathcal{V}_{B_i}$ we have that $f = \chi_C f$ and $g = \chi_{D_i} g$ and $\langle f, g \rangle = \langle \chi_C f, \chi_{D_i} g \rangle$ that can be rewritten as $\langle \chi_C \chi_{D_i} f, g \rangle$. Since $C \cap D_i = \emptyset$, then $\chi_C \chi_{D_i} = 0$ from which the orthogonality of $f$ and $g$ follows.

Moreover, since $\chi_{B_i} = \chi_C + \chi_{D_i}$, each function $f \in \mathcal{V}_{B_i}$ can be written as $f = \chi_{B_i} f = \chi_C f + \chi_{D_i} f$, that is, the sum of a function from $\chi_C \mathcal{V}_{B_i}$ with a function from $\chi_{D_i} \mathcal{V}_{B_i}$. Such a fact, together with the orthogonality property and the hypothesis that $\chi_C \mathcal{V}_{B_i}$ is a subspace of $\mathcal{V}_{B_i}$, implies that $\mathcal{V}_{B_i} = \chi_C \mathcal{V}_{B_i} \oplus \chi_{D_i} \mathcal{V}_{B_i}$. Similar reasoning holds for $\mathcal{V}_{B_j}$. It is clear that the question about orthogonality of $\mathcal{V}_{B_i}$ and $\mathcal{V}_{B_j}$ becomes about orthogonality of $\chi_C \mathcal{V}_{B_i}$ and $\chi_C \mathcal{V}_{B_j}$.

In the one-dimensional case $\chi_{[\alpha_0 - \varepsilon_0, \alpha_0 + \varepsilon_0]} \mathcal{V}[\alpha_0 - \varepsilon_0, \alpha_1 + \varepsilon_1]$ is the vector space of functions $f = wS$ with $S$ having support $C = [\alpha_0 - \varepsilon_0, \alpha_0 + \varepsilon_0]$ and symmetric or antisymmetric with respect to $\alpha_0$, depending on the sign chosen in (3). Moreover, $w$ restricted to $[\alpha_0 - \varepsilon_0, \alpha_0 + \varepsilon_0]$ enjoys the property

$$w^2(x) + w^2(2\alpha - x) = 1, \quad \forall x \in [\alpha_0 - \varepsilon_0, \alpha_0 + \varepsilon_0]. \tag{7}$$

for each $x$ belonging to $[\alpha_0 - \varepsilon_0, \alpha_0 + \varepsilon_0]$. In a more general case we can have more than one symmetry. For example, in the two-dimensional case we can have the symmetry $u_x$ with respect to the $x$ axis, the symmetry $u_y$ with respect to the $y$ axis and they can be combined to obtain $u_{xy}$ that is, the symmetry with respect to the origin. Note that $\Gamma = \{\mathcal{I}, u_x, u_y, u_{xy}\}$, where $\mathcal{I}(x) = x$, is a group and each element is an involution (symmetry). To give the type of the symmetry (if even or odd) one can assign to each symmetry $u$ a sign $s(u) = \pm 1$. It can be readily seen that, in order to have coherence between the types of symmetries, if $u, v \in \Gamma$ then it must be $s(uv) = s(u)s(v)$, that is, $s$ must be a representation of $\Gamma$. Note that

[3] For this to hold we need a fairly technical hypothesis which can be found in [**5**].

in the one-dimensional case $\Gamma$ has only two elements. With such a point of view, property (7) can be rewritten as

$$\sum_{u\in\Gamma} w^2\big(u(x)\big) = 1.$$

Inspired by these observations we can impose the following structure on the vector space $\chi_C\mathcal{V}_{B_i}$:

DEFINITION 2.1. Let us make the following assumptions:

- Let $\mathcal{V}_C$ be the subspace of $\mathcal{V}_X$ formed by the functions of $\mathcal{V}_X$ with support on $C$.
- Let $\Gamma$ be a group of involutions on $C$. Each involution induces a transformation on functions of $\mathcal{V}_C$ as $f_u \triangleq f\big(u(x)\big)$.
- Suppose that $f \in \mathcal{V}_C \implies f_u \in \mathcal{V}_C$ and that $\langle a, b\rangle = \langle a_u, b_u\rangle$.
- Let $s$ be a function mapping $\Gamma$ onto $\{+1, -1\}$ such that $s(uv) = s(u)s(v)$, for all $u, v \in \Gamma$, that is, $s$ is a representation of order two of $\Gamma$.
- Finally, suppose that $w$ is a function with support $C$ such that

(8) $$\sum_{u\in\Gamma} w^2\big(u(x)\big) = 1.$$

Then we define the vector space $\mathcal{V}_C(\Gamma, s, w)$ as the space of the functions $f$ of $\mathcal{V}_C$ such that $f$ can be written as $wS$, where $S$ is a function with support $C$ satisfying

(9) $$S_u(x) = S\big(u(x)\big) = S(u)S,$$

that is, $S$ possesses certain symmetry properties.

It is possible to prove that the orthogonal projection from $\mathcal{V}_X$ to $\mathcal{V}_C(\Gamma, s, w)$ is

(10) $$P(\Gamma, s, w)f \triangleq \sum_{u\in\Gamma} s(u)ww_uf_u.$$

Note that all the proofs can be found in [**5**]. It is worth noting that the problems of the orthogonality and completeness can be rephrased in term of the corresponding projections, as we will do. If we impose on $\chi_C\mathcal{V}_{B_i}$ the structure described by Definition 1 (we can, since we are designing $\mathcal{V}_{B_i}$), the solution to the problem of orthogonality is included in the following property:

PROPERTY 2.1. *Let $P_1 \triangleq P(\Gamma, s, w)$ and $P_2 \triangleq P(\Gamma, t, w_v)$ where $v \in \Gamma$, $t(v) = -s(v)$. Then $P_1 \perp P_2$.*

In the one-dimensional case, for $C = [\alpha_0 - \varepsilon_0, \alpha_0 + \varepsilon_0]$, there is only one nontrivial symmetry (that is, $u_0(x) = 2\alpha_0 - x$) and Property 2.1 asserts that the right tail of the window of $[\alpha_{-1} - \varepsilon_{-1}, \alpha_{-1+1} + \varepsilon_{-1+1}]$ must be obtained from the left one of $[\alpha_0 - \varepsilon_0, \alpha_1 + \varepsilon_1]$ with a symmetry around $\alpha_0$ and that the signs of the two projections must be different, as given in Proposition 1 of [**1**]. Moreover, Property 2.1 holds in the multidimensional case as well.

2.2.2. *Structure of the sum of two vector spaces.* In order to solve the problem of completeness suppose that a given set $C$ is contained in sets $B_1, B_2, \ldots, B_M$ and it has empty intersection with the others. For the relationship (6) to hold it must be

(11) $$\chi_C\mathcal{V}_X = \oplus_{i=1}^{M}\chi_C\mathcal{V}_{B_i}$$

or, in other words, by summing functions in $\chi_C \mathcal{V}_{B_i}$, $i = 1, \ldots, M$ one must be able to obtain every function of $\mathcal{V}_X$ with support $C$. If $\chi_C \mathcal{V}_{B_i}$ has the structure described by Definition 1, the following property gives a solution to the problem of completeness in the specific case when there is only one symmetry beside the trivial one:

PROPERTY 2.2. *Suppose* $\Gamma = \{\mathcal{I}, v\}$ *and let* $P_1 \triangleq P(\Gamma, s, w)$ *and* $P_2 \triangleq P(\Gamma, t, w_v)$ *where* $t(v) = -s(v)$. *Then* $P_1 \perp P_2$ *and* $P_1 f + P_2 f = f$, *for all* $f \in \chi_C \mathcal{V}_X$.

First part of Property 2.2 is a particular case of Property 2.1 and ensures the orthogonality of $\mathcal{V}_1 = \mathcal{V}_C(\Gamma, s, w)$ and $\mathcal{V}_2 = \mathcal{V}_C(\Gamma, t, w_v)$, that is, the sum of $\mathcal{V}_1$ and $\mathcal{V}_2$ is direct. The second part claims that in this particular case, where $\Gamma$ has only one nontrivial symmetry, the sum $\mathcal{V}_1 \oplus \mathcal{V}_2$ is also complete, that is, $\mathcal{V}_1 \oplus \mathcal{V}_2 = \chi_C \mathcal{V}_X$ or, in other words, each function of $\mathcal{V}_X$ having support $C$ can be expressed as the sum of a function of $\mathcal{V}_1$ and a function of $\mathcal{V}_2$.

Since in the one-dimensional case each transition zone $[\alpha_0 - \varepsilon_0, \alpha_0 + \varepsilon_0]$ has only one symmetry, Property 2.2 is sufficient to study the orthogonality and completeness of a given decomposition in the one-dimensional case. In the spirit of Property 2.2 let us give the following definition:

DEFINITION 2.2. Two projections satisfying the hypothesis of Property 2.2 will be called *complementary* to one another. The complementary projection to a projection $P$ will be denoted as $P^*$ (not to be confused with the adjoint).

In order to handle the case in which $\Gamma$ has more than one symmetry, we need a tool to "split" a projection as a chain of simpler projections; such a tool is furnished by the two properties shown below; they can be seen as inverses of each other. These properties have no counterpart in the one-dimensional case and to give an example we must resort to multiple dimensions. The first property says, in a formal way, an intuitively appealing fact: if we perform, in sequence, first a projection with a certain symmetry (for example, even symmetry around $x$ axis) then another with another symmetry (for example, odd symmetry around $y$ axis) we obtain a signal with a combination of the two symmetries; in this case we would expect a signal that is even symmetric with respect to the $x$ axis, odd symmetric with respect to the $y$ axis and odd symmetric with respect to the origin, since this symmetry is the combination of an odd symmetry and an even one.

PROPERTY 2.3. *Let* $P_1 \triangleq P(\Gamma_1, s, w)$ *and* $P_2 \triangleq P(\Gamma_2, t, y)$. *Suppose that* $\Gamma_1 \cap \Gamma_2 = \mathcal{I}$, *that* $\langle \Gamma_1 \Gamma_2 \rangle$ *is a group of involutions and that for each* $v \in \Gamma_2$ *and* $u \in \Gamma_1$ *the following holds*

$$w_v = w, \quad y_u = y. \tag{12}$$

*Then, operator* $P_1 P_2$ *is a projection and, moreover*

$$P_1 P_2 = P\big(\langle \Gamma_1, \Gamma_2 \rangle, s \oplus t, wy\big). \tag{13}$$

Finally, the last property is the inverse of Property 2.3 and it tells us when a given projection can be expressed as a composition of two simpler projections. It can be easily seen that (13) imposes some structure on the window of the resulting projections. It is clear that in order to invert Property 2.3 we need some hypothesis on the window.

PROPERTY 2.4. *Let $P(\Gamma, s, w)$ be a projection and suppose that $\Gamma_1$ and $\Gamma_2$ are two subgroups of $\Gamma$ such that $\Gamma = \langle \Gamma_1, \Gamma_2 \rangle$. Also let $r$ and $t$ be two representations of $\Gamma_1$ and $\Gamma_2$, respectively, such that $s = r \oplus t$. Suppose that window $w$ is such that $ww_{uv} = w_u w_v$, for each $u \in \Gamma_1$, $v \in \Gamma_2$. Then, projection $P(\Gamma, s, w)$ can be written as*

$$P(\Gamma, s, w) = P(\Gamma_1, r, \theta) P(\Gamma_2, t, \tau) \tag{14}$$

*where $\theta$ and $\tau$ are defined as*

$$\theta = \frac{w}{(\sum_{u \in \Gamma_1} {w_u}^2)^{1/2}}, \qquad \tau = \frac{w}{(\sum_{v \in \Gamma_2} {w_v}^2)^{1/2}}. \tag{15}$$

In order to complete the construction of basis (4), one has to find an orthonormal basis for $\mathcal{V}_{B_i}$. Note that $\mathcal{V}_{B_i}$ can be written as the direct sum of its subspaces $\mathcal{V}_{C_{i,j}}$ relative to the subsets $C_{i,j}$ forming $B_i$. Using such a fact, an orthonormal basis for $\mathcal{V}_{B_i}$ can be found by finding an orthonormal basis $b_{i,j,k}$, $k \in \mathbb{N}$, for each $\mathcal{V}_{C_{i,j}}$; an orthonormal basis for $B_i$ will be $\{b_{i,1,k_1}, b_{i,2,k_2}, \dots\}$ with $k_i \in \mathbb{Z}$. Thus, we can limit ourselves to the study of orthonormal bases of $\mathcal{V}_{C_{i,j}}$, as it is done in the next section. For convenience we will call $C_{i,j} = C$ and $\mathcal{V}_{C_{i,j}} = \mathcal{V}_C$.

2.2.3. *Finding a basis.* Finally, we address Point 3 in Section 2.2. The problem of finding an orthonormal basis for $\mathcal{V}_C(\Gamma, s, w)$ is simplified by the property presented below. The intuitive meaning is that $\mathcal{V}_C(\Gamma, s, w)$ is constructed by multiplying functions having certain symmetries by the window $w$ and, for this reason, one can expect that a basis for $\mathcal{V}_C(\Gamma, s, w)$ could be obtained by multiplying by $w$ a basis of the space of functions opportunely symmetric on $C$. Note that such a space can not be written as $\mathcal{V}_C(\Gamma, s, \chi_C)$ since $\chi_C$ does not satisfy the condition of normalization of the power; instead one has to write $\mathcal{V}_C(\Gamma, s, \chi_C/\sqrt{|\Gamma|})$, that is, with a scaled window which we will call $w_C = \chi_C/\sqrt{|\Gamma|}$, where $|\Gamma|$ is the cardinality of $\Gamma$, that is, the number of elements of $\Gamma$.

PROPERTY 2.5. *Let $b_i$ be an orthonormal basis of $\mathcal{V}_C(\Gamma, s, w_C)$. Then $\sqrt{|\Gamma|} w b_i$ is an orthonormal basis of $\mathcal{V}_C(\Gamma, s, w)$.*

For simplicity, here we discuss the problem of finding an orthonormal basis for $\mathcal{V}_C(\Gamma, s, w_C)$ only in the case of $C$ discrete and finite, that is, when $\mathcal{V}_C$ has a finite dimension. Similar reasoning could be used in the continuous-time case. What we need to find out now, is how many free parameters we have in the system. Thus, we will first examine the dependencies that exist.

If $f \in \mathcal{V}_C(\Gamma, s, w_C)$, then, by definition, $f\big(u(x)\big) = s(u) f(x)$, that is, the values assumed by $f$ on a given orbit $O(x)$ are not independent, rather they are determined by the value assumed by $f$ on any element of $O(x)$. It is worth noting that if for some $x \in C$ there is an $u \in \Gamma$ such that $u(x) = x$ and $s(u) = -1$, then $f(x) = f\big(u(x)\big) = s(u) f(x) = -f(x) = 0$, that is, $f$ is zero on each point of $O(x)$. By partitioning $C$ in orbits and choosing one representative from each orbit, a set of representatives $\check{C}$, can be made. Note that each function belonging to $\mathcal{V}_C(\Gamma, s, w_C)$ is uniquely determined by the values assumed on $\check{C}$.

For example, in the one-dimensional case it is worth noting that a function of $\mathcal{V}_C(\Gamma, s, w)$, where $C$ is the transition zone $[\alpha_0 - \varepsilon_0, \alpha_0 + \varepsilon_0]$, is uniquely determined by its values on $\check{C} = [\alpha_0, \alpha_0 + \varepsilon_0]$, since the other values can be obtained by symmetry. Since a similar reasoning holds also for $[\alpha_1 - \varepsilon_1, \alpha_1 + \varepsilon_1]$ it is easy to see that a function belonging to $\mathcal{V}[\alpha_0 - \varepsilon_0, \alpha_1 + \varepsilon_1]$ is determined by its values on $[\alpha_0 - \varepsilon_0, \alpha_1 + \varepsilon_1] = [\alpha_0, \alpha_1]$.

By discarding all $x$ on which $f$ is identically zero (we can do so, since such points do not add information to $f$ and we can "restore" them whenever we wish), we can form a vector by ordering the points of $C$ "by orbits", that is, by putting first the points of one orbit, then the points of another orbit, and so on. Only one value of those belonging to a "subvector" corresponding to a particular orbit is free. We will choose as the free parameter the first one, the others will be obtained by multiplying it by $\pm 1$, according to the sign of $s(u)$. Then, one can show that each $f \in \mathcal{V}_C$ can be written as

$$f = \boldsymbol{W}\boldsymbol{K}g = \boldsymbol{W}(\boldsymbol{I}_{N_O} \oplus K)g \tag{16}$$

where $N_O$ is the number of orbits of $\Gamma$ on which $f$ is not zero, $g$ is a vector without constraints, $\boldsymbol{K}$ is the block diagonal matrix diag $(K, K, \ldots, K)$, where $K$ is a column vector with entries $\pm 1$ and $\boldsymbol{W}$ is a diagonal matrix which contains the values of the window. It is easy to see that $(\boldsymbol{W}\boldsymbol{K})^T\boldsymbol{W}\boldsymbol{K} = \boldsymbol{I}$ as long as the window satisfies the power-complementarity conditions.

Suppose now that vectors $c_i$, $i = 1, \ldots, M$, are an orthonormal basis of $\mathcal{V}_C$. Let $\boldsymbol{C}$ be the matrix having as columns the vectors $c_i$. Observe that each $c_i$ is a linear combination of the columns of $\boldsymbol{W}\boldsymbol{K}$, therefore $\boldsymbol{C} = \boldsymbol{W}\boldsymbol{K}\boldsymbol{G}$, for some $M \times M$ matrix $\boldsymbol{G}$. From the orthogonality of $c_i$ it follows

$$\boldsymbol{I} = \boldsymbol{C}^T\boldsymbol{C} = \boldsymbol{G}^T\boldsymbol{K}^T\boldsymbol{W}\boldsymbol{W}\boldsymbol{K}\boldsymbol{G} = \boldsymbol{G}^T\boldsymbol{G} \tag{17}$$

that is, the columns of $\boldsymbol{G}$ form an orthonormal basis of $\mathbb{R}^M$. It is clear that the inverse is also true, more precisely, if the columns of $\boldsymbol{G}$ are an orthonormal basis of $\mathbb{R}^M$, then the columns of $\boldsymbol{W}\boldsymbol{K}\boldsymbol{G}$ are an orthonormal basis of $\mathcal{V}_C$.

Moreover, if $B = \cup_i C_i$, a vector $f$ of $\mathcal{V}_B$ can always be obtained by "stacking" vectors of $\mathcal{V}_{C_i}$. The resulting structure of $f$ is, therefore, $f = \mathcal{W}\mathcal{K}g$, where $\mathcal{K}$ and $\mathcal{W}$ are defined as

$$\mathcal{K} \triangleq \operatorname{diag}\{\boldsymbol{K}_1, \ldots, \boldsymbol{K}_i, \ldots\}, \quad \mathcal{W} \triangleq \operatorname{diag}\{\boldsymbol{W}_1, \ldots, \boldsymbol{W}_i, \ldots\} \tag{18}$$

where $\boldsymbol{K}_i$ and $\boldsymbol{W}_i$ are, respectively, matrices $\boldsymbol{K}$ and $\boldsymbol{W}$ relative to the vector space $\mathcal{V}_{C_i}$. It is also clear that for (18) one can write $(\mathcal{W}\mathcal{K})^T\mathcal{W}\mathcal{K} = \boldsymbol{I}$; therefore, all arguments used for the construction of a basis for $\mathcal{V}_C$ still hold.

This way of constructing an orthonormal basis of $\mathcal{V}_B$ can be easily seen as the general extension of the technique to extend a basis by symmetry or antisymmetry, as presented in [**27**]. The general structure of an orthonormal basis of $\mathcal{V}_B$ is, therefore,

$$\mathcal{W}\mathcal{K}\boldsymbol{G} \tag{19}$$

where $\boldsymbol{G}$ is a unitary matrix, $\mathcal{K}$ "starting basis", and $\mathcal{W}$ is a diagonal matrix having as elements the values that window $w$ assumes on $B$. Note that $\operatorname{Tr}(\mathcal{W}^2) = \text{const}$, that is, the sum of all the elements of $\mathcal{W}$ squared ($\mathcal{W}$ is diagonal) equals a constant.

It is interesting to see how the coefficients of a given function $f \in \mathcal{V}_X$ can be computed. First, observe that $\mathcal{V}_B$ is a subspace of $\chi_B\mathcal{V}_X$. Therefore, we can first project $f$ on $\chi_B\mathcal{V}_X$ by obtaining a vector with $|B|$ components; note that such a step does not involve any operations. Next, multiply $\chi_B f$ by $(\mathcal{W}\mathcal{K}\boldsymbol{G})^T = \boldsymbol{G}^T\mathcal{K}^T\mathcal{W}$ to obtain the components of $f$. Since $\boldsymbol{W}$ is a $|\check{B}| \times |\check{B}|$ diagonal matrix, the first product costs only $|\check{B}|$ products, therefore it has linear complexity. The cost of the product by $\mathcal{K}$ depends on the symmetries.

Suppose, for simplicity's sake, that, for each $B_i$ and each $x \in B_i$ the only $u \in \Gamma$ such that $u(x) = x$ is $u = \mathcal{I}$. Note that in this case $\text{Stab}(x) = \{\mathcal{I}\}$. Such a fact implies that each orbit of $B_i$ has $|\Gamma|$ points and that $|\check{B}_i| = |B_i|/|\Gamma|$. Therefore, the block of $\mathcal{K}$ corresponding to $B_i$ will have $|B_i|/|\Gamma|$ rows and $|\Gamma|$ columns with values $\pm 1$ and will cost $\big(|B_i|/|\Gamma|\big)\big(|\Gamma - 1|\big) \approx |B_i|$ sums. The total cost of $\mathcal{K}$ will be, therefore, less then $|B|$, that is, linear with the number of point of $B$.

The product with $\boldsymbol{G}^T$, instead, will cost in general approximately $2|\check{B}|^2$ operations, that is a quadratic cost that can be responsible for the greatest part of computational weight. If we want an orthonormal basis that allows for a fast algorithm we must choose a matrix $\boldsymbol{G}$ that can be computed with a small number of operations. Some examples of such fast orthonormal bases can be the DCT, the DFT (although it requires complex arithmetic), the discrete Hartley transform (DHT) or the Walsh transform. It is worth noting that a fast algorithm for $\boldsymbol{G}$ can be immediately transformed into a fast algorithm for the local orthogonal system.

Let us observe that in a filter-bank interpretation of this system, the columns of matrix (19) are the impulse responses of the filters.

**2.3. Window design.** Given a lattice and support for the basis functions, designing local orthogonal bases consists basically of two steps:

(a) one needs to find the "starting basis" **G** and
(b) one has to design a window.

The starting basis is free and when designing it we have to bear in mind that it has to be suitable for fast implementation. As for the window, the procedures for designing it are different in continuous and discrete time. In this short section, we will just outline those procedures. For more details, the reader is referred to [**6**].

2.3.1. *Discrete time.* We will here show how to design a window in discrete time. Due to space limitations, we will show how to do that in a specific case and show an example of window design. More details on the general case can be found in [**6**].

We first choose a sampling lattice. The one we will consider here will be rectangular sampling by 8 in two dimensions. The next step is to find support $B$ having a finite number of points such that $B$ and its copies $B_i$ on the lattice we chose cover the whole space. We chose for a window with support $B$ to be of size $16 \times 16$, that is, we have overlaps between adjacent windows. We then decompose $B$ into nonoverlaping supports $C_{i,j}$ which in our case are of size $8 \times 8$ each and correspond to four corners of the $16 \times 16$ window. For each $C_{i,j}$ we then chose a set of symmetries $\Gamma$, which in this case are identical and are $\{I, u_x, u_y, u_{xy}\}$, that is, identity, symmetry along the horizontal axis, symmetry along the vertical axis, and symmetry along the middle point, respectively. We can now find the window bearing in mind that it has to satisfy the constraint of power-complementarity. An example is given in Figure 1(a) where a two-dimensional window has been designed by using as cost function the power in the band outside of the square $[-1/16\pi, 1/16\pi]^2$.

2.3.2. *Continuous Time.* The algorithm in continuous time is somewhat different. It can be summarized as follows: choose functions $\mathring{h}(k, y)$ such that they are smooth in $y$, zero for $y \notin [0, 1]$, and symmetric with respect to $y = 1/2$. Orthogonalize them using the Gram-Schmidt procedure and obtain $h(k, y)$. Compute

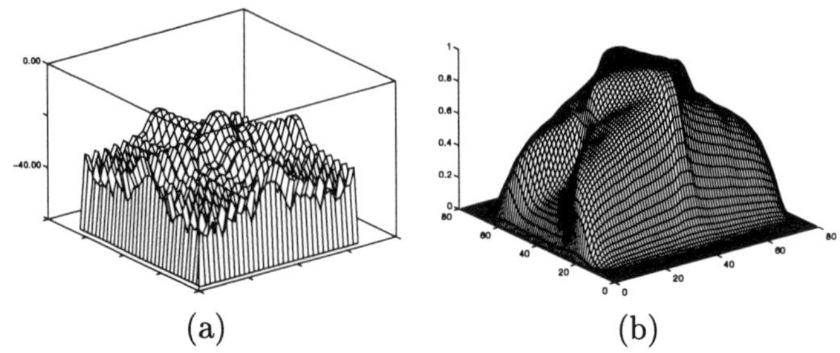


FIGURE 1. (a) Example of a window design in discrete time. Magnitude frequency response is given. (b) Example of a window design in continuous time.

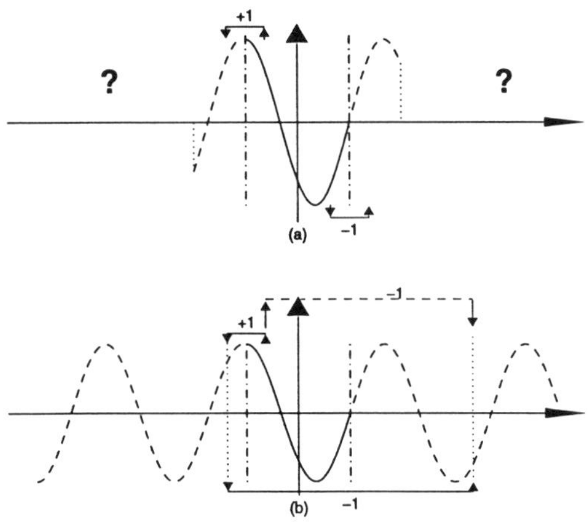


FIGURE 2. Effect of the characteristic functions in (3). (a) Function $S(x)$ satisfying (3). (b) If the characteristic functions are not used, the two symmetries interfere with one another and $S(x)$ is not free anymore outside $[\alpha-\varepsilon_1, \beta+\varepsilon_2]$.

the autocorrelation of $h(k, y)$ and obtain $g(n, y)$. Use $g(n, y)$ as Fourier coefficients in the discrete-time Fourier transform of $q(x, y)$ and find $q(x, y)$. Integrate $q(x, y)$ with respect to y to obtain $Q(x, y)$. Compute the semi-window $p(x, y)$ as $p(x, y) = \sin\bigl(\pi Q(x, y)/2\bigr)$. Finally, use $p(x, y)$ in $w_h(x, y) = p(x, y)p(x, 2-y)$ to get the horizontal window $w_h(x, y)$. Apply again this procedure to obtain the vertical window $w_v(x, y)$. Find the total window $w$ by

$$w(x, y) = w_h(x, y)w_v(x, y).$$

An example is given in Figure 1(b). More details on the described procedure can be found in [**6**].

**2.4. More than one overlap.** In the one-dimensional case, there are solutions to the multiple overlap problem [**10, 16, 18, 22**]. The theory presented in [**5, 6, 1**] and in the previous section does not work in the case of multiple overlapping (for example, in the one-dimensional case we cannot have more than two

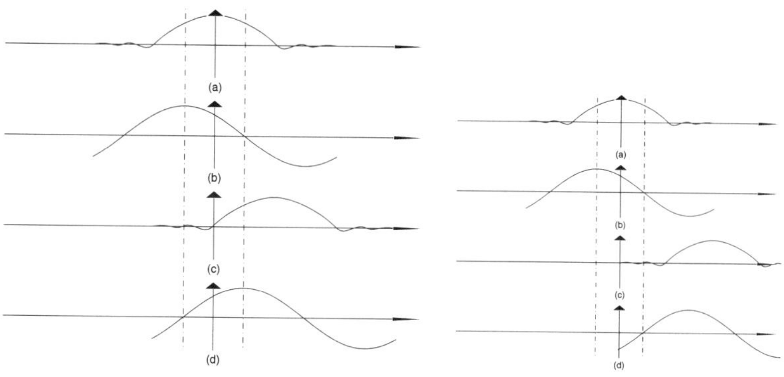


(a) (b)

FIGURE 3. Left side: Multiple overlapping and odd translation. (a) Window $w(x)$. (b) Symmetric signal $S(x)$. (c) and (d) Signals in (a) and (b) translated by one step. Right side: Multiple overlapping and even translation. (a) Window $w(x)$. (b) Symmetric signal $S(x)$. (c) and (d) Signals in (a) and (b) translated by two steps.

windows overlapping). The problem is that the characteristic functions used in (3) limits the "influence zone" of the symmetries, leaving undetermined the function $S(x)$ outside $[\alpha - \varepsilon_1, \beta + \varepsilon_2]$. In Figure 2(a) one can see a function $S(x)$ verifying (3). The part depicted with the solid line can be arbitrarily chosen, while the part drawn with dashed line is determined by the symmetries around $\alpha$ and $\beta$. Nothing is known about $S(x)$ outside the interval $[\alpha - \varepsilon_1, \beta + \varepsilon_2]$, which does not matter in the case of single overlapping because $w(x)$ is zero outside $[\alpha - \varepsilon_1, \beta + \varepsilon_2]$.

The observation that allows one to avoid such a limitation is that the cosines used to modulate window $w(x)$ enjoy symmetries (3) even if the multiplication by the characteristic functions is dropped[4]. Such an "extended" symmetry is, of course, inherited also by their linear combinations and in the spirit of [**5**], one could define the corresponding vector space in a way similar to (3), but without characteristic functions.

To show how orthogonality works in this case, we present, as an example, a graphical study of continuous-time local cosine bases with *multiple overlapping.* In Figure 3(a) (left side) one can see a symmetric window $w(x)$ designed for multiple overlapping, while in Part (b) function $S(x)$ satisfying (3) is displayed, but without the characteristic functions. Parts (c) and (d) show the functions in (a) and (b) translated by one step. Multiplying the windows in Parts (a) and (c) gives a symmetric function, while multiplying the signals in (b) and (d) gives an antisymmetric one. The overall product is therefore antisymmetric and its area is zero. It is possible to show that the same reasoning holds for every *odd* translation, that is, orthogonality for odd translations is granted by window symmetry $w(x) = w(-x)$.

[4]It is worth nothing that without the characteristic functions, symmetries (3) "interfere" with one another, giving rise to an infinite set of symmetries. As an example, in Figure 2(b) one can see the effect of dropping the characteristic function in (3). The dashed line shows a new "symmetry" obtained by applying both the symmetries around $\alpha$ and $\beta$.

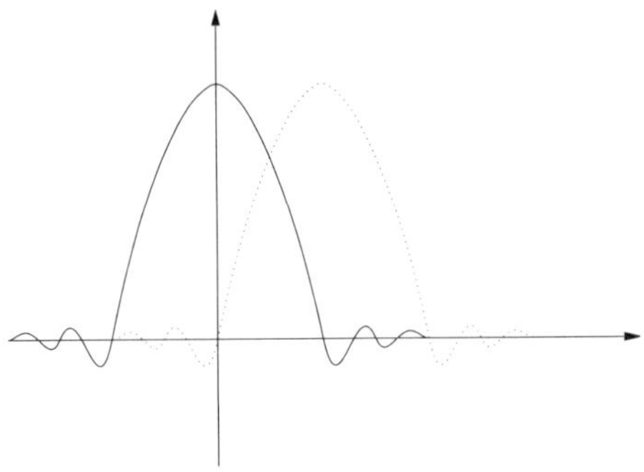

FIGURE 4. Solid line: example of a continuous-time smooth (derivable) window designed for multiple overlapping. Dotted line: first translation of the designed window.

The case of even translations is depicted in the same figure on the right side. Parts (a) to (d) show, respectively, the window $w(x)$, the corresponding symmetric function $S(x)$, the translated window $w(x-2)$ and the corresponding function $S_2(x)$. Since both $S(x)$ and $S_2(x)$ are antisymmetric around $1/2$, their product is symmetric and a cancellation similar to the one on the left side is not possible. It is, however, possible to prove that orthogonality is granted if the window satisfies *self-orthogonality* conditions

$$\sum_{k\in\mathbb{Z}} w(x+2k)w(x+2k+2n) + w\big((2\beta-x)+2k\big)w\big((2\beta-x)+2k+2n\big) = \delta(n). \tag{20}$$

By defining the following sequence:

$$v_x[2k+i] = \begin{cases} w(x+2k) & \text{if } i=0 \\ w\big((2\beta-x)+2k\big) & \text{if } i=1, \end{cases} \tag{21}$$

one can restate (20) by saying that sequence $v_x[n]$ must be sequence orthogonal with respect to its even translations.

It is interesting to observe that, even in the continuous-time case, orthogonality is expressed by discrete-time conditions and one can exploit that for designing an arbitrarily smooth window. The idea is that, for every $x$, sequence $v_x[n]$ can be parameterized with a set of angles $a_1, \ldots, a_N$ [**18, 25**]. In order to obtain a continuous time window one can consider every $a_i$ as a function of $x$ and it is possible to prove that conditions on smoothness of $w(x)$ can be transformed into boundary conditions upon functions $a_i(x)$ and their derivatives.

An example of a continuous and differentiable window for multiple overlapping, designed with such a technique, is shown in Figure 4.

## 3. Arbitrary Tilings of the Time-Frequency Plane Using Local Bases

One of the main goals of signal analysis in recent years has been to develop mixed signal representations in terms of some elementary blocks well localized in time and frequency, where these blocks are known as time-frequency atoms [**19**]. Each one of these blocks would reside mostly[5] in a well-defined area (usually a

[5]By mostly we mean that most of its energy lies within a prescribed region.

rectangle) in the time-frequency plane. An appropriate analogy is that of a musical score. Knowing that there exists a $C$ within a first few bars of a musical piece will not help us unless we know its position in time and its duration. Thus, we can think of notes as time-frequency atoms [**19**].

Currently, two classes of time-frequency atoms are in use, corresponding to two types of analyses that Ville proposed: The first, wavelet packets [**8**], splits the signal first into frequency bands and then in time slices, while the second one, local cosine bases [**7, 1**], does the opposite, that is, it slices first in time and then in frequency. Our approach in [**2**] produces a mixture of these two classes leading to more general tilings.

Here, we show how one can construct time-frequency atoms leading to fairly arbitrary tilings of the time-frequency plane using these local bases. It is also possible to obtain time-varying tilings, since the transitions between various tilings depend only on the windows used. For more details, see [**2**].

**3.1. Construction of arbitrary tilings.** Our goal is to construct a basis such that its basis functions are localized in a certain area of the time-frequency plane. For simplicity and due to practical importance, consider the one-dimensional discrete-time case (the same analysis could be performed in continuous time as well as in multiple dimensions).

In order to express localization in frequency, let us write the Fourier transform of the signal $\boldsymbol{h}$

$$\boldsymbol{H}(\omega) = \sum_{n=0}^{N-1} \boldsymbol{h}[n] \exp(-j\omega n) = \boldsymbol{W}^T(\omega)\boldsymbol{h}, \tag{22}$$

where $\boldsymbol{W}(\omega)$ is a column vector containing functions $\exp(j\omega n)$ organized in a way compatible with the ordering of the samples of $\boldsymbol{h}$. Note that we will use the transpose sign$^T$ to denote transposition and conjugation when the vector is complex.

The power of $\boldsymbol{H}$ at a given frequency $\omega$ can be expressed as

$$\boldsymbol{H}^T(\omega)\boldsymbol{H}(\omega) = \boldsymbol{h}^T\boldsymbol{W}(\omega)\boldsymbol{W}^T(\omega)\boldsymbol{h}, \tag{23}$$

and, we will want to minimize its power in a certain area $A_F$, that is,

$$P_F = \int_{\omega\in A_F} \boldsymbol{h}^T\boldsymbol{W}(\omega)\boldsymbol{W}^T(\omega)\boldsymbol{h}\, d\omega = \boldsymbol{h}^T\left(\int_{\omega\in A_F} \boldsymbol{W}(\omega)\boldsymbol{W}^T(\omega)\, d\omega\right)\boldsymbol{h}. \tag{24}$$

The integral in (24) depends on the filter support (via the definition of functions in $\boldsymbol{W}$) and on the frequency area, both design specifications, but not on vector $\boldsymbol{h}$. Therefore, cost function (24) is of quadratic type and can be expressed as

$$P_F = \boldsymbol{h}^T\boldsymbol{C}_F\boldsymbol{h} \tag{25}$$

where $\boldsymbol{C}_F$ is the matrix having the integral entries as in (24).

A slightly more general form of (24) can sometimes be useful. More precisely, replace the integral in (24) with

$$\int_\omega \rho(\omega)\boldsymbol{W}(\omega)\boldsymbol{W}^T(\omega)\, d\omega \tag{26}$$

where the integral is extended to the whole frequency space and $\rho(\omega)$ is a weight function. Note that the integral in (24) can be written as in (26) with $\rho(\omega) = \chi_{A_F}$.

Since the frequency power distribution is symmetric with respect to the origin, it is a reasonable assumption that $\rho(\omega)$ is symmetric as well.

So far we saw that frequency localization can be expressed as a quadratic function of vector $\boldsymbol{h}$. Of course, the same reasonings can be repeated for time localization. For example, we could ask for good time localization of vector $\mathbf{h}_1$ between 0 and 4. This can be expressed as the minimization of the power between 5 and 8

$$P_T = \sum_{n=5}^{8} \mathbf{h}_1^2[n] = \mathbf{h}_1^T \begin{bmatrix} \mathbf{0}_4 & \\ & \mathbf{I}_4 \end{bmatrix} \mathbf{h}_1 \tag{27}$$

where $\mathbf{I}_4$ and $\mathbf{0}_4$ are, the identity and the $4 \times 4$ null matrices respectively.

In general, we could express time localization in a way similar to (24), as

$$\sum_n \rho[n]\boldsymbol{h}^2[n] \tag{28}$$

where $\rho$ is a weight function whose meaning is similar to $\rho$ in (26). Then, (27) becomes

$$P_T = \boldsymbol{h}^T \boldsymbol{C}_T \boldsymbol{h}, \tag{29}$$

with $\boldsymbol{C}_T$ a diagonal matrix formed with the values of $\rho$.

If we want to perform joint time-frequency localization optimization we can sum the two cost functions (24) and (29) to obtain

$$\boldsymbol{h}^T(\boldsymbol{C}_F + \boldsymbol{C}_T)\boldsymbol{h} = \boldsymbol{h}^T\boldsymbol{C}\boldsymbol{h} \tag{30}$$

that is still a quadratic cost function. It is worth observing that the two costs (temporal and frequency) can be weighted in different ways.

This reasoning can be repeated to obtain, for each basis vector, a corresponding cost function represented by its matrix $\boldsymbol{C}_i$. Summarizing, our cost function is

$$C_i(\boldsymbol{h}_i) = \boldsymbol{h}_i^T \boldsymbol{C}_i \boldsymbol{h}_i \tag{31}$$

where $\boldsymbol{C}_i$ is the "total" cost matrix, accounting for frequency and temporal localizations, each one with its own weight.

Function (31) is the cost function relative to a *single* vector. If we want a global optimization for the whole basis we can simply sum all of the cost functions in order to obtain the overall cost function. Thus, our problem becomes that of finding a basis $\{\boldsymbol{h}_1, \ldots \boldsymbol{h}_N\}$, such that

$$\sum_{i=1}^{N} \boldsymbol{h}_i^T \boldsymbol{C}_i \boldsymbol{h}_i \tag{32}$$

is minimized. This can be solved in several ways, described in the next section.

**3.2. Solving for the basis.** Here, we will describe ways in which one can solve (32). A key observation is that if vectors $\boldsymbol{h}_i$ are orthogonal to one another, they can be organized as columns of a unitary matrix $\boldsymbol{U} \triangleq [\boldsymbol{h}_1\ \boldsymbol{h}_2\ \ldots\ \boldsymbol{h}_N]$. Such a fact is important because matrix $\boldsymbol{U}$, even if not square, can be written as a function of some free parameters. Thus, we can parameterize this matrix and then extract the vectors $\boldsymbol{h}_i$, now given as functions of free parameters. A unitary matrix $\boldsymbol{U}$ can be written either using Givens' rotations

$$\boldsymbol{U} = \boldsymbol{R}_1\boldsymbol{R}_2 \ldots \boldsymbol{R}_M\mathcal{D}_G, \tag{33}$$

where $\boldsymbol{R}_k$ are suitable rotation matrices and $\mathcal{D}_G$ is a diagonal matrix having $d_{ii} = \pm 1$, or using Householder building blocks

$$\boldsymbol{U} = \boldsymbol{P}_1\boldsymbol{P}_2 \ldots \boldsymbol{P}_{N-1}\mathcal{D}_H, \tag{34}$$

where each matrix $\boldsymbol{P}_k$ is defined as

$$\boldsymbol{P}_k \stackrel{\Delta}{=} \boldsymbol{I} - 2 \cdot \boldsymbol{u}_k\boldsymbol{u}_k^T, \tag{35}$$

with $\boldsymbol{u}_k$ a vector having the first $k-1$ components equal to zero such that $\|\boldsymbol{u}_k\| = 1$. It is worth noting that matrices $\mathcal{D}_G$ and $\mathcal{D}_H$ have no effect on the value of (32) because they simply correspond to a change of sign for some vector $\boldsymbol{h}_i$; therefore, one can assume $\mathcal{D}_G = \mathcal{D}_H = \boldsymbol{I}$.

3.2.1. *Closed-form solution.* This method, although completely general, is feasible only for small vector spaces and maybe with the help of a program capable of doing computation with polynomials. The idea is to use decomposition (34). By observing (35) it is clear that the entries of each matrix $\boldsymbol{P}_k$ in (34) can be written as polynomials of degree two in the nonzero components of $\boldsymbol{u}_k$. Therefore, $\boldsymbol{U}$, being a product of polynomial matrices, is a polynomial matrix and cost function (32) can be expressed as a polynomial of the free parameters. Bearing this in mind, our problem becomes that of finding the minimum value assumed by a given polynomial on a multidimensional interval and it can be solved by classical analytical methods. Such a polynomial will have $N(N-1)/2$ free variables and its degree will be $4(N-1)$. For example, if $N = 10$ we have a polynomial of degree 36 with 45 variables. The gradient of such a polynomial is a vector having as components 45 polynomials of degree 35. Because of this, the closed-form solution is feasible only for small values of $N$. The fact that we can look at $\boldsymbol{U}$ as a polynomial matrix, however, can have theoretical relevance since it gives us strong information about the structure of (32).

3.2.2. *Iterative solution.* If a closed-form solution is not feasible one can always exploit decompositions (33) and (34) to minimize (32) by using a program for unconstrained optimization. This solution is applicable to problems that are too big for the closed-form approach. It is worth remembering that the number of free variables is always $N(N-1)/2$, that is, quadratic with the problem size. This means that if $N = 100$ we have a problem with almost $5,000$ variables.

3.2.3. *A simpler solution.* A simpler, even if suboptimal, approach that works also for bigger problems is to minimize each term of (32) separately, which in turn is obtained by finding the maximum eigenvalue of the matrix $\boldsymbol{C}_i$ and its corresponding eigenvector.

The resulting set of vectors is not guaranteed to be an orthonormal basis, but can be orthogonalized. An interesting way to do so is to write the vectors obtained from the minimization of each term of (32) as the columns of a matrix $\boldsymbol{A}$ and decompose it, with the singular values decomposition, as $\boldsymbol{A} = \boldsymbol{OSQ}$ where $\boldsymbol{O}$ and $\boldsymbol{Q}$ are orthogonal matrices and $\boldsymbol{S}$ is a diagonal matrix. The columns of $\boldsymbol{OQ}$ are orthogonal to one another and enjoy the property that they are the orthonormal basis having the minimum distance from the original set of vectors. However, this procedure does not guarantee that the result will be optimum. Note, though, that it is reasonable to suppose that if the given tiles are disjoint, the vectors obtained in the first step should be nearly orthogonal. Indeed, in all the examples we tried, the two solutions, optimum and orthogonal, are usually very close to one another. Note that such a distance can be estimated from the singular values of $\boldsymbol{A}$.

**3.3. Critical-band basis for audio coding.** The problem we want to attack now is complex and will be governed by the needs of an audio coder. In [**14**], Johnston lists a few of the interactions between the human auditory system and the filter banks that need to be used, among them: time-domain effects such as pre-echo and stereo imaging effects, frequency-domain artifacts such as unmasked aliasing and large overcoding requirements due to the filter bank bandwidth and coding gain problems, such as lack of sufficient coding gain as well as loss of coding gain due to pre-echo or nonstationarity issues. Many of these issues have been addressed in current filter banks designs. However, many are still left open, and we will try to address a few of them. "The primary need in audio coding is the ability to flexibly change the time-frequency tiling of the filter banks, in a way dictated by the joint understanding of the signal and the human auditory system, so that a joint minimum in bit rate can be reached for each kind of audio signal ... The key in such a filter bank is the flexibility of time and frequency tiling, in the sense that frequency portions of the signal that are highly varying may require a time resolution matching that of the critical band, and portions that are highly stationary may require very, very high frequency resolution, both in the same frame of the coded signal" [**14**]. Johnston further states that a filter bank that switches between the critical-band and the high-resolution states has to have a certain signal and frequency meaning. Thus, an allpass filter bank would not do the job. Moreover, an orthonormal filter bank is highly desirable. Finally, the resulting filter bank has to be of sufficiently low complexity so that it can be implemented.

Our technique lends itself well to the above requirements. First, since our scheme is window based, it is easy to switch from the uniform mode to a critical-band mode (since these transitions depend only on the window). This further implies that the transition filter bank will keep its frequency meaning. Due to the fact that our approach is based upon ((19)) and if we design $\mathbf{G}$ so that it has a fast algorithm for implementation, the resulting filter bank will not be overly complex. In the rest of this section we will describe a way in which we designed a critical-band filter bank (the uniform filter bank is the local cosine basis filter bank, or, the MDCT).

3.3.1. *Reducing complexity by using translations.* We want to design a critical-band filter bank that will follow (or precede) a 1024 uniform bank. Therefore, the size of our basis is $N = 1024$. This means that we will have 1024 basis functions tiling our given space in the time-frequency plane. Such a problem is inherently big and even the approximate method would require too much computational time. Thus, we use the eigenvalue approach described in the previous section.

Figure 5 shows the time-frequency tiling used in this case. We have 25 frequency bands. Intuitively, it is clear that each filter in the same frequency band has to be some kind of a translation of the first one. Thus, in order to reduce the computational complexity of the corresponding filter bank, the filters corresponding to the same frequency band are constrained to be circular translation one another.

Therefore, as a first step, we can impose that these filters be obtained by linear transformations of the first filter, that is, for band 8, for example, $\boldsymbol{h}_{i_8} = \boldsymbol{L}_{i_8}\boldsymbol{h}_8$,

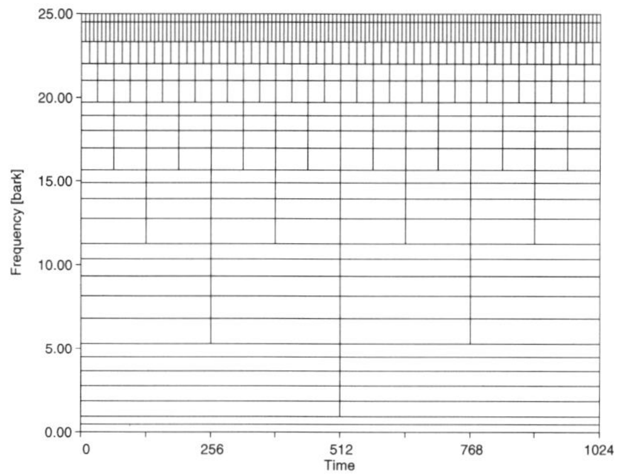


FIGURE 5. Tiling of the time–frequency plane in the case of audio coding. Frequency axis is given in barks where $B = 13.0 \arctan(0.76f/1000) + 3.5 \arctan(f/7500)^2$, with $B$ being bark frequency and $f$ frequency in Hz.

where $\boldsymbol{L}_{i_8}$ is a linear transformation describing some kind of a translation. This can be used in (32) to obtain a simpler form. Indeed, by rewriting (32) one obtains

$$\sum_i \boldsymbol{h}_{i_8}^T \boldsymbol{C}_{i_8} \boldsymbol{h}_{i_8} = \boldsymbol{h}_8^T \Big( \sum_i \boldsymbol{L}_{i_8}^T \boldsymbol{C}_{i_8} \boldsymbol{L}_{i_8} \Big) \boldsymbol{h}_8. \tag{36}$$

By repeating this process, cost function (32) can be rewritten as a function of the "primitive" vectors (25 in our case), leading to a simpler problem

$$\sum_{j=1}^{25} \boldsymbol{h}_j^T \Big( \sum_i \boldsymbol{L}_{i_j}^T \boldsymbol{C}_{i_j} \boldsymbol{L}_{i_j} \Big) \boldsymbol{h}_j. \tag{37}$$

Care has to be exercised when imposing linear dependencies because too difficult a problem can result. For example, in the case of circular translations discontinuities appear (see [**2**] for more details).

Because of the requirement on frequency localization, the cost function is high when the impulse response is smooth. Since these filters belong to a local orthogonal bank given by ((19)), the vectors obtained by the minimization of (32) are extended by symmetry and modulated with a window. Since at the right symmetry point the function is extended by anti-symmetry, the requirement on the final basis function to be smooth implies the requirement on the original vector to be almost zero in the vicinity of such a point. However, if this is true, when the filter is translated we obtain a discontinuity. By the same token, if the circular translation is smooth we have a discontinuity in the first filter. Therefore, the optimization procedure has to accommodate two incompatible constraints, introducing the discontinuities as mentioned before.

A possible solution to this problem is to search for some kind of "pseudo-translations" with the following requirements: they are approximate translations, they are periodic and they do not introduce discontinuities as we mentioned before. A way of constructing these "pseudo-translations" is described in [**2**]. It is based on the fact that we are using local bases for our construction, that is, the final basis function is always obtained by extending by symmetry and windowing the starting basis function. Due to this, the continuity of the final basis function descends from the continuity of the starting basis function and from it being zero at the right symmetry point. Therefore, we will ask for our basis functions to be 0 at that point. Then, by looking at the translation operator expressed as a matrix, we conclude that we have to "truncate" the matrix, so that we obtain a diagonal block matrix. For each block $\boldsymbol{B}$ we then do the following: we diagonalize it as $\boldsymbol{B} = \boldsymbol{A}^{-1}\boldsymbol{D}\boldsymbol{A}$. We approximate each diagonal element of $\boldsymbol{D}$ by a matrix $\widehat{\boldsymbol{D}}$ whose eigenvalues are all $N$th roots of unity (this is needed so that the operator has period $N$). Finally, we reconstruct $\widehat{\boldsymbol{B}}$ as $\widehat{\boldsymbol{B}} \triangleq \boldsymbol{A}^{-1}\widehat{\boldsymbol{D}}\boldsymbol{A}$. For more details, see [**2**].

3.3.2. *Other design considerations.* For the highest-frequency filters time localization was far more important than frequency response. In order to achieve good time localization we weighted differently the frequency and the time error (time error was given a weight of 25, as opposed to 1 for the frequency error).

The first problem is related to pre-echo effects due to the filter symmetry. For example, if the original vector has one burst, by translating it, filters obtain a second burst, "reflection" of the original one. The second burst is particularly annoying in audio coding because it gives rise to pre-echo artifacts, since the listener perceives a copy of the signal before its due time. Unfortunately, such a fact cannot be avoided by choosing the basis because it depends on the symmetries imposed on the filter (theoretically necessary in order to achieve the orthogonality) and on the window shape. Because of this, in order to solve the pre-echo problem, we had to choose a sharper window.

The second problem is that the vectors obtained are not necessarily orthogonal; indeed, because of the strong requirements imposed, they are nonorthogonal. Moreover, even if weighting the cost function allowed us to obtain filters better localized in time, such localization was not enough for audio applications. In order to overcome the orthogonality problem, the basis has been orthogonalized via the singular value decomposition that gives us the orthonormal basis having a minimum distance from the original one. In order to improve time localization, the resulting vectors have been forced to zero outside the desired interval in order to obtain a filter having the desired support. The result of such an operation can be seen in Figure 6, where all of the 25 filters are displayed in a $5 \times 5$ matrix. It is worth observing that they are exactly zero outside the desired intervals.

## 4. Designing Local Orthogonal Bases on General Groups with Applications to Evaluating Image Quality

Many image processing tasks rely on measures of image quality. Until now, objective evaluations used mainly the mean-squared error, while subjective evaluations were used to get a more precise idea about the perceptual quality. Recently, this topic has attracted some attention, including [**23, 15, 26, 9**]. Several approaches to subjective distance of images rely on a weighted measure of the error in frequency, with the justification that the visual system is lowpass and high-frequency errors will be less noticeable (thus lowpass parts will have more weight). Sometimes the error is processed with a log-like nonlinear function. However, this approach works better for audio than for images since one could argue that, while perception of audio is in the "frequency domain", perception of images is in the "spatial domain".

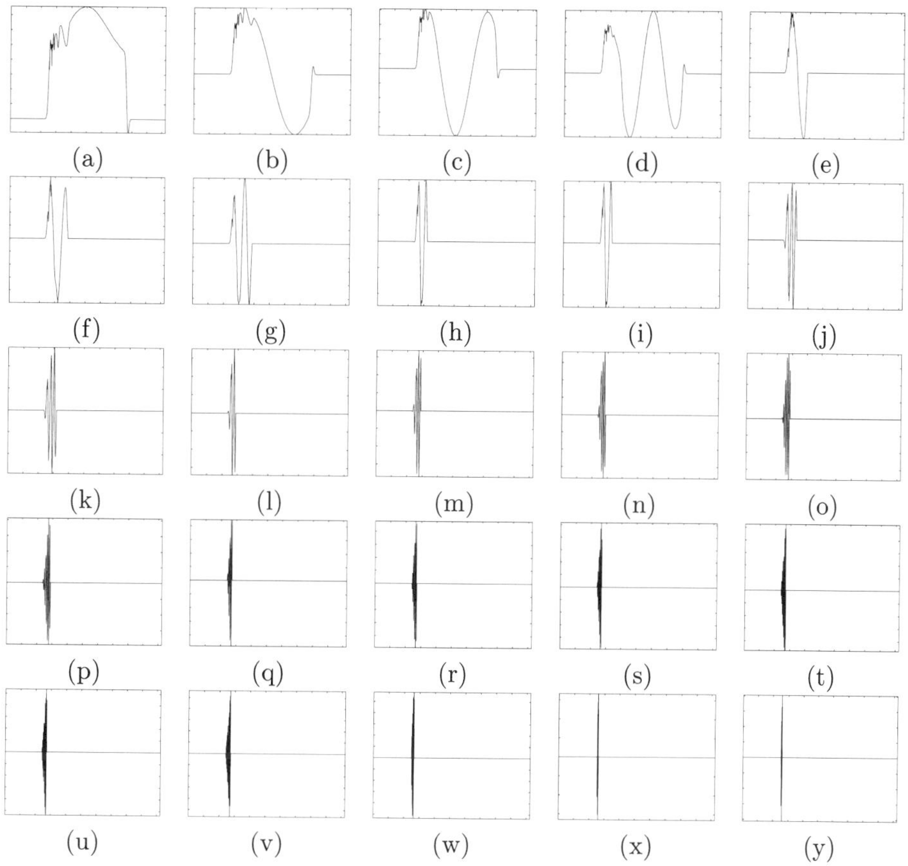


FIGURE 6. Filters forming the critical-band filter banks (25 of them). The rest of the basis functions $(1024 - 25)$ are obtained by pseudo-translations and circular translations (in the case of higher-frequency filters) of the 25 critical-band filters. To solve the problem of pre-echo, a very sharp window was used. The filters were then orthogonalized and truncated to allow for the desired time localization.

Indeed, people are so aware of the spatial structure of the noise that sometimes they recognize it and give it a name such as "blocking effect", "ringing", "blurring" and so on. Another important point is that some types of noise are more disturbing than other. For example, a slight blurring or a slight scaling is less noticeable and less annoying than the blocking effect. A subjective measure of image quality has to account for this and such a goal can be accomplished only by considering the image from a more "high-level" perspective.

The main idea we want to pursue is that in an image we can recognize some "structures". It is widely believed that the more "structured" noise is the more annoying it is. Some physiological results point out that the early stage of the human visual system (HVS) works like a filter bank on the retinal image. The filters of such a bank can be seen as obtained by rotation and modulation from an original prototype filter. Other theories suggest that, starting from the response of the first low-level stage, the visual system performs some higher-level processing by recognizing "regularities" in the filter bank output.

One could find a measure of subjective quality of an image in a "structure domain". This problem is very important and would give a means of comparing coders and other visual products. This involves the following steps: By using the technique developed for the design of local orthogonal bases, project a filter bank made of modulations and rotations of some prototype filter. With such a filter bank analyze some images and figure out how the structures reflect themselves in the output. Figure out the characteristics of "bad" structures, that is, structures associated with noise and/or artifacts. Write a software capable of recognizing the image structures from the filter bank response. By comparing the output of such a software with the results of psychophysical experiments, figure out a function mapping the "structure content" of an image into a "physiological quality score".

The work presented in this section and [**3, 4**], is just the first step and it concentrates on the design of local bases obtained by modulations and rotations. This involves extending the relation:

$$\langle f(t), f(t-n)\rangle = \delta_n \iff \sum_{k=-\infty}^{+\infty} \left|F(\omega + 2\pi k)\right|^2 = 1, \tag{38}$$

which allows us both to test the orthogonality of $f(t)$ with respect to its integer translations and to produce functions satisfying such a property. Mostly, this relation has been used to orthogonalize existent nonorthogonal bases. This is performed by dividing the basis functions in the frequency domain by the normalizing factor, that is, square root of the sum given above. A more general case is if instead of the group of integer translations, we could have a similar relation for more general groups. Our particular interest is in the group of rotations and modulations.

**4.1. Problem definition.** Our aim here is to design the starting basis $\boldsymbol{G}$ consisting of modulations and rotations. To that end, we want to extend (38) to a more general case.

Consider a vector space $\mathcal{V}$, with an inner product $\langle\cdot,\cdot\rangle$, and let $\Gamma$ be a group of unitary transforms of $\mathcal{V}$, that is, $\Gamma$ is a set of linear transforms such that for each $U \in \Gamma$ and $\boldsymbol{b}, \boldsymbol{c} \in \mathcal{V}$, $\langle U\boldsymbol{b}, U\boldsymbol{c}\rangle = \langle \boldsymbol{b}, \boldsymbol{c}\rangle$.

Our aim is, given a vector $\boldsymbol{b} \in \mathcal{V}$, to find some necessary and sufficient condition that are to be satisfied in order for the set $\mathcal{O} = \{\boldsymbol{b}\colon \boldsymbol{b} = U\boldsymbol{b}, U \in \Gamma\}$ to be an orthonormal set of vectors. It is worth observing that we are not asking if such a set is a basis, rather we just want to know whether it is an orthonormal set; if vector space $\mathcal{V}$ has finite dimension, then we know that an orthonormal set is a basis if and only if its cardinality equals the dimensionality of $\mathcal{V}$.

We can put the condition of the orthogonality of $\mathcal{O}$ in a form closer to (38) as

$$\langle \boldsymbol{b}, U\boldsymbol{b}\rangle = \begin{cases} 1 & U = \mathcal{I}, \\ 0 & \text{otherwise.} \end{cases}$$

This can be written in a more compact form by defining, analogously to the usual Kronecker delta,

$$\delta_U = \begin{cases} 1 & U = \mathcal{I}, \\ 0 & \text{otherwise.} \end{cases}$$

With such a notation the left-hand side of (38) becomes

(39) $$\langle \boldsymbol{b}, U\boldsymbol{b}\rangle = \delta_U, \quad U \in \Gamma.$$

We will say that the vector $\boldsymbol{b}$ is *orthogonal with respect to the action of* $\Gamma$ if it satisfies (39).

Our aim is to prove something of the type

$$\langle \boldsymbol{b}, U\boldsymbol{b}\rangle = \delta_U \iff ??? = 1$$

where "???" will be a (still to be found) function of $\boldsymbol{b}$. Since the couple $(\delta, 1)$ usually forms a Fourier transform pair, one can think of using some sort of a Fourier transform to find "???".

Note that here, we will be using concepts from the theory of groups, representations and harmonic analysis on groups. For results and details on representations on finite groups, consult the excellent text by Serre [**21**], or the Appendix in [**3**]. For results on how to use harmonic analysis on fairly general groups, see [**11**] or the Appendix in [**3**].

**4.2. How it works on Abelian groups?** If $\Gamma$ is an abelian group, it is well-known that every irreducible representation $\pi_\omega$ is one dimensional, or, equivalently, it is a homomorphism mapping each $U \in \Gamma$ into a complex number with a unitary modulus, $\pi_\omega(U)$. Then, the Fourier transform of a function $h(U)$, defined on $\Gamma$, is a function $H$ defined on the set of irreducible representations of $\Gamma$, usually denoted as $\widehat{\Gamma}$. Function $H(\omega)$ is

(40) $$H(\omega) = \sum_{U\in\Gamma} h(U)\pi_\omega(U).$$

We can now find the Fourier transform of signal $\varphi(U) = \langle \boldsymbol{b}, U\boldsymbol{b}\rangle$ as

$$\Phi(\omega) = |\Gamma|\langle \boldsymbol{b}, \mathcal{U}_\omega \boldsymbol{b}\rangle,$$

where matrices $\mathcal{U}_\omega$ are defined as

$$\mathcal{U}_\omega \triangleq \frac{1}{|\Gamma|}\sum_{U\in\Gamma} U\pi_\omega(U).$$

We can also find the Fourier transform pair of $\delta_U$ as 1, yielding finally the orthogonality condition in the frequency domain, and consequently, the counterpart of (38)

(41) $$\langle \boldsymbol{b}, U\boldsymbol{b}\rangle \iff |\Gamma|\langle \boldsymbol{b}, \mathcal{U}_k \boldsymbol{b}\rangle = 1.$$

It is important to note here that matrices $\mathcal{U}_\omega$ are projections orthogonal to one another, that is,

(42) $$\begin{aligned} \mathcal{U}_\omega\mathcal{U}_{\omega_1} &= 0, \quad \text{if } \omega \neq \omega_1, \\ 11\ \mathcal{U}_\omega\mathcal{U}_\omega &= \mathcal{U}_\omega \\ \mathcal{U}_\omega^T &= \mathcal{U}_\omega. \end{aligned}$$

Due to this, if

(43) $$\{\mathcal{U}\boldsymbol{b} \mid U \in \Gamma\},$$

is not orthonormal, it can be orthogonalized using

(44) $$\hat{\boldsymbol{b}} \triangleq \sum_{\omega\in\widehat{\Gamma}} \mathcal{U}_\omega \boldsymbol{b}\frac{1}{\sqrt{\Phi(\omega)}}.$$

It is easy to show [**3**] that the set

$$\{\mathcal{U}\hat{\boldsymbol{b}} \mid U \in \Gamma\}$$

is an orthonormal set. Moreover, the resulting vector $\hat{\boldsymbol{b}}$ is the vector orthogonal with respect to the action of $\Gamma$ such that $\|\boldsymbol{b} - \hat{\boldsymbol{b}}\|$ is minimum.

In [**3**] we also show that the vector space spanned by vectors $\{\mathcal{U}\hat{\boldsymbol{b}} \mid U \in \Gamma\}$ is equal to the vector space spanned by $\{\mathcal{U}\boldsymbol{b} \mid U \in \Gamma\}$. Another interesting point is that the space associated with $\mathcal{U}_\omega$ is $\Gamma$-invariant. With this property we can rephrase (44) as

> In order to orthogonalize vector $\boldsymbol{b}$ with respect to the action of $\Gamma$, project $\boldsymbol{b}$ on the $\Gamma$-invariant subspaces (left product with $\mathcal{U}_\omega$), orthogonalize the resulting vectors (right product with $1/\sqrt{\Phi(\omega)}$) and recombine them.

Another way to state the same fact is to say that each system that is orthogonal with respect to the action of $\Gamma$ can be decomposed as the sum of orthonormal systems belonging to the invariant subspaces of $\Gamma$.

Note that (44) is based on the fact that $\Phi(\omega)$ is nonnegative. If $\Phi(\omega)$ is zero, one can show that this is equivalent to $\boldsymbol{b}$ belonging to one of the auto-spaces and thus $U\boldsymbol{b}$ cannot be orthogonalized to span the other auto-spaces as well.

Note also that one could use more than one prototype vector. This will give rise to vector sets explained in [**3**]. All the results can be extended to that case as well. For more details, see [**3**].

**4.3. Orthonormal sets obtained from nonabelian groups of linear transformations.** Our discussion in the previous section on how to build orthonormal sets of vectors obtained by applying abelian groups of unitary linear transformations is far away from being completely general. Indeed, the group of rotations and modulations that triggered this work is not abelian; thus, the previous theory has to be modified in order to be able to work with such a group. The main problem that arises with noncommutative groups is that the representations are no longer one dimensional and this will give rise to some technical difficulties. Let us now see how the reasonings in the previous section have to be modified for nonabelian groups. If the group is not commutative, an irreducible representation, in general, will map an element $U \in \Gamma$ into a matrix $\pi_\omega(U)$; we will denote the element $(i,j)$ of the matrix $\pi_\omega(U)$ as $\pi_\omega^{ij}(U)$. Because of this, the Fourier transform becomes a function with three indices

$$\Phi(\omega, i, j) \triangleq \sum_{U \in \Gamma} \varphi(U)\pi_\omega^{ij}(U), \tag{45}$$

where $\omega$ determines the representation and $i$ and $j$ are the row and column indices of $\pi_\omega^{ij}(U)$. We want to apply (45) to the condition of $\{U\boldsymbol{B} \mid U \in \Gamma\}$ being an orthonormal system, that is, on

$$\langle \boldsymbol{B}, U\boldsymbol{B} \rangle = \delta_U \boldsymbol{I}. \tag{46}$$

Thus, multiply both sides of the above equation by $\pi_\omega^{ij}(U)$ and sum over $U \in \Gamma$, yielding

$$\Phi(\omega, i, j) \triangleq \left\langle \boldsymbol{B}, \sum_{U \in \Gamma} U\pi_\omega^{ij}(U)\boldsymbol{B} \right\rangle = \sum_{U \in \Gamma} \boldsymbol{I}\pi_\omega^{ij}(U)\delta_U. \tag{47}$$

Define

$$\mathcal{U}_\omega^{ij} \triangleq \frac{\dim(\pi_\omega)}{|\Gamma|} \sum_{U \in \Gamma} U \pi_\omega^{ij}(U). \tag{48}$$

Remember that while $\pi_\omega^{ij}$ are scalars, $\mathcal{U}_\omega^{ij}$ are matrices, of the same degree as $U$. By using (48) in (47), and remembering that $\pi_\omega(\mathcal{I}) = \boldsymbol{I}$ for each $\omega$, one obtains

$$\frac{|\Gamma|}{\dim(\pi_\omega)} \langle \boldsymbol{B}, \mathcal{U}_\omega^{ij} \boldsymbol{B} \rangle = \overline{\boldsymbol{I} \pi_\omega^{ij}(\mathcal{I})} = \delta_{i-j} \boldsymbol{I}, \tag{49}$$

that is,

$$\langle \boldsymbol{B}, \mathcal{U}_\omega^{ij} \boldsymbol{B} \rangle = \frac{\dim(\pi_\omega)}{|\Gamma|} \boldsymbol{I} \delta_{i-j}. \tag{50}$$

As in the commutative case (50) is our "analysis tool"; now we want to find the corresponding "orthonormalization tool". The noncommutative case does not have big conceptual differences from the commutative one and we will limit ourselves to giving the algorithm and verifying its correctness.

A major role in the proof will be played by the following property, generalization of (42), which we give here without proof:

PROPERTY 4.1. *If matrices $\mathcal{U}_\omega^{ij}$ are as in* (48), *then*

$$\mathcal{U}_{\omega_1}^{i_1 j_1} \mathcal{U}_{\omega_2}^{i_2 j_2} = \delta_{j_1 - i_2} \delta_{\omega_1 - \omega_2} \mathcal{U}_{\omega_1}^{i_1 j_2}, \quad (\mathcal{U}_\omega^{ij})^T = \overline{\mathcal{U}_\omega^{ji}}. \tag{51}$$

From (51) it is easy to see that $\mathcal{U}_\omega^{ii}$ are projections, therefore it makes sense to speak about a basis of the space $\mathcal{V}_\omega^i$ associated with $\mathcal{U}_\omega^{ii}$.

To obtain a vector set $\widehat{\boldsymbol{B}}$ with $N$ vectors which are orthogonal with respect to the action of $\Gamma$, proceed as follows:

ALGORITHM 1.

1. For each $\omega$, find an orthonormal basis of $\mathcal{V}_\omega^1$ and extract from it $\dim(\pi_\omega)$ blocks, $\bar{B}_{\omega,1}, \bar{B}_{\omega,2}, \ldots, \bar{B}_{\omega,\dim(\pi_\omega)}$, each one consisting of $N$ vectors. Of course, the dimension of $\mathcal{V}_\omega^1$ must be greater or equal than $\dim(\pi_\omega)N$. (This might seem an unnecessary restriction at this point, but is perfectly reasonable, as will be demonstrated shortly).
2. Compute

$$\widehat{\boldsymbol{B}} - \sum_{\omega} \sum_{j=1}^{\dim(\pi_\omega)} \mathcal{U}_\omega^{j1} \bar{B}_{\omega,j}. \tag{52}$$

3. Vector set $\widehat{\boldsymbol{B}}$ is orthonormal with respect to the action of $\Gamma$.

The proof can be found in [**4**].

In the first step of Algorithm 1 we required that the dimension of $\mathcal{V}_\omega^1$ be greater than $\dim(\pi_\omega)N$. Let us think about this requirement for a moment:

1. As is shown in [**4**], vector spaces $\mathcal{V}_\omega^j$ have the same dimension that has to be greater that $\dim(\pi_\omega)N$.
2. Therefore, the vector space $\mathcal{V}_\omega$ associated with $\mathcal{P}_\omega + \sum_j \mathcal{U}_\omega^{jj}$, being direct sum of $\mathcal{V}_\omega^j$, has dimension at least $\dim(\pi_\omega)^2 N$.

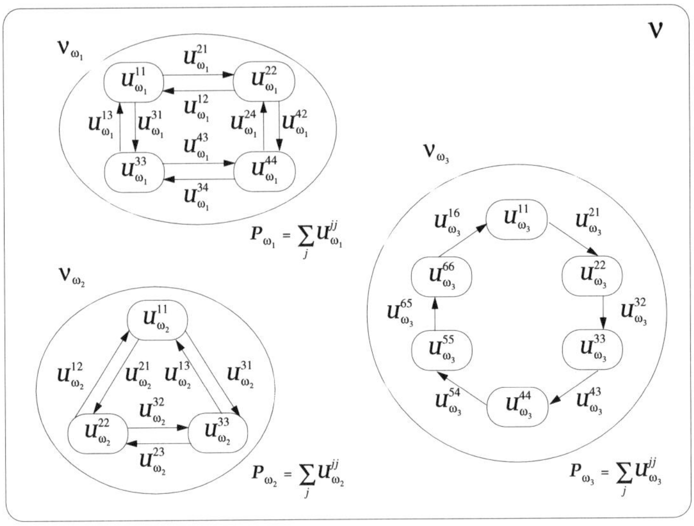


FIGURE 7. Subdivision of the vector space $\mathcal{V}$. Note that not all the branches are shown in order not to clog the figure.

3. Finally, vector space $\mathcal{V}$, being direct sum of the vector spaces $\mathcal{V}_\omega$, has dimension at least

$$\sum_\omega \dim(\pi_\omega)^2 N = N \sum_\omega \dim(\pi_\omega)^2. \tag{53}$$

4. A theorem of group theory claims that the sum on the right-hand side of (53) is equal to the cardinality of $\Gamma$, that is, the dimension of $\mathcal{V}$ has to be at least

$$\dim \mathcal{V} \geq |\Gamma| N. \tag{54}$$

5. If we apply $\Gamma$ to a vector set $\boldsymbol{B}$ of dimension $N$, we obtain $|\Gamma|N$ vectors. Furthermore, if we want these vectors being orthogonal to one another (which implies linear independence) condition (51) is obvious.

The first step of Algorithm 1 requires one to choose a set of vectors belonging to $\mathcal{V}_\omega^1$ and orthogonal to one another. Since $\mathcal{U}_\omega^{11}$ is the projection associated with $\mathcal{V}_\omega^1$ one can simply extract some linearly independent columns from $\mathcal{U}_\omega^{11}$ and orthogonalize them. We can see the composition of the vector space $\mathcal{V}$ in Figure 7.

**4.4. Filter design.** The theory presented in the previous sections concerns a technique to find a set orthonormal with respect to a transformation group $\Gamma$. In particular, we will be searching for bases with local orthogonal structure [**5, 6**].

In the introduction we suggested that a filter bank "invariant" under the action of a given group can be useful to analyze an image and to find its local symmetries and regularities. The local orthogonal basis technique allows us to design a multidimensional filter bank starting from an orthonormal basis. If such a basis is designed invariant with respect to a group action we can obtain a set of filters that can be used for this purpose.

Therefore, we will ask of $\boldsymbol{G}$ to be a matrix whose columns are obtained by applying a transformation group $\Gamma$ to a single vector $\boldsymbol{g}$, that is,

$$\boldsymbol{G} = [\boldsymbol{g}\ U_1\boldsymbol{g}\ U_2\boldsymbol{g}\ \ldots] \tag{55}$$

with $U_n \in \Gamma$. More generally we could ask that the column of $\boldsymbol{G}$ are obtained with the action of $\Gamma$ on two or more vectors. For the sake of simplicity, we will limit ourselves to the case with only one vector. With this position, the filters of the designed filter bank will have impulse responses

$$\mathcal{L}\boldsymbol{g},\ \mathcal{L}U_1\boldsymbol{g},\ \mathcal{L}U_2\boldsymbol{g}, \ldots,$$

with $U_n \in \Gamma$ and $\mathcal{L} = \boldsymbol{W}\boldsymbol{K}$.

It is common practice to design filters optimizing some measure of frequential and/or temporal error. Similarly to Section 3.1 this kind of error can be expressed as a quadratic function of each filter $\mathcal{L}U\boldsymbol{g}$, that is, the cost function has the form

$$(\mathcal{L}U\boldsymbol{g})^T\boldsymbol{C}_U\mathcal{L}U\boldsymbol{g}, \tag{56}$$

where $\boldsymbol{C}_U$ is an opportune square matrix, eventually dependent on $U$, that we can assume is symmetric (otherwise we can replace it with $(\boldsymbol{C}_U + \boldsymbol{C}_U^T)/2$) and positive semidefinite (otherwise the error could assume a negative value and this would not make sense).

So far we have the set of cost functions (56), but we need a single cost function to carry out the optimization. To that end, sum functions (56) to obtain

$$\sum_{U\in\Gamma}(\mathcal{L}U_{\boldsymbol{g}})^T\boldsymbol{C}_U\mathcal{L}U\boldsymbol{g} = \left(\sum_{U\in\Gamma}U^T\mathcal{L}^T\boldsymbol{C}_U\mathcal{L}U\right)\boldsymbol{g} = \boldsymbol{g}^T\boldsymbol{C}\boldsymbol{g}, \tag{57}$$

where $\boldsymbol{C} \triangleq \sum_{U\in\Gamma}U^T\mathcal{L}^T\boldsymbol{C}_U\mathcal{L}U$. Note that (57) could become a weighted sum by multiplying each term by a real number. Such a number can successively be "absorbed" in the corresponding cost matrix, as we will assume in the following. It also shows that the global cost function can be expressed as a quadratic cost of the single vector $\boldsymbol{g}$, yielding a simpler problem.

As seen in the previous sections, vector $\boldsymbol{g}$ is not free since the request that it be orthogonal with respect to the action of $\Gamma$ gives to $\boldsymbol{g}$ a particular structure. Here comes in handy the parameterization that allows us to obtain $\boldsymbol{g}$ as a function of certain free parameters that can be interpreted as certain Givens' rotations. Now we have a simple problem of unconstrained optimization with a quadratic cost function that can be easily solved with numerical methods.

Often the use of a quadratic cost is motivated more by theoretical convenience than by a real cost with a quadratic characteristic. Bearing this in mind, a solution that maybe does not achieve the minimum cost, but is close to the optimal solution can be interesting if, for example, it is simpler to compute. Indeed, to obtain vector $\boldsymbol{g}$ from Givens' rotations is theoretically simple, but it can be computationally intensive and when included in an optimization loop it can give rise to long design times.

To this end we exploit properties telling us that, if we choose our vector in the particular way described earlier, the resulting vector has a minimum distance from the original one.

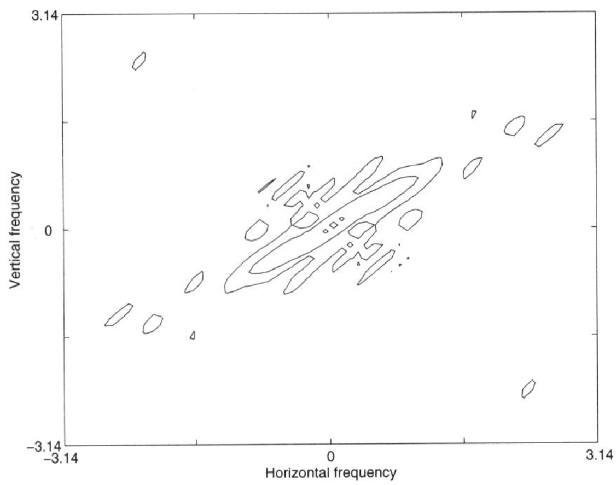


FIGURE 8. Magnitude frequency response of the filter designed which can be used to generate an orthonormal set by rotations and translations of itself. Levels are at 20, 15 and 12 dB.

The idea is to minimize the cost function with the only constraint of $\|\boldsymbol{g}\| = 1$. Such a constraint is necessary because otherwise we could achieve arbitrarily small values by simply scaling $\boldsymbol{g}$. Such a problem has a well-known solution, that is, $\boldsymbol{g}$ is the eigenvector of $\boldsymbol{C}$ having the minimum eigenvalue.

4.4.1. *Example of design.* As an example of filter design in the nonabelian case, we will present briefly the case of rotations and modulations of a two-dimensional filter. For more details, see [**4**].

Let $s(\boldsymbol{n})$, $\boldsymbol{n} \in \mathbb{Z}^2$, be a two-dimensional filter. Its modulated and rotated versions are defined as

$$\begin{aligned} \text{modulated version} &= \exp(-j2\pi\boldsymbol{\omega}_0^T\boldsymbol{n})s(\boldsymbol{n}), \\ \text{rotated version} &= s(\boldsymbol{R}\boldsymbol{n}), \end{aligned} \tag{58}$$

where $\boldsymbol{R}$ is a $2 \times 2$ integer matrix. Note that in (58) one cannot use the usual rotation matrix

$$\boldsymbol{R} = \begin{bmatrix} \cos\alpha_0 & \sin\alpha_0 \\ -\sin\alpha_0 & \cos\alpha_0 \end{bmatrix} \tag{59}$$

because it does not have integer entries. In [**4**] an "approximate rotation" matrix

$$\boldsymbol{R} = \begin{bmatrix} 0 & 1 \\ -1 & 1 \end{bmatrix} \tag{60}$$

is used. Note that $\boldsymbol{R}^6 = \boldsymbol{I}$, so $\boldsymbol{R}$ can thought of as an approximate rotation of $\pi/3$.

It is more convenient to express (58) in frequency domain. By defining the Fourier transform of $s(\boldsymbol{n})$ as

$$S(\boldsymbol{\omega}) \triangleq \sum_{\boldsymbol{n}\in\mathbb{Z}^2} \exp(-j2\pi\boldsymbol{\omega}^T\boldsymbol{n})s(\boldsymbol{n}), \tag{61}$$

(58) can be rewritten as

$$\begin{aligned}\text{Fourier transform of the modulated version} &= S(\boldsymbol{\omega}_0 + \boldsymbol{\omega}),\\ \text{Fourier transform of the rotated version} &= S(\boldsymbol{R}^{-T}\boldsymbol{\omega}).\end{aligned} \tag{62}$$

The resulting group can therefore be described as the group $\Gamma$ of affine transformations mapping $\boldsymbol{\omega}$ into

$$\widehat{\boldsymbol{\omega}} = (\boldsymbol{R}^{-T})^i \boldsymbol{\omega} + \boldsymbol{\omega}_0, \tag{63}$$

where $\boldsymbol{\omega}_0$ belongs to a translation lattice $\Lambda(\boldsymbol{Q}) \supset \mathbb{Z}^2$. Some technical assumptions that are necessary in order to be sure that (63) is a group can be found in [**4**] where the matrix $\boldsymbol{Q} = \mathrm{diag}(1/8, 1/8)$ is used.

An example of filter orthogonal with respect to such a group can be seen in Figure 8.

## 5. Conclusions

We gave an overview of the developments concerning local orthogonal bases in the past couple of years. These bases were constructed trying to obtain more general solutions that would keep all the advantageous properties that their one-dimensional counterparts—local cosine bases—have. These include: construction from a single prototype window, fast implementation algorithm, possibility of constructing time-varying bases.

We gave a general framework for construction and ways to design the prototype window both in discrete and continuous time. As examples we showed how to construct arbitrary tilings of the time-frequency plane using such bases as well as how to use them to construct bases on arbitrary groups such as the group of rotations and modulations which can be used in systems for evaluating image quality.

We believe that the local orthogonal bases are a very flexible tool for building custom bases for many applications in signal processing and that the next few years will see further developments in this area.

## References


1. P. Auscher, G. Weiss, and M. V. Wickerhauser, *Local sine and cosine bases of Coifman and Meyer and the construction of smooth wavelets*, Wavelets, Anal. Appl., vol. 2, Academic Press, Boston, MA, 1992, pp. 237–256.
2. R. Bernardini and J. Kovačević, *Local bases yielding arbitrary tilings of the time-frequency plane*, IEEE Trans. Signal Proc., 1995 (submitted).
3. ———, *Designing local orthogonal bases on groups* I. *Abelian case*, J. Fourier Anal. Appl. (1997) (submitted)
4. ———, *Designing local orthogonal bases on groups* II. *Nonabelian case*, J. Fourier Anal. Appl. (1996) (submitted).
5. ———, *Local orthogonal bases* I. *Construction*, Multidimens. Systems Signal Process. **7** (1996), nos. 3–4, 331–369.
6. ———, *Local orthogonal bases* II. *Window design*, Multidimens. Systems Signal Process. **7** (1996), nos. 3-4, 371–399.
7. R. R. Coifman and Y. Meyer, *Remarques sur l'analyse de Fourier à fenêtre*, C. R. Acad. Sci. Paris **I**, Math. **312** (1991), no. 3, 259–261.
8. R. R. Coifman, Y. Meyer, S. Quake, and M. V. Wickerhauser, *Signal processing and compression with wavelet packets*, Progress in Wavelet Analysis and Applications (Toulouse, 1992), Frontires, Gif-sur-Yvette, 1993, pp. 77–93.
9. S. Daly, *Visual factors in electronic image communications*, An Algorithm for the Assessment of Image Fidelity, MIT Press, 1993.

10. R. L. de Queiroz, T. Q. Nguyen, and K. R. Rao, *The genLOT: Generalized linear phase lapped orthogonal transform*, IEEE Trans. Signal Proc. **41** (1993), no. 12 (special issue on Wavelets and Signal Processing, to appear).
11. G. B. Folland, *A course in abstract harmonic analysis*, Studies in Advanced Mathematics, CRC Press, Boca Raton, FL, 1995.
12. C. Herley, J. Kovačević, K. Ramchandran, and M. Vetterli, *Tilings of the time-frequency plane: Construction of arbitrary orthogonal bases and fast tiling algorithms*, IEEE Trans. Signal Processing **41** (1993), no. 12, 3341–3359 (special issue on Wavelets and Signal Processing).
13. A. W. Johnson, J. Princen, and M. H. Chan, *Frequency scalable video coding using the* MDCT, Proc. IEEE Intern. Conf. on Acoustics, Speech, and Signal Processing (Adelaide, Australia), 1994.
14. J. D. Johnston, *Subband and wavelet transforms: Designs and applications*, Filterbank Interactions with Audio Coding (A. Akansu and M. J. T. Smith, eds.), Kluwer, 1995.
15. S. A. Karunasekera and N. G. Kingsbury, *A distortion measure for blocking artifacts in images based on human visual sensitivity*, IEEE Trans. Image Processing **4** (1995), no. 6, 713–724.
16. R. D. Koilpillai and P. P. Vaidyanathan, *New results on cosine-modulated* FIR *filter banks satisfying perfect reconstruction*, IEEE Trans. Signal Processing **40** (1992), no. 4, 770–783.
17. J. Kovačević, D. J. LeGall, and M. Vetterli, *Image coding with windowed modulated filter banks*, Proc. IEEE Intern. Conf. on Acoustics, Speech, and Signal Processing (Glasgow, 1989), pp. 1949–1952.
18. H. S. Malvar, *Signal processing with lapped transforms*, Artech House, Norwood, MA, 1992.
19. Y. Meyer, *Wavelets: Algorithms and applications*, SIAM, Philadelphia, PA, 1993 (Transl. and revised by R. D. Ryan).
20. J. Princen, A. Johnson, and A. Bradley, *Subband transform coding using filter bank designs based on time domain aliasing cancellation*, Proc. IEEE Intern. Conf. and Acoustics, Speech, and Signal Processing (Dallas, 1987), pp. 2161–2164.
21. J.-P Serre, *Linear representations of finite groups*, Springer-Verlag, New York, 1971.
22. T. A. Ramstad and J. P. Tanem, *Cosine-modulated analysis-synthesis filterbank with critical sampling and perfect reconstruction*, Proc. IEEE Intern. Conf. Acoustics, Speech, and Signal Processing (Toronto, 1991), pp. 1789–1792.
23. P. C. Teo and D. J. Heeger, *Perceptual image distortion*, Proc. IEEE Conf. on Image Proc. II (Austin, TX), 1994, pp. 982–986.
24. C. C. Todd, G. A. Davidson, M. F. Davis, L. D. Fielder, B. D. Link, and S. Vernon, AC-3: *Flexible preceptual coding for audio transmission and storage*, Convention of the AES (Amsterdam, 1994).
25. M. Vetterli and J. Kovačević, *Wavelets and subband coding*, Signal Processing, Prentice-Hall, Englewood Cliffs, NJ, 1995.
26. A. B. Watson, *Visually optimal* DCT *quantization matrices for individual images*, Proc. Data Compr. Conf. (Snowbird, 1993), pp. 178–187.
27. Xian-Gen Xia and Bruce W. Suter, *A family of two dimensional nonseparable* Malvar *wavelets*, J. Appl. Comput. Harmonic Analysis **2** (1995), 243–256.
28. ______, *Construction of Malvar wavelets on hexagons*, J. Appl. Comput. Harmonic Analysis **3** (1996), 65–71.

Bell Laboratories, Murray Hill, NJ 07974, USA
*E-mail address*: `jelena@bell-labs.com`

Universita di Padova, I-35131 Padova, Italy
*E-mail address*: `bernardi@dei.unipd.it`

Centre de Recherches Mathématiques
CRM Proceedings and Lecture Notes
Volume **18**, 1999

# On the $L^p$-Lipschitz Exponents of the Scaling Functions

Ka-Sing Lau and Mang-Fai Ma

ABSTRACT. Based on a necessary and sufficient condition for the existence of $L^p$-scaling function as solution of a dilation equation [**LW**], we derive a formula for the $L^p$-Lipschitz exponent. For $p$ an even integer (or for all $p \geq 1$ in some special cases), the formula can be simplified to a computationally efficient matrix form.

## 1. Introduction

A nonzero function $f(x)$ is called a *scaling function* if it satisfies the *two-scale dilation equation*

$$f(x) = \sum_{n=0}^{N} c_n f(2x-n). \tag{1.1}$$

Such functions play significant roles in wavelet theory, constructive approximation theory and fractal geometry. Their existence and regularity, treated as compactly supported $L^2$ or continuous solutions of (1.1), have been studied in great detail (e.g. [**CD,CH,D,DL1,DL2,H,V**] and the references there). In this paper we will continue such investigation for the compactly supported $L^p$-solutions $1 \leq p < \infty$ (notation: $L^p_c$-solutions).

For the $L^2$-case the analysis depends heavily on the Fourier transformation of (1.1): $\hat{f}(\xi) = m_0(\xi/2)\hat{f}(\xi/2)$ where $m_0(\xi) = 1/2 \sum c_n e^{in\xi}$. For the $L^p$-case we rely on the following linear algebraic set-up used by Daubechies and Lagarias [**DL2**], Collela and Heil [**CH**], and independently by Micchelli and Prautzsch [**MP**]: For any $g$ supported by $[0, N]$, we convert $g$ into the vector-valued function

$$\mathbf{g}(x) = \big[g(x), g(x+1), \ldots, g\big(x+(N-1)\big)\big]^t, \quad x \in [0,1] \tag{1.2}$$

and use the right hand side of the dilation equation to construct two $N \times N$ matrices $T_0$ and $T_1$ (see Section 2). Let $\mathbf{v}$ be a $N$-vector, let $\mathbf{f}_0(x) = \mathbf{v}$ be the initial function, and let $\{\mathbf{f}_k(x)\}_{k=0}^{\infty}$ be obtained by iterating $\mathbf{f}_0(x)$ with the matrices $T_0$ and $T_1$ inductively (the *Cascade algorithm*). The limit, if it exists, will be the solution of the dilation equation. A natural question is: is there any difference in choosing the

---

1991 *Mathematics Subject Classification.* Primary: 42C15; Secondary: 41A25.
This is the final form of the paper.

initial vector? Although there are many choices of $\mathbf{v}$ for the iterated sequence to converge, we find that the most special one is the right-eigenvector corresponding to the eigenvalue 2 (notation: 2-eigenvector) of the matrix $T_0+T_1$. We can use this eigenvector to obtain a necessary and sufficient condition for the existence of $L^p_c$-solution ([**LW**] and Theorem 2.2 in the following) and also the smoothness of the solution. In [**J**], Jia studied the same existence question by using a "hat" function as initial function and obtained a slightly different criterion.

To study the regularity of the $L^p_c$-scaling function $f$, we use the *$L^p$-Lipschitz exponent* defined by

$$\mathrm{Lip}_p(f) = \liminf_{h\to 0^+} \frac{\ln \|\Delta_h f\|_p}{\ln h}$$

where $\Delta_h f(x) = f(x+h) - f(x)$, equivalently,

$$\mathrm{Lip}_p(f) = \inf\left\{ s\colon 0 < \limsup_{h\to 0^+} h^{-s}\|\Delta_h f\|_p \right\}.$$

When $p = 2$, $\alpha$ equals the Sobolev exponent of $f$. Recently there is a great interest on the $L^p$-Lipschitz exponent ($L^p$-dimension if measures are concerned) in connection to the multifractal theory (e.g. [**FP,HJ,DL3,J1,2,LN**]). The function $\tau(p) = p\,\mathrm{Lip}_p(f)$ (called the *moment scaling exponent* of $f$) has a very elegant heuristic relationship with $\gamma(\alpha) = \dim_H\{x\colon \lim_{h\to 0^+}(\ln|\Delta_h f(x)|/\ln h) = \alpha\}$, the Hausdorff dimension of the set of all $x$ with local Lipschitz exponent $\alpha$, namely the *multifractal formalism*: the Legendre transformation (equivalently, the concave conjugate) of $\tau(p)$ is $\gamma(\alpha)$. This elegant relationship was proposed by physicists [**FP,HJ**] and attracts a lot of attention from mathematicians. Daubechies and Lagarias [**DL3**] showed that the formalism does not hold for the scaling function $D_4$. Nevertheless it was observed that the failure is due to the nondifferentiability of $\tau(p)$ at $p = 2$ (see Figure 2e) and it raises more interesting questions on the $L^p$-Lipschitz exponent.

For the scaling functions, the $L^2$-Lipschitz exponent has been studied in detail in [**CD,H**] and [**V**] and sharpened in [**LMW**]. In this note we will consider the $L^p$-case. We note that in the proof of existence, the iteration of $\mathbf{f}_k$, starting with the 2-eigenvector $\mathbf{v}$, converges to the solution $\mathbf{f}$ at a geometric rate. Amazingly, this rate actually gives us the $L^p$-Lipschitz exponent of $f$ (Theorem 3.1). The rate is given by a series of matrix products and is therefore not computationally efficient. For some special cases (including those considered in [**DL3**]), we can reduce it to simple formulas. For the more general case we can still calculate the $L^p$-Lipschitz exponent when $p$ is a positive integer. Note that in our consideration we only treat the Lipschitz exponent of order not greater that 1. For higher order of smoothness we need to consider the higher order difference $\Delta_h^k f$.

In the following we give the results and sketch some of the proofs. The detail will appear in [**LM**].

## 2. Existence of $L^p_c$-Scaling Functions

It is known that if (1.1) has an integrable solution, then $\sum c_n = 2^m$, $m > 0$ an integer ($m$ is the order of zero of $\hat{f}(\xi)$ at 0, and $m = 1$ if $\int f \neq 0$) [**DL1**]. We will hence use the natural assumption $\sum c_n = 2$ on the coefficients unless otherwise specified. Let $T_0 = [c_{2i-j-1}]_{1\le i,j\le N}$ and $T_1 = [c_{2i-j}]_{1\le i,j\le N}$, i.e.,

$$T_0 = \begin{pmatrix} c_0 & 0 & 0 & \cdots & 0 \\ c_2 & c_1 & c_0 & \cdots & 0 \\ c_4 & c_3 & c_2 & \cdots & 0 \\ \vdots & \vdots & \vdots & \ddots & \vdots \\ 0 & 0 & 0 & \cdots & c_{N-1} \end{pmatrix}, \quad T_1 = \begin{pmatrix} c_1 & c_0 & 0 & \cdots & 0 \\ c_3 & c_2 & c_1 & \cdots & 0 \\ c_5 & c_4 & c_3 & \cdots & 0 \\ \vdots & \vdots & \vdots & \ddots & \vdots \\ 0 & 0 & 0 & \cdots & c_N \end{pmatrix}.$$

For any $g \in L^p(\mathbb{R})$ with support in $[0, N]$, let $\mathbf{g}(x)$ be the vector-valued function representing $g$ as in (1.2) and let

$$\mathbf{T}g(x) = \begin{cases} T_0\mathbf{g}(2x) & \text{if } x \in [0, 1/2), \\ T_1\mathbf{g}(2x-1) & \text{if } x \in [1/2, 1). \end{cases}$$

It is easy to show that $f$ is a solution of (1.1) if and only if $\mathbf{f} = \mathbf{T}f$. With no confusion, we use $\|\cdot\|$ to denote the $L^p$-norm of $g$ and also the vector-valued function $\mathbf{g}$. For any $k \ge 1$, we let $J = (j_1, \dots, j_k)$, $j_i = 0$ or $1$, denote the multi-index ($J = \emptyset$ if $k = 0$), and let $|J|$ denote the length of $J$. The matrix $T_J$ represents the product $T_{j_1} \cdots T_{j_k}$; $I_J = I_{(j_1,\dots,j_k)} = [a, b)$ denotes the dyadic interval where

$$a = \frac{j_1}{2} + \frac{j_2}{2^2} + \dots + \frac{j_k}{2^k} \quad \text{and} \quad b = a + \frac{1}{2^k},$$

and $g_I$ means the average $|I|^{-1}\int_I g(x)\,dx$ of $g$ on an interval $I$. We first give a necessary condition for (1.1) to have an $L^p_c$-solution [**LW**].

PROPOSITION 2.1. *Assume $\sum_{n=0}^N c_n = 2$. For $1 \le p < \infty$, let $f$ be an $L^p_c$-solution of* (1.1) *and let $\mathbf{v} = [f_{[0,1]}, \dots, f_{[N-1,N]}]^t$ be the vector defined by the average of $f$ on the $N$ subintervals. Then*

(i) *$\mathbf{v}$ is a 2-eigenvector of $(T_0 + T_1)$.*

(ii) *Let $\mathbf{f}_0(x) = \mathbf{v}$, $x \in [0, 1)$, and let $\mathbf{f}_{n+1} = \mathbf{T}f_n$, $n = 0, 1, \dots$, then $\mathbf{f}_n(x) = \sum_{|J|=n}(T_J\mathbf{v})\chi_{I_J}(x)$, $x \in [0, 1)$, and $\mathbf{f}_n \to \mathbf{f}$ in $L^p([0, 1], \mathbb{R}^N)$.*

To look for a criterion for the existence of $L^p_c$-solution, the above proposition suggests that we should concentrate on the 2-eigenvector of $T_0 + T_1$. Let $\mathbf{v}$ be such a vector, then $(T_0 - I)\mathbf{v} = -(T_1 - I)\mathbf{v}$. Let $\tilde{\mathbf{v}} = (T_0 - I)\mathbf{v}$ and let $H(\tilde{\mathbf{v}})$ be the subspace in $\mathbb{R}^N$ spanned by $\{T_J\tilde{\mathbf{v}} \colon J \text{ a multi-index}\}$. Now, let $\{\mathbf{f}_k\}_{k=0}^\infty$ be defined as in the proposition, then $\mathbf{f}_n = \mathbf{f}_0 + \sum_{k=0}^{n-1}(\mathbf{f}_{k+1} - \mathbf{f}_k)$ and

$$\begin{aligned} \|\mathbf{f}_{k+1} - \mathbf{f}_k\|^p &= \frac{1}{2^{k+1}} \sum_{|J|=k} \big\|T_J(T_0 - I)\mathbf{v}\big\|^p + \big\|T_J(T_1 - I)\mathbf{v}\big\|^p \\ &= \frac{1}{2^k} \sum_{|J|=k} \|T_J\tilde{\mathbf{v}}\|^p. \end{aligned} \tag{2.1}$$

If $2^{-k}\sum_{|J|=k}\|T_J\tilde{\mathbf{v}}\|^p \to 0$, it can be shown that it actually converges at a geometric rate, hence $\{\mathbf{f}_n\}$ converges and the limit is the $L^p_c$-solution. Indeed we have the following stronger result [**LW**]:

THEOREM 2.2. *Suppose $\sum_{n=0}^N c_n = 2$ and $1 \le p < \infty$. Then the following statements are equivalent:*

(i) *Equation* (1.1) *has a nonzero $L^p_c$-solution.*

(ii) *There exists a 2-eigenvector* $\mathbf{v}$ *of* $(T_0 + T_1)$ *satisfying*

$$\lim_{n\to\infty} \frac{1}{2^n} \sum_{|J|=n} \|T_J \tilde{\mathbf{v}}\|^p = 0.$$

(iii) *There exist a 2-eigenvector* $\mathbf{v}$ *of* $(T_0 + T_1)$ *and an integer* $\ell \geq 1$ *such that*

$$\frac{1}{2^\ell} \sum_{|J|=\ell} \|T_J \mathbf{u}\|^p < 1 \textit{ for all } \mathbf{u} \in H(\tilde{\mathbf{v}}), \quad \|\mathbf{u}\| \leq 1.$$

We remark that in [**J**], Jia used the initial function $f_0$ defined by

$$f_0(x) = \begin{cases} 1 + x & \text{if } x \in [-1, 0), \\ 1 - x & \text{if } x \in [0, 1), \\ 0 & \text{otherwise} \end{cases}$$

and showed that the necessary and sufficient condition for the $L^p$-convergence of the iteration is the existence of $\ell$ such that

$$\frac{1}{2^\ell} \sum_{|J|=\ell} \|T_J \mathbf{u}\|^p < 1 \text{ for all } \mathbf{u} \in H := \left\{ \mathbf{u} \in \mathbb{R}^N : \sum_{i=1}^N u_i = 0 \right\}, \quad \|\mathbf{u}\| \leq 1.$$

Note that (iii) is a weaker condition and $H = H(\tilde{\mathbf{v}})$ in many cases.

## 3. Regularity of $L^p_c$-Scaling Functions

In this section we assume that $f$ is an $L^p_c$-solution of (1.1). It follows from Theorem 2.2 that $\lim_{n\to\infty} 2^{-n} \sum_{|J|=n} \|T_J \tilde{\mathbf{v}}\|^p = 0$. In fact the convergence is at a geometric rate [**LW**]. Under some slightly stronger conditions, this rate actually gives us the $L^p$-Lipschitz exponent of $f$.

THEOREM 3.1. *Assume* $\sum c_{2n} = \sum c_{2n+1} = 1$ *and* 1 *is a simple eigenvalue of* $T_0$ *and* $T_1$. *Then for* $1 \leq p < \infty$,

$$\operatorname{Lip}_p(f) = \liminf_{n\to\infty} \frac{\ln(2^{-n} \sum_{|J|=n} \|T_J \tilde{\mathbf{v}}\|^p)}{p \ln(2^{-n})}. \tag{3.1}$$

We will give a sketch of the proof. First we need some basic facts concerning the eigenvectors of $T_0$ and $T_1$. Some of them can be found in [**H**]. Let $H = \{\mathbf{u} \in \mathbb{R}^N : \sum_{i=1}^N u_i = 0\}$.

LEMMA 3.2. *Suppose* $\sum c_{2n} = \sum c_{2n+1} = 1$. *Let* $\mathbf{v}$ *be a 2-eigenvector of* $(T_0 + T_1)$, $\tilde{\mathbf{v}} = (T_0 - I)\mathbf{v}$ *and* $H(\tilde{\mathbf{v}})$ *be the subspace spanned by the* $T_J \tilde{\mathbf{v}}$*'s. Then*

(i) $H(\tilde{\mathbf{v}})$ *is a subspace of* $H$.

(ii) 1 *is an eigenvalue of* $T_0$ *and* $T_1$; *let* $\mathbf{v}_0$ *and* $\mathbf{v}_1$ *be the corresponding eigenvectors, then* $T_0 \mathbf{v}_1 = T_1 \mathbf{v}_0$.

(iii) $\mathbf{v}$ *can be decomposed as* $\mathbf{v} = c\mathbf{v}_0 + \mathbf{h}_0 = c\mathbf{v}_1 + \mathbf{h}_1$ *where* $c > 0$ *and* $\mathbf{h}_0$, $\mathbf{h}_1 \in H$; *furthermore if* 1 *is a simple eigenvalue of* $T_0$ *and* $T_1$, *then* $\mathbf{h}_0$, $\mathbf{h}_1 \in H(\tilde{\mathbf{v}})$.

It is easy to see that the assumption $\sum c_{2n} = \sum c_{2n+1} = 1$ implies $\tilde{\mathbf{v}} \in H$ and $H$ is invariant under $T_0$ and $T_1$. Hence (i) follows immediately. Note that 1 is an eigenvalue of the common submatrix $M = [c_{2i-j}]_{1\le i,j\le N-1}$ of $T_0$ and $T_1$. If $\mathbf{a} = [a_1, \ldots, a_{N-1}]^t$ is an 1-eigenvector of $M$, then $\mathbf{v}_0 := [0, a_1, \ldots, a_{N-1}]^t$ and $\mathbf{v}_1 := [a_1, \ldots, a_{N-1}, 0]^t$ are 1-eigenvectors of $T_0$ and $T_1$ respectively, and a direct calculation yields (ii). For (iii) we need only choose $c = \sum_{i=1}^N v_i / \sum_{i=1}^{N-1} a_i$ where $v_i$ are the coordinates of $\mathbf{v}$. To prove the last statement, we note that the assumption implies that $T_0 - I$ restricted on $H$ is bijective; it is hence also bijective on the $(T_0 - I)$-invariant subspace $H(\tilde{\mathbf{v}})$. From the decomposition we have $(T_0 - I)\mathbf{h}_0 = \tilde{\mathbf{v}}$ and hence $\mathbf{h}_0$ must be in $H(\tilde{\mathbf{v}})$. The case for $\mathbf{h}_1$ is the same.

To prove the theorem, we let $\mathbf{v}$ be the 2-eigenvector of $T_0 + T_1$ satisfying condition (ii) of Theorem 2.2, $\mathbf{f}_n(x) = \sum_{|J|=n} (T_J \mathbf{v}) \chi_{I_J}(x)$, and $f_n$ the corresponding function on $\mathbb{R}$ supported by $[0, N]$. For fixed $0 < h < 1/2$, let $n$ be such that $2^{-(n+1)} \le h < 2^{-n}$. We claim that $\|\Delta_h f_n\|$ has order $2^{-n} \sum_{|J|=n} \|T_J \tilde{\mathbf{v}}\|^p$. Indeed, let $E_n = \bigcup_{j=0}^{N-1} [j,\ j+1-2^{-n})$ and $\widetilde{E}_n = [-2^{-n}, N] \setminus E_n$. Since $f_n$ is supported by $[0,\ N]$, we have

$$
\begin{aligned}
\|\Delta_h f_n\|^p &= \int_{-2^{-n}}^N \left|\Delta_h f_n(x)\right|^p dx \\
&= \int_{E_n} \left|\Delta_h f_n(x)\right|^p dx + \int_{\widetilde{E}_n} \left|\Delta_h f_n(x)\right|^p dx \\
&= \int_0^{1-2^{-n}} \left\|\mathbf{f}_n(x+h) - \mathbf{f}_n(x)\right\|^p dx + h \sum_{k=0}^N \left| f_{[k,k+2^{-n})} - f_{[k-2^{-n},k)} \right|^p \\
&:= I_1 + I_2
\end{aligned}
$$

where $f_{[a,b)}$ is the average of $f$ on the interval $[a, b)$. By Lemma 3.2(ii), (iii) and using $\mathbf{f}_n(x) = \sum_{|J|=n} (T_J \mathbf{v}) \chi_{I_J}(x)$, we have

$$
I_1 \le C' \frac{1}{2^n} \left[ \sum_{k=0}^{n-1} \sum_{|J|=n-k-1} \|T_J T_1 T_0^k \mathbf{h}_0\|^p + \sum_{k=0}^{n-1} \sum_{|J|=n-k-1} \|T_J T_0 T_1^k \mathbf{h}_1\|^p \right]
$$

and

$$
I_2 \le C'' \frac{1}{2^n} \left( \Big\| T_{\underbrace{0\cdots 0}_{n \text{ terms}}} \mathbf{h}_0 \Big\|^p + \Big\| T_{\underbrace{1\cdots 1}_{n \text{ terms}}} \mathbf{h}_1 \Big\|^p \right),
$$

where $C'$ and $C''$ are constants independent of $n$. Hence we have

$$
\|\Delta_h f_n\|^p \le C_1 \frac{1}{2^n} \left( \sum_{|J|=n} \|T_J \mathbf{h}_0\|^p + \sum_{|J|=n} \|T_J \mathbf{h}_1\|^p \right)
$$

for some constant $C_1$. The claim follows from observing that the terms on the right hand side have the desired order. Now note that $\|\Delta_h f\| \le 2\|f - f_n\| + \|\Delta_h f_n\|$ and that $\|f - f_n\|^p$ has order $2^{-n} \sum_{|J|=n} \|T_J \tilde{\mathbf{v}}\|^p$ as $n \to \infty$ (by (2.1) and Theorem 2.2). This together with the claim implies "$\ge$" of (3.1).

The reverse inequality follows from the following inequality which can be proved by the similar argument as above: there exists $C_2$ such that for $n, \ell \geq 1$,

$$C_2\bigg(\frac{1}{2^{n-1}} \sum_{|J|=n-1} \|T_J\tilde{\mathbf{v}}\|^p\bigg) \leq \|\Delta_{2^{-n}} f_{n+\ell}\|^p.$$

## 4. Reduction of the Criterion

In the following we consider solutions $f$ of some special four-coefficient dilation equations with $c_0+c_2 = c_1+c_3 = 1$. We first calculate the $\mathrm{Lip}_p(f)$ for some special cases.

EXAMPLE 1. $c_3 = 0$. This is actually a 3-coefficient case with $c_0 + c_2 = 1$ and $c_1 = 1$. By a direct calculation, $\tilde{\mathbf{v}} = \big[c_0(c_0 - 1), -c_0(c_0 - 1)\big]^t$ is a common eigenvector of $T_0$ and $T_1$ with eigenvalues $c_0$ and $1 - c_0$ respectively. From this we show that $2^{-n}\sum_{|J|=n} \|T_J\tilde{\mathbf{v}}\|^p = 2^{-n}\big(|c_0|^p + |1 - c_0|^p\big)^n \|\tilde{\mathbf{v}}\|^p$. It follows that

$$\mathrm{Lip}_p(f) = \frac{\ln(2^{-1}(|c_0|^p + |1 - c_0|^p))}{-p\ln 2}.$$

EXAMPLE 2. $c_0 + c_3 = 1$. In this case $\tilde{\mathbf{v}} = [-c_0^2c_3, c_0^2c_3 - c_0c_3^2, c_0c_3^2]^t$ is again a common eigenvector of $T_0$ and $T_1$ with eigenvalues $c_0$ and $1 - c_0$ respectively. It follows that $\mathrm{Lip}_p(f)$ is the same as in Example 1 by using an identical calculation. (See Appendix, Figure 1 for some graphs.)

EXAMPLE 3. $c_0+c_3 = 1/2$. (This includes the Daubechies scaling function $D_4$ where $c_0 = (1 + \sqrt{3})/4, c_3 = (1 - \sqrt{3})/4)$. Note that $\mathbf{u} = [0, 1, -1]^t$, $\mathbf{h} = [1, -2, 1]^t$ are eigenvectors of $T_0$ corresponding to the eigenvalues $1/2$ and $c_0$; also $T_1\mathbf{u} = 1/2\mathbf{u} + c_0\mathbf{h}$ and $T_1\mathbf{h} = (1/2 - c_0)\mathbf{h}$. By using $\{\mathbf{u}, \mathbf{h}\}$ as a basis of the subspace $H(\tilde{\mathbf{v}}) = H$, we can rewrite $T_0$ and $T_1$ on $H$ as

$$\begin{pmatrix} \frac{1}{2} & 0 \\ 0 & c_0 \end{pmatrix}, \quad \begin{pmatrix} \frac{1}{2} & 0 \\ c_0 & \frac{1}{2} - c_0 \end{pmatrix}.$$

Let $\beta_0 = c_0$ and $\beta_1 = \frac{1}{2} - c_0$. Then the corresponding matrix for $T_J$ is

$$\begin{pmatrix} 2^{-n} & 0 \\ \lambda_J & \mu_J \end{pmatrix},$$

where $\lambda_J = \beta_0(j_1 2^{-(n-1)} + j_2 2^{-(n-2)}\beta_{j_1} + \cdots + j_n\beta_{j_1}\cdots\beta_{j_{n-1}})$, $\mu_J = \beta_{j_1}\beta_{j_2}\cdots\beta_{j_n}$, $j_i = 0$ or $1$. A direct estimation yields

$$\mathrm{Lip}_p(f) = \min\bigg\{1, \frac{\ln(2^{-1}(|c_0|^p + |1/2 - c_0|^p))}{-p\ln 2}\bigg\}.$$

This formula was also proved by Daubechies and Lagarias [**DL3**] with another method and the additional assumption that $1/2 < c_0 < 3/4$. (See Appendix, Figure 2 for some graphs.)

In all the three examples the computation depends on the existence of a common eigenvector of $T_0$ and $T_1$ (which may be associated with different eigenvalues). This technique cannot be used for the general case. However we show in the following that if $p$ is a positive even integer, then $\mathrm{Lip}_p(f)$ is related to the spectral radius of a matrix $W_p$ whose entries are induced from the coefficients of the dilation equation. For simplicity, we only give the construction of $W_p$ for the 4-coefficient dilation equation. It is not hard to extend this to the case with more coefficients.

We will first develop a simple expression for the sum $2^{-n}\sum_{|J|=n}\|T_J\mathbf{v}\|^p$ for $p$ a positive even integer. Let $\mathbf{e}_1 = [0,1,-1]^t$ and $\mathbf{e}_2 = [1,-1,0]^t$ be a basis of $H = \{\mathbf{u}\in\mathbb{C}^3 : \sum u_i = 0\}$. Then

$$T_0 = \begin{pmatrix} 1-c_0-c_3 & c_3 \\ 0 & c_0 \end{pmatrix}, \quad T_1 = \begin{pmatrix} c_3 & 0 \\ c_0 & 1-c_0-c_3 \end{pmatrix} \text{ and } \tilde{\mathbf{v}} = \begin{pmatrix} a \\ b \end{pmatrix}.$$

We define the vector $\mathbf{a}_n$ with the $i$-th entry by

$$(\mathbf{a}_n)_i = \sum_{|J|=n} (\mathbf{e}_0 T_J\mathbf{u})^{p-i}(\mathbf{e}_1 T_J \mathbf{u})^i, \quad i = 0,\ldots,p.$$

If $p$ is an even integer, then

$$\begin{aligned}\sum_{|J|=n}\|T_J\mathbf{u}\|^p &= \sum_{|J|=n}\left(|\mathbf{e}_0T_J\mathbf{u}|^p + |\mathbf{e}_1T_J\mathbf{u}|^p\right) \\ &= (\mathbf{a}_n)_0 + (\mathbf{a}_n)_p = \left|(\mathbf{a}_n)_0\right| + \left|(\mathbf{a}_n)_p\right|.\end{aligned} \tag{4.1}$$

Note that $\mathbf{a}_0 = [\alpha^p, \alpha^{p-1}\beta, \ldots, \alpha\beta^{p-1}, \beta^p]^t$ for some $\alpha, \beta$. If we let $d = 1 - c_0 - c_3$, we have, in view of the expression of $T_0$ and $T_1$ on $H$ and by using simple iteration we have

PROPOSITION 4.1. *For any integer $p \geq 1$, we have*

$$\mathbf{a}_{n+1} = W_p\mathbf{a}_n = W_p^{n+1}\mathbf{a}_0$$

*where $W_p$ is a $(p+1)\times(p+1)$ matrix defined by*

$$(W_p)_{ij} = \begin{cases} \binom{p-i}{j-i} c_0^i c_3^{j-i} d^{p-j} & \text{for } 0\leq i<j\leq p \\ c_0^i d^{p-i} + c_3^{p-i}d^i & \text{for } i=j \\ \binom{i}{j} c_0^{i-j} c_3^{p-i} d^j & \text{for } 0\leq j<i\leq p \end{cases}$$

*where $d = 1-c_0-c_3$. In particular if $p$ is an even integer, then*

$$\sum_{|J|=n}\|T_J\mathbf{u}\|^p = [1,0,0,\ldots,0,1]W_p^n\mathbf{a}_0.$$

The matrix $W_p$ can be written as $W_p = W_p^{(L)} + W_p^{(U)}$, where $W_p^{(L)}$ and $W_p^{(U)}$ are the lower and upper triangular part of $W_p$, in a very symmetric manner. For example,

$$W_2^{(L)} = \begin{pmatrix} \binom{0}{0}c_3^2 & 0 & 0 \\ \binom{1}{0}c_0c_3 & \binom{1}{1}c_3 d & 0 \\ \binom{2}{0}c_0^2 & \binom{2}{1}c_0 d & \binom{2}{2}d^2 \end{pmatrix}, \quad W_2^{(U)} = \begin{pmatrix} \binom{2}{0}d^2 & \binom{2}{1}c_3 d & \binom{2}{2}c_3^2 \\ 0 & \binom{1}{0}c_0 d & \binom{1}{1}c_0c_3 \\ 0 & 0 & \binom{0}{0}c_0^2 \end{pmatrix};$$

$$W_4^{(L)} = \begin{pmatrix} \binom{0}{0}c_3^4 & 0 & 0 & 0 & 0 \\ \binom{1}{0}c_0c_3^3 & \binom{1}{1}c_3^3d & 0 & 0 & 0 \\ \binom{2}{0}c_0^2c_3^2 & \binom{2}{1}c_0c_3^2d & \binom{2}{2}c_3^2d^2 & 0 & 0 \\ \binom{3}{0}c_0^3c_3 & \binom{3}{1}c_0^2c_3d & \binom{3}{2}c_0c_3d^2 & \binom{3}{3}c_3d^3 & 0 \\ \binom{4}{0}c_0^4 & \binom{4}{1}c_0^3d & \binom{4}{2}c_0^2d^2 & \binom{4}{3}c_0d^3 & \binom{4}{4}d^4 \end{pmatrix},$$

$$W_4^{(U)} = \begin{pmatrix} \binom{4}{0}d^4 & \binom{4}{1}c_3d^3 & \binom{4}{2}c_3^2d^2 & \binom{4}{3}c_3^3d & \binom{4}{4}c_3^4 \\ 0 & \binom{3}{0}c_0d^3 & \binom{3}{1}c_0c_3d^2 & \binom{3}{2}c_0c_3^2d & \binom{3}{3}c_0c_3^3 \\ 0 & 0 & \binom{3}{0}c_0^2d^2 & \binom{2}{1}c_0^2c_3d & \binom{2}{2}c_0^2c_3^2 \\ 0 & 0 & 0 & \binom{1}{0}c_0^3d & \binom{1}{1}c_0^3c_3 \\ 0 & 0 & 0 & 0 & \binom{0}{0}c_0^4 \end{pmatrix}.$$

For the 4-coefficient dilation equation, it is easy to check that $\dim H(\tilde{\mathbf{v}}) = 0$ if and only if $(c_0, c_3) \in \{(0,0), (1,0), (0,1), (1,1)\}$, and the solutions are characteristic functions ([**LW**], Lemma 3.3]). Hence $\mathrm{Lip}_p(f) = 1/p$. Also $\dim H(\tilde{\mathbf{v}}) = 1$ if and only if $c_0 + c_3 = 1$, and in Example 2 we have given a formula of $\mathrm{Lip}_p(f)$ for this case. It remains to consider the case $\dim H(\tilde{\mathbf{v}}) = 2$, which will complete all the cases for all 4-coefficient scaling functions.

THEOREM 4.2. *Consider the 4-coefficient dilation equation with the assumption that* $\dim H(\tilde{\mathbf{v}}) = 2$. *For $p$ a positive even integer, the equation has a (unique) $L_c^p$-solution $f$ if and only if $\rho(W_p)/2 < 1$, and in this case*

$$\mathrm{Lip}_p(f) = \frac{\ln(\rho(W_p)/2)}{-p\ln 2}.$$

The proof follows easily from Proposition 4.1, together with Theorems 2.2 and 3.1. We remark that when $p = 2$, the matrix $W_2$ is the same as the transition matrix obtained in [**DL3,H,LMW**] and [**V**] and the $L^2$-Lipschitz exponent and the Sobolev exponent coincide. If $c_0 > 0$, $c_3 > 0$, and $1 - c_0 - c_3 > 0$, then $T_0$ and $T_1$ are non-negative matrices. Hence (4.1) still holds if $p$ is a positive odd integer. Consequently, we have

COROLLARY 4.3. *Consider the 4-coefficient dilation equation with $c_0 > 0$, $c_3 > 0$, and $1 - c_0 - c_3 > 0$. Suppose* $\dim H(\tilde{\mathbf{v}}) = 2$, *then the conclusion of Theorem 4.2 holds for $p$ a positive integer.*

For the four coefficient case with $c_0 + c_2 = 1$, $c_1 + c_3 = 1$, we use $(c_0, c_3)$ as the two free parameters. Figure 3 shows the domain of $(c_0, c_3)$ for the existence of $L_c^p$-solutions for even integers using the above criterion $\rho(W_p)/2 < 1$. The curves are $\rho(W_p)/2 = 1$ corresponds to $p = 2, 4, 6, 10, 20$, and $40$. Note that when $p \to \infty$ the limit is the triangular region which is the approximate region plotted in [**H**] for the existence of continuous 4-coefficient scaling functions using the joint spectral radius.

APPENDIX. We illustrated some of our results by the graphs. Figure 1 gives the graphs of the scaling functions together with the curve of $L^p$-Lipschitz exponent in Example 2, and Figure 2 is for Example 3.

Figure 4 is the graph of $\text{Lip}_4(f)$ plotted against the $(c_0, c_3)$-plane, using Theorem 4.2. It shows the overall picture of $\text{Lip}_4(f)$ for the 4-coefficient case. It looks similar to the graph of $\text{Lip}_2(f)$ plotted in [**LMW**].

## References


[CD] A. Cohen and I. Daubechies, *A stability criterion for biorthogonal wavelet bases and their related subband coding scheme*, Duke Math. J. **68** (1992), 313–335.

[CH] D. Colella and C. Heil, *Characterizations of scaling functions: Continuous solutions*, SIAM J. Matrix Anal. Appl. **15** (1994), 496–518.

[D] I. Daubechies, *Ten lectures on wavelets*, CBMS-NSF Regional Conf. Ser. in Appl. Math., vol. 61, 1992.

[DL1] I. Daubechies and J. C. Lagarias, *Two-scale difference equation.* I. *Global regularity of solutions*, SIAM J. Math. Anal. **22** (1991), 1388–1410.

[DL2] ———, *Two-scale difference equation.* II. *Local regularity, infinite products and fractals*, SIAM J. Math. Anal. **23** (1991), 1031–1079.

[DL3] ———, *On the thermodynamic formalism for multifractal*, Rev. Math. Phys. **6**, 1033–1070.

[E] T. Eirola, *Sobolev characterization of solutions of dilation equations*, SIAM J. Math. Anal. **23** (1992), 1015-1079.

[FP] U. Frisch and G. Parisi, *On the singularity structure of fully developed turbulence*, Proc. Int. Sch. Phys., North Holland, Amsterdam, 1985, pp. 84–88.

[H] C. Heil, *Method of solving dilation equations*, Prob. and Stoch. Methods in Analysis with Applications, vol. 372, NATO Adv. Sci. Inst. Ser. C Math. Phys. Sci., 1992, pp. 15–45.

[HJ] T. C. Halsey, M. H. Jensen, L. P. Kadanoff, I. Procaccia, and B. J. Shraiman, *Fractal measures and there singularities: The characterization of strange sets*, Phys. Rev. A **33** (1986), 1144–1151.

[J1] S. Jaffard, *Multifractal formalism for functions.* I. *Results valid for all functions*, SIAM J. Math. Anal. **28** (1977), 944–970.

[J2] ———, *Multifractal formalism for functions.* II. *Self-similar functions*, SIAM J. Math. Anal. **28** (1977), 971–998.

[J] R. Jia, *Subdivision schemes in $L^p$ spaces*, Adv. Comput. Math. **3** (1995), 309–341.

[LM] K. S. Lau and M. F. Ma, *The regularity of $L^p$-scaling functions*, Asian J. Math. **1** (1997), 272–292.

[LMW] K. S. Lau, M. F. Ma, and J. Wang, *On some sharp regularity estimations of $L^2$-scaling functions*, SIAM J. Math. Anal. **27** (1996), 835-864.

[LN] K. S. Lau and S. M. Ngai, *Multifractal formalism and a weak separation condition*, Adv. Math. (to appear).

[LW] K. S. Lau and J. Wang, *Characterization of $L^p$-solution for the two scale dilation equations*, SIAM J. Math. Anal. **26** (1995), 1018-1046.

[MP] C. A. Micchelli and H. Prautzsch, *Uniform refinement of curves*, Linear Algebra Appl. **114/115** (1989), 841–870.

[V] L. Villemoes, *Energy moments in time and frequency for two-scale difference equation solutions and wavelets*, SIAM J. Math. Anal. **23** (1992), 1519–1543.



DEPARTMENT OF MATHEMATICS, THE CHINESE UNIVERSITY OF HONG KONG, HONG KONG.
*E-mail address*: `kslau@euler.math.cuhk.edu.hk`

DEPARTMENT OF MATHEMATICS, UNIVERSITY OF PITTSBURGH, PITTSBURGH, PA 15260, USA
*E-mail address*: `mfma@vmo.cis.pitt.edu`

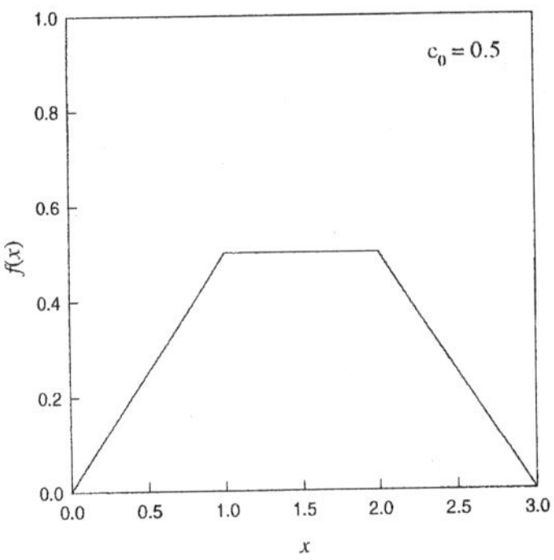


**Figure 1a**

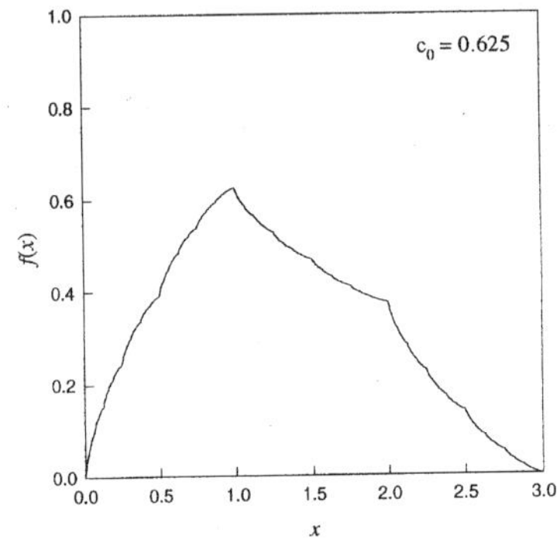


**Figure 1b**

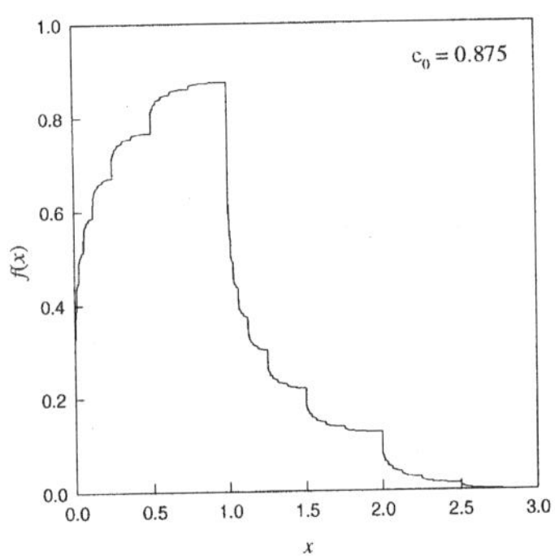


**Figure 1c**

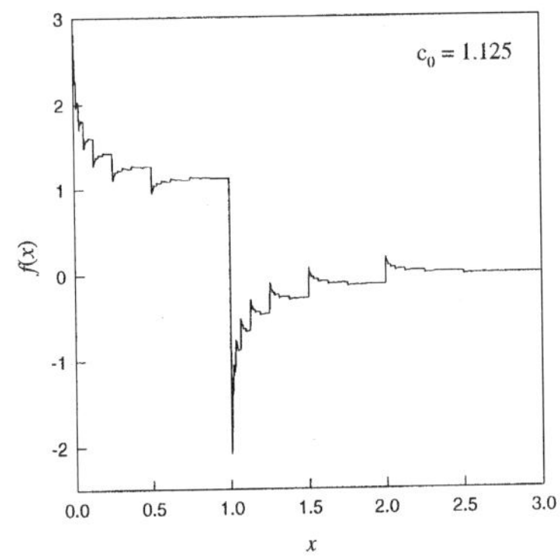


**Figure 1d**

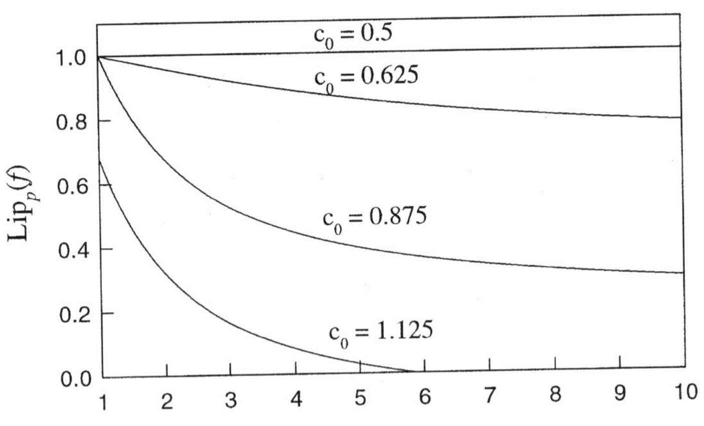


**Figure 1e**

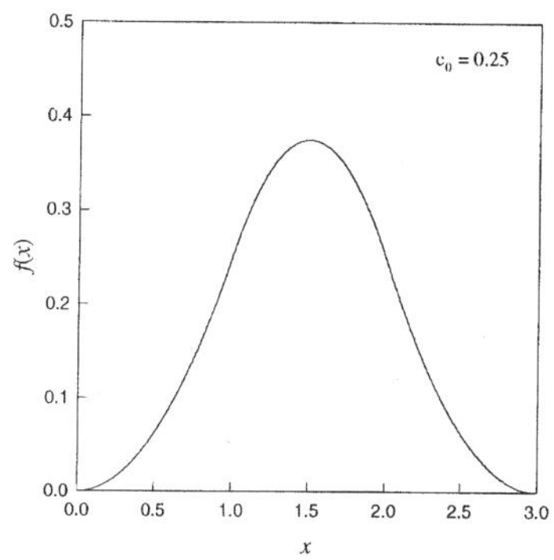


**Figure 2a**

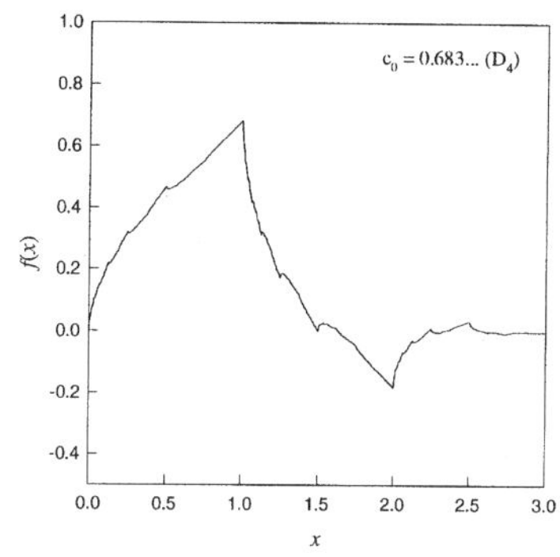


**Figure 2b**

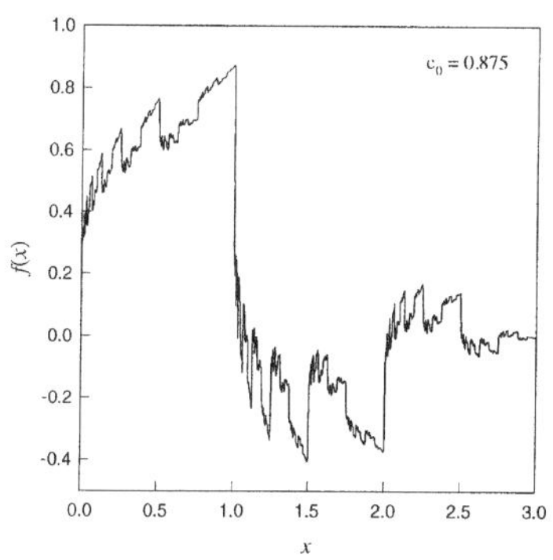


**Figure 2c**

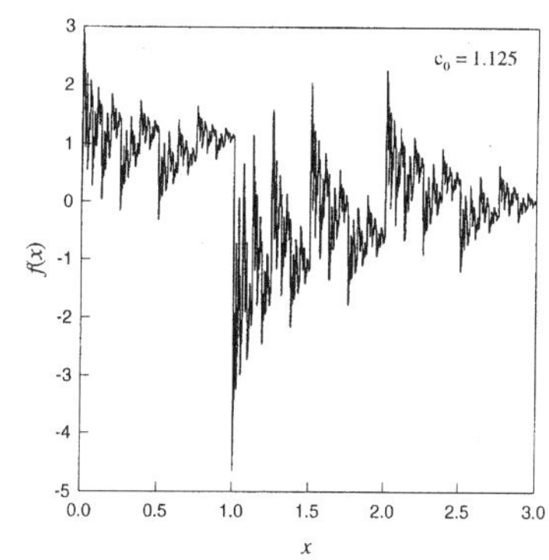


**Figure 2d**

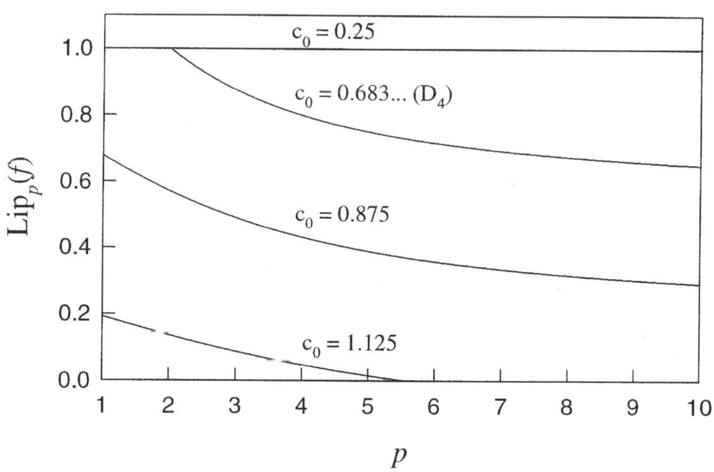


**Figure 2e**

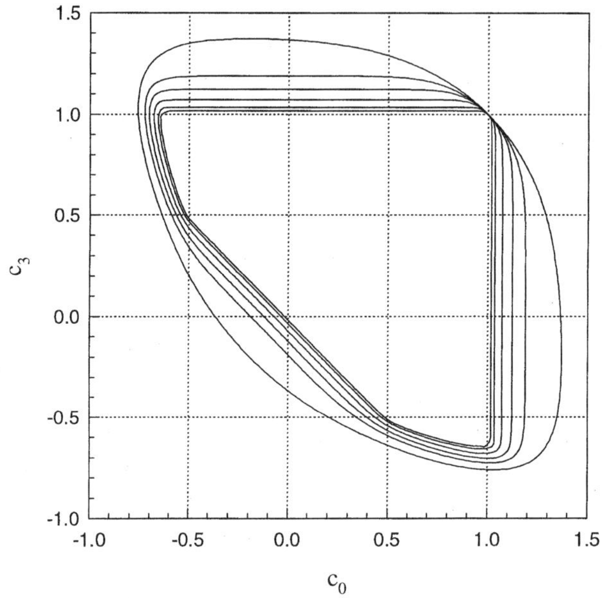


**Figure 3**

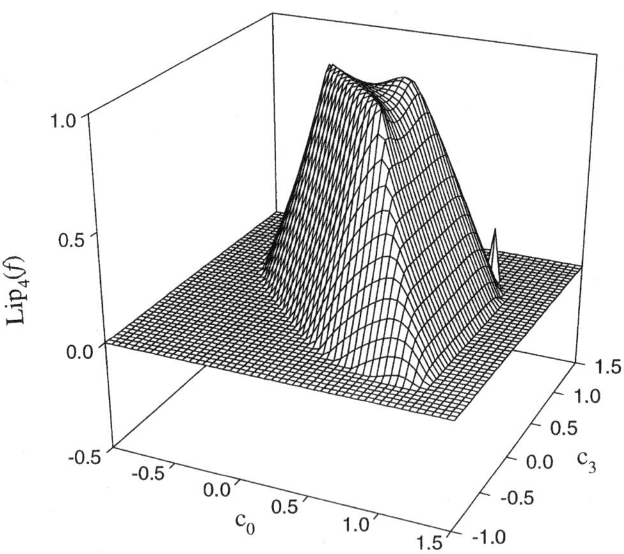


**Figure 4**

Centre de Recherches Mathématiques
CRM Proceedings and Lecture Notes
Volume **18**, 1999

# Robust Speech and Speaker Recognition Using Instantaneous Frequencies and Amplitudes Obtained with Wavelet-Derived Synchrosqueezing Measures

Stéphane H. Maes

Abstract. Speech production models often lead to the introduction of *modulation models* to represent speech signals with primary components which are amplitude-and-phase-modulated sine functions.

Since their early investigations as continuous transforms and the development of *ridge-skeleton* representations, wavelets have proved well adapted to extract the modulation laws of isolated or distinct enough primary components. However, the approximations intrinsic to these approaches limit their applications when it comes to real life problems like speech analysis. Parallelisms between properties of the wavelet transform of primary components and algorithmic representations of speech signals derived from auditory nerve models like the EIH lead to the introduction of *synchrosqueezing measures.*

This paper presents a new transformation of the time-scale plane obtained by a quasi-continuous wavelet transform into a time-frequency plane: the *synchrosqueezed plane representation.* The resulting analyses have a broad scope of interest not limited to speech applications or wavelet analyses. Furthermore, these approaches provide representations essentially independent of the generating analysis wavelets.

Our analysis is "phase-oriented" and it formalizes naturally auditory nerve models like the IFD, SBS and EIH models. It is also based on overcomplete (frame) analysis and synthesis of signals. Robustness naturally result from these analyses.

The analysis can be inverted to extract the primary components of speech signals

## 1. INTRODUCTION

In general, in speech coding as well as in speech recognition, the parameters and the tools which are commonly used rely on power spectrum approaches. This is the case for the spectrum representation, with its Mel or Bark variations, the cepstrum,

---

1991 *Mathematics Subject Classification.* 94A12, 94A14.

This work was performed when Dr. S. Maes was a Research Assistant of the Belgian National Fund for Scientific Research (FNRS). he author was affiliated to CAIP Center, Rutgers University, Piscataway, NJ 08855, USA and Unités TELE and FYMA, Université Catholique de Louvain, B-1348 Louvain-la-Neuve, Belgium.

This work was done in collaboration with Prof. I. Daubechies, Princeton University.

This is the final form of the paper.

FFT-derived (Fast Fourier Transform) or LPC-derived (Linear Predictive Coding), all the LPC-derived sets of features, the autocorrelation, the energy content and all the associated delta and delta-delta coefficients [**24, 23, 25, 68**]. It is also the case of the classical FMMM (Formant Mode Modulation Model) [**39, 14, 69, 4**] and the Sinusoidal representation [**61**].

Only a few models have been proposed in the literature to characterize speech with phase-derived features. In the framework of our method the most relevant models are: EIH (Ensemble Interval Histogram) [**27, 28, 68**], SBS (in-Synchrony Bands Spectrum) [**26**] and IFD (Instantaneous-Frequency Distribution) [**22**]. These three models are derived from (nonplace/temporal) auditory nerve models of the human auditory system [**27, 28**]. Note that other models have been proposed using wavelets to model the auditory nerve system [**9**].

Starting from the classical source-filter model of speech production, the modulation model (MM) is briefly motivated. A short review of the wavelet transform and its application to MM primary components illustrate the shortcomings of direct time-scale representations. By introducing the EIH approach, we discover possible ways to handle similar issues. These principles are transposed to WT analyses, leading to the concept of *synchrosqueezed plane.* Incidentally, it is also essentially independent of the generating analysis wavelet.

The use of wavelet analyses results from different considerations. The WT (Wavelet Transform) performed with a complex analysis wavelet is known to carry relevant information in its modulus as well as in its phase [**18, 17, 12, 19, 35, 58**]. The information contained in the modulus is similar to power-spectrum derived parameters. On the other hand, the phase is partially independent of the amplitude level of the input signal. Also, the filter banks which are often used to simulate the first stage of the peripheral human auditory system are very close to constant Q systems [1] [**27, 68**]. As the wavelet transform is also, up to a normalization factor, implemented with a constant Q filter bank, its interest in speech processing is obvious; indeed the resulting scale invariance of the representation matches expectations based on physiological considerations. Eventually the dilation applied on the generating analysis wavelet provides an adaptive time-frequency support adequate to study phenomena existing over a large range of scales. It is typically the case of speech signals where short-time and long-time spectrograms extract different relevant aspects of the signal.

## 2. The Modulation Model

A source-filter model, slightly different from what is usually presented in the literature [**24, 65**], is depicted in Figure 1. It represents a linear equivalent circuit model of speech production mechanisms. Its limitations are discussed in [**21, 24, 16**].

For voiced sounds, a periodic impulse sequence is filtered with a glottal filter which produces the periodic part of the excitation signal. The non-periodicity and the turbulent content are obtained by adding an additional white noise source. This last source is usually omitted in the literature. For unvoiced sounds, only the white

[1] The quality factor of a filter is defined as the ratio of its center-frequency to its bandwidth.

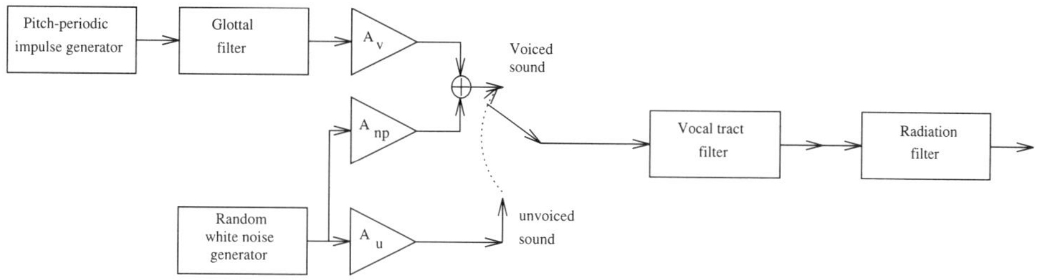


FIGURE 1. Source filter model for speech production. $A_v$ is the voiced gain, $A_u$ is the unvoiced part and $A_{np}$ is the unvoiced contribution present in a voiced signal.

noise source is present. The resulting signal excites the vocal tract characterized by a transfer function $H_{vt}(z)$ [2].

An additional term $R(z)$ models the radiation effect at the lips. $R(z)$ always has a highpass behavior [**10, 65, 21, 20, 63, 64**].

Such a system has dominant contributions in pressure or velocity at resonances and anti-resonances of the vocal-tracts. These dominant contributions behave as eigenmodes of the transfer function and the model leads to the MM. This interpretation is especially valid for voiced sounds, where some primary components are closely related to the definitions of formants. However, unvoiced utterances still satisfy the model.

The (MM) considers that a speech signal can be written as a linear combination of amplitude and phase modulated components: the primary components;

$$f(t) = \sum_{k=1}^{N_{pc}} A_k(t) \cos\big(\Phi_k(t)\big) + \eta(t) \tag{2.1}$$

where $A_k(t)$ denotes the instantaneous amplitude of the $k^{th}$ primary component and $\Phi_k(t)$ denotes its instantaneous phase. $\eta(t)$ takes into account additive noise and errors of modeling, especially errors due to the finite summation. This model has also been proposed in the literature as the elementary waveform speech model [**39, 14**].

In speech signals, the first primary components are the pitch component along with some of its first harmonics. They are disregarded by most of the classical algorithms which pre-emphasize the signal [**51, 4**]. By definition, they are not considered as formants. The other primary components encompass naturally the formants of a speech signal. The formants characterize rather well the different voiced phones. However, besides the formants, the pitch and its first harmonics, the MM often introduces other primary components which result from the interferences among all the different harmonics and from their interactions within the vocal tract. Depending on the application, more or less primary components are considered relevant. In speaker identification, the relevant components are usually identified with the classical formants [**4**]. But for coding purposes, synthesis-by-analysis coders based on the sinusoidal representation use a quite larger number of

[2]Strictly speaking, $z$ should be the Laplace variable as the signal is continuous. However, the relationship between the Laplace transform and the z-transform is well known [**11**]. Especially as the speech signals are always sampled before processing.

primary components, even for unvoiced sounds [**61, 36, 59, 60, 67, 2, 26**]. The possibility to model unvoiced sounds with a large set of modulated sine waves offers additional justifications for the approaches discussed in this paper which uses the MM without distinction between voiced and unvoiced sounds. For unvoiced utterances, most of the primary components seem to disappear. The remaining primary components have wider instantaneous bandwidths, overlapping their neighboring primary components. The center frequencies become tricky to extract. The extraction of the primary components usually relies on cocurvilinearity considerations with the neighboring voiced frames. By definition of an unvoiced sound, the first primary components associated with the pitch and first harmonics are usually quite attenuated, if not absent. But the MM remains valid as described in [**5, 14**]. Basically it can be modeled as if the excitation was different for each primary component. This is visible on extracted instantaneous amplitudes as illustrated in [**48, 52, 54**].

For a given signal $f(t)$, the task addressed by our work is to minimize the Euclidean norm in $L^2(\mathbb{R}, dx)$ of $\eta(t)$. Similarly to the classical problem of analytic representation of a signal, the solution to our problem is not unique.

## 3. Wavelet Transform

It is useful to briefly present the notations for the WT which are used in this paper. The WT of a signal $f(t)$ using a generating analysis wavelet $g(t)$ is defined as:

$$\mathcal{WT}_f(a, b, g) = \langle g_{(a,b)} \mid f \rangle \tag{3.1}$$

where,

$$g_{(a,b)}(t) = \frac{1}{\sqrt{a}} g\left(\frac{t-b}{a}\right) \tag{3.2}$$

$a \in \mathbb{R}_+ \setminus \{0\}$ is the scale or dilation parameter and $b \in \mathbb{R}$ is the translation parameter[3]. $a$ is often related to the inverse of the frequency while $b$ can be considered as the time variable. $\langle \cdot \mid \cdot \rangle$ is the classical functional inner product. The WT at a given scale level $a$ appears as the projection of $f(t)$ over the set of translated versions of $g_{(a,0)}(t)$[4].

Any signal $f(t)$ in a satisfying functional Hilbert space $\mathcal{H}$ can be reconstructed from its WT, using a generating analysis wavelet $h(t)$. This operation is called *resolution of the identity,*

$$\forall f(t) \in \mathcal{H} \colon f(t) = \frac{1}{C_{g,h}} \int_{-\infty}^{+\infty} \int_0^{+\infty} h_{(a,b)}(t) \frac{db\, da}{a^2} \mathcal{WT}_f(a, b, g) \tag{3.4}$$

[3] In the literature, another convention is also often used where $a \in \mathbb{R} \setminus \{0\}$.

[4] The projection of $f(t)$ on the space spanned by the set $\{e_k(t)\}_k$ is defined as: $\{P_{\{e_k(t)\}}\{f(t)\}\}_k \triangleq \{\langle e_k \mid f \rangle\}_k$. When the term "projection" is used instead of orthogonal projection, the reader should always use this definition. It is the series of lengths of the orthogonal projections of $f(t)$ on $e_k(t)$, multiplied by the length of $e_k(t)$. The absence of normalization by the length of $e_k(t)$ allows to select different normalizations for the wavelet transform. For example, some authors use another normalization for $g_{(a,b)}(t)$:

$$\breve{\mathcal{WT}}_f(a, b, g) = \frac{1}{\sqrt{a}} \langle g_{(a,b)} \mid f \rangle. \tag{3.3}$$

and it is subject to the following *admissibility condition* [5]:

$$0 < \left| C_{g,h} = \int_0^{+\infty} \frac{\hat{h}(\omega)\hat{g}^*(\omega)\,d\omega}{\omega} = \int_0^{+\infty} \frac{\hat{h}(-\omega)\hat{g}^*(-\omega)\,d\omega}{\omega} \right| < +\infty. \tag{3.5}$$

Note that these equations are written for the continuous biorthogonal case: the generating analysis wavelet $g(t)$ can be different from the generating synthesis wavelet [**37**].

## 4. The Wavelet Transform Applied to the Modulation Model

**4.1. Techniques of extraction of simple curves.** One of the main application of the CWT (continuous WT) is curve extraction, detection and manipulation. In this section, the application of the WT to the analysis of the MM is briefly described.

The generating analysis wavelet, used in this paper, is assumed to be analytic:

$$\omega \leq 0 \rightarrow \hat{g}(\omega) = 0. \tag{4.1}$$

**4.2. Single complex exponential characterization.** The input signal is assumed to be an isolated complex exponential signal,

$$f(t) = A_f\ e^{j(\omega_f t + \Phi_f)}. \tag{4.2}$$

Its WT is given by

$$\mathcal{W}T_f(a,b,g) = \sqrt{a} A_f \hat{g}^*(a\omega_f) e^{j(\omega_f b + \Phi_f)}. \tag{4.3}$$

Let the WT be written as a phasor by

$$\mathcal{W}T_f(a,b,g) \triangleq \left|\mathcal{W}T_f(a,b,g)\right| e^{j\Phi_{\mathcal{W}T_f}(a,b,g)}. \tag{4.4}$$

A pseudo-frequency is defined by

$$\omega_{\mathcal{W}T_f}(a,b,g) = \frac{d\Phi_{\mathcal{W}T_f}}{db}(a,b,g). \tag{4.5}$$

Although associated with a single spectral line, the WT appears as a ribbon of essential width, in the scale direction, given by $\left[(\omega_l)/(\omega_f), (\omega_h)(\omega_f)\right]$ where $[\omega_l, \omega_h]$ denotes the essential frequency support of $\hat{g}(\omega)$. Figure 2 presents the modulus of the WT of signal (4.2).

In other words, the useful information, desired to be concentrated on a single straight line, is spread over a ribbon which is only a function of the generating analysis wavelet. This illustrates, in a striking way, the claim that the time-scale representation obtained by a direct application of the WT has no real physical meaning. It motivates the search for an alternative representation, which would not be conditioned by the generating analysis wavelet. For the complex exponential,

---

[5] $\hat{g}(\omega)$ stands for the Fourier transform of $g(t)$. It is normalized as: $\hat{g}(\omega) \triangleq \int_{-\infty}^{+\infty} g(t)(e^{-j\omega t})/(\sqrt{2\pi})\,dt$

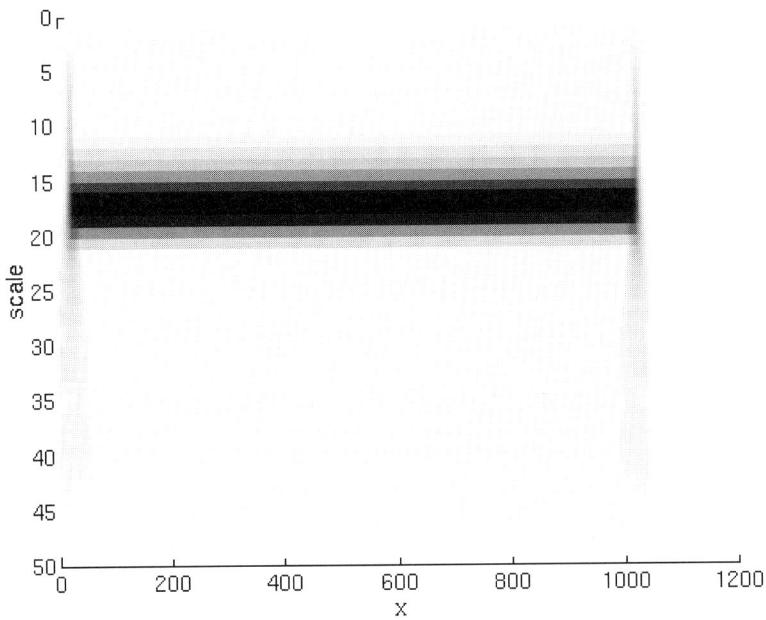


FIGURE 2. Modulus of the wavelet transform of the single complex exponential. The horizontal axis is sampled at $8\,kHz$. The vertical axis represents the indexes of the different subbands, arranged as explained in [**48, 44, 47**]. There are 5 octaves, with 7 equally spaced sub-octaves per octave and an upper-octave interpolation. The low indexes are associated to the higher frequencies. A boundary treatment was applied but the parasitic contributions at the extremities have not been eliminated in this plot [**48, 46, 44**].

the following relation is satisfied [6]:

$$\omega_{\mathcal{W}T_f}(a,b,g) = \omega_f. \tag{4.7}$$

More complex cases are studied in details in [**48, 53, 58, 34, 31, 35**].

Figure 3 illustrates the modulus of the WT of the speech utterance $/\underbrace{\cdots - a - a}\, \underbrace{- i - i - \cdots}/$. Each primary component of this voiced utterance presents a behavior similar to the isolated complex exponential.

**4.3. The ridge-skeleton method.** The *ridge-skeleton* methods are described in details in [**18, 17, 19, 70, 29, 32, 33**]. The theory of asymptotic expansions,

---

[6]Equation (4.7) is always satisfied to the second order by an isolated phase modulated exponential. It becomes:

$$\frac{d\Phi_{\mathcal{W}T_f}}{db}(a,b,g) = \omega_f(b) = \frac{d\Phi_t}{db}(b). \tag{4.6}$$

It can be verified by using the *steepest descent method* [**13, 38**]. It justifies the extension of our analysis to any phase modulation law.

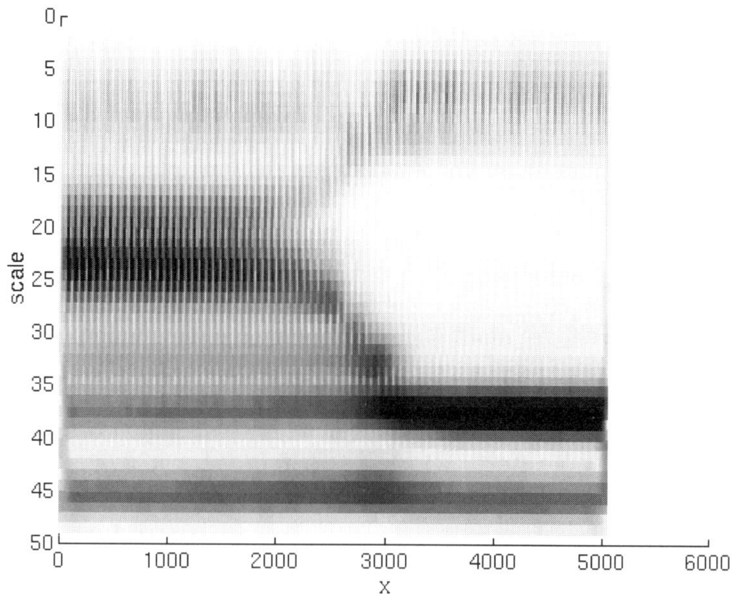


FIGURE 3. Modulus of the wavelet transform of the $/\underbrace{\cdots - a - a}\;\underbrace{- i - i - \cdots}/$. A colored noise is present with SNR $= 15\,dB$. The horizontal axis is sampled at $8\,kHz$. The vertical axis represents the indexes of the different subbands, arranged as explained in [**53, 48, 44, 47**]. There are 5 octaves, with 7 equally spaced sub-octaves per octave and an upper-octave interpolation. The low indexes are associated to the higher frequencies. The analysis generating wavelet is given by (17). No boundary treatment has been applied.

in particular the *steepest descent method*, applied to rapidly oscillating integrals, implies that the dominant contributions of the integral of the WT is provided by the stationary points $t_s$ of the phase of the integrand [**13, 38**]. The *ridges* are the curves associated in the time-scale plane with the stationary points of the integrand of the WT. *Ridges* yield direct reading of the instantaneous frequencies of the signal in the time-scale plane.

The *skeleton* is defined as the restriction, to the *ridges*, of the WT of the signal. It can be approximated by analytical equations.

For analytic asymptotic generating analysis wavelets and for analytic wavelet with normal Gaussian amplitudes, the *ridges* and the *skeleton* are independent of the generating analysis wavelet. However, the errors of approximation depend on the signal and on the generating analysis wavelet.

The knowledge of the *skeleton* and the *ridge* enables the extraction of the instantaneous amplitudes. Algorithms exist to extract the *ridges* without computing the WT over the whole time-scale plane.

As long as the different primary components are not too close to each other, the dominant contributions originate from the regions of the time-scale plane where the wavelet oscillates at the same frequency as the closest component. Hence, for all the values of $b$ where there is no overlapping, no crossing, nor any component coming too close to another, $N_{pc}$ *ridges* can be extracted. As long as no interference occurs, the resulting curves describe the frequency modulation laws of the different components. When some primary components come too close to each others, at least two *ridges* are blurred into one and, because of the associated beat phenomenon, the *steepest descent method* does not lead to the dominant contributions to the WT. Hence the *ridges* are not only blurred, they also start wandering around, with no link to the modulation laws of the associated primary components.

It is possible to use perturbation techniques to forbid crossing of *ridges*. This perturbation technique results from the observation that most of the information about a primary component is not only present along the *ridge*, which is merely a curve drawn in a middle of the time-scale plane, but it is also present in a ribbon, which extends around the *ridge*. Hence by drawing two perturbed curves around the *ridges*, enough information is extracted to obtain an estimation of the associated frequency and amplitude modulation laws.

## 5. The Ensemble Interval Histogram

Detailed descriptions of the human peripheral auditory system can be found in [**68, 24, 21, 10, 28, 1, 30**]. The EIH representation results from an attempt to exploit the in-synchrony phenomena observed in neuron firing patterns which contain all the information processed by the higher auditory system stages [7]. In other words, it is expected that for speech analysis, at least with a recognition purpose, it is sufficient to retain only the characteristics of the signal that determine the firing patterns.

As illustrated in Figure 4, the frequency selectivity of the basilar membrane is modeled by a cochlear filter bank. The filter are essentially Q constant except for the first $1000\,Hz$. Details can be found in [**68, 27, 28**]. The next stages model the non-linearities of the neural firing patterns along the auditory nerve: the level-crossings simulates the transduction from motion to neural activity of the inner hair cells. The model is considered nonplace/temporal as it is assume that no information is retained about the position of the inner hair cells along the basilar membrane, but only about the relative instants of firing. This is implemented by the interval histogram. The summation stage measures the extent of the coherent neural activity along the auditory nerve.

The global algorithm is the following. The signal is split up into subbands with a generalized hybrid wavelet filter bank [8]. In each subband, dominant frequencies are extracted. At each time $t$ and for each extracted frequency, a measure accumulates the contributions of each subbands for which that frequency is detected at time $t$.

[7]Similar models have been proposed earlier on: the instantaneous-frequency distribution (IFD) [**22**] and the in-synchrony bands spectrum (SBS) [**26**].

[8]In our context, the terminology Generalized hybrid wavelet filter bank means that the generating filters can change from level to level in a hierarchical implementation [**45**].

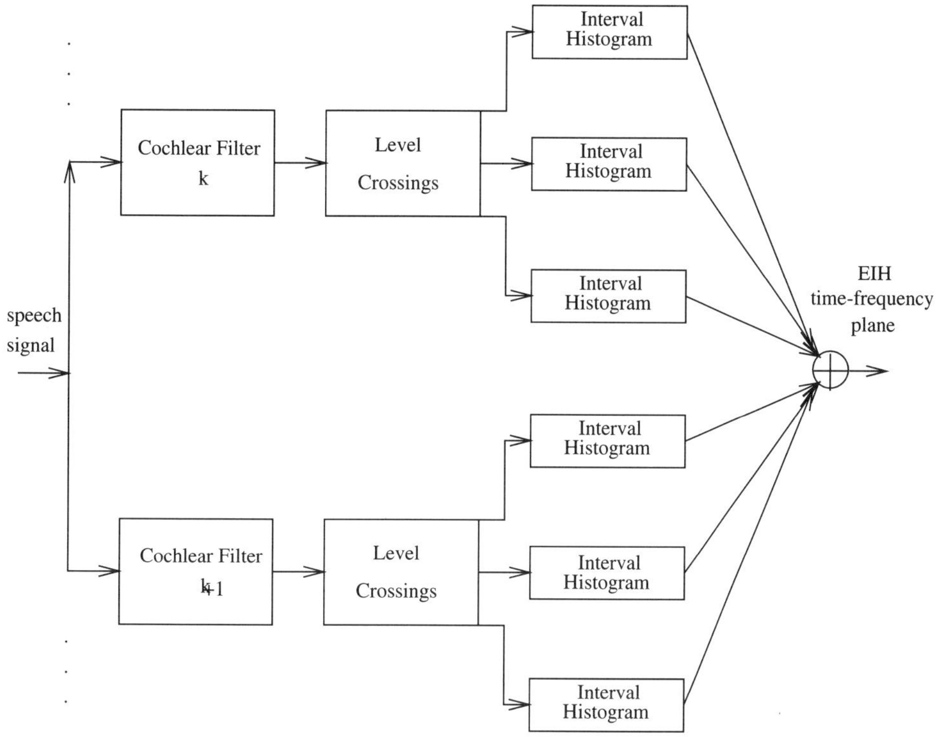


FIGURE 4. It presents a block diagram of the EIH model. The filter bank models the frequency response of the inner hair cells. The two following blocks implement the auditory nerve processing according to the nonplace/temporal model. At each time, the interval histogram are from each level in all the subbands are added to produce the EIH representation.

## 6. The Synchrosqueezed Place Representation

As already mentioned, WT provide generating-analysis-wavelet-dependent representations of a signal. However, it is possible to extract from this representation features which depend essentially only on the signal. Existing methods draw dominant curves in the time-scale plane from where they extract the only information used to characterize the signal. The extracted features are not conditioned by the generating analysis wavelet, at least explicitly.

The most important lesson is probably that the information about the phase modulation laws is totally contained in the phase representation of the wavelet transform. This information can easily be separated from the influence of the amplitudes, while it is merged with the information about the amplitude modulation laws in the modulus representation.

Unfortunately, the *ridge-skeleton* method is not robust enough to noise and its accuracy suffers when the primary components interfere.

Also, it should be clear at this stage that the information about a primary components has been spread over its whole associated ribbon. This is one of the consequences of equation (4.3). Hence, instead of selecting curves from the ribbon,

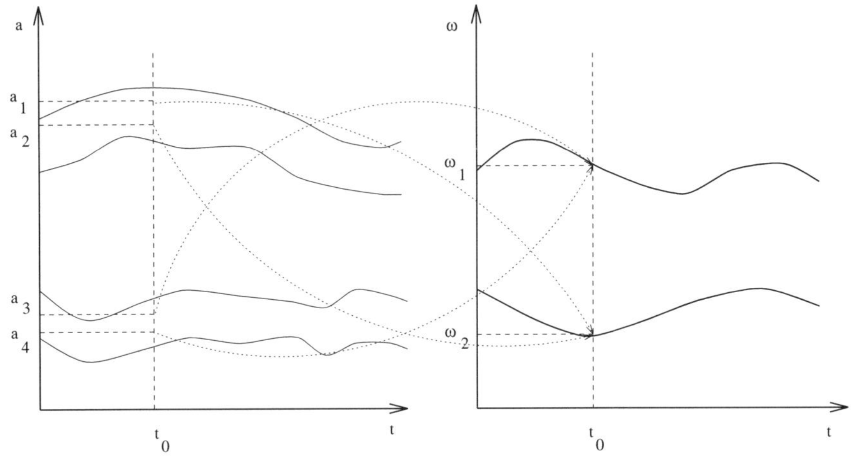


FIGURE 5. If each ribbon is associated to only one pseudo-frequency, then the *squeezed ribbons* degenerate into two curves. This figure sketches the *synchrosqueezing process.*

an interesting alternative is to *squeeze* the ribbon on a curve. It guarantees that no useful information is lost, while at the same time, the representation is not conditioned by the analysis generating wavelet any more. Furthermore, because a *squeezing process* is performed instead of simply drawing a curve, the robustness to noise is dramatically increased.

The concept of regrouping information from subbands which have the similar behavior is similar to the EIH approach. We just have to define our frequency extractor and the cumulated measure of information.

The new representations proposed in this paper are applicable to the WT, and to the Gabor transform, or, in general, to any time-frequency distribution which contains enough informations about the frequency modulation laws of the signal. Generalized cases are presented in [**48, 46, 52, 53**].

The *synchrosqueezing measure* associated with the continuous WT is introduced [**48, 53, 15**] [9],

$$\mu_{\text{Synmod}}(\omega, b, f, g) = \left| \int^{SIS}_{(b,\omega_{\mathrm{WT}_f}(a,b,g))\in \mho(b,\delta_b,\omega,\delta_\omega)} \mathrm{WT}_f(a,b,g)w_p(a)\,da \right| + \int^{SIS}_{(b,\omega_{\mathcal{WT}_f}(a,b,g))\in \mho(b,\delta_b,\omega,\delta_\omega)} \left|\mathcal{WT}_f(a,b,g)\right| w_m(a)\,da \tag{6.1}$$

where

$$\mho(b,\delta_b,\omega,\delta_\omega) = \left[b-\delta_b(\omega), b+\delta_b(\omega)\right] \otimes \left[\omega-\delta_\omega, \omega+\delta_\omega\right] \tag{6.2}$$

$\delta_\omega$ is half of the resolution in frequency which is achievable by this approach and $\delta_b(\omega)$ is half the length of the interval used to average the measure in $b$ direction

[9]The boundary notation of the integral is symbolic as an isolated frequency value has a zero measure. It must always be understood as a given set of non-zero measure parameterized by $\omega$.

[10]. It is related to the time reassignment described in [**6, 7**]. $w_p(a)$ and $w_m(a)$ are weighting factors.

In the case of a (Quasi-Continuous Wavelet Transform) (QCWT) [11], the distribution is implemented with finite summations [**53, 48, 41, 57, 56, 43, 42, 47, 15**].

As a complex exponential has always its contribution spread over the essential frequency support of the accordingly dilated wavelet, only one main connex interval of $a$ values should be taken into account in the integral of equation (6.1). All the other contributions should be rejected. This technique is called the selective increment strategy (SIS). It reduces the perturbation of the *squeezed plane* due to noise and increases the robustness and stability of any method based on it. The notation $\int^{SIS}$ denotes the compatibility of the associated measure with a SIS: the integral is subject to the SIS.

Complex values are used so that when two primary components come close together, as they present different phase behaviors in their WT, they interfere incoherently and a clear separation automatically appears at the edge between the two components. This formulation is also practical for selective reconstructions based on the selective fusion algorithm described later on and the direct summation strategy for QCWT synthesis described in [**48, 49**].

## 7. Synchrosqueezed Place Representation of Speech

Figures 6 to 8 illustrate respectively the *squeezed plane* and the *synchrosqueezed plane* obtained without SIS for three speech signals. The horizontal axis represents the $b$ axis. The vertical axis plots the frequency.

The fast QCWT algorithm that we use is described in details in [**44, 48, 46, 49, 55**]. 101 uniformly spaced levels in frequency have been obtained from a QCWT analysis with 5 octave levels, with 7 sub-octaves between each octave and an upper-octave interpolation. The generating analysis wavelet is defined by:

$$\psi(t) = e^{-t^2/16}(e^{j\,0.69\pi t} - 4.89098\,10^{-4}). \tag{7.1}$$

The projection on the lower *scaling function* has been disregarded. The speech signals are sampled with $\nu_s = 8\,kHz$ The *synchrosqueezed plane* are downsampled with a factor 20, the contribution of each of the 20 samples being accumulated at the same abscissa $b$.

---

[10]The most obvious choice is to use $\delta_b(\omega) \propto 1/\omega$.

[11]The QCWT is a discrete time transform with no downsampling along the time axis and the possibility to choose any sampling grid along the scale axis. Actually, a very useful sampling grid for analysis purposes is defined by

$$a = a_0^{-m-\varepsilon/(n_{so}+1)}, \quad a_0 > 0,\ a_0 \neq 1,\ m \in \mathbb{Z},\ \varepsilon = 0, \ldots, n_{so} \tag{6.3}$$

where $n_{so} \in \mathbb{N}\backslash\{0\}$.

In the QCWT case, $a_0$ is usually chosen equal to 2. In this particular case, the different levels, numbered by the parameter $m$, are arranged in octaves and the sub-levels are arranged in fractions of octaves and called sub-octaves or sub-subbands. For other choices of the value of $a_0$, the same denomination is maintained. However these choices are of very little interest, as the sampling grid (6.3) should always give enough resolution in scale.

If $a_0 \neq 2$,the generating analysis wavelets are usually not dilated version of a same generating analysis wavelet. However, it is not an issue as we develop a representation essentially independent of the generating analysis wavelet.

In this paper all the experiments were done with $a_0 = 2$. Fast QCWT algorithms for analysis and synthesis are described in [**49, 55, 48**].

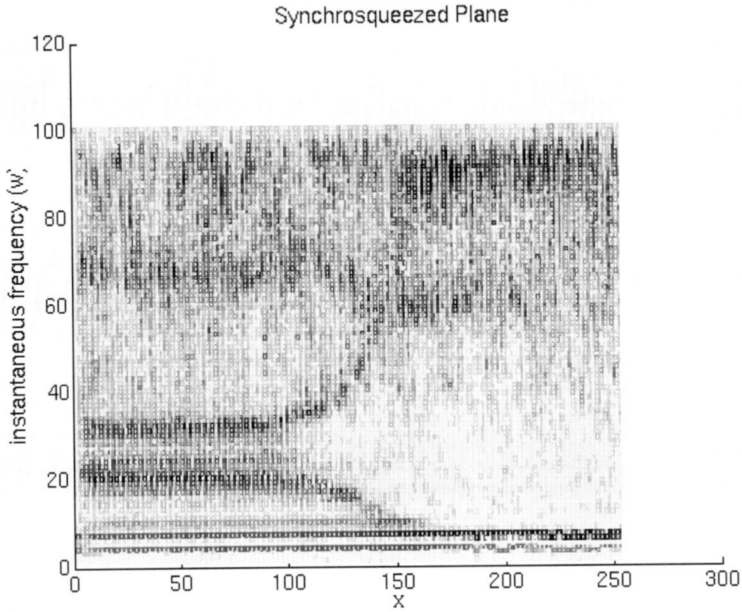


FIGURE 6. *Synchrosqueezed plane representation* for $/\underbrace{\cdots - a - a} - \underbrace{i - i - \cdots}/$. A colored noise is present with SNR $= 15\,dB$.

The first signal is a transition from the sustained French vowel $/a/$ to the sustained French vowel $/i/$, pronounced by a male voice in a noisy environment. It is denoted by $/\underbrace{\cdots - a - a} - \underbrace{i - i - \cdots}/$. The corrupting noise has been measured in situ. It produces a colored noise with SNR $= 15\,dB$. Starting from the lower frequencies, the first three primary components met in the $/a/$ portion represent the pitch and the first two harmonics of the pitch. It is particularly well visible as the signal has not been pre-emphasized. Then, the two first dominant *squeezed ribbons* are associated with the first two formants. The third and fourth formants are also clearly apparent. The transition between a characteristic $/a/$ and a characteristic $/i/$ is particularly striking, and the associated formant behavior is in perfect agreement with what is described in the literature. [**21, 10, 68**].

Notice the oscillation of the whole *synchrosqueezed plane* at the pitch frequency. Note also that the boundary effects have been completely canceled with the VCE ($V_0$ Compatible Extrapolation) strategy described in [**48, 46, 44**].

The second signal is the sentence $/h - \delta - w - a - r - j - u\ ?/$ (How are you?) pronounced by a male American voice, in the New Jersey dialect, with the same SNR. The behavior of the different formants clearly appears, still in perfect agreement with the theory.

The third example is $/ow - g - \Lambda - s - t/$ (August), which is partially voiced except for the last part $/st/$ which is unvoiced. The disappearance of the pitch and its harmonics is particularly striking.

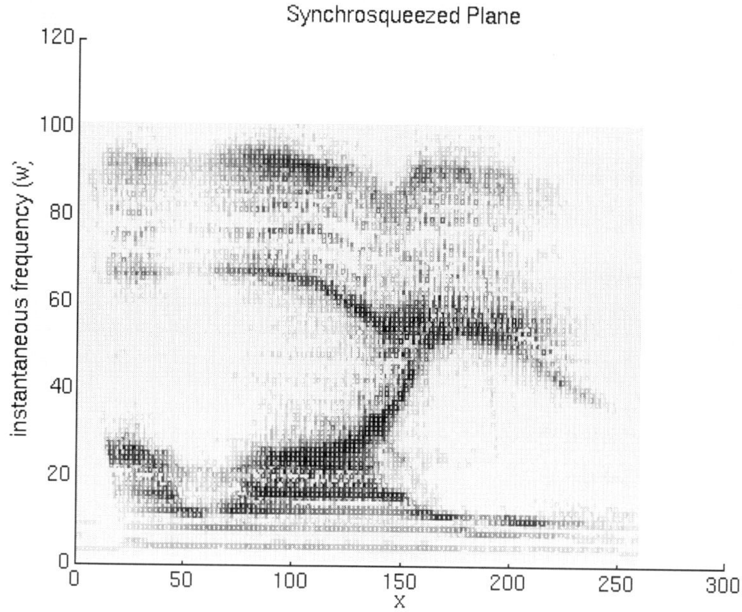


FIGURE 7. *Synchrosqueezed plane representation* for $/h-\delta-w-a-r-j-u\ ?/$. A colored noise is present with $\mathrm{SNR}=15\,dB$.

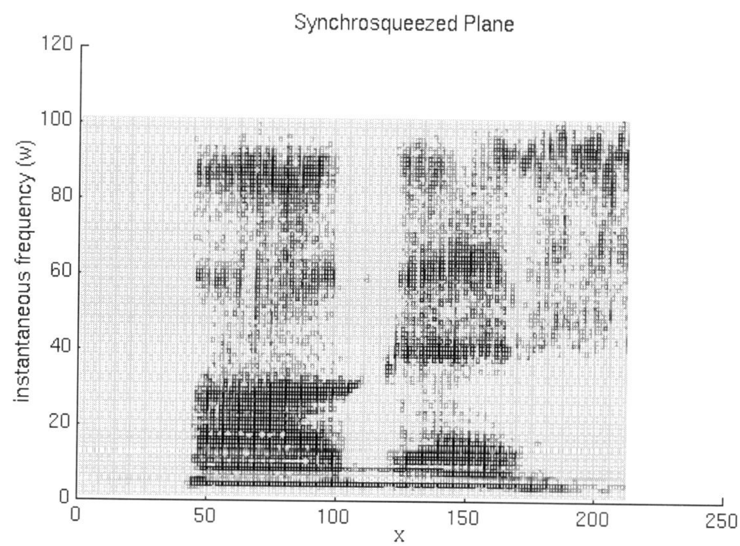


FIGURE 8. *Synchrosqueezed plane representation* for $/ow-g-\lambda-s-t\ /$. A colored noise is present with $\mathrm{SNR}=15\,dB$.

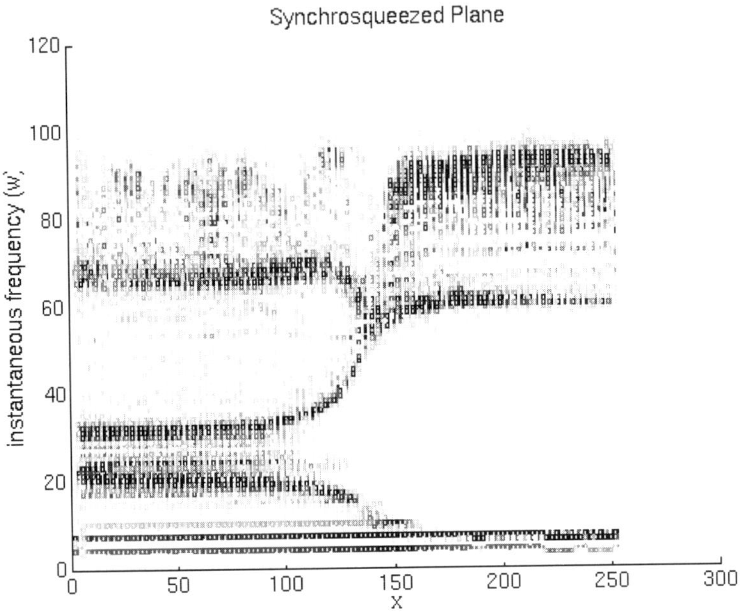


FIGURE 9. *Synchrosqueezed plane representation* for $/\underbrace{\cdots - a - a}\ - \underbrace{i - i - \cdots}/$. A colored noise is present with SNR $= 15\,dB$. An additional white noise is added with SNR $= 11\,dB$.

In Figure 9 the *synchrosqueezed plane representation* of $/\underbrace{\cdots - a - a}-\underbrace{i - i - \cdots}/$ is presented for an additional white noise with SNR $= 11\,dB$.

A detailed discussion of the robustness of features derived from the *synchrosqueezed plane* can be found in [**53, 54**]. The over-complete nature of the analysis and synthesis presented in this paper is a key contributor to the robustness of our approaches [**8**].

## 8. The Selective Fusion Algorithm

The *synchrosqueezed plane* representation provides immediate readings of the instantaneous frequencies of all the unmasked primary components. However, it is rather difficult to use this representation for a direct extraction of the instantaneous amplitudes of the primary components. Also, it would be interesting to invert the representation and reconstruct portions or the totality of the *synchrosqueezed plane.*

**8.1. The SFA synchrosqueezing measure.** The selective fusion algorithm SFA provides such a tool [**41, 57, 56, 43, 42**].

A SFA is another particular type of *synchrosqueezing measure* which, in theory, provides directly the frequency modulation laws and the amplitude modulation laws of each primary component. However, in practice it is usually split into two successive stages:

- Determination of the frequency modulation law with a first *synchrosqueezing measure*,

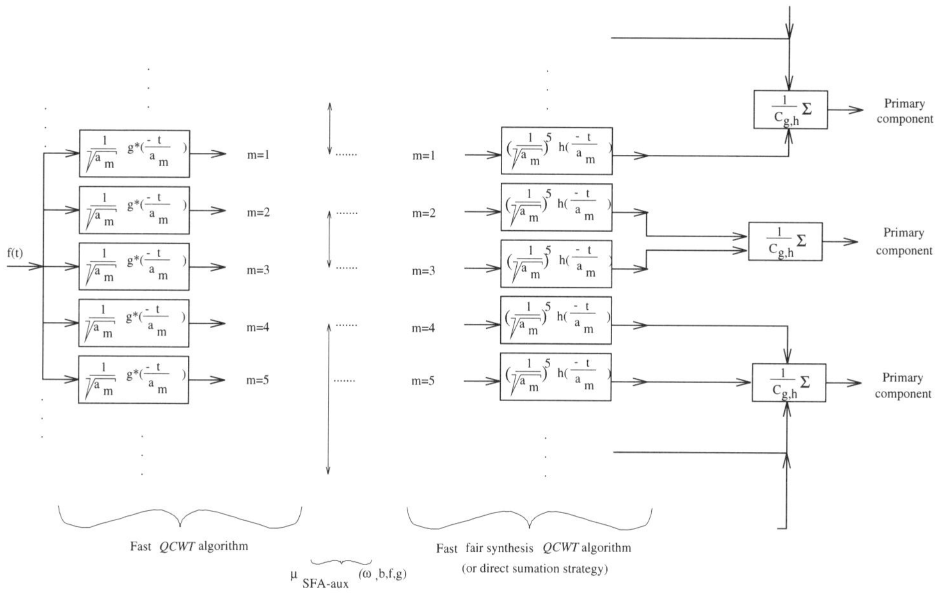


FIGURE 10. Selective fusion algorithm.

- Extraction of the instantaneous amplitude and the initial phase constant, using the SFA. This stage is usually not presented as a *synchrosqueezing measure*.

The non-necessarily positive *synchrosqueezing measure* associated with the SFA is

(8.1)

$$\mu_{\mathrm{SFA-aux}}(\omega, b, f, g) = \int_{\omega_{\mathrm{WT}_f}(a,b,g)=\omega}^{\mathrm{SIS}} \int_{-\infty}^{+\infty} \mathrm{WT}_f(a,t,g)\frac{1}{\sqrt{a}} g\left(\frac{b-t}{a}\right)\frac{dt\, da}{C_{g,g}a^2}$$

and

(8.2)

(8.3) $$\mu_{SFA}(\omega, b, f, g) = \sum_k \delta(\omega - \omega_{c_k}) \int_{\omega \in \mathrm{SqueezedRibbon}(\omega_{c_k})} d\mu_{\mathrm{SFA-aux}}(\omega, b, f, g)$$

where $\omega_{c_k}$ is the center frequency of the associated $k^{th}$ *squeezed ribbon*.

The noise robustness is increased by implementing the integral along $a$, in equation (8.1), with a SIS.

The SFA provides a time-frequency plane representation where the center frequency of each *squeezed ribbon* supports the reconstructed primary components. For these components, the *synchrosqueezing measure*, $\mu_{\mathrm{SFA-aux}}(\omega, b, f, g)$ already provides a direct reading of their frequency modulation laws.

**8.2. Practical implementation.** The general principle underlying the SFA algorithm is the observation of all the scale levels of the WT of the signal and the grouping of the scale levels which present a similar behavior in the time evolution of a property [**41, 57, 56, 43, 42**]. All the levels which belong to a same group are reconstructed with the formula of *reconstruction of the identity*.

The output, for signal satisfying the MM, are the different primary components, and the difference between the input signal and the combination of the different outputs provides $\eta(t)$, as described by equation (2.1),

$$f(t) - \int_0^{+\infty} \mu_{\mathrm{SFA}}(\omega, t, f, g)\, d\omega = \eta(t). \tag{8.4}$$

Hence, if some primary components are present but masked by other dominant primary components, they can be extracted from the remainder after subtraction of the dominant contributions.

Also, the estimation of $\eta(t)$ enables the test of validity of the MM for a given application or family of signal, or the evaluation of the level of additive noise present in the signal.

For speech applications, where the MM is valid, $\eta(t)$ can be considered as noise, at least when the pitch is present, which means when the signal is voiced. In fact, this remains valid for short unvoiced parts located between voiced signals. Therefore, it is possible to use the SFA as a denoising algorithm which cleans the signal from all non-formant-like components.

It can also be used as a classical formant tracker with numerous potential applications, including coding with the *analysis-synthesis* approach [**24, 65, 16**].

Nothing imposes to determine which components must be grouped together based on the measure proposed in equation (8.1). Any other *synchrosqueezing measure* is acceptable.

In that more general case, each level is tagged with the group to which it belongs, including the possibility of belonging to no group at all. For each group, at each time, the levels which are tagged as part of the group are recombined together. This is illustrated in Figure 10.

The synthesis algorithm, used to implement the *resolution of the identity* over each group, must fairly use the contribution coming from each level. In a QCWT, it is often possible to simply use the direct summation technique described in [**48, 46**]. In that case, equation (8.1) becomes

$$\mu_{\mathrm{SFA-ds}}(\omega, b, f, g) = \int_{\omega_{\mathrm{WT}_f}(a,b,g)=\omega}^{\mathrm{SIS}} \mathrm{WT}_f(a, b, g) \frac{dt\, da}{C_{g,\delta} a^{3/2}}. \tag{8.5}$$

The total amplitude and phase determination algorithm is presented in Figure 11.

Different techniques can be used to extract the instantaneous amplitude of each extracted primary component $f_k(t)$. Of course, it is possible to directly use the Hilbert transform, but it does not use the knowledge of the instantaneous frequency.

Hence a more efficient approach fits an amplitude function to minimize a distance $d(;)$ [12]:

$$\tilde{A}_k(t) = \operatorname{argmin}_{B(t)} d\big(f_k(t), B(t)\cos(\Phi_k(t))\big). \tag{8.6}$$

[12] Of course, this assumes that the phase constant is also known. Otherwise, the phase constant must also be adjusted to minimize the distances.

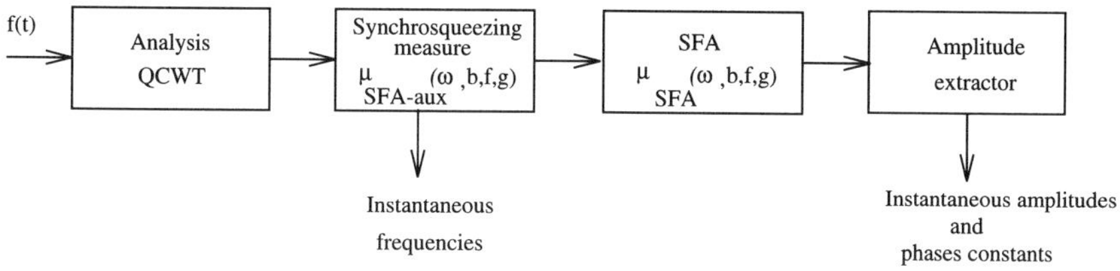


(a)

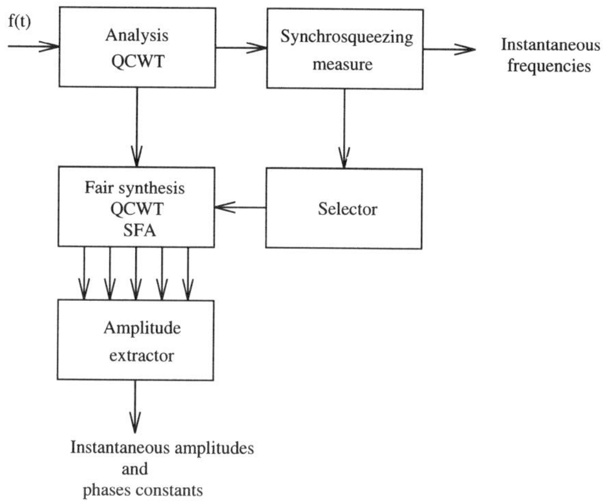


(b)

FIGURE 11. Instantaneous phase and amplitude extraction algorithm. (a) Theoretical approach (b) Practical implementation.

Eventually, it is also possible to adopt the point of view of the sinusoidal model [**61, 36, 59, 60, 67, 2, 26**]. Each primary component can be resolved in a set of sine waves.

**8.3. A primary component tracker.** The numerous oscillations present in *synchrosqueezed plane representations* make it extremely difficult to correctly, efficiently and rapidly track the primary components. A summary of the shortcomings of the classical image processing methods of extraction of saliencies can be found in [**48**].

An *automatic* approach is proposed. It is a fitting technique based on a statistical model which allows an accurate extraction of the *squeezed ribbons* [**48, 46, 54**].

The method proposed in this paper is called the MLE-based living cubic spline extractor (MLE–LCSE) [13]. MLE stands for maximum likelihood estimation [**66, 71, 11, 3**]. MLE are notorious for being slowly converging techniques. This approach avoids that problem, by using clustering estimations which select initial guesses very close to the final solution.

In order to correctly track the *squeezed ribbons* on the basis of dominant amplitudes in the *synchrosqueezed plane* ($\mathcal{V}_{(b,\omega)}$), the *synchrosqueezed* measure is interpreted as a probability density function in $\mathcal{V}_{(b,\omega)}$. The problem becomes a statistical

[13] "living" results from the possibility to kill or create the splines when needed.

problem where this probability density function is simply estimated by the statistical frequency [**66**]. Hence the sampling density along the time axis is given by

$$\rho_s(t) = \int_{\omega \in \mathcal{V}_{(b,\omega)}} d\mu(t,\omega). \tag{8.7}$$

Initially, the model supposes that $N_{pc}$ primary components are present in the *synchrosqueezed plane.* The plane $(\omega \mid t, N_{pc})$, with the notation which means that the values of $\omega$ are conditioned by the position $t$, is modeled as a mixture of Gaussians,

$$P(\omega \mid t, N_{pc}) = \sum_{k=1}^{N_{pc}} \frac{\pi_k}{(\sigma_k(t))^2} \exp -(\omega - \omega_{c_k}(t))^2/2(\sigma_k(t))^2 \tag{8.8}$$

where:

- $P(\omega \mid t, N_{pc})$ is the conditional probability to observe $\omega$ at the time $t$, knowing that there are $N_{pc}$ primary components.
- $\omega_{c_k}(t)$ is the value of the center frequency of the $k^{th}$ *squeezed ribbon*, at the time $t$.
- $(\sigma_k(t))^2$ is the variance of the $k^{th}$ *squeezed ribbon*, at the time $t$. It can be interpreted as the bandwidth of the associated primary component.
- $\pi_k$ is the probability to be associated with the $k^{th}$ *squeezed ribbon.*
- $1/((\sigma_k(t))^2 \exp -(\omega - \omega_{c_k}(t))^2/2(\sigma_k(t))^2$ is the probability to observe the value $\omega$, knowing that it is associated with the $k^{th}$ *squeezed ribbon.*

The center frequencies $\omega_{c_k}(t)$ are described by a cubic spline representation,

$$\omega_{c_k}(t) = \omega_{c_k}(t, k, \theta_k) \tag{8.9}$$

where $\theta_k$ stands for all the parameters which define the $k^{th}$ cubic spline. In order to be coherent with the definition of a formant, which are never allowed to cross each other, non-crossing of the components is enforced. The parameters $\pi_k$, $\theta_k$ and $(\sigma_k(t))^2$ are evaluated by maximizing the likelihood

$$\int_{t \in \mathcal{V}_{(b,\omega)}} \rho(t)\, dt \log_2\big(P(\omega \mid t, k, \pi_k, \theta_k, (\sigma_k(t)))\big). \tag{8.10}$$

The optimization problem (8.10) is solved as described [**48, 46, 54**].

For the first pass through the whole $\mathcal{V}_{(b,\omega)}$, the initial guesses required to solve the equations are obtained by clustering.

Equation (8.10) is solved by writing each center frequency as

$$\omega_{c_k}(t) = \sum_{i=1}^{J_i} \omega_{c_k}(t_i)\xi_i(t) \tag{8.11}$$

where $\xi_i(t)$ are basis functions for the cubic spline interpolations. $\xi_i(t)$ is obtained by imposing a value 1 at $t_i$ and zero at the other $t_k$, and performing a cubic spline interpolation. This trick guarantees a fast optimization of equation (8.10), with a small number of parameters $\theta_k$. After determination of the center frequencies, the variances are straightforwardly determined.

After each estimation of center frequencies and variances, the *synchrosqueezed measures* of each extracted ribbon are computed. This value is compared with the values of its closest neighbor, of the measure of a neighbor region which does not belong to any extracted ribbon, and to the previous value along the ribbon. Based on these three values, a decision is taken to keep the ribbon or to kill it.

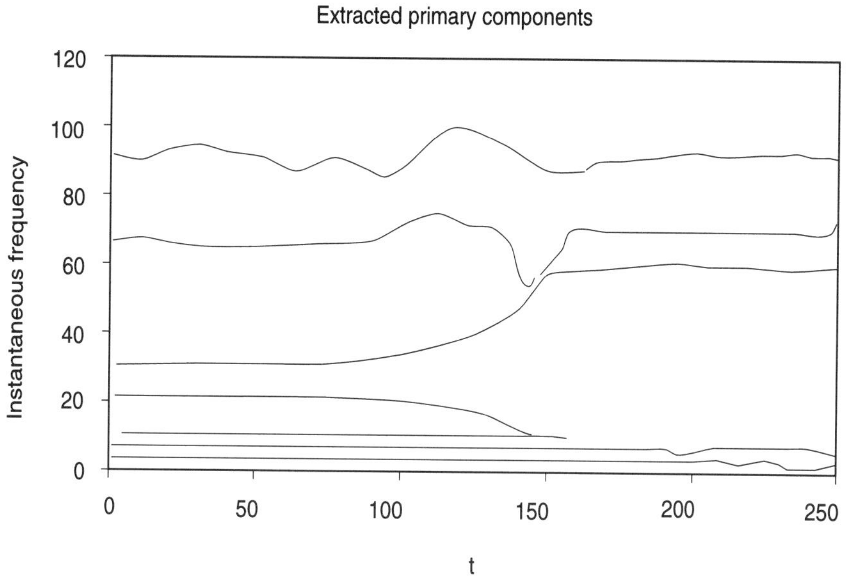


FIGURE 12. Extracted primary components for $/\underbrace{\cdots - a - a}\ -\underbrace{i - i - \cdots}/$. A colored noise is present with SNR $= 15\,dB$. A white noise is added with: SNR $= 10\,dB$.

Figure 12 presents the center frequency obtained for $/\underbrace{\cdots - a - a} - \underbrace{i - i - \cdots}/$ under noisy conditions.

Once tracked down with this algorithm, the SFA algorithm can be applied to select and reconstruct the different primary component of the signal.

Applications of the extracted ribbons to extract the amplitude modulation laws can be found in [**48, 46, 47, 53, 54, 52**]. The reader will find that *synchrosqueezed*-derived cepstral parameters give are more robust to acoustic distortions. Also in these papers, the representation has been applied for closed-set speaker identification in the context of the *modulation model*. This leads to the introduction of a "phase-oriented" cepstral parameter: the *wastrum*.

The excellent robustness of the *synchrosqueezed plane representations* or wastrum-type of parameters is discussed in [**53, 54, 50, 40**].

## 9. Conclusion

An important conclusion of this paper is that *synchrosqueezing measures* appear as natural tools to trim time-scale or time-frequency representations; especially as it is through *synchrosqueezing* that the representations lose most of the perturbations introduced by the analyzing tools.

Besides, because of the *synchrosqueezing process*, the resulting representation is robust to noise, especially when it is implemented with a SIS.

With the SFA, it is easy to extract the frequency, phase and amplitude modulation laws of the primary components and to reconstruct selectively the different primary components.

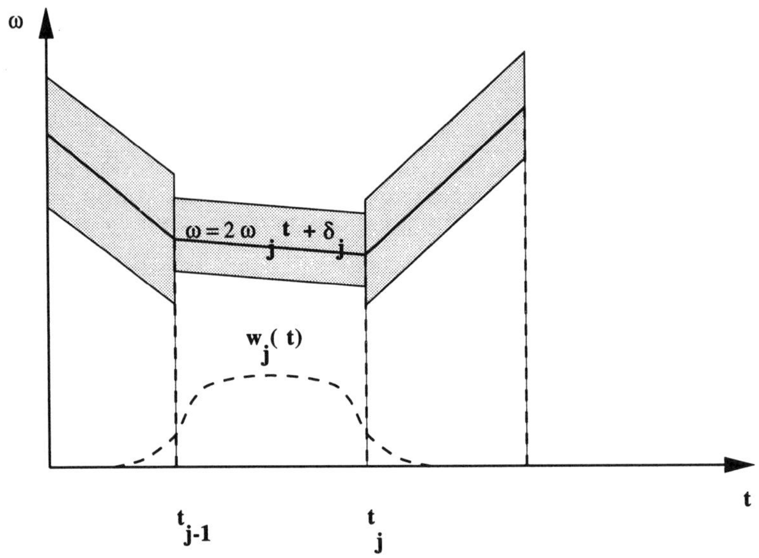


FIGURE 13. It illustrates the time-frequency essential support of the windows and the linear frequency modulation laws which are imposed by Meyer's approach. The continuity of the center frequency is important.

The formalism can be adapted to representations based on auditory nerve models. As a result, the SFA approach invert these representations. It is another interesting contribution of this work.

At this stage, we can conclude that the phase-derived features, and in particular the *synchrosqueezed representation* offers promising perspectives for speech recognition. They are worth further investigation, especially in order to avoid the need to extract the center frequencies and bandwidths [**48, 46, 40, 50**].

## 10. A Digression: Wigner-Ville Wavelets à la Meyer

Recently, Professor Y. Meyer introduced new orthonormal bases in order to decompose signals into series of orthonormal chirp-type of functions [**62**]. Furthermore, this analysis is closely related to Wigner-Ville distributions with cancellation of the cross-terms. We refer the reader to [**62**] for detailed discussions.

Considering a set of windows $\{W_k\}_k$ such that:

$$\sum_{k=-\infty}^{+\infty} W_k^2(t) = 1 \tag{10.1}$$

$$W_k(t_k + s) = W_{k-1}(t_k - s) \quad |s| < \eta_k \tag{10.2}$$

$$\eta_k + \eta_{k+1} \leq l_k = t_{k+1} - t_k. \tag{10.3}$$

Meyers shows that the set $\{g(k,n)\}_{k\in\mathbb{Z},n\in\mathbb{Z}}$ forms an orthonormal basis of $L^2(\mathbb{R})$, when:

$$g(k,n) = \sqrt{\frac{2}{l_n}}\sin(\pi k + \frac{1}{2})\left(\frac{t - t_n}{l_n}\right)e^{j(\omega_n t^2 + \theta_n t)}W_n(t). \tag{10.4}$$

The meaning of the parameters $\omega_n$ and $\theta_n$ is presented in Figure 13.

The difficulty of the approach resides in the selection of these parameters for a given signal. As *synchrosqueezed analysis* are essentially independent of the analyzing tool, it can be used together with an overcomplete analysis to extract acceptable values for the parameters $\omega_n$ and $\theta_n$.

Consider the over-complete set:

$$g(k,n) = \sqrt{\frac{2}{l_n}} \sin\left(\pi(\underline{\delta(k)} + \frac{1}{2})\right)\left(\frac{t-t_n}{l_n}\right) e^{j(\omega_n t^2 + \theta_n t)} W_{\underline{\delta_n}}(t) \tag{10.5}$$

$\underline{\delta(k)}$ and $\underline{\delta_n}$ denote the steps of the over-complete set. A Wigner-Ville transform à la Meyer (WVAM) is introduced as

$$T_f(k,n,g) = \langle g(k,n) \mid f \rangle = \|T_f(k,n,g)\| e^{j\Phi_{T_f(k,n,g)}}. \tag{10.6}$$

The pseudo-frequency is defined as

$$\omega_{T_f}(k,n,g) = \frac{\partial \Phi_{T_f(k,n,g)}}{\partial \delta_n}. \tag{10.7}$$

A *synchrosqueezing measure* automatically results:

$$\mu(\omega,n,g,f) = \sum_{\omega_{T_f}(k,n,g)\in[\omega-\delta_\omega,\omega+\delta_\omega]}^{SIS} T_f(k,n,g).w(k). \tag{10.8}$$

The desired parameters are obtained by iteration over the following scheme:

$$\begin{array}{l} g_0 \Rightarrow T_f(k,n,g_0) \Rightarrow \mu(\omega,n,g_0,f) \\ \Downarrow \\ g_1 \Longrightarrow T_f f(k,n,g_1) \Rightarrow \mu(\omega,n,g_1,f) \end{array} \tag{10.9}$$

## 11. Acknowledgments

The author's pre-eminent thanks to his research advisor, Professor I. Daubechies, for her help, encouragement and guidance during this research. Special thanks to Professor R. Mammone and Professor J. L. Flanagan for their comments and suggestions and for welcoming the author in the speech recognition team at CAIP, Rutgers University, where most of this research was performed. The author is also grateful to Dr. O. Ghitza and Dr. T. Hastie, with whom he had fruitful discussions. Eventually the author wants to give thanks, many thanks, to his thesis advisors, Professor J-P. Antoine and Professor P. Delogne.

The author is also grateful to the members of the departments TELE and FYMA at UCL, CAIP at Rutgers University, the speech research department at AT&T Bell Laboratories and the department of human language technologies at IBM T. J. Watson Research Center for their continuous assistance and friendship.

## References

1. J. B. Allen, *Cochlear modeling*, IEEE Trans. Acoust. Speech Signal Process. **2** (1985), no. 1, 3–29.
2. L. B. Almeida and F. M. Silva, *Variable-frequency synthesis: An improved harmonic coding scheme*, IEEE Proc. ICASSP, San Diego, CA, 1984, p. 27.5.1.
3. T. Anderson, *An introduction to multivariate statistical analysis*, Wiley, New York, NY, 1958.
4. K. Assaleh, *Robust features for speaker identification*, CAIP Center, Rutgers University, New Brunswick, NJ, 1993.

5. B. S. Atal and J. R. Remde, *A new model of LPC excitation for producing natural-sounding speech at low bit rates*, IEEE Proc. ICASSP, Paris, France, 1982, pp. 614–617.
6. F. Auger and P. Flandrin, *The why and how of time-frequency reassignment*, Internat. Symposium on Time-frequency and Time Scale Analysis, Philadelphia, PA, 1994, pp. 197–200.
7. ______, *Improving the readability of time-frequency and time-scale representations by the reassignment method*, IEEE Trans. Acoust. Speech Signal Process. **2** (1985), no. 1, 1995.
8. J. J. Benedetto, *Noise reduction in terms of the theory of frames*, Spline Functions and the Theory of Wavelets (Montréal, 1996), Centre de recherches mathématiques, Univ. de Montréal, Montreal, Canada, 1996 (talk).
9. J. J. Benedetto and A. Teolist, *A wavelet auditory model and data compression*, ACHA **1** (1993), no. 1, 3–28.
10. P. Calliope, *La parole et son traitement automatique*, Masson, 1989.
11. , J. V. Candy, *Signal processing: the modern approach*, Mc Graw-Hill, 1988.
12. J. M. Combes, A. Grossmann, and Ph. Tchamitchian (eds.), *Wavelets, time-frequency methods and phase space*, Proc. of the Int. Conf. (Marseille, 1987), Springer-Verlag, 1989.
13. E. Copson, *Asymptotic expansions*, Cambridge Univ. Press, Cambridge, MA, 1965.
14. C. D'Alessandro, *Time-frequency speech transformation based on an elementary waveform representation*, Speech Comm. **9** (1990), 419–431.
15. I. Daubechies and S. Maes, *A nonlinear squeezing of the continuous wavelet analysis based on auditory nerve models* Wavelets in Medicine and Biology (A. Alroubi and M. Unser), CRC Press, 1996.
16. J. R. Deller, J. G. Proakis and J. H. L. Hansen, *Discrete-time processing of speech signals*, MacMillan, New York, NY, 1993.
17. N. Delprat, *Extraction of frequency modulation laws in sound synthesis*, Res. Notes Appl. Math., Masson and Springer-Verlag, 1991.
18. N. Delprat, B. Escudié, P. Guillemain, R. Kronland-Martinet, Ph. Tchamitchian, and B. Torresani, *Asymptotic wavelet and Gabor analysis: Extraction of instantaneous frequencies*, IEEE Trans. Inform. Theory **38** (1992), no. 2-II, 644–664.
19. B. Escudié and B. Torresani, *Wavelet analysis as asymptotic signals.* I. *Ridge and skeleton of the transform*, RMA Res. Notes Appl. Math., vol. 10, Masson, Paris; Springer-Verlag, Berlin, 1992.
20. G. Fant, *Acoustic theory of speech production*, Mouton's Co, Hague, 1960.
21. J. L. Flanagan, *Speech analysis synthesis and perception*, Springer-Verlag, 1972.
22. D. H. Friedman, *Instantaneous-frequency distribution vs. time: An interpretation of the phase structure of speech*, IEEE Proc. ICASSP, 1985, pp. 1121-1124.
23. S. Furui, *Cepstral analysis technique for automatic speaker verification*, IEEE Trans. Acoust. Speech Signal Process. **29** (1981), 254–272.
24. ______, *Digital speech processing, synthesis and recognition*, Marcel Dekker, New York, NY, 1989.
25. S. Furui and M. Sondhi (eds.), *Advances in speech signal processing*, Marcel Dekker, New York, NY, 1991.
26. O. Ghitza, *Auditory nerve representation criteria for speech analysis/synthesis*, IEEE Trans. Acoust. Speech Signal Process. **6** (1987), no. 35, 736-740.
27. ______, *Advances in speech signal processing*, Advances in Speech Signal Processing, Marcel Dekker, New York, NY, 1991.
28. O. Ghitza ______, *Auditory models and human performances in tasks related to speech coding and speech recognition*, IEEE Trans. Speech and Audio Proc. **2** (1994), no. 1, 115–132.
29. C. Gonnet and B. Torresani, *Local frequency analysis with two-dimensional wavelet transform*, Tech. report no. CPT-92/P.2870, Centre de Physique Théorique, CNRS, (Luminy, 1993).
30. S. Greenberg (ed.), *Representation of speech in the auditory periphery*, J. Phonetics **16** (1988), no. 1, 1–151.
31. P. Guillemain, *Analyse et modélisation de signaux acoustiques*, Mémoire de D.E.A. Acoustique et Dynamique des vibrations, Laboratoire de Mécanique et d'Acoustique, CNRS (Marseille, 1990).
32. ______, *Estimation of frequency modulation laws associated to a sum of asymptotic signals with continuous wavelet and Gabor transform*, Tech. report, Centre de Physique Théorique, CNRS (Luminy, 1992).

33. ———, *Ridges associated to isolated frequential and temporal singularities*, Tech. report, Centre de Physique Théorique, CNRS, (Luminy, 1992).
34. P. Guillemain, R. Kronland-Martinet, and B. Martens, *Application de la transformée en ondelettes à la spectroscopie* NMR, Note no. 112, Laboratoire de Mécanique et d'Acoustique, CNRS (Marseille, 1989).
35. ———, *Estimation of spectral lines with the help of the wavelet transform, application in* N.M.R *spectroscopy*, Wavelets and Applications (Marseille, 1989) RMA Res. Notes Appl. Math. vol 20, Paris, 1992, pp. 38–60.
36. P. Hedelin, *A tone-oriented voice-excited vocoder*, IEEE Proc. ICASSP, Atlanta, GA, 1981, p. 205.
37. M. Holschneider, *General inversion formulas for wavelet transforms*, J. Math. Phys. **34** (1993), no. 9, 4190–4198.
38. M. Lavrentiev and B. Chabat, *Méthodes de la théorie des fonctions d'une variable complexe*, Mir, Moscow, 1977.
39. J. S. Liénard, *Speech analysis and reconstruction using short-time, elementary, waveforms*, IEEE Proc. ICASSP (Dallas, 1987), pp. 948–951.
40. S. Maes, *Impact of filtering distorsions on acoustic speech parameters*, NSA, (Fort Meade, MD, 1995) (viewgraphs).
41. ———, *Quasi-continuous wavelets used to extract parameters in speech processing speech*, Summer School on Wavelet Techniques and Applications, (Bordeaux, 1993) (viewgraphs).
42. ———, *Non-linear techniques for parameter extraction from quasi-continuous wavelet transform, with application to speech*, Neural Networks and Artificial Intelligence, Proc. EOS-SPIE of the International Symposium on Substance Identification Technologies (Innsbrück, 1993), paper no. 2093-03, EOS-SPIE.
43. ———, *Application of the wavelet transform for speech processing*, Proc. of the 2nd Annual Intern Space Univ. Alumni Conf. (L. Johnson and P. Robinson, eds.) (Huntville, 1994) NASA Conf. publ. no. 3253; NASA, G.C.M.S.F.C., Marshal Space Flight Center, Alabama.
44. ———, *A fast quasi-continuous wavelet transform algorithm*, Proc. Colloque Temps-Fréquence, Ondelettes et Multirésolution: Théorie, Modèles et Applications, Inst. Nat. Sci. Appl. Lyon et CNRS, Lyon, France, 1994.
45. ———, *The generalized hybrid wavelet transform yields a new reading of subband coding decompositions*, Annex PhD Dissertation, Université Catholique de Louvain, Louvain-la-Neuve, Belgium, 1994.
46. ———, *The synchrosqueezed plane representation, a new time frequency representation applied to speaker identification*, CAIP Center, Rutgers Univ., New Brunswick, NJ., 1994.
47. ———, *Wastrum: Wavelet derived generalized cepstrum, using the synchrosqueezed plane*, Proc. Colloque Temps-fréquence, Ondelettes et Multirésolution: Théorie, Modèles et Applications (Lyon, 1994), Inst. Nat. Sci. Appl. de Lyon and CNRS.
48. ———, *The wavelet transform in signal processing, with application to the extraction of the speech modulation model features*, Ph.D. thesis, Univ. Catholique de Louvain, Louvain-la-Neuve, Belgium, 1994.
49. ———, *Fast quasi-continuous wavelet algorithms for analysis and synthesis of* 1-D *signals*, SIAM J. Appl. Math. (1995) (to appear).
50. ———, *Investigation of new acoustic features for robust automatic speech recognition*, Techn. Memorandum, AT&T Bell Lab., Speech Res. Depart., Murray Hill, NJ, 1995.
51. ———, *LPC Techniques and some new ear-model-based front-ends*, Lecture Notes of internal course on Speech Recognition, IBM T. J. Watson Research Center, Human Language Technology, Hawthorne, NY., 1995.
52. ———, *The synchrosqueezed representation yields a new reading of the wavelet transform*, Proc. SPIE 1995 on OE/Aerospace Sensing and Dual Use Photonics—Wavelet Applications for Dual Use—Session on Acoustic and Signal Processing, Wavelet Applications II (Orlando, FL, 1995) (H. H. Szu, ed.), Part I, vol. 2491, pp. 532-559.
53. ———, *The wavelet-derived synchrosqueezed plane representation yields a new time-frequency analysis of* 1-D *signals, with application to speech*, ACHA (1995) (submitted).
54. ———, *The wavelet-derived synchrosqueezed plane representation yields new front-ends for automatic speech recognition*, IEEE Trans. Speech Audio Process. (1995) (submitted).
55. *Signal analysis and synthesis with* 1-D *quasi-continuous wavelet transform*, Proc. 12th. Intern. Conf. on Analysis and Optimization of Systems, IRSIA, Paris, 1996.

56. S. Maes and I. Daubechies, *An application of wavelets to the characterization of formants parameters in speech*, AT&T Bell (Holmdel, USA, 1993) (viewgraphs).
57. ———, *Wavelets and the modulation model*, Speech Processing Using the Modulation Model (Washington DC, 1992); Tech. report, CAIP Center, Rutgers Univ., New Brunswick, NJ, 1992.
58. B. Martens, *Application de l'analyse en ondelettes à la spectroscopie* RMN, Mémoire de Licence en Physique, Univ. Catholique de Louvain, Louvain-la-Neuve, Belgium, 1989.
59. R. J. McAulay and T. F. Quatieri, *Magnitude-only reconstruction using a sinusoidal speech model*, IEEE Proc. ICASSP, San Diego, CA, 1984, p. 27.6.1.
60. ———, *Mid-rate coding based on a sinusoidal representation of speech*, IEEE Proc. ICASSP, Tampa, FL, 1985, p. 945.
61. ———, *Speech analysis/synthesis based on a sinusoidal representation*, IEEE Trans. Acoust. Speech Signal Process. **34** (1986), no. 4, 744–754.
62. Y. Meyer, *Time-scale and time-frequency analysis in signal or image processing*, Program on Spline Functions and the Theory of Wavelets (Chaire Aisenstadt), Centre de Recherches Mathématiques, Univ. de Montréal, Montreal, Canada, 1996.
63. P. M. Morse, *Vibrations and sounds*, Mc Graw-Hill, New York, NY, 1948.
64. P. M. Morse and K. U. Ingard, *Theoretical acoustics*, Mc Graw-Hill, New York, NY, 1968.
65. F. J. Owens, *Signal processing of speech*, Macmillan, UK, 1993.
66. V. Pugachev, *Théorie des probabilités et statistique mathématique*, Mir, Moscow, 1982.
67. T. F. Quatieri and R. J. McAulay, *Speech transformations based on a sinusoidal representation*, IEEE Proc. ICASSP (Tampa, 1985), p. 489.
68. L. Rabiner and B-H. Juang, *Fundamentals of speech recognition*, Prentice Hall, Englewoods Cliffs, NJ, 1993.
69. M. A. Ramalho, *The pitch modulation model with applications in speech processing*, CAIP Center, Rutgers University, New Brunswick, NJ, 1994.
70. G. Sarraco, G. Sessarego, J. Sageloli, P. Guillemain, and R. Kronland-Martinet, *Extraction of modulation laws of elastic shells by the use of the wavelet transform*, Res. Notes Appl. Math., Masson and Springer-Verlag, 1991.
71. L. R. Shenton, *Maximum likelihood estimation in small samples*, Macmillan, New York, NY, 1977.
72. B. Torresani, *Wavelet analysis as asymptotic signals*. II. *A tentative model for bat sonar receiver*, RMA Res. Notes Appl. Math., vol. 10, Masson, Paris; Springer-Verlag, Berlin, 1992.


IBM, T. J. Watson Research Center, Human Language Technologies, Acoustic Processing Department, Room J2–N52, P.O. Box 704, Yorktown Heights, NY 10598, USA.

*E-mail address*: smaes@watson.ibm.com

Centre de Recherches Mathématiques
CRM Proceedings and Lecture Notes
Volume **18**, 1999

# Extensions of the Heisenberg Group and Wavelet Analysis in the Plane

Eckart Schulz and Keith F. Taylor

## 1. Introduction

In this lecture, we describe a class of groups formed as semidirect products with the positive real numbers acting on the Heisenberg group as dilations. Special cases from this class have been studied by Torresani [**5**] and Hogan and Lakey [**4**]. Our purpose here is to demonstrate how to take advantage of square-integrability when it is present. All except one of the groups in the class we consider have square-integrable representations and each of these square-integrable representations leads to continuous wavelet transforms of elements of $L^2(\mathbb{R}^2)$. We compute the reconstruction formulae explicitly in these cases. The techniques that we use are based on those developed in [**1**]. We begin by briefly presenting the continuous wavelet transform on $L^2(\mathbb{R})$ that arises from the affine group.

## 2. The Affine Wavelet Transform

Let

$$\mathcal{H}^2_+ = \{g \in L^2(\mathbb{R})\colon\ \operatorname{supp}\hat{g} \subseteq (0,\infty)\, a.e.\}$$

and

$$\mathcal{H}^2_- = \{g \in L^2(\mathbb{R})\colon\ \operatorname{supp}\hat{g} \subseteq (-\infty,0)\, a.e.\}.$$

Then, $L^2(\mathbb{R}) = \mathcal{H}^2_+ \oplus \mathcal{H}^2_-$.

For $a > 0$ and $b \in \mathbb{R}$ and any $g \in L^2(\mathbb{R})$, let

$$\left[\rho(b,a)g\right](t) = \frac{1}{\sqrt{a}}\, g\left(\frac{t-b}{a}\right), \quad \text{for all } t \in \mathbb{R}.$$

The function $\rho(b,a)g$ is a translated and dilated copy of $g$.

An affine wavelet for $\mathcal{H}^2_+$ is any $\omega \in \mathcal{H}^2_+$ such that

$$\int_0^\infty \frac{|\hat{\omega}(\gamma)|^2}{\gamma}\, d\gamma = 1.$$

---

1991 *Mathematics Subject Classification.* Primary: 22E27; Secondary: 43A80, 22E70.
This is the final form of the paper.

With such $\omega$ fixed, let $\omega_{b,a} = \rho(b,a)\omega$, for all $b \in \mathbb{R}, a > 0$. The continuous wavelet transform of $f \in \mathcal{H}_+^2$ using the wavelet $\omega$ is the function $T_\omega f$ of $(b,a)$ given by

$$[T_\omega f](b,a) = \langle f, \omega_{b,a}\rangle = \int_{-\infty}^{\infty} f(t)\overline{\omega_{b,a}(t)}\,dt.$$

The reconstruction formula for $f$ is

$$f = \int_0^\infty \int_{-\infty}^{\infty} [T_\omega f](b,a)\omega_{b,a}\frac{db\ da}{a},$$

with the integral on the right converging in the weak sense in the Hilbert space $\mathcal{H}_+^2$. Of course, a properly modified version of this holds on $\mathcal{H}_-^2$.

In terms of square-integrable group representations, the group here is the affine group $G_{\text{aff}} = \{(b,a)\colon b \in \mathbb{R}, a \in \mathbb{R}^+\}$ which has the group product of $(b,a)$ with $(b',a')$ given by $(b,a)(b',a') = (b+ab', aa')$. For each $(b,a)$ in $G_{\text{aff}}, \rho(b,a)$ is a unitary operator on $L^2(\mathbb{R})$ and $(b,a) \to \rho(b,a)$ is a unitary representation of $G_{\text{aff}}$. The representation $\rho$ is irreducible and square-integrable.

## 3. A Generalization

The group $G_{\text{aff}}$ can be viewed as a semidirect product formed by the multiplicative group of positive real numbers acting on $\mathbb{R}$ by dilations. This situation was generalized in **[1]** to the following: Let $H$ be a closed subgroup of $GL_n(\mathbb{R})$, the general linear group of invertible $n \times n$ real matrices. Consider $\mathbb{R}^n$ as the space of $n$-dimensional column vectors and let $h \in H$ act on $\mathbf{x} \in \mathbb{R}^n$ by $\mathbf{x} \to h\mathbf{x}$, with simple matrix multiplication. Let

$$G = \mathbb{R}^n \rtimes H = \{(\mathbf{x},h)\colon \mathbf{x} \in \mathbb{R}^n, h \in H\}$$

with the product of $(\mathbf{x},h)$ and $(\mathbf{y},k)$ in $G$ given by

$$(\mathbf{x},h)(\mathbf{y},k) = (\mathbf{x}+h\mathbf{y}, hk).$$

The dual group to $\mathbb{R}^n$ is isomorphic to $\mathbb{R}^n$ but we prefer to distinguish it by considering it as $\widehat{\mathbb{R}^n} = \{\gamma = (\gamma_1,\dots,\gamma_n)\colon \gamma_j \in \mathbb{R}, 1 \le j \le n\}$, the space of $n$ dimensional row vectors. Each $\gamma \in \widehat{\mathbb{R}^n}$ acts as a character on $\mathbb{R}^n$ by $(\gamma,\mathbf{x}) = e^{2\pi i\gamma\mathbf{x}}$, for all $\mathbf{x} \in \mathbb{R}^n$. The group $H$ acts on $\widehat{\mathbb{R}^n}$ with the action written on the right. For $h \in H$ and $\gamma \in \widehat{\mathbb{R}^n}, \gamma\cdot h$ is defined by

$$(\gamma\cdot h,\mathbf{x}) = (\gamma, h\mathbf{x}), \text{ for all } \mathbf{x} \in \mathbb{R}^n.$$

Clearly, $\gamma\cdot h = \gamma h$, simple matrix multiplication, so we will simply write $\gamma h$ for $\gamma\cdot h$.

For $g \in L^1(\mathbb{R}^n)$, the Fourier transform of $g$ is denoted $\hat{g}$ and is given by

$$\hat{g}(\gamma) = \int_{\mathbb{R}^n} g(\mathbf{x})(\gamma,\mathbf{x})\,d\mathbf{x}.$$

If $g \in L^1(\mathbb{R}^n) \cap L^2(\mathbb{R}^n)$, then $\hat{g} \in L^2(\widehat{\mathbb{R}^n})$ and $\|\hat{g}\|_2 = \|g\|_2$. Consequently, there exists a unitary map $\mathcal{P}$ of $L^2(\mathbb{R}^n)$ onto $L^2(\widehat{\mathbb{R}^n})$, called the Plancherel transform, such that $\mathcal{P}_g = \hat{g}$, for $g \in L^1(\mathbb{R}^n) \cap L^2(\mathbb{R}^n)$.

There is a natural unitary representation $\rho$ of $G = \mathbb{R}^n \rtimes H$ on $L^2(\mathbb{R}^n)$ given, for $(\mathbf{x},h) \in G$ and $g \in L^2(\mathbb{R}^n)$, by

$$\rho(\mathbf{x},h)g(\mathbf{y}) = \delta(h)^{-1/2}g\big(h^{-1}(\mathbf{y}-\mathbf{x})\big),$$

for all $\mathbf{y} \in \mathbb{R}^n$, where $\delta(h) = |\det(h)|$.

Let $\pi$ denote the unitary representation of $G$ that is equivalent to $\rho$ via $\mathcal{P}$. That is, $\pi$ is a representation on $L^2(\widehat{\mathbb{R}^n})$ such that

$$\pi(\mathbf{x}, h) = \mathcal{P}\rho(\mathbf{x}, h)\mathcal{P}^{-1}, \text{ for all } (\mathbf{x}, h) \in G.$$

A simple calculation (see Proposition 1, [**1**]) shows that, for $\xi \in L^2(\widehat{\mathbb{R}^n})$,

$$\pi(\mathbf{x}, h)\xi(\gamma) = \delta(h)^{1/2}(\gamma, \mathbf{x})\xi(\gamma h), \text{ for all } \gamma \in \widehat{\mathbb{R}^n}.$$

Generally, $\rho$ (or $\pi$) is not an irreducible representation of $G$. To see this, suppose $U$ is a measurable $H$-invariant subset of $\widehat{\mathbb{R}^n}$. Let

$$L^2(U) = \{\xi \in L^2(\widehat{\mathbb{R}^n}) \colon \chi_U\xi = \xi\},$$

where $\chi_U$ is the characteristic function of $U$. Let

$$\mathcal{H}^2_U = \{g \in L^2(\mathbb{R}^n) \colon \mathcal{P}g \in L^2(U)\}.$$

Then, it is easy to check that $L^2(U)$ is a $\pi$-invariant subspace of $L^2(\widehat{\mathbb{R}^n})$ and, so, $\mathcal{H}^2_U$ is a $\rho$-invariant subspace of $L^2(\mathbb{R}^n)$. If $U$ is such that both $U$ and $\widehat{\mathbb{R}^n}\backslash U$ have positive measure, then $L^2(U)$ and $\mathcal{H}^2_U$ are proper subspaces and $\pi$ and $\rho$ are reducible.

Now, suppose that there exists a $\gamma \in \widehat{\mathbb{R}^n}$ such that the $H$-orbit $\gamma H = \{\gamma h \colon h \in H\}$ is open in $\widehat{\mathbb{R}^n}$ and that the map $h \to \gamma h$ is a bijection of $H$ onto $\gamma H$. Let $U = \gamma H$ and let $\rho_U$ denote the subrepresentation of $\rho$ corresponding to the subspace $\mathcal{H}^2_U$ of $L^2(\mathbb{R}^n)$. In Theorem 1 of [**1**], it was shown that $\rho_U$ is an irreducible square-integrable representation of $G$.

There is a beautiful theory of generalized orthogonality relations for square-integrable representations due to Duflo and Moore [**2**]. In the situation at hand, the relevant aspects are as follows (see Theorem 2 of [**1**] for full details):

Let $\Psi$ denote the Radon-Nikodym derivative of the left Haar measure of $H$, transfered to the orbit $U$, with respect to the Lebesgue measure of $\widehat{\mathbb{R}^n}$, restricted to $U$. Define an operator $K$ whose domain is $\operatorname{dom} K = \{g \in \mathcal{H}^2_U \colon \mathcal{P}g\Psi \in L^2(U)\}$ and, for $g \in \operatorname{dom} K$, define

$$Kg = \mathcal{P}^{-1}(\mathcal{P}g/\Psi).$$

The operator $K$ is (generally) unbounded, self-adjoint and positive. It is a generalization of the "dimension" of $\rho_U$. The operator $K^{-1/2}$ has domain $\{g \in \mathcal{H}^2_U \colon \hat{g}\Psi^{1/2} \in L^2(U)\}$ and $K^{-1/2}g = \mathcal{P}^{-1}\big((\mathcal{P}g)\Psi^{1/2}\big)$. For $f, g \in \mathcal{H}^2_U$, let $V_{g,f}(\mathbf{x}, h) = \langle f, \rho_U(\mathbf{x}, h)g\rangle$, for all $(\mathbf{x}, h) \in G$. If $g \in \operatorname{dom} K^{-1/2}$, then $V_{g,f} \in L^2(G)$, for all $f \in \mathcal{H}^2_U$ and, for $f_1, f_2 \in \mathcal{H}^2_U$,

$$\langle V_{g,f_1}, V_{g,f_2}\rangle_{L^2(G)} = \langle f_1, f_2\rangle_{\mathcal{H}^2_U} \|K^{-1/2}g\|^2_{\mathcal{H}^2_U}.$$

Thus, if we fix $\omega \in \mathcal{H}^2_U$ such that $\|\hat{\omega}\Psi^{1/2}\|_{L^2(U)} = 1$, then $f \to V_{\omega,f}$ is an isometry of $\mathcal{H}^2_U$ into $L^2(G)$ and we get the weakly convergent reconstruction formula, for $f \in \mathcal{H}^2_U$,

$$f = \int_H \int_{\mathbb{R}^n} \langle f, \omega_{\mathbf{x},h}\rangle \omega_{\mathbf{x},h} \frac{d\mathbf{x}\, dh}{\delta(h)}, \tag{3.1}$$

where $\omega_{\mathbf{x},h}(\mathbf{y}) = \rho(\mathbf{x}, h)\omega(\mathbf{y})$, for all $\mathbf{y} \in \mathbb{R}^n$ and $(\mathbf{x}, h) \in G$.

For fixed $\omega$, let $T_\omega f(\mathbf{x}, h) = \langle f, \omega_{\mathbf{x},h}\rangle = \langle f, \rho(\mathbf{x}, h)\omega\rangle$, for all $f \in \mathcal{H}^2_U$. Then $T_\omega$, from $\mathcal{H}^2_U$ into $L^2(G)$, can be considered as a continuous wavelet transform. Any closed subgroup $H$ of $GL_n(\mathbb{R})$, which is such that it gives rise to a free open orbit $U$

in $\widehat{\mathbb{R}^n}$, gives us another form of continuous wavelet decompositions on $\mathbb{R}^n$. We will investigate a family of such decompositions on $\mathbb{R}^2$ that arise by adding dilations to the Heisenberg group.

## 4. Extensions of the Heisenberg Group

Let $\mathbb{H}$ denote the three dimensional, simply connected, two step, nilpotent Lie group which we prefer to realize by three by three matrices as follows:

$$\mathbb{H} = \left\{ \begin{pmatrix} 1 & x & z \\ 0 & 1 & y \\ 0 & 0 & 1 \end{pmatrix} : x, y, z \in \mathbb{R} \right\}.$$

To compress notation a bit, denote

$$\begin{pmatrix} 1 & x & z \\ 0 & 1 & y \\ 0 & 0 & 1 \end{pmatrix}$$

by $[x, y, z]$. Thus, the group product in $\mathbb{H}$ is given by

$$[x_1, y_1, z_1][x_2, y_2, z_2] = [x_1 + x_2, y_1 + y_2, z_1 + z_2 + x_1 y_2].$$

See Chapter 1 of [**3**] for a discussion of the Heisenberg groups—$\mathbb{H}$ is one realization of the 3-dimensional Heisenberg group—and their elegant representation theory.

We will say that "dilations of $\mathbb{H}$" means an action of $\mathbb{R}^+$ on $\mathbb{H}$ that is diagonal with respect to the $[x, y, z]$ parametrization. Since the action must respect the group structure of $\mathbb{H}$, a dilation is specified by two parameters $\alpha$ and $\beta$, say. For each $a \in \mathbb{R}^+$, the dilation by $a$ is the map $\delta_{\alpha,\beta}(a)\colon \mathbb{H} \to \mathbb{H}$ given by

$$\delta_{\alpha,\beta}(a)[x, y, z] = [a^\alpha x, a^\beta y, a^{\alpha+\beta} z],$$

for all $[x, y, z] \in \mathbb{H}$. Let

$$G_{\alpha,\beta} = \mathbb{H} \rtimes_{\alpha,\beta} \mathbb{R}^+ = \{([x, y, z], a)\colon [x, y, z] \in \mathbb{H}, a \in \mathbb{R}^+\}.$$

The group product in $G_{\alpha,\beta}$ is given by

$$([x, y, z], a)([x', y', z'], a') = ([x + a^\alpha x', y + a^\beta y', z + a^{\alpha+\beta} z' + x a^\beta y'], aa').$$

When we refer to an extension of the Heisenberg group by dilations, we mean a group which is isomorphic to $G_{\alpha,\beta}$, for some $(\alpha, \beta) \in \mathbb{R}^2 \backslash (0, 0)$. The question naturally arises of whether the $G_{\alpha,\beta}$ are distinct and it turns out that they are not. Our first task is to find a convenient cross-section of the isomorphism classes of the collection of $G_{\alpha,\beta}$'s .

It is instructive to realize $G_{\alpha,\beta}$ as a group of $3 \times 3$ matrices. The map

$$([x, y, z], a) \to \begin{pmatrix} a^{\alpha+\beta} & a^\beta x & z \\ 0 & a^\beta & y \\ 0 & 0 & 1 \end{pmatrix}$$

is an isomorphism of $G_{\alpha,\beta}$ onto a closed subgroup of $GL_3(\mathbb{R})$. It is clear that each $G_{\alpha,\beta}$ is a four dimensional simply connected solvable Lie group. Let $\mathbf{g}_{\alpha,\beta}$ denote its Lie algebra. Since each $G_{\alpha,\beta}$ is simply connected, $G_{\alpha_1,\beta_1}$ is isomorphic to $G_{\alpha_2,\beta_2}$ if and only if $\mathbf{g}_{\alpha_1,\beta_1}$ is isomorphic to $\mathbf{g}_{\alpha_2,\beta_2}$.

Now $\{([x, 0, 0], 1)\colon x \in \mathbb{R}\}$, $\{([0, y, 0], 1)\colon y \in \mathbb{R}\}$, $\{([0, 0, z], 1)\colon z \in \mathbb{R}\}$ and $\{([0, 0, 0], a)\colon a \in \mathbb{R}^+\}$ are four one-parameter subgroups of $G_{\alpha,\beta}$. Let $X, Y, Z$, and $A_{\alpha,\beta}$ denote the corresponding generators in $\mathbf{g}_{\alpha,\beta}$. With the above $3 \times 3$-matrix realization of $G_{\alpha,\beta}$, it is easy to realize these basic elements of $\mathbf{g}_{\alpha,\beta}$ by matrices. This gives

$$X = \begin{pmatrix} 0 & 1 & 0 \\ 0 & 0 & 0 \\ 0 & 0 & 0 \end{pmatrix}, \quad Y = \begin{pmatrix} 0 & 0 & 0 \\ 0 & 0 & 1 \\ 0 & 0 & 0 \end{pmatrix}, \quad Z = \begin{pmatrix} 0 & 0 & 1 \\ 0 & 0 & 0 \\ 0 & 0 & 0 \end{pmatrix},$$

and

$$A_{\alpha,\beta} = \begin{pmatrix} \alpha+\beta & 0 & 0 \\ 0 & \beta & 0 \\ 0 & 0 & 0 \end{pmatrix}.$$

So we can realize $\mathbf{g}_{\alpha,\beta}$ as the linear span of $\{X, Y, Z, A_{\alpha,\beta}\}$. Now it is clear that $\mathbf{g}_{\alpha,\beta} = \mathbf{g}_{\alpha/\beta,1}$, if $\beta \neq 0$ and $\mathbf{g}_{\alpha,0} = \mathbf{g}_{1,0}$, for all $(\alpha,\beta) \neq (0,0)$. Thus, we reparametrize. Let

$$A_p = \begin{pmatrix} p+1 & 0 & 0 \\ 0 & 1 & 0 \\ 0 & 0 & 0 \end{pmatrix},$$

for $p \in \mathbb{R}$ and

$$A_\infty = \begin{pmatrix} 1 & 0 & 0 \\ 0 & 0 & 0 \\ 0 & 0 & 0 \end{pmatrix}$$

and let $\mathbf{g}_p = \langle X, Y, Z, A_p \rangle$, for $p \in \mathbb{R} \cup \{\infty\}$. Let $G_p$ be the simply connected Lie group

$$\exp(\mathbf{g}_p) = \left\{ \begin{pmatrix} a^{p+1} & x & z \\ 0 & a & y \\ 0 & 0 & 1 \end{pmatrix} : x, y, z \in \mathbb{R}, a \in \mathbb{R}^+ \right\}.$$

Each extension of the Heisenberg group is isomorphic to one of these $G_p$'s; however, the $G_p$'s are still not mutually nonisomorphic.

THEOREM 4.1. *Let $p, q \in \mathbb{R} \cup \{\infty\}, p \neq q$. Then $\mathbf{g}_p$ is isomorphic to $\mathbf{g}_q$ if and only if $p = 1/q$. If $q \neq \infty$ and $p = 1/q$, then an isomorphism $\Phi\colon \mathbf{g}_p \to \mathbf{g}_q$ is given by $\Phi(X) = -Y, \Phi(Y) = X, \Phi(Z) = Z$, and $\Phi(A_p) = pA_q$. An isomorphism $\Phi\colon \mathbf{g}_0 \to \mathbf{g}_\infty$ is the same on $X, Y$, and $Z$ but $\Phi(A_0) = A_\infty$.*

PROOF. One directly checks that $\Phi$ as given provides an isomorphism in the indicated cases.

Conversely, suppose $p, q \in \mathbb{R}$ and $\Phi\colon \mathbf{g}_p \to \mathbf{g}_q$ is an isomorphism of Lie algebras. Now $\{A_p, X, Y, Z\}$ and $\{A_q, X, Y, Z\}$ are vector bases of $\mathbf{g}_p$ and $\mathbf{g}_q$, respectively. The only nontrivial Lie brackets among the basis elements of $\mathbf{g}_p$ are

$$[X, Y] = Z, [A_p, X] = pX, [A_p, Y] = Y, \text{ and } [A_p, Z] = (p+1)Z.$$

Thus, $[\mathbf{g}_p, \mathbf{g}_p] = \langle X, Y, Z \rangle$, if $p \neq 0$ and $[\mathbf{g}_0, \mathbf{g}_0] = \langle Y, Z \rangle$. Hence, $\mathbf{g}_0$ cannot be isomorphic to any $\mathbf{g}_p$ with $p \neq 0$ or $\infty$. So we may assume $p, q \in \mathbb{R}\backslash\{0\}$.

Furthermore, the centers of $[\mathbf{g}_p, \mathbf{g}_p]$ and $[\mathbf{g}_q, \mathbf{g}_q]$ are then both $\mathbb{R}Z$. Thus, an isomorphism $\Phi\colon \mathbf{g}_p \to \mathbf{g}_q$ must map $\langle X, Y, Z \rangle$ to $\langle X, Y, Z \rangle$ and $\mathbb{R}Z$ to $\mathbb{R}Z$. These observations mean that the matrix of $\Phi$ with respect to the two bases $\{A_p, X, Y, Z\}$ and $\{A_q, X, Y, Z\}$ has the form:

$$\begin{pmatrix} \varphi_{11} & 0 & 0 & 0 \\ \varphi_{21} & \varphi_{22} & \varphi_{23} & 0 \\ \varphi_{31} & \varphi_{32} & \varphi_{33} & 0 \\ \varphi_{41} & \varphi_{42} & \varphi_{43} & \varphi_{44} \end{pmatrix}.$$

Since this matrix must be full rank, $\varphi_{11} \neq 0$ and $\varphi_{44} \neq 0$. The relation $[A_p, Z] = (p+1)Z$ implies $[\Phi A_p, \Phi Z] = (p+1)\Phi Z$. That is,

$$[\varphi_{11}A_q + \varphi_{21}X + \varphi_{31}Y + \varphi_{41}Z, \varphi_{44}Z] = (p+1)\varphi_{44}Z$$

which reduces to $(q+1)\varphi_{11}\varphi_{44} = (p+1)\varphi_{44}$. Since $\varphi_{11}$ and $\varphi_{44}$ are nonzero, either $q = -1$ and $p = -1$ or $\varphi_{11} = p+1/q+1$. In the case that either $p = -1$ or $q = -1$, the other is forced to be $-1$ also. Thus, we may assume $p \neq -1$ and $q \neq -1$ from now on. Thus,

$$\varphi_{11} = \frac{p+1}{q+1}. \tag{4.1}$$

Similarly, $[\Phi A_p, \Phi X] = p\Phi X$ implies

$$q\varphi_{11}\varphi_{22} = p\varphi_{22} \tag{4.2}$$

and

$$\varphi_{11}\varphi_{32} = p\varphi_{32}. \tag{4.3}$$

The relation $[\Phi A_p, \Phi Y] = \Phi Y$ implies that

$$q\varphi_{11}\varphi_{23} = \varphi_{23} \tag{4.4}$$

and

$$\varphi_{11}\varphi_{33} = \varphi_{33}. \tag{4.5}$$

Finally, $[\Phi X, \Phi Y] = \Phi Z$ gives

$$\varphi_{22}\varphi_{33} - \varphi_{32}\varphi_{23} = \varphi_{44}. \tag{4.6}$$

Using (4.6), we see that $\varphi_{22}\varphi_{33}$ and $\varphi_{32}\varphi_{23}$ cannot both be 0. Now $\varphi_{22}\varphi_{33} \neq 0$ forces $\varphi_{11} = p/q$ from (4.2) and $\varphi_{11} = 1$ from (4.5). But then we would have $p = q$, contrary to our assumptions.

Thus $\varphi_{32}\varphi_{23} \neq 0$. Then (4.3) implies $\varphi_{11} = p$ and (4.4) implies $\varphi_{11} = 1/q$. Hence, $p = 1/q$. Thus, if $p \neq q$ and $\mathbf{g}_p \cong \mathbf{g}_q$, then $p = 1/q$. □

COROLLARY 4.1. *$\{G_p\colon -1 \leq p \leq 1\}$ is a complete cross-section of the isomorphism classes of $\{G_{\alpha,\beta}\colon (\alpha,\beta) \in \mathbb{R}^2\backslash(0,0)\}$.*

When we refer to an extension of the Heisenburg group by dilations, from now on, we will mean one of the $G_p$ for $-1 \leq p \leq 1$.

## 5. Square-Integrable Representations of $G_p$

Recall that

$$G_p = \left\{ \begin{pmatrix} a^{p+1} & x & z \\ 0 & a & y \\ 0 & 0 & 1 \end{pmatrix} : x, y, z \in \mathbb{R}, a \in \mathbb{R}^+ \right\}.$$

Noting that

$$\begin{pmatrix} a^{p+1} & x & z \\ 0 & a & y \\ 0 & 0 & 1 \end{pmatrix} = \begin{pmatrix} 1 & 0 & z \\ 0 & 1 & y \\ 0 & 0 & 1 \end{pmatrix} \begin{pmatrix} a^{p+1} & x & 0 \\ 0 & a & 0 \\ 0 & 0 & 1 \end{pmatrix}$$

leads us to realizing $G_p$ in the form $\mathbb{R}^2 \rtimes H_p$. These are the kinds of groups that were studied in Section 3. Let

$$H_p = \left\{ \begin{pmatrix} a^{p+1} & x \\ 0 & a \end{pmatrix} : x \in \mathbb{R}, a \in \mathbb{R}^+ \right\},$$

which is a closed subgroup of $GL_2(\mathbb{R})$. Then

$$\mathbb{R}^2 \rtimes H_p = \left\{ \left( \begin{pmatrix} z \\ y \end{pmatrix}, \begin{pmatrix} a^{p+1} & x \\ 0 & a \end{pmatrix} \right) : \begin{pmatrix} z \\ y \end{pmatrix} \in \mathbb{R}^2, \begin{pmatrix} a^{p+1} & x \\ 0 & a \end{pmatrix} \in H_p \right\}$$

is isomorphic to $G_p$ via the map

$$\left( \begin{pmatrix} z \\ y \end{pmatrix}, \begin{pmatrix} a^{p+1} & x \\ 0 & a \end{pmatrix} \right) \to \begin{pmatrix} a^{p+1} & x & z \\ 0 & a & y \\ 0 & 0 & 1 \end{pmatrix}.$$

Following the analysis of Section 3, we consider the action of $H_p$ on $\widehat{\mathbb{R}^2} = \{(\gamma, \mu) : \gamma, \mu \in \mathbb{R}\}$. That is, for $(\gamma, \mu) \in \widehat{\mathbb{R}^2}$

$$(\gamma, \mu) \begin{pmatrix} a^{p+1} & x \\ 0 & a \end{pmatrix} = (a^{p+1}\gamma, a\mu + x\gamma) \tag{5.1}$$

for

$$\begin{pmatrix} a^{p+1} & x \\ 0 & a \end{pmatrix} \in H_p.$$

If $p = -1$, then there are three kinds of orbits in $\widehat{\mathbb{R}^2}$;

(1) for $\gamma \neq 0$,

$$(\gamma, 0)H_{-1} = \{(\gamma, \mu) : \mu \in \mathbb{R}\}$$

(2) for $\gamma = 0$,

$$(0, 1)H_{-1} = \{(0, \mu) : \mu > 0\}$$

and

$$(0, -1)H_{-1} = \{(0, \mu) : \mu < 0\}$$

(3) finally, $\{(0,0)\}$ is an orbit itself.

Thus, there are no open $H_{-1}$-orbits in $\widehat{\mathbb{R}^2}$. In fact, $G_{-1}$ has no square-integrable representations and does not fit under the general analysis developed in [**1**]. However, it is interesting that $G_{-1}$ has been studied in detail, from the point of view of generating frames, in [**4**], where it is called the one dimensional UT group.

Therefore, from now on, assume $-1 < p \leq 1$. By (5.1), we see that there are two open $H_p$-orbits in $\widehat{\mathbb{R}^2}$. They are

$$U_+ - (1, 0)H_p = \{(\gamma, \mu) : \gamma > 0\}$$

and

$$U_- = (-1, 0)H_p = \{(\gamma, \mu) : \gamma < 0\}.$$

Since $U_+ \cup U_-$ is co-null in $\widehat{\mathbb{R}^2}$, we have $L^2(\widehat{\mathbb{R}^2}) = L^2(U_+) \oplus L^2(U_-)$ and $L^2(\mathbb{R}^2) = \mathcal{H}^2_{U_+} \oplus \mathcal{H}^2_{U_-}$.

Let us apply the theory of Section 3 to determine the form that the continuous wavelet transform and reconstruction formula (3.1) takes in $\mathcal{H}^2_{U_+}$. We get a different analysis for each $p$, $-1 < p \leq 1$.

As a manifold, $G_p = \mathbb{R} \times \mathbb{R} \times \mathbb{R} \times \mathbb{R}^+$. Let us write $(x, y, z, a)$ for the element

$$\begin{pmatrix} a^{p+1} & x & z \\ 0 & a & y \\ 0 & 0 & 1 \end{pmatrix}$$

of $G_p$. Also, write the elements of $\mathbb{R}^2$ in the form $(s, t)$ instead of as a column. Then, the square-integrable representation of $G_p$ associated with $U_+$ is, for $g \in \mathcal{H}^2_{U_+}$ and $(x, y, z, a) \in G_p$,

$$\rho_{U_+}(x, y, z, a)g(s, t) = \frac{1}{a^{(p+2)/2}} g\left( \frac{as - az - xt + xy}{a^{p+2}}, \frac{t - y}{a} \right)$$

for all $(s, t) \in \mathbb{R}^2$.

To find the operator $K$, we need the Radon-Nikodym derivative $\Psi$. That is found as follows: Fix an element in $U_+$, say $(1, 0)$. Let $\xi \in C_{00}(U_+)$. Then, using left Haar integration on $H_p$

$$\begin{aligned} \int_{H_p} \xi\big((1,0)h\big)\, dh &= \int_0^\infty \int_{-\infty}^\infty \xi(a^{p+1}, x) \frac{dx\, da}{a^{p+1}} \\ &= \int_0^\infty \int_{-\infty}^\infty \xi(\gamma, \mu) \frac{1}{(p+1)\gamma^{(2p+1)/(p+1)}}\, d\mu\, d\gamma. \end{aligned}$$

Thus,

$$\Psi(\gamma, \mu) = \frac{1}{(p+1)\gamma^{(2p+1)/(p+1)}}, \text{ for } (\gamma, \mu) \in U_+.$$

Note that when $p = 0$, $\Psi(\gamma, \mu) = \gamma^{-1}$ and, when $p = 1$, $\Psi(\gamma, \mu) = 1/2\gamma^{-3/2}$.

Now, we fix any $\omega \in \mathcal{H}^2_{U_+}$ such that

$$(5.2) \qquad (p+1) \int_0^\infty \int_{-\infty}^\infty \left|\widehat{\omega}(\gamma, \mu)\right|^2 \gamma^{(2p+1)/(p+1)}\, d\mu\, d\gamma = 1.$$

Let, for $(x, y, z, a) \in \mathbb{R} \times \mathbb{R} \times \mathbb{R} \times \mathbb{R}^+$,

$$(5.3) \qquad \omega_{(x,y,z,a)}(s, t) = \frac{1}{a^{(p+2)/2)}} \omega\left( \frac{as - az - xt + xy}{a^{p+2}}, \frac{t - y}{a} \right)$$

for all $(s, t) \in \mathbb{R}^2$. For $f \in \mathcal{H}^2_{U_+}$, define $T_\omega f$ on $\mathbb{R} \times \mathbb{R} \times \mathbb{R} \times \mathbb{R}^+$ by

$$(5.4) \qquad T_\omega(x, y, z, a) = \int_{-\infty}^\infty \int_{-\infty}^\infty f(s, t) \overline{\omega_{(x,y,z,a)}(s, t)}\, ds\, dt.$$

Then the reconstruction formula (1) becomes,

$$(5.5) \qquad f = \int_0^\infty \int_{-\infty}^\infty \int_{-\infty}^\infty \int_{-\infty}^\infty T_\omega f(x, y, z, a) \omega_{(x,y,z,a)} \frac{dx\, dy\, dz\, da}{a^{2p+3}},$$

weakly in $\mathcal{H}^2_{U_+}$.

We will leave the calculations for $\mathcal{H}^2_{U_-}$ to the interested reader.

## 6. Conclusions

We looked at those groups that arise by adding dilations to the Heisenberg group structure of $\mathbb{R}^3$, where we considered that the dilation must respect the group structure. This led us to the class of groups $G_p$, $-1 \leq p \leq 1$. Each $G_p$ could be alternately realized in the form $\mathbb{R}^2 \rtimes H_p$ for an appropriate 2-dimensional subgroup $H_p$ of $GL_2(\mathbb{R})$. Using the general theory of [**1**] for groups of this form, we found that $G_p$ has square-integrable representations when $-1 < p \leq 1$.

For each $p$, $-1 < p \leq 1$, formulas (5.2), (5.3), (5.4) and (5.5) constitute a form of continuous wavelet analysis on $\mathbb{R}^2$ that was discovered via representation theory.

## References

1. D. Bernier and K. F. Taylor, *Wavelets from square-integrable representations*, SIAM J. Math. Anal. **27** (1996), 594–608.
2. M. Duflo and C. C. Moore, *On the regular representation of a nonunimodular locally compact group*, J. Funct. Anal. **21** (1976), 209–243.
3. G. B. Folland, *Harmonic analysis in phase space*, Annals of Mathematics Studies, no. 122, Princeton Univ. Press (1989), Princeton.
4. J. A. Hogan and J. D. Lakey, *Extensions of the Heisenberg group by dilations and frames*, Appl. Comput. Harmon. Anal. **2** (1995), 174–199.
5. B. Torresani, *Wavelets and representations of the affine Weyl-Heisenberg group*, J. Math. Phys. **32** (1991), 1273–1279.

School of Mathematics, Suranaree University of Technology, Nakhon Ratchasim, 3000, Thailand

*E-mail address*: `eckart@math.sut.ac.th`

Department of Mathematics and Statistics, University of Saskatchewan, 106 Wiggins Road, Saskatoon, SK S7N 5E6

*E-mail address*: `taylor@snoopy.usask.ca`

# Part 3

# Wavelets in Physics

# Introduction and Summary

Before jumping right into this part, let us give a guided tour through the various contributions concerning the application of wavelets in Physics, to whet the reader's taste.

Arneodo *et al.* in Section 1 with an extension of the multifractal formalism to fractal functions involving oscillating singularities. The singular behavior of functions is generally characterized by the so-called Hölder exponent $h$. But this exponent poorly characterizes oscillating singularities which require the introduction of a second exponent $\beta$ that accounts for the singular behavior of the instantaneous frequency of the function. One of the main point reported in this paper is that the continuous wavelet transform is a natural tool to characterize pointwise regularity. In particular it can be used to distinguish between cusps and oscillating singularities through the computation of both exponents $h$ and $\beta$. From wavelet analysis one can thus describe statistically the fluctuations of both the Hölder and the oscillation exponents through the computation of the singularity spectrum $D(h, \beta)$ which corresponds to the Hausdorff dimension of the set of points $x$ associated to the same Hölder and oscillation exponents $\big(h(x) = h$ and $\beta(x) = \beta\big)$. For a large class of fractal functions whose cusps and oscillating singularities lie on Cantor sets, an extended wavelet based multifractal formalism is proposed and proved to give access to $D(h, \beta)$ singularity spectrum. Whereas the partition function used in the classical multifractal formalism is indexed by a single parameter (conjugated to the Hölder exponent $h$), this new description is based on a partition function involving two intensive parameters (associated with the exponents $h$ and $\beta$). In that sense, it is the analog of a "grand canonical" multifractal formalism, whereas the classical formalism can be identified with a "canonical" description.

In Section 2, Ali briefly reviews a method of quantization of classical systems based on the theory of coherent states. Coherent states arise as special families of vectors in the representation spaces of locally compact groups. When properly chosen, they display the geometrical properties of complex line bundles which, in the framework of geometric quantization, provide the first link in the chain going from a classical to a quantum system. The method proposed by Ali consists in using families of coherent states with symplectic geometry inherited from the projective space of a complex Hilbert space, to induce a quantization of the underlying manifold parametrizing the states. This method is illustrated on an example based on the Poincaré group. Even though one can desperately look for the word "wavelets" in this paper, one should not forget that wavelets, as originally defined by Morlet and Grossmann, can be seen as affine coherent states (i.e. irreducible and square-integrable representations of the "$ax + b$" group).

The next two sections bring the reader to applications of wavelets in the domains of atomic, molecular and condensed-matter physics. In Section 3, Antoine *et al.* start with the case of the emission of light by an atom when hit by a very intense, ultrashort, laser pulse. Odd harmonics of the laser frequency are emitted. In order to understand the emission mechanisms better, one would like to know for instance when the harmonics are emitted during the optical cycle and what the time evolution is during the laser pulse. These questions are out of reach of the traditional Fourier spectral methods. For atomic hydrogen, the emission is investigated by both the Gabor transform (windowed Fourier transform) and the wavelet transform, yielding time profiles of each individual harmonic. Analysis of these profiles leads to the conclusion that harmonic emission takes place only when the electron is close to the nucleus. Special emphasis is made on the complementarity of Gabor and wavelet analysis in this type of study. For the case of multi-electronic atoms, the authors propose the use orthogonal wavelet bases for the description of the radial part of the wave functions. Here (discrete) wavelet bases adequately replace traditional plane waves or LCAO (Linear Combination of Atomic Orbitals) bases, allowing improvements over more conventional Hartree-Foch methods. As further emphasized by Handy and Murenzi in Section 5, a combination of wavelet transforms and conventional techniques is likely to allow a better calculation of energy levels in atoms.

As in atomic physics, wavelets have found so far relatively few applications in molecular and condensed-matter physics. Indeed, as in the case self-consistent atomic methods, what is needed here are mostly orthogonal bases with good localization properties, in order to minimize the number of terms required for an accurate calculation of wave functions and related observable quantities. In Section 4, Antoine reports on three different applications of wavelet analysis in dimensions three, two and one respectively. The first one concerns the improvement of standard numerical methods like the LCAO or plane wave basis methods to computing the wave function shape or the energy levels of atoms or dimer molecules like $H_2$ or $O_2$. Let us note that both non-orthogonal and orthogonal wavelet bases have been applied successfully. In a second part, the author addresses the issue of the Fractional quantum Hall effect, i.e. a (quasi) planar gas of electrons submitted to a strong magnetic field perpendicular to the plane. The main point is that 2D orthonormal wavelets yields various adapted bases for the lowest Landau level. The last part of this paper is dedicated to the study of electron propagation in 1D crystals. Antoine shows that the Wannier functions associated with a particle in a periodic potential satisfy two-scale relations similar to those defining multiresolution analysis. This equivalence becomes an equality in the case of a free particle for which Wannier functions are shown to coincide with Littlewood-Paley wavelets.

Recent attempts to applying wavelet analysis to solve linear differential (ODE) and partial differential (PDE) equations, such as the Schrödinger equation with a rational fraction potential, have not proven too successful partly because many of these equations do not admit a tractable wavelet transform. In Section 5, Handy and Murenzi elaborate on a way to circumvent these difficulties which consists in calculating the continuous wavelet transform $\omega_t(b,a)$ via the computation of dilated and translated (generalized) moments $\mu_{b,1/2a^2}(p)$. Using the Eigenvalue Moment Method of Handy and Bessis, one can first solve the infinite scale ($a = \infty$) problem to arbitrary precision and then use standard integration to calculate the moments for arbitrary $b$ and $a$ values. Since many important wavelet transforms can be expressed as linear combinations of $\mu_{b,a}$, the $\omega_t(b,a)$ transform is then trivially obtained. This method is illustrated on two particular examples, namely the quartic anharmonic oscillator and a rational fraction potential $\big(\nu(x) = gx^6/(I + \lambda x^2)\big)$. Let us note that although the analysis is mainly focussed on the ground state, it is readily extendable to excited states.

In Section 6, Battle reviews some background knowledge on the renormalization group theory with the specific goal to demonstrate that wavelets can advantageously be used in this context. Indeed, in his pioneering work, Wilson postulated the existence of wavelets for his renormalization group analysis more than ten years before Morlet, Grossmann and the mathematics community began to construct them. His own motivation was to develop a hierarchical approximation for the non-perturbative calculations of the renormalization group transformation. On the other hand, the development of perturbative calculations was based on imposing a standard ultraviolet (UV) cutoff for a scalar field theory, rather than on a wavelet cutoff or even a lattice cutoff. The $\varepsilon$-expansion was introduced in the UV formalism. The advantage in working with the standard UV cutoff on a continuum field is that the kinetic contribution to the Hamiltonian is a fixed point which makes easier the RG procedure. Indeed these calculations simply reduce to the evaluation of Feynman diagrams. But this cutoff is only qualitatively similar to the lattice cutoff, since the integration of high-momentum amplitudes is only qualitatively similar to discrete averaging. Actually this sharp cutoff also destroys the notion of a critical manifold because no model can have a finite correlation length in the absence of real analyticity in momentum space. The aim of Battle's lecture is to advertise the merit of developing the $\varepsilon$-expansion in the wavelet formalism. Using wavelet cutoff to prevent UV divergencies presents several advantages: while this cutoff shares the advantage of the UV cutoff, it has also a lattice picture with an averaging transformation; moreover it admits the exponential decay of correlations. "Wavelet diagrams" can then be defined as analogs of Feynman diagrams. Finally, it is shown how the notion of fractional dimension ($d = 4 - \varepsilon$) can be naturally introduced by a scale-dependent omission of some wavelet coefficients.

First introduced in the context of critical phenomena, the renormalization group analysis was later extended to non equilibrium systems such as turbulent flows. The common goal of any RG procedure is the reduction of the number of relevant degrees of freedom. Namely, starting from some partial differential equation, any RG method generates a family of "renormalized" differential equations by successive eliminations of small scale modes. This set of equations is supposed to converge to a fixed point which yields a simpler PDE at the critical point. Section 7, Elezgaray *et al.* discuss how wavelet decompositions can be used to obtain such coarse-grained equations in fluids dominated by coherent structures, with no need to introduce special assumptions about the behaviour of the small scales. Inspired by the Karhunen-Loeve decomposition method, they report on numerical simulations ot the Kuramoto-Sivashinsky equation that illustrate how Galerkin projections onto some orthogonal wavelet basis naturally yield a set of effective equations for the large scales of the system which preserve the statistics of the original equations, and in particular the energy transfer between large and small scales. This study clearly reveals that a good model for the behaviour of the small scales is nothing but the dynamics of the original KS equation in a small domain. This confirms that no ad-hoc modelling is needed to parametrize the net effect of small scales upon large scales. Very much like the RN methods, the approach advocated by

Elezgaray *et al.* is not concerned about the approximation of particular solutions but rather aims at understanding their statistical properties. It has to be considered as a very attractive alternative to approximation methods based on the notion of inertial manifold that actually preserve the asymptotic behavior of almost any particular solution.

A. Arneodo

Centre de Recherches Mathématiques
CRM Proceedings and Lecture Notes
Volume **18**, 1999

# Coherent States and Quantization

S. Twareque Ali

ABSTRACT. We outline a method of quantization of classical systems based on the theory of coherent states for locally compact groups. The method exploits the symplectic geometry inherent in the projective space of a separable complex Hilbert space. This geometry is inherited by appropriate families of coherent states and their natural embeddings into the projective Hilbert space. We also point out the mathematical link between the theory of coherent states, and hence of this technique of quantization, and the continuous wavelet transform.

## 1. Introduction

Coherent states and quantization are two time honoured areas of research in the current mathematical physics literature, and there is an interesting link between the two and the theory of generalized continuous wavelet transforms. Coherent states arise (see, for example, [**4, 10**]) as special families of vectors in the representation spaces of locally compact groups, and when properly chosen, display the geometrical properties of complex line bundles [**9**]. In the scheme of geometric quantization, on the other hand, complex line bundles provide the first link in the chain going from a classical to a quantum system (see, for example, [**11**]). Using the fact that the projective space of a complex Hilbert space is an analytic manifold with a Kähler structure, it is possible to use families of coherent states to induce a quantization of the underlying manifold parametrizing the states. We briefly explain this procedure in this paper and illustrate it with an example based on the Poincaré group. Looked upon from the point of view of coherent states, the continuous wavelet transform is the natural representation of signals in terms of their time-frequency content using the coherent states of the affine group. Analogous transforms, using the coherent states of other groups, thus lead to generalized wavelet transforms which are appropriate for the analysis of signals having more complex structural content. Detailed expositions may be found in [**3, 4, 9**].

## 2. Quantization

*Quantization* is the mathematical procedure for making the transition:

$$\textit{Classical Mechanics} \longrightarrow \textit{Quantum Mechanics}.$$

1991 *Mathematics Subject Classification.* Primary: 81S10; Secondary: 81R30.
This is the final version of the paper.

In other words, given a classical system (evolving with some specific Hamiltonian), the procedure arrives at the corresponding quantum system (and its quantum mechanical evolution). Classical mechanics is formulated on phase space, which very generally, is a *symplectic manifold* $(\Gamma, \Omega)$, i.e., a $2n$-dimensional real manifold $\Gamma$, equipped with a non-degenerate two-form $\Omega$, which is closed, $d\Omega = 0$, and which in some preferred local coordinates $q^i$, $p_j$, $i, j = 1, 2, \ldots, n$, has the form

$$\Omega = \sum_{i=1}^{n} dq^i \wedge dp_i. \tag{2.1}$$

More generally, one can find coordinates $\eta^i$, $i = 1, 2, \ldots, 2n$ such that locally

$$\Omega = \frac{1}{2} \sum_{i,j=1}^{2n} \omega_{ij}\, d\eta^i \wedge \eta^j \tag{2.2}$$

where at each point $x \in \Gamma$,

$$\omega_{ij}(x) = -\omega_{ji}(x) \quad \text{and} \quad \det\|\omega_{ij}(x)\| \neq 0. \tag{2.3}$$

Typically, $\Gamma = T^*M$, i.e., it is the cotangent bundle of a manifold $M$ of dimension $n$. In that case there exists a *symplectic potential* $\theta$, which locally can be brought into the form $\theta = \sum_{i=1}^{n} p_i\, dq^i$ and for which $\Omega = d\theta$. Using $\Omega$, one defines the *Hamiltonian vector fields* $X_u$, corresponding to any smooth function $u \colon \Gamma \to \mathbb{R}$:

$$du = -X_u \rfloor \Omega = -2\Omega(X_u, \cdot), \tag{2.4}$$

and the Poisson bracket,

$$\{u, v\} = 2\Omega(X_u, X_v). \tag{2.5}$$

Furthermore, the closedness of $\Omega$ implies the *Jacobi identity*:

$$\{u, \{v, w\}\} + \{v, \{w, u\}\} + \{w, \{u, v\}\} = 0. \tag{2.6}$$

Thus, very generally, a classical mechanics consists of a symplectic manifold along with a Poisson bracket emanating from its associated two-form.

*States* in classical mechanics are probability measures $\mu$ on $\Gamma$ with

$$\int_\Gamma d\mu(x) = 1, \tag{2.7}$$

and a pure state is a delta-measure:

$$\delta_x(f) = f(x), \quad f \in C^\infty(\Gamma). \tag{2.8}$$

*Observables* in classical mechanics are real-valued functions $f$ on $\Gamma$, and the expectation value of an observable $f$ in the state $\mu$ is the quantity

$$\langle \mu; f \rangle = \int_\Gamma f(x)\, d\mu(x). \tag{2.9}$$

Quantum mechanics, on the other hand, is based on a separable, complex Hilbert space $\mathfrak{H}$. Pure states are rays,

$$\mathbb{C}\phi, \ \phi \in \mathfrak{H}, \quad \phi \neq 0, \tag{2.10}$$

or alternatively, one-dimensional projection operators,

$$\Phi = \frac{|\phi\rangle\langle\phi|}{\|\phi\|^2}, \quad \phi \neq 0. \tag{2.11}$$

The set of all these projection operators constitutes an infinite-dimensional symplectic manifold, $\mathbb{CP}(\mathfrak{H})$, so that the situation is analogous to classical mechanics where the pure states constitute a finite-dimensional symplectic manifold. *Mixed states* in quantum mechanics are density matrices,

$$\rho = \sum_{j=1}^{\infty} \alpha_j |e_j\rangle\langle e_j|, \quad \sum_{j=1}^{\infty} \alpha_j = 1, \qquad \alpha_j \geq 0, \tag{2.12}$$

where the $e_j$ constitute an orthonormal basis in $\mathfrak{H}$. Observables in quantum mechanics are self-adjoint operators $A$ on $\mathfrak{H}$, and the expectation value of an observable $A$ in the state $\rho$ is

$$\langle \Phi; A\rangle = \mathrm{Tr}[\rho A]. \tag{2.13}$$

The set of classical observables forms a non-associative algebra with respect to the product

$$f \circ_{\mathrm{cl}} g = \{f, g\}, \quad \text{(Poisson bracket)}, \tag{2.14}$$

while the set of quantum observables is a non-associative algebra with respect to the product

$$A \circ_{\mathrm{qu}} B = \frac{1}{i\hbar}[A, B] = \frac{1}{i\hbar}(AB - BA), \quad \text{(commutator bracket)}. \tag{2.15}$$

Let $\mathcal{S}(\mathfrak{H})$ be some suitably large algebra of self-adjoint operators on $\mathfrak{H}$ with respect to the $\circ_{\mathrm{qu}}$ product and defined on a common dense domain, and $\mathcal{A}(\Gamma)$ some suitably large subalgebra of $\mathcal{C}^{\infty}(\Gamma)$ with respect to the $\circ_{\mathrm{cl}}$ product. Then a quantization would imply finding an algebraic homomorphism between $\mathcal{A}(\Gamma)$ and $\mathcal{S}(\mathfrak{H})$. More precisely, in the *Dirac scheme* quantization involves constructing a linear map $f \mapsto \hat{f}$ of $\mathcal{A}(\Gamma)$ into $\mathcal{S}(\mathfrak{H})$, satisfying,

(1) the set $\{\hat{f}\}$ acts irreducibly on $\mathfrak{H}$;
(2) $\mathbb{I} \mapsto I$, where $\forall x \in \Gamma, \mathbb{I}(x) = 1$, and $I$ is the identity operator on $\mathfrak{H}$;
(3) for a chosen set of functions $f, g$ (which cannot be made to include all of $\mathcal{A}(\Gamma)$ without relaxing condition 1. above [**7**]),

$$\{f, g\} \to \frac{1}{i\hbar}[\hat{f}, \hat{g}]. \tag{2.16}$$

There exist several well-known methods for implementing the Dirac scheme—geometric quantization [**11**], Berezin quantization [**5**], to mention two. We describe here a quantization method based on *coherent states*, but for that we first need to describe briefly the theory of coherent states [**4**].

## 3. Coherent States

Let $G$ be a locally compact group, $\mu$ the left Haar measure on it, $H$ a closed subgroup of $G$ and $X = G/H$ the corresponding left coset space. For a global Borel section $\sigma\colon X \to G$ of the group (note that if $\pi\colon G \to X$ is the canonical projection map then $\pi(g) = gH$ and $\pi(\sigma(x)) = x$) let $\nu_\sigma$ be the quasi-invariant measure defined by:

$$d\nu_\sigma(x) = \lambda(\sigma(x), x)\, d\nu(x), \quad \text{where}, \quad \lambda(g, x)\, d\nu(x) = d\nu(g^{-1}x), \ (\forall g \in G). \tag{3.1}$$

Let $U$ be a unitary representation of $G$ in a Hilbert space $\mathfrak{H}$ and $\eta^i$, $i = 1, 2, \ldots, n$, a set of linearly independent vectors in $\mathfrak{H}$. Then, if the set of vectors

$$\mathfrak{S}_\sigma = \{\eta^i_{\sigma(x)} = U(\sigma(x))\eta^i \mid i = 1, 2, \ldots, n, x \in X\} \tag{3.2}$$

is total in $\mathfrak{H}$, we call it a family of covariant coherent states for $U$. Let $F = \sum_{i=1}^n |\eta^i\rangle\langle\eta^i|$, $F_\sigma(x) = U(\sigma(x))FU(\sigma(x))^*$. If

$$\int_X F_\sigma(x)\, d\nu_\sigma(x) = A_\sigma, \quad A_\sigma, A_\sigma^{-1} \text{ bounded operators on } \mathfrak{H}, \tag{3.3}$$

the integral converging weakly, the coherent states $\mathfrak{S}_\sigma$ are called square integrable and the representation $U$ is said to be square integrable mod$(H, \sigma)$. In this case, the vectors $\eta^i$, $i = 1, 2, \ldots, n$, are called admissible mod$(H, \sigma)$.

Given a family of covariant coherent states, there is an associated positive definite kernel

$$K_\sigma(x, y) = F_\sigma(x)^{1/2} F_\sigma(y)^{1/2}, \tag{3.4}$$

with the properties:

(1) $K_\sigma(x, x)$ is a strictly positive operator on $\mathfrak{K}_x$, the subspace of $\mathfrak{H}$ generated by the vectors $\eta^i_{\sigma(x)}$, $i = 1, 2, \ldots, n$;
(2) $K_\sigma(x, y)$ is a linear operator mapping $\mathfrak{K}_y$ to $\mathfrak{K}_x$;
(3) if $N$ is a finite integer $x_i \in X$, $v_i \in \mathfrak{K}_{x_i}$, $i = 1, 2, \ldots, N$, then

$$\sum_{i,j=1}^N \langle v_i \mid K_\sigma(x_i, x_j) v_j\rangle \geq 0. \tag{3.5}$$

The quantities

$$\xi_x^{v_x} = K_\sigma(\cdot, x) v_x, \quad \forall v_x \in \mathfrak{K}_x, \tag{3.6}$$

are measurable vector fields $X(y) = \xi_x^{v_x}(y) = K_\sigma(y, x) v_x$. On their linear span we define a scalar product

$$\langle \xi_x^{v_x} \mid \xi_y^{v_y}\rangle_{K_\sigma} = \langle v_x \mid K_\sigma(x, y) v_y\rangle_{\mathfrak{K}_x}. \tag{3.7}$$

Closing the linear span in this scalar product we get a *reproducing kernel Hilbert space* $\mathfrak{H}_{K_\sigma}$. The distinctive feature of this Hilbert space is the *reproducing property*, that for any $\Phi \in \mathfrak{H}_{K_\sigma}$ and $v_x \in \mathfrak{K}_x$,

$$\langle v_x \mid \Phi(x)\rangle_{\mathfrak{K}_x} = \langle K_\sigma(\cdot, x) v_x \mid \Phi\rangle_{K_\sigma}, \tag{3.8}$$

which also translates into the continuity of the *evaluation map*,

$$E(x)\colon \mathfrak{H}_{K_\sigma} \to \mathfrak{K}_x, \quad E(x)\Phi = \Phi(x). \tag{3.9}$$

There also exists a natural unitary map $W\colon \mathfrak{H} \to \mathfrak{H}_{K_\sigma}$,

$$(W\phi)(x) = \Phi(x) = F_\sigma^{1/2}(x)\phi. \tag{3.10}$$

Examples of coherent states are easily obtained. Suppose that $U$ is a representation of the locally compact group $G$ on the Hilbert space $\mathfrak{H}$, admitting a cyclic vector $\eta$. In this case, the set of vectors $\eta_g = U(g)\eta$, $g \in G$, is dense in $\mathfrak{H}$ and hence taking $H = \{e\}$, the trivial subgroup consisting of the identity element of $G$, (so that $X = G$), the trivial section $\sigma(g) = g$ and $n = 1$, we see that the $\eta_g$ are a set of coherent states. Thus the orbit of any cyclic vector is a family of coherent states, which do not in general have to be square integrable. Furthermore, square integrability in this case would imply that

$$\int_G |\eta_g\rangle\langle\eta_g|\, d\mu(g) = \lambda I, \tag{3.11}$$

for some real number $\lambda$.

On the other hand, consider the closed subgroup $H$ of $G$ which stabilizes $\eta$ up to a multiplicative factor, i.e.,

$$U(h)\eta = \exp\bigl[i\omega(h)\bigr]\eta, \quad h \in H, \tag{3.12}$$

where $\omega$ is a real-valued function on $\mathfrak{H}$. It follows that $\chi(h) = \exp\bigl[i\omega(h)\bigr]$ is a character of $H$. Letting $X = G/H$ and taking any Borel section $\sigma\colon X \to G$, we may define another family of covariant CS

$$\eta_{\sigma(x)} = U\bigl(\sigma(x)\bigr)\eta, \quad x \in X, \tag{3.13}$$

since this set of vectors is also total in $\mathfrak{H}$. These are the *Gilmore-Perelomov coherent states* [**6, 10**].

The unitary map in (3.10) is also the group theoretical basis for the existence of the *continuous wavelet transform.* In the case where the group $G$ is the affine group of transformations of $\mathbb{R}$, which map points $x \in \mathbb{R}$ to $ax + b \in \mathbb{R}$, for arbitrary $a > 0$ and $b \in \mathbb{R}$, the map $W$ in (3.10) leads to an isometric embedding of the"space of signals" $L^2(\mathbb{R}, dx)$ into a Hilbert space of continuous functions in the two variables $a$ and $b$, identified with the time and frequency, respectively. The map $W$ is now called the continuous wavelet transform. More generally, if signal features other than just the time and frequency need to be analyzed, the use of more complex groups lead to what are called *generalized continuous wavelet transforms.*

## 4. Quantization Using Coherent States

Let $N = \dim \mathfrak{H}$, which could be (countably) infinite or finite and consider $\mathbb{CP}(\mathfrak{H})$. This is a complex analytic manifold with a *Kähler structure.* An analytic atlas of $\mathbb{CP}(\mathfrak{H})$ is given by the coordinate charts

$$\bigl\{(V_\Phi, \mathbf{h}_\Phi, \mathfrak{H}_\Phi) \mid \Phi \in \mathbb{CP}(\mathfrak{H})\bigr\}, \tag{4.1}$$

where

$$V_\Phi = \bigl\{\Psi \in \mathbb{CP}(\mathfrak{H}) \mid \langle\phi \mid \psi\rangle \neq 0\bigr\} \tag{4.2}$$

is an open, dense set in $\mathbb{CP}(\mathfrak{H})$ and

$$\mathfrak{H}_\Phi = (I - \Phi)\mathfrak{H} = \langle\phi\rangle^\perp \tag{4.3}$$

is the subspace of $\mathfrak{H}$ orthogonal to the range of $\Phi$ and $\mathbf{h}_\Phi\colon V_\Phi \to \mathfrak{H}_\Phi$ is the diffeomorphism

$$\mathbf{h}_\Phi(\Psi) = \frac{1}{\langle\hat{\phi} \mid \psi\rangle}(I - \Phi)\psi, \quad \hat{\phi} = \frac{\phi}{\|\phi\|}, \tag{4.4}$$

where $\hat{\phi}$ is a fixed unit vector in the range of $\Phi$. Since $V_\Phi$ is dense in $\mathfrak{H}$, we consider only one coordinate chart, setting $e_0 = \hat{\phi}$ and choosing an orthonormal basis $\{e_i\}_{i=1}^{N-1}$ of $\mathfrak{H}_\Phi$ to obtain a basis of $\mathfrak{H}$. For arbitrary $\Psi \in \mathfrak{H}$ we have

$$z_i = \langle e_i \mid \psi\rangle, \quad i = 1, 2, \ldots, N-1, \tag{4.5}$$

and the coordinates of $\Psi \in \mathbb{CP}(\mathfrak{H})$ are the homogeneous coordinates

$$Z_i = \frac{z_i}{z_0}, \quad i = 1, 2, \ldots, N-1. \tag{4.6}$$

The projection map $\pi\colon \mathfrak{H} \setminus \{0\} \to \mathbb{CP}(\mathfrak{H})$ that assigns to each $\phi \in \mathfrak{H} \setminus \{0\}$ the corresponding projector $\Phi \in \mathbb{CP}(\mathfrak{H})$ is holomorphic in these coordinates. For $\Phi \in \mathbb{CP}(\mathfrak{H})$ we have $\pi^{-1}(\Phi = \mathbb{C}^*\phi$ where $\mathbb{C}^* = \mathbb{C} \setminus \{0\}$, and so $\pi\colon \mathfrak{H} \setminus \{0\} \to \mathbb{CP}(\mathfrak{H})$ is a $GL(1, \mathbb{C})$ principal bundle, sometimes called the *canonical line bundle* over $\mathbb{CP}(\mathfrak{H})$. We will denote the associated holomorphic line bundle by $\Xi(\mathfrak{H})$ and write elements in it as $(\Psi, \psi)$, where $\psi \in \Psi(\mathfrak{H})$. A local trivialization of $\Xi(\mathfrak{H})$ over $V_\Phi$ is given by the (holomorphic) unit section $\hat{s}$ of $\Xi(\mathfrak{H})$:

$$\hat{s}(\Psi) = \left(\Psi, \frac{\psi}{\langle \hat{\phi} \mid \psi\rangle}\right), \tag{4.7}$$

and any other section $s\colon V_\Phi \to \Xi(\mathfrak{H})$ is given by

$$s(\Psi) = \big(\Psi, \kappa(\Psi)\big) \tag{4.8}$$

where $\kappa\colon \mathbb{CP}(\mathfrak{H}) \to \mathfrak{H}\setminus\{0\}$ is a holomorphic map with $\big(|\kappa(\Psi)\rangle \langle\kappa(\Psi)|\big)/\big(\|\kappa(\Psi)\|^2\big) = \Psi$. Denote by $s_0$ the zero-section of $\Xi(\mathfrak{H})$. The identification map $\imath_\Xi\colon \Xi(\mathfrak{H}) \setminus s_0\big(\mathbb{CP}(\mathfrak{H})\big) \to \mathfrak{H}$ given as

$$\imath_\Xi(\Psi, \psi) = \psi, \tag{4.9}$$

yields a global coordinatization of $\Xi(\mathfrak{H}) \setminus s_0\big(\mathbb{CP}(\mathfrak{H})\big)$. For any $\psi \in \mathfrak{H}$ let $\langle\psi|$ be its dual element. The restriction of $\langle\psi|$ to the fibre $\pi^{-1}(\Psi')$ in $\Xi(\mathfrak{H})$, for arbitrary $\Psi' \in \mathbb{CP}(\mathfrak{H})$, then yields a section $s^*_{\Psi'}$ of the (anti-holomorphic) dual bundle $\Xi(\mathfrak{H})^*$.

The tangent space $T_\Psi\mathbb{CP}(\mathfrak{H})$ to $\mathbb{CP}(\mathfrak{H})$ at the point $\Psi$ has a natural identification with $\mathfrak{H}_\Psi$, so that the complex structure of $\mathfrak{H}_\Psi$ endows the tangent space with an integrable complex structure $J_\Psi$, making $\mathbb{CP}(\mathfrak{H})$ into a Kähler manifold. The corresponding canonical 2-form $\Omega_{FS}$, called the *Fubini-Study 2-form*, is given pointwise by

$$\Omega_{FS}(X_\Psi, Y_\Psi) = \frac{1}{2i}\big(\langle\xi \mid \zeta\rangle - \langle\zeta \mid \xi\rangle\big), \tag{4.10}$$

where $\xi, \zeta \in \mathfrak{H}_\Psi$ correspond to the tangent vectors $X_\Psi$, $Y_\Psi$ respectively. In the local coordinates $Z_i$, defined in (4.6), $\Omega_{FS}$ assumes the form

$$\Omega_{FS} = \frac{1}{1 + \|\mathbf{Z}\|^2} \sum_{i,j=1}^{N-1} \left[\delta_{ij} - \frac{Z_i \overline{Z}_j}{1 + \|\mathbf{Z}\|^2}\right] d\overline{Z}_i \wedge dZ_j, \qquad \mathbf{Z} = (Z_1, Z_2, \ldots, Z_{N-1}). \tag{4.11}$$

Thus, clearly, $d\Omega_{FS} = 0$, implying that $\Omega_{FS}$ is a *closed* 2-form, derivable from the real Kähler potential

$$\Phi(\mathbf{Z}, \overline{\mathbf{Z}}) = \log\left[1 + \|\mathbf{Z}\|^2\right]. \tag{4.12}$$

A hermitian metric $H_{FS}$ and a connection $\nabla_{FS}$ on $\Xi(\mathfrak{H})$ can be defined pointwise by

$$H_{FS}\big((\Psi, \psi), (\Psi, \psi')\big) = \langle\psi \mid \psi'\rangle \tag{4.13}$$

for all $(\Psi, \psi)$, $(\Psi, \psi') \in \pi^{-1}(\Psi)$. Using the identification map $\imath_\Xi$ defined in (4.9) we now construct a connection on $\Xi(\mathfrak{H})$. Define the 1-form $\alpha$ on $\mathfrak{H}$ by

$$\alpha(\psi) = \frac{\langle d\psi \mid \psi \rangle}{\|\psi\|^2}. \tag{4.14}$$

Then the pullback

$$\alpha_{FS} = \imath_\Xi^* \alpha \tag{4.15}$$

defines a $\mathbb{C}^*$-invariant 1-form on $\Xi(\mathfrak{H})$. For an arbitrary section $s\colon V_\Phi \to \Xi(\mathfrak{H}) \setminus s_0\big(\mathbb{CP}(\mathfrak{H})\big)$, as in (4.8), the pullback

$$\theta_{FS} = s^* \alpha_{FS} \tag{4.16}$$

defines a local 1-form on $\mathbb{CP}(\mathfrak{H})$. Pointwise,

$$\theta_{FS}(\Psi) = \frac{\langle d\kappa(\Psi) \mid \kappa(\Psi) \rangle}{\|\kappa(\Psi)\|^2} = \overline{\partial} \log \big\|\kappa(\Psi)\big\|^2, \tag{4.17}$$

where $\overline{\partial}$ denotes exterior differentiation with respect to the *anti-holomorphic* variables. In terms of the coordinatization introduced in (4.5), with $f$ as the holomorphic function representing $\kappa$, we have

$$\theta_{FS}(\mathbf{Z}) = \frac{\overline{df(\mathbf{Z})}}{\overline{f(\mathbf{Z})}} + \frac{\sum Z_i d\overline{Z}_i}{1 + \|\mathbf{Z}\|^2}. \tag{4.18}$$

Furthermore, $\theta_{FS}$ locally defines a metric connection $\nabla_{FS}$

$$\nabla_{FS} s = \theta_{FS} \otimes s, \tag{4.19}$$

and it is easy to verify that

$$\Omega_{FS} = i\partial\theta_{FS} = i \operatorname{curv} \nabla_{FS}, \tag{4.20}$$

where $\partial$ denotes exterior differentiation with respect to the holomorphic variables and $\operatorname{curv} \nabla_{FS}$ is the curvature of the line bundle $\Xi(\mathfrak{H})$.

Thus the hermitian line bundle $(\Xi(\mathfrak{H}),\ H_{FS},\ \nabla_{FS})$ is a prequantization of $(\mathbb{CP}(\mathfrak{H}),\ \Omega_{FS})$ in the sense of geometric quantization [**8, 11**].

The prequantization of $(\mathbb{CP}(\mathfrak{H}),\ \Omega_{FS})$ can be exploited to obtain a prequantization of an arbitrary symplectic manifold $(X, \Omega)$ whenever there exists a symplectomorphism $\mathrm{Coh}\colon X \to \mathbb{CP}(\mathfrak{H})$. (This is the first step in the quantization process.) In this case, $\Omega = \mathrm{Coh}^* \Omega_{FS}$ and $L = (\mathrm{Coh}^* \Xi(\mathfrak{H}_\sigma),\ \mathrm{Coh}^* H_{FS},\ \mathrm{Coh}^* \nabla_{FS})$ is a prequantization of $X$, i.e. in particular, $\Omega = i \operatorname{curv}(\mathrm{Coh}^* \nabla_{FS})$. To make the connection with coherent states, consider those of the Gilmore-Perelomov type in (3.13) and suppose that $X = G/H = \Gamma$ is a symplectic manifold. Let $V \subset X$ be an open dense set on which the section $\sigma$ is smooth and such that the restriction of $L$ to $V$ is trivial. For a properly chosen $\eta$, the map

$$\mathrm{Coh}(x) = \frac{|\eta_{\sigma(x)}\rangle \langle \eta_{\sigma(x)}|}{\|\eta_{\sigma(x)}\|^2} \tag{4.21}$$

can then be made into a symplectomorphism [**9**], and we again have here a prequantization of $\Gamma$.

In terms of the reproducing kernel $K_\sigma(x, y) = \langle \eta_{\sigma(x)} \mid \eta_{\sigma(y)} \rangle$ and locally on $V$, the expression

$$\theta_K(x) = d_1 \log K(x_1, x_2)\big|_{x_1 = x_2 = x} = (\mathrm{Coh}^* \theta_{FS})(x), \tag{4.22}$$

($d_1$ denoting exterior differentiation with respect to $x_1$) defines a 1-form on $L$, for which $\Omega = i\, d\theta_K$. The hermitian metric $H_K = \text{Coh}^*\, H_{FS}$ and the metric connection $\nabla_K$, are given by

$$H_K\big((x,\psi),(x,\psi')\big) = \langle \psi \mid \psi' \rangle \tag{4.23}$$

$$\nabla_K\, \text{coh} = \theta_K \otimes \text{coh}, \tag{4.24}$$

and $i\, \text{curv}\, \nabla_K = \Omega$.

A quantization of an arbitrary $f \in \mathcal{A}(\Gamma)$ is now obtained by the prescription

$$f \to \hat{f} = \int_\Gamma f(x) \mid \eta_{\sigma(x)}\rangle\langle \eta_{\sigma(x)} \mid d\nu(x). \tag{4.25}$$

The Poisson brackets of a certain set of classical observables $f$ will now go over to commutator brackets for the corresponding quantized observables $\hat{f}$; the determination of the exact set is analogous to that of finding a polarization in geometric quantization.

## 5. An Example

We illustrate the above method for quantization using the coherent states of the Poincaré group $\mathcal{P}_+^\uparrow(1,1)$ in 1-space and 1-time dimensions. This group is the semidirect product $\text{SO}(1,1) \otimes_s M^2$, where $M^2$ denotes the two dimensional Minkowski space and $\text{SO}(1,1)$ the (identity component of the) corresponding special orthogonal group. The elements of $\mathcal{P}_+^\uparrow(1,1)$ are of the form

$$g = (a, \Lambda_p) = \begin{pmatrix} \Lambda_p & a \\ 0 & 1 \end{pmatrix} \tag{5.1}$$

where $\Lambda_p$ is the Lorentz boost parameterized by the 2-vector $p = (p_0, \mathbf{p}) \in \mathcal{V}_m^+$ for some $m > 0$:

$$\Lambda_p = \begin{pmatrix} p_0 & \mathbf{p} \\ \mathbf{p} & p_0 \end{pmatrix}, \tag{5.2}$$

$\mathcal{V}_m^+$ denotes the forward mass hyperbola:

$$\mathcal{V}_m^+ = \big\{(p_0, \mathbf{p}) \in \mathbb{R}^2 \mid p_0 > 0, p_0^2 - \mathbf{p}^2 = m^2\big\}, \tag{5.3}$$

and $a = (a_0, \mathbf{a}) \in \mathbb{R}^2$. Let $T$ denote the time-translation subgroup of $\mathcal{P}_+^\uparrow(1,1)$, consisting of all elements of the type $\big((a_0, \mathbf{0}), \mathbb{I}_2\big)$, where $\mathbb{I}_2$ is the $2 \times 2$ unit matrix. We take $\Gamma = \mathcal{P}_+^\uparrow(1,1)/T$ and note that it has the structure of a symplectic manifold [**3**]. We consider the Wigner representation $U_W \equiv U_W^{(m)}$ of mass $m \geq 0$:

$$\big(U_W(a,\Lambda_p)\phi\big)(k) = e^{ik\cdot a}\phi(\Lambda_p^{-1}k), \quad \phi \in \mathfrak{H}_m = L^2(\mathcal{V}_m, d\mathbf{k}/k_o). \tag{5.4}$$

In this representation, the energy and momentum operators read, respectively:

$$P_o\psi(k) = k_o\phi(k), \quad \mathbf{P}\phi(k) = \mathbf{k}\phi(k). \tag{5.5}$$

The phase space $\Gamma = \mathcal{P}_+^\uparrow(1,1)/T$ has a global coordinatization $(\mathbf{q}, \mathbf{p})$ in terms of which the $\mathcal{P}_+^\uparrow(1,1)$-invariant measure is $d\mathbf{q}\, d\mathbf{p}$. In the principal bundle $\mathcal{P}_+^\uparrow(1,1) \to \Gamma$ a section $\sigma\colon \Gamma \to \mathcal{P}_+^\uparrow(1,1)$ reads:

$$\sigma(\mathbf{q},\mathbf{p}) = \sigma_o(\mathbf{q},\mathbf{p})\big((f(\mathbf{q},\mathbf{p}), \mathbf{0}), \mathbb{I}_2\big) \equiv (\hat{q}, \Lambda_p), \tag{5.6}$$

where $\sigma_o(\mathbf{q},\mathbf{p}) = \big((0,\mathbf{q}), \Lambda_p\big)$ is called the Galilean section, and the expression multiplying $\sigma_o(\mathbf{q},\mathbf{p})$ is an arbitrary element of $T$, indexed by the function $f$. The section $\sigma$ is said to be *affine* if $f$ is an affine function of $\mathbf{q}$: $f(\mathbf{q},\mathbf{p}) = \mathbf{q}\cdot\vartheta(\mathbf{p})$. Alternatively, the section $\sigma$ may be parametrized in terms of a momentum-dependent "speed" $\boldsymbol{\beta}$:

$$\hat{q}_o = \boldsymbol{\beta}(\mathbf{p})\cdot\hat{\mathbf{q}}. \tag{5.7}$$

Geometrically, (5.7) implies that for fixed $\mathbf{p}$, the choice of $\boldsymbol{\beta}$ determines a particular reference frame in $\hat{q}$-space [**1, 2**]. In particular, we see that the vector $\hat{q}$ is time-like, light-like or space-like, whenever $|\boldsymbol{\beta}| > 1$, $|\boldsymbol{\beta}| = 1$ or $|\boldsymbol{\beta}| < 1$, respectively, and this classification is also $\mathcal{P}_+^\uparrow(1,1)$-invariant. Notice that $\hat{q}$ being space-like, i.e., $\hat{q}^2 < 0$ or $|\boldsymbol{\beta}| < 1$, is equivalent to the condition:

$$\left|\boldsymbol{\vartheta}(\mathbf{p}) - \frac{\mathbf{p}}{m}\right| < \frac{p_o}{m}. \tag{5.8}$$

We introduce also the dual speed

$$\boldsymbol{\beta}^*(\mathbf{p}) = \frac{\mathbf{p} - p_o\boldsymbol{\beta}(\mathbf{p})}{p_o - \mathbf{p}\cdot\boldsymbol{\beta}(\mathbf{p})}. \tag{5.9}$$

The functions $\boldsymbol{\beta}$, $\boldsymbol{\beta}^*$ satisfy a number of interesting identities. In particular, $\boldsymbol{\beta}^{**} = \boldsymbol{\beta}$, which means that the sections come in dual pairs, and moreover the duality $\boldsymbol{\beta} \Leftrightarrow \boldsymbol{\beta}^*$ preserves the space-like (respectively, time-like or light-like) character of the vector $\hat{q}$.

Square-integrable coherent states may now be obtained as a consequence of the following result.

THEOREM 5.1. *Let* $\sigma\colon \Gamma = \mathcal{P}_+^\uparrow(1,1)/T \to \mathcal{P}_+^\uparrow(1,1)$ *be any affine space-like section. Then a vector* $\eta$ *is admissible* mod$(T,\sigma)$ *if and only if* $\eta \in \mathcal{D}(P_o^{1/2})$ *(the domain of* $P_o^{1/2}$*), and every such vector generates a family of square-integrable coherent states indexed by points in the phase space* $\Gamma$:

$$\mathfrak{S}_\sigma = \{\eta_{\sigma(\mathbf{q},\mathbf{p})} = U_W(\sigma(\mathbf{q},\mathbf{p}))\eta \mid (\mathbf{q},\mathbf{p}) \in \Gamma\}. \tag{5.10}$$

*The operator*

$$A_\sigma^\eta = \int_\Gamma |\eta_{\sigma(\mathbf{q},\mathbf{p})}\rangle\langle\eta_{\sigma(\mathbf{q},\mathbf{p})}|\, d\mathbf{q}\, d\mathbf{p} \tag{5.11}$$

*is a positive bounded multiplication operator with bounded inverse:* $(A_\sigma^\eta\psi)(k) = \mathcal{A}_\sigma^\eta(k)\psi(k)$, *where*

$$\mathcal{A}_\sigma^\eta(k) = \int_{\mathcal{V}_m^+} \mathcal{A}_\sigma(k,p)|\eta(p)|^2\frac{d\mathbf{p}}{p_o}. \tag{5.12}$$

*The kernel* $\mathcal{A}_\sigma(k,p)$ *is defined in terms of the function* $\boldsymbol{\vartheta}$ *or the speed* $\boldsymbol{\beta}$, *defining the section* $\sigma$, *as follows:*

$$\begin{aligned}\mathcal{A}_\sigma(k,p) &= \frac{2\pi}{m}\frac{k\cdot p}{k_o - \mathbf{p}\cdot\boldsymbol{\vartheta}(\underline{\Lambda}_p^{-1}\underline{k})} \\ &= 2\pi m\left(p_o + \mathbf{p}\cdot\boldsymbol{\beta}^*(\underline{\Lambda}_p^{-1}\underline{k})\right)^{-1}\end{aligned} \tag{5.13}$$

*where* $\underline{z} \equiv \mathbf{z}$ *denotes the space component of a* 2*-vector* $z = (z_o, \mathbf{z})$.

Assuming now that $A_\sigma^\eta$ in (5.11) is a multiple of the identity, we have

$$\Omega_{K_\sigma} = F_\sigma(\mathbf{p})\, d\mathbf{p}\wedge d\mathbf{q} \tag{5.14}$$

where

$$F_\sigma(\mathbf{p}) = \frac{1}{\|\eta_{\mathbf{q},\mathbf{p}}\|^2} \int_{\mathcal{V}_m^+} \left[ \frac{\mathbf{k}}{k_o p_o} \cdot \left( \frac{\mathbf{p}}{m} - \boldsymbol{\vartheta}(\mathbf{p}) \right) + \frac{1}{m} \right] |\eta|^2(k)\, d\mathbf{k}. \tag{5.15}$$

The one-form $\theta_{K_\sigma}$ is, in this case, given by

$$\theta_{K_\sigma} = \frac{i}{\|\eta_{\mathbf{q},\mathbf{p}}\|^2} (G_1(\mathbf{p})\, d\mathbf{q} + \mathbf{q} G_2(\mathbf{p})\, d\mathbf{p}) \tag{5.16}$$

with

$$\begin{aligned} G_1(\mathbf{p}) &= \int_{\mathcal{V}_m^+} \left( k_o \boldsymbol{\vartheta}(\mathbf{p}) - \frac{\mathbf{k} p_o + \mathbf{p} k_o}{m} \right) |\eta|^2(k) \frac{d\mathbf{k}}{k_o}, \\ G_2(\mathbf{p}) &= \int_{\mathcal{V}_m^+} \left( k_o \boldsymbol{\vartheta}'(\mathbf{p}) - \frac{\mathbf{k}}{p_o} \cdot \boldsymbol{\vartheta}(\mathbf{p}) \right) |\eta|^2(k) \frac{d\mathbf{k}}{k_o}. \end{aligned} \tag{5.17}$$

In order for the condition (4.21) to be satisfied in this case, the function $\eta$ has to be an analytic vector of $\mathfrak{H}$ (i.e., infinitely differentiable as a function of $k$) and (for all $(\mathbf{q}, \mathbf{p}) \in \mathbb{R}^2$)

$$\|\eta_{\mathbf{q},\mathbf{p}}\|^2 = \lambda \int_{\mathcal{V}_m^+} \left[ \frac{\mathbf{k}}{k_o p_o} \cdot \left( \frac{\mathbf{p}}{m} - \boldsymbol{\vartheta}(\mathbf{p}) \right) + \frac{1}{m} \right] |\eta|^2(k)\, d\mathbf{k}, \tag{5.18}$$

where $\lambda$ is a fixed constant.

As shown in [**3**] depending on the section chosen, the quantization yields either a configuration space, momentum space or a phase space representation of a one-particle relativistic system.

## 6. Conclusion

Quantization using coherent states is often more versatile than geometric quantization and it is closer in spirit to Berezin's technique [**5**]. It is often possible to build different families of coherent states on the same phase space, depending on the symmetry group or complex structure exploited, and these in general lead to different quantizations. Furthermore, the existence of a family of coherent states induces, under appropriate conditions, a symplectic structure on the underlying parameter space, and when there is more than one such structure available, it is interesting to study the corresponding classical systems. There is also the possibility of using coherent state techniques for infinite dimensional groups and to arrive at quantizations of the homogeneous spaces of such groups. These homogeneous spaces are usually not locally compact and the problem becomes much more subtle. Finally, as pointed out, continuous wavelet transforms share the same mathematical origin as the unitary map $W$ (see (3.10)) linking coherent states to the Hilbert space of the quantized system.

## References

1. S. T. Ali, J.-P. Antoine, and J.-P. Gazeau, *Square integrability of group representations on homogeneous spaces.* II. *Coherent and quasi-coherent states, The case of the Poincaré group,* Ann. Inst. H. Poincaré Phys. Théor. **55** (1991), 857–890.
2. S. T. Ali, J.-P. Antoine, and J.-P. Gazeau, *Relativistic quantum frames,* Ann. Physics **222** (1993), 38–88.
3. S. T. Ali and U.-A. Mueller, *Quantization of a classical system on a coadjoint orbit of the Poincaré group in* $1+1$ *dimensions,* J. Math. Phys. **35** (1994), 4405–4422.
4. S. T. Ali, J.-P. Antoine, J.-P. Gazeau, and U.-A. Mueller, *Coherent states and their generalizations: A mathematical overview,* Rev. Math. Phys. **7** (1995), 1013–1104.

5. F. A. Berezin, *Quantization*, Izv. Akad. Nauk SSSR Ser. Mat. **38** (1974), 1116–1175 (Russian).
6. R. Gilmore, *Geometry of symmetrized states*, Ann. Physics **74** (1972), 391–463; *On properties of coherent states*, Rev. Mexicana Fís. **23** (1974), 143–187.
7. L. van Hove, *Sur le problème des relations entre les transformations unitaires de la mécanique quantique et les transformations canoniques de la mécanique classique*, Acad. Roy. Belgique. Bull. Cl. Sci. (5) **37**, (1951), 610–620.
8. B. Kostant, *Quantization and unitary representations*, Lecture Notes in Math., vol. 170, Springer-Verlag, Berlin, 1970, 87–208.
9. A. Odzijewicz, *Coherent states and geometric quantization*, Commun. Math. Phys. **150** (1992), 385–413.
10. A. Perelomov, *Generalized coherent states and their applications*, Springer-Verlag, Berlin, 1986.
11. N. J. M. Woodhouse, *Geometric quantization*, 2nd ed., Clarendon Press, Oxford, 1992.

Department of Mathematics and Statistics, Concordia University, Montréal, Québec, Canada H4B 1R6

*E-mail address*: `stali@neumann.concordia.ca`

Centre de Recherches Mathématiques
CRM Proceedings and Lecture Notes
Volume **18**, 1999

# Wavelets in Molecular and Condensed-Matter Physics

J.-P. Antoine

## 1. Introduction

As in atomic physics [**1**], wavelets have found so far relatively few applications in molecular and condensed-matter physics, and all of them use only discrete multiresolution wavelets [**13**]. Indeed, as in the case of self-consistent atomic methods, what is needed here are mostly orthogonal bases with good localization properties, in order to minimize the number of terms required for an accurate calculation of wave functions and related observable quantities.

We will describe in this paper three very different applications, in dimensions three, two and one, respectively.

(1) The most active field of research is that of *ab initio* electronic structure calculations, in atoms, molecules and crystals; the crucial feature is the narrow support of the wavelets constituting an orthonormal basis (or pseudobasis).
(2) An electron gas submitted to a strong magnetic field may be considered as a two-dimensional physical system. This is, for instance, the set-up of the quantum Hall effect (QHE). Here wavelets yield various bases for the lowest Landau level, a necessary step for the description of the Fractional QHE.
(3) In certain cases, the Wannier functions of a 1-D crystal obey two-scale relations similar to those of the multiresolution wavelets, and this analogy may bring new physical insight.

## 2. Electronic Structure Calculations

**2.1. Principle.** A basic problem in condensed-matter physics is the *ab initio* calculation of the electronic structure of a given material (ground state energy, wave function, etc.) Now, since the crystal is a 3 D periodic structure, it suffices to describe the electronic structure around a single lattice site and apply a Bloch transformation. Thus one comes back to the study of the electronic structure of a single atom or molecule. As explained in [**1**], the key is to find a good orthonormal basis, consisting of functions well adapted to the problem. Two standard methods are popular among chemists:

1991 *Mathematics Subject Classification.* Primary. 81R30, Secondary: 81V55, 82D20.
This is the final version of the paper.

- *LCAO bases* (Linear Combination of Atomic Orbitals), based for instance on Slater or Gaussian orbitals. The method yields a good description of the electronic structure with relatively few terms, but it is difficult to improve it systematically (the LCAO method is sometimes described as a art rather than a science!). In addition the expression for the forces is extremely complicated.
- *Plane wave bases*: These two difficulties disappear, but *a priori* plane waves are not suited for describing localized objects, since all information on space localization is lost in Fourier transforms. Yet the electronic structure of an atom is highly inhomogeneous in space, the wave function oscillates much more rapidly close to the nucleus. As a consequence a large number of terms is needed for describing the small inner region, but this increased precision is not necessary elsewhere.

Combining the advantages of the two methods would require an orthogonal basis of localized functions, allowing to vary the precision according to the local electronic density. That suggests a wavelet basis, since wavelets are well localized, and adapt automatically to the scale of the object to represent (the so-called automatic zoom effect). Several research groups have tried such an analysis, with different types of discrete wavelets, and we shall quickly review them.

**2.2. A non-orthogonal wavelet basis.** The first attempt, by Cho *et al.* [**10**] was based on a non-orthogonal multiresolution basis and its approximation by Gaussians and Mexican hats. The mathematical background is the following. One starts with the familiar multiresolution scale in three dimensions

$$\cdots \subset V_{-1} \subset V_0 \subset V_1 \subset \cdots \subset L^2(\mathbb{R}^3), \tag{2.1}$$

but *without* the assumption of orthogonality

$$V_j = V_{j-1} \dot{+} W_{j-1}, \tag{2.2}$$

where, as usual, $W_{j-1}$ describes the additional details needed for passing from resolution $2^{j-1}$ to resolution $2^j$, and $\dot{+}$ denotes a direct sum, not necessary orthogonal (hence the decomposition in (2.2) is still unique). Thus one gets

$$L^2(\mathbb{R}^3) = \dot{\sum_{j\in\mathbb{Z}}} W_j = V_{j_o} \dot{+} \Big(\dot{\sum_{j\geq j_o}} W_j\Big), \tag{2.3}$$

where $j_o$ corresponds to the lowest resolution considered and $\dot{\sum}$ denotes again a direct sum. Then one chooses two orthogonal bases, localized around the nodes of a fixed 3-D lattice:

$$\{\phi_{j\mathbf{n}}(\mathbf{r})\} \in V_j \text{ and } \{\psi_{j\mathbf{n}}(\mathbf{r})\} \in W_j, \tag{2.4}$$

where $\mathbf{n}$ denotes a lattice point and $j \in \mathbb{Z}$. From (2.3), the practical expansion of a general wave function into these bases reads:

$$f(\mathbf{r}) = \sum_{\mathbf{n}} a_{j_o\mathbf{n}}\phi_{j_o\mathbf{n}}(\mathbf{r}) + \sum_{j=j_o}^{j_{\max}} \sum_{\mathbf{n}} d_{j\mathbf{n}}\psi_{j\mathbf{n}}(\mathbf{r}). \tag{2.5}$$

In this truncated expansion, $\phi_{j_o\mathbf{n}}$ is a scaling function, $\psi_{j\mathbf{n}}$ is a wavelet, both centered around the lattice point $\mathbf{n}$, and $j_{\max}$ is the finest resolution, corresponding to the desired precision. This expansion still has an infinite number of terms, since $\mathbf{n}$ runs over the whole lattice. However, both $\phi_{j\mathbf{n}}$ and $\psi_{j\mathbf{n}}$ are supposed to be well localized, so one has to keep only those functions which have significantly large coefficients $a_{j_o\mathbf{n}}$, resp. $d_{j\mathbf{n}}$, for the problem at hand. This means that one allows different resolutions for different localized regions. For instance, we know that the electronic wave function oscillates more rapidly in the atomic core region. Then one should add higher resolution scales $j$ in the core region, where the precision must be higher, but only there. Globally the number of terms is thus reduced.

The technique introduced in [**10**] for practical calculations runs as follows.

- One starts at resolution $j_o = 0$ with basis functions centered on the nodes of a simple cubic lattice $L_0$, with lattice spacing $d_o$ small enough to give a sufficiently tight frame.
- For each resolution $j = 1, 2 \ldots, j_{\max}$, the basis functions are localized on the nodes of the refined lattice $L_j$ of spacing $2^{-j}d_o$, excluding the nodes of the previous lattice $L_{j-1}$. Thus all the basis functions taken together are localized on the lattice $L_{j_{\max}}$ without overlap.
- In order to adapt the resolution to the electronic density, one draws around each nucleus a sequence of concentric spheres $S_j$ of radii $r_j = 2^{-j}r_0, j = 0, \ldots, j_{\max}$. Then one keeps as basis functions only those scaling functions that are localized inside $S_0$ and, for each $j = 0, \ldots, j_{\max}$, wavelets $\psi_{j\mathbf{n}}$ localized inside $S_{j+1}$. For each nuclear center, the resulting finite basis consists of the functions

$$\{\phi_{0\mathbf{n}}(\mathbf{n} \in L_0 \cap S_0); \psi_{0\mathbf{n}}(\mathbf{n} \in L_0 \cap S_1); \psi_{1\mathbf{n}}(\mathbf{n} \in (L_1 \setminus L_0) \cap S_2); \\ \ldots; \psi_{j_{\max}\mathbf{n}}(\mathbf{n} \in (L_{j_{\max}} \setminus L_{j_{\max}-1}) \cap S_{j_{\max}+1})\}.$$

  This arrangement is schematized in Figure 1 in a 2-D version. Notice that the lattice, hence the basis, is fixed, it is *not* correlated to the position of the atomic nuclei.

The net result is a highly flexible finite basis, with higher resolutions introduced only where they are needed, namely close to the nuclear core.

The next problem is to choose basis functions which are both efficient and simple to work with. The authors of [**10**] use Gaussians for $\phi_{j,\mathbf{n}}$ and Mexican hat wavelets for $\psi_{j,\mathbf{n}}$. Now, strictly speaking, this choice is not allowed, since these functions do not generate an orthogonal basis. But they give an extreme good approximation to such a basis.

The argument runs as follows [**6**]. First a criterion for ensuring a unique decomposition (2.2) (without orthogonality in general) is that (see [**12**, Theor.5.16])

$$\begin{vmatrix} P(z) & Q(z) \\ P(-z) & Q(-z) \end{vmatrix} \neq 0, \text{ for all } |z| = 1, \tag{2.6}$$

where $P(z)$, $Q(z)$ are the familiar functions defining the two-scale relations:

$$\widehat{\phi}(2\omega) = P(e^{i\omega})\widehat{\phi}(\omega) \equiv \widetilde{P}(\omega)\widehat{\phi}(\omega) \text{ (scaling function)},$$
$$\widehat{\psi}(2\omega) = Q(e^{i\omega})\widehat{\psi}(\omega) \equiv \widetilde{Q}(\omega)\widehat{\psi}(\omega) \text{ (wavelet)},$$

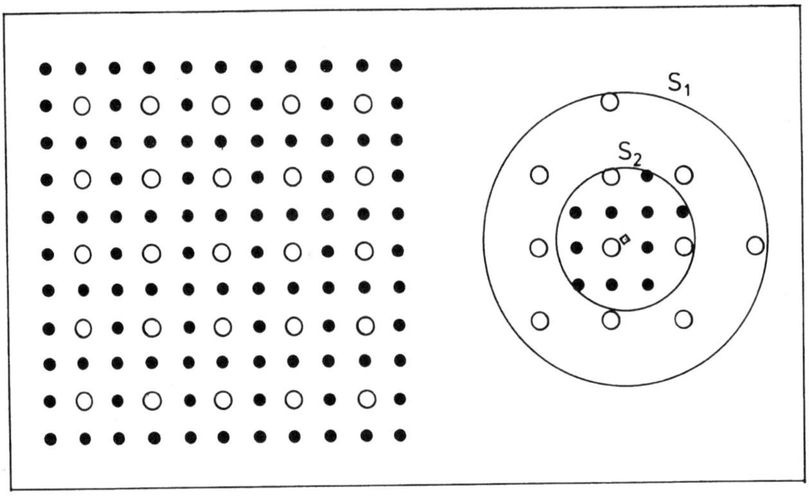


FIGURE 1. The geometrical arrangement of [**10**] in a 2-D version: (left) Centers of the basis functions with $j = 0$ (open circles) and $j = 1$ (black circles); (right) The spheres $S_1$ and $S_2$, both centered on an atomic nucleus (lozenge), and containing centers of wavelets with $j = 0$ and $j = 0, 1$, respectively.

with $\widetilde{P}$, $\widetilde{Q}$ two $2\pi$-periodic functions (denoted $m_0$, $m_1$ in [**13**]). Then the authors of [**10**] choose the following functions:

$$\widetilde{P}(\omega) = \sum_{j\in\mathbb{Z}} \exp\Big(-\frac{3}{2}\sigma^2(\omega - 2\pi j)^2\Big),$$

$$\widetilde{Q}(\omega) = -e^{i\omega}\left(\omega - 2\pi\left[\frac{\omega+\pi}{2\pi}\right]\right)^2 \widetilde{P}(\omega),$$

where $[\,.\,]$ denotes the integer part function. It is readily seen by direct computation (noting that $P(-z)$ corresponds to $\widetilde{P}(\omega+\pi)$) that these functions satisfy the criterion (2.6), thus guaranteeing a good decomposition (2.2). Furthermore, and this is the interesting point, for $\sigma = 1.35$, $\widetilde{P}$ and $\widetilde{Q}$ generate a scaling function $\phi(x)$ and a "wavelet" $\psi(x)$ that match, respectively, a Gaussian $G(x)$ and a Mexican hat $-G''(x)$ within an absolute error of $10^{-10}$. This is indeed a good approximation!

Using this method, the following results are obtained in [**10**]:

(1) *Ground state wave function of hydrogen atom.*
Using a basis of 25 functions (7 scaling functions for $j = 0$ and 6 wavelets for each of $j = 0, 1, 2$ ($r_0 = 1$ atomic unit (a.u.)), the (known) ground state radial wave function of hydrogen is reproduced with a 2% error; the latter may be reduced to 0.3% with an 85 function basis ($j_{\max} = 1$, $r_o = 2$ a.u.).

(2) *Ground state energies of arbitrary atoms.*
By the same technique, the ground state energies of all elements, from hydrogen to uranium, may be estimated within 3% using a 67 function basis, with $j_{\max} = 10$.

(3) *Energy of the hydrogen molecular ion $H_2^+$.*
The next example is the calculation of the total energy of a hydrogen molecular ion $H_2^+$ as a function of the separation $R$ between the two protons. With a basis of 141 to 167 functions (depending on $R$), the exact values are reproduced within 1%.

The conclusion is that a (quasi-)wavelet basis offers a very good alternative to standard methods for calculating the electronic structure of atoms or molecules, because it allows to vary the spatial resolution with space. The analysis extends immediately to periodic systems by introducing a Bloch transformation. But, of course, several problems remain open.

One of them is the conflict between the spherical geometry of the atom and the Cartesian geometry of the lattice used in the wavelet expansion. Recent work of Arias *et al.* [**5**] has made progress in this direction, by introducing a more isotropic scheme based on the so-called *interpolets*, which combines wavelets with finite element methods [**11**]. Actually this rather suggests using a (discretized version of) a 3-D continuous wavelet transform, for which the resolving power (precision) is easy to control [**3, 4**].

Another open issue is the comparison with *multigrid* methods, which also allow to vary the spatial resolution. Preliminary results [**6,11**] indicate that the interpolet method is more efficient.

**2.3. Orthogonal wavelet bases.** Instead of the non-orthogonal scheme of Cho *et al.* [**10**], several groups have considered genuine orthonormal wavelet bases, e.g. Daubechies or Meyer bases.

1. Wei and Chou [**23**] choose a Daubechies D6 wavelet basis of compact support to generate a 3-D orthonormal basis by a threefold tensor product. Exploiting fast wavelet transform algorithms, they use the resulting basis for calculating the physical quantities in the so-called local density approximation. This requires evaluating matrix elements of the Hamiltonian, $\int \psi_l H \psi_m d^3x$, where $\psi_l$ stands for the wavelet basis vectors, and in particular potential terms $\int \psi_l U \psi_m d^3x$, where $U$ may be a local potential (ionic, Hartree or exchange-correlation) or a nonlocal separable pseudopotential. The grid is the standard multiresolution grid, and its position relative to the centers of the atoms is essentially irrelevant. As compared to [**10**], no artificial cut-off is put by hand, the rapid convergence of the calculation follows from the smallness of the compact support of the D6 wavelet, which yields sparse matrices.

With this technique, Wei and Chou compute the total energies for a (fictitious) hydrogen atom and for a dimer molecule, $H_2$ or $O_2$ (as a function of the bond length). In all cases, the method requires relatively few basis functions (1000 or 2000) as compared to the plane wave method ($\sim 8700$), with similar results. Thus it offers hope for a serious improving of standard calculations.

2. Tymczak and Wang [**22**] also use a Daubechies wavelet basis for performing electronic structure calculations in a local density approximation. The latter consists in starting from the Kohn-Sham equation (instead of the Schrödinger equation):

$$\left(-\frac{\hbar^2}{2m}\Delta + V\big[\mathbf{r};\rho(\mathbf{r})\big]\right)\psi_n(\mathbf{r}) = \varepsilon_n \psi_n(\mathbf{r}), \quad \rho(\mathbf{r}) = \sum_n f_n \big|\psi_n(\mathbf{r})\big|^2,$$

and transform it with the WT. These authors report good convergence results for the total energy (ground state) of an harmonic oscillator and the hydrogen atom, and announce further work on the helium atom and diatomic molecules.

3. Yamaguchi and Mukoyama [**24**] use instead a Meyer wavelet basis ($C^\infty$, all moments vanishing, but noncompact support) for solving the *radial* Schrödinger equation by a variational (Galerkin) method. By this technique they compute energy eigenvalues and wave functions for hydrogen, neon and argon atoms, and continuum wavefunctions for the argon atom. As a test of the numerical quality of their wave functions, they calculate, respectively, radiative transition rates in neon and argon atoms, and partial photoionization rates for an argon atom in its 1s, 2s and 3s state. The results are in excellent agreement with those obtained by traditional methods. An interesting remark is that each type of atomic state (lower bound state, Rydberg state, continuum state) has a characteristic distribution in the wavelet parameter space. This permits to choose for each kind of physical process an adequate trial function for the variational procedure.

## 3. Wavelet-Like Orthonormal Bases for Lowest Landau Level

**3.1. The fractional quantum Hall effect setup.** The system considered in the Fractional Quantum Hall Effect (FQHE) is a (quasi)-planar gas of electrons in a strong magnetic field perpendicular to the plane (see [**20**] for a review and the original references). The first problem to tackle is to find the ground state of the system. Two main methods have been proposed in the literature to that effect, depending on the electron density, which is measured by the so-called filling factor $\nu$. The first one is a Hartree-Fock approach to a system of $N$ two-dimensional electrons [**9**], that is, a Slater determinant of one-electron wave functions [**1**]. This picture gives good energy values for small ($0 < \nu < \frac{1}{5}$) or high ($4/5 < \nu < 1$) electron densities. In the intermediate range ($1/5 < \nu < 4/5$), however, the best results are obtained with the Laughlin wave function, which is derived by a variational technique based on a non mean-field approach to the same two-dimensional gas of electrons. We will consider here the first method only.

The first step is to select an adequate wave function for a single electron in the magnetic field. As it is well-known [**20**], the energy levels, the so-called Landau levels, are infinitely degenerate, and there arises the problem of finding a good basis in the corresponding Hilbert subspace. This is crucial for an easy computation of the energy levels of the whole system, in the presence of perturbations. In particular, the ground state belongs to the lowest Landau level (LLL). In a joint work with Bagarello [**2,7,8**], we have proposed a general method for constructing an orthogonal basis for the LLL, starting from standard 1-D orthogonal wavelet bases. We will describe this construction below, first recalling the physical background of the problem.

Consider a single electron confined in the $xy$-plane and subjected to a strong magnetic field in the $z$-direction. In the symmetric gauge, the Hamiltonian reads

$$H_o = \frac{1}{2}(p_x - y/2)^2 + \frac{1}{2}(p_y + x/2)^2. \tag{3.1}$$

Introducing the canonical variables

$$P' = p_x - y/2, \quad Q' = p_y + x/2, \tag{3.2}$$

this can be written in the form of the Hamiltonian of a harmonic oscillator:

$$H_o = \frac{1}{2}(Q'^2 + P'^2) \tag{3.3}$$

(we use units such that $\hbar = M = e|\vec{H}|/c = 1$).

The eigenstates of the Hamiltonian (3.1) can be found explicitly, and they have the following form:

$$\Phi_{mn}(x,y) \sim e^{(x^2+y^2)/4}(\partial_x + i\partial_y)^m(\partial_x - i\partial_y)^n e^{-(x^2+y^2)/2}, \quad m,n = 0,1,2,\ldots, \tag{3.4}$$

corresponding to the eigenvalues

$$E_{mn} \equiv E_n = n + 1/2. \tag{3.5}$$

Thus the energy levels are all degenerate in $m$, so that the ground level (LLL) is spanned by the set $\{\Phi_{m0}(x,y)\}$, which forms an orthonormal basis. However these wave functions are not very well localized, since the mean value of the distance from the origin, $r \equiv \sqrt{x^2+y^2}$, increases with $m$. Yet the physics of the problem requires that the wave functions be fairly well localized, since the system tends, as $T \to 0$, to the configuration of the Wigner crystal, that is a triangular lattice [**9**]. Thus the basis $\{\Phi_{m0}(x,y)\}$ of the LLL is inadequate for our purposes.

**3.2. The LLL basis problem.** In order to find another basis of eigenfunctions, orthogonal or not, spanning the same energy level, it is convenient to use the method introduced in [**9**], which is based on a technique introduced in [**21**]. The transformation (3.2) can be seen as a part of a canonical transformation from the variables $x$, $y$, $p_x$, $p_y$ into the new ones $Q$, $P$, $Q'$, $P'$, where

$$P = p_y - x/2, \quad Q = p_x + y/2. \tag{3.6}$$

These operators satisfy the following commutation relations:

$$[Q,P] = [Q',P'] = i, \tag{3.7}$$

$$[Q,P'] = [Q',P] = [Q,Q'] = [P,P'] = 0. \tag{3.8}$$

Then a wave function in the $(x,y)$-space is related to its $PP'$-expression by the formula

$$\Psi(x,y) = \frac{e^{ixy/2}}{2\pi}\int\!\!\int_{\mathbb{R}^2} e^{i(xP'+yP+PP')}\Psi(P,P')\,dP\,dP'. \tag{3.9}$$

In virtue of the expression (3.3) of $H_o$, the Schrödinger equation $H_o\Psi = \frac{1}{2}(Q'^2 + P'^2)\Psi = E\Psi$ admits factorized solutions $\Psi(P,P') = f(P')h(P)$. Thus the ground state wave function of (3.3) must have the form

$$\Psi_0(P,P') = f_0(P')h(P), \tag{3.10}$$

where $f_0(P') = \pi^{-1/4}e^{-P'^2/2}$, and the function $h(P)$ is arbitrary, which manifests the infinite degeneracy of the LLL.

Depending on the choice of $h(P)$, several types of bases for the LLL may be obtained, according to the following general scheme. Inserting (3.10) into the integral (3.9), the Gaussian integration on $P'$ can be performed exactly. Next, taking a wave function $\Psi_n(P,P') = f_0(P')h_n(P)$, where $\{h_n(P)\}$ is an arbitrary basis in $L^2(\mathbb{R})$, we define:

$$h_n^{(2)}(x,y) = \frac{e^{ixy/2}}{\sqrt{2}\pi^{3/4}} \int_{-\infty}^{\infty} e^{iyP} e^{-(x+P)^2/2} h_n(P)\, dP. \tag{3.11}$$

Then the set $\{h_n^{(2)}(x,y)\}$ is a basis for the LLL, and it is orthonormal iff $\{h_n(P)\}$ is orthonormal in $L^2(\mathbb{R})$. This follows from the canonicity of the change of variables given in Eqs. (3.2), (3.6) or simply by an explicit calculation of the matrix element $\langle h_n \mid h_m\rangle$, using the integral (3.11). Several examples of this construction have been presented in the literature.

1. Bagarello *et al.* [**9**] choose for the ground state a Gaussian $h_0(P) = f_0(P)$, which yields, by (3.11), $h_0^{(2)}(x,y) = \Phi_{00}(x,y)$ as defined in (3.4). Furthermore they construct a complete set of basis functions of the LLL by acting on $\Phi_{00}(x,y)$ with the so-called magnetic translation operators

$$T(\vec{a_i}) = \exp i(Qa_{ix} + Pa_{iy}), \quad i = 1,2,$$

where $\vec{a_i} = (a_{ix}, a_{iy})$ are the lattice basis vectors. The set obtained in this way is still in the LLL since the operators $T(\vec{a_i})$ commute with $H_o$, in virtue of the commutation relations (3.8). Moreover $[T(\vec{a_1}), T(\vec{a_2})] = 0$ if the area of the cell of the lattice is such that $a_{1x}a_{2y} - a_{1y}a_{2x} = 2N\pi$, $N \in \mathbb{Z}$, a condition assumed henceforth. Thus the basis vectors in $L^2(\mathbb{R})$ have the form

$$h_{kj}(P) = T(\vec{a_1})^k T(\vec{a_2})^j f_o(P), \quad k,j \in \mathbb{Z},$$

where $k\vec{a_1} + j\vec{a_2}$ is an arbitrary lattice point. Completeness of the set results from its unitary equivalence with the set of canonical coherent states associated to $Q, P$. These basis vectors have Gaussian localization around the sites of a regular two-dimensional lattice, and thus the basis is lattice-translation invariant. However, each vector has a well-defined, fixed (essential) support, so that there is no possibility of modifying the mutual overlap for fixed electron density. In addition, this basis is not orthogonal, since coherent states are in general not mutually orthogonal. Enforcing orthogonality (e.g. Gram-Schmidt) spoils much of the simplicity of the basis functions, and in particular the localization properties for intermediate fillings and the lattice-translation invariance property.

2. In order to keep the good localization properties and some sort of translation invariance, Ferrari [**14**] has constructed an orthonormal basis for the LLL by taking infinite superpositions of the above (coherent) states. The resulting basis vectors are Bloch functions, which may be made translation invariant over the nodes of a given lattice, typically triangular or hexagonal (remember that the Wigner crystal is a triangular lattice). Clearly this basis describes very well the two-dimensional low-density system of electrons of the FQHE, but its construction is rather involved and ad hoc.

3. Since one wants basis wave functions which are both well localized *and* orthogonal, obvious candidates are orthogonal wavelets. They do enjoy good localization properties, and the latter are easily controlled by varying the scale parameter, in contrast to the Gaussian-like functions of [**9**]. In addition, wavelets seem well adapted to a physical problem which has an intrinsic hierarchical structure [**16,17**]. In particular, the relevant parameter, namely the filling factor, may take arbitrary rational values, and this suggests a fractal behavior, as pointed out recently [**19**].

This again points to wavelets. Several examples of this construction have been proposed in [**2, 7, 8**] and we will review them in the next section.

### 3.3. Wavelet-like bases.

3.3.1. *The Haar basis.* Let us look first at the LLL basis generated by the Haar wavelet basis. The mother wavelet reads:

$$h(x) = \begin{cases} 1, & \text{if } 0 \leqslant x < 1/2 \\ -1, & \text{if } 1/2 \leqslant x < 1 \\ 0, & \text{otherwise} \end{cases} \tag{3.12}$$

and the corresponding wavelet orthonormal basis in $L^2(\mathbb{R})$ is obtained via the usual multiresolution analysis (MRA) formula

$$h_{mn}(x) = 2^{-m/2} h(2^{-m}x - n). \tag{3.13}$$

Of course, since $h(x)$ is a discontinuous function, its localization in frequency space is poor, it decays as $\omega^{-1}$. However, the transformation (3.11) is not a Fourier transform, hence it is not clear *a priori* that the corresponding functions $\{h^{(2)}_{mn}(x,y)\}$ will also have a poor localization in both variables. In fact it is *not* the case, as can be see by investigating the asymptotic behavior of the basis functions.

Using standard results on Gaussian integrals, one finds for the ground state wave function:

$$H_{00}(x,y) = \frac{e^{-ixy/2} e^{-y^2/2}}{2\pi^{1/4}} \{2\Xi(x - iy + 1/2) - \Xi(x - iy) - \Xi(x - iy + 1)\} \tag{3.14}$$

where

$$\Xi(z) = \operatorname{erf}\big(z/\sqrt{2}\big) = \frac{2}{\sqrt{\pi}} \int_0^{z/\sqrt{2}} e^{-t^2}\, dt, \quad z \in \mathbb{C}.$$

Using the asymptotic expansion of the error function, we obtain, for $|x|$, $|y| \gg 1$:

$$H_{00}(x,y) \simeq \frac{e^{ixy/2} e^{-x^2/2}}{2\pi^{1/4}} \sqrt{\frac{2}{\pi}} \left( \frac{1}{x - iy} + \frac{e^{-1/2 - x + iy}}{x - iy + 1} - 2\frac{e^{-1/8 - (x-iy)/2}}{x - iy + 1/2} \right), \tag{3.15}$$

which displays the Gaussian localization of the wave function in the variable $x$ and the rather poor one in $y$ (decay as $y^{-1}$). This behavior is confirmed in Figure 2, which shows the modulus of $H_{00}(x,y)$, first in 3-D perspective (a), then along the $x$-axis (b), finally along the $y$-axis (c). Clearly the function $H_{00}(x,y)$ is much better localized in the $x$ variable than in $y$.

An analogous behavior can be obtained for the generic function $H_{mn}(x,y)$, which may also be calculated exactly. Here, according to (3.13), $n$ indexes the center of the original mother wavelet and $m \in \mathbb{Z}$ is the scale parameter. Using (3.11), it is easily seen that the asymptotic behavior of $h^{(2)}_n(x,y)$ in $x$ is governed by the asymptotic behavior of $h_n(P)$, and the one in $y$ by that of the Fourier transform of $h_n(P)$. Since in the present case, $h_n(P)$ has compact support (increasing monotonically with $m$), we expect $H_{mn}(x,y)$ to be strongly localized in $x$ and delocalized in $y$, and that its decay in $x$ gets faster for smaller $m$. This is indeed the case, as may be seen on the figures presented in [**7**] consider for instance the function $H_{m0}(x,y)$. For $m = 1, 2, 3\ldots$, the support of $h_{m0}$ increases with $m$, so that the decay of $H_{m0}(x,y)$ in $x$ gets slower, whereas, for $m = -1, -2, -3, \ldots$, the support of $h_{m0}$ decreases with $m$, so that the decay of $H_{m0}(x,y)$ in $x$ gets faster.

3.3.2. *The Littlewood-Paley basis.* Another simple example of an orthonormal wavelet basis in $L^2(\mathbb{R})$, also coming from MRA, is the Littlewood-Paley basis, generated from the mother wavelet

$$\Psi(x) = (\pi x)^{-1}(\sin 2\pi x - \sin \pi x). \tag{3.16}$$

The behavior of this function is complementary to that of the Haar wavelet: it is very well localized in frequency space (it has a compact support):

$$\widehat{\Psi}(\omega) = \begin{cases} (2\pi)^{-1/2}, & \text{if } \pi \leqslant |\omega| \leqslant 2\pi \\ 0, & \text{otherwise,} \end{cases} \tag{3.17}$$

while, as one can see from (3.16), it decays like $1/x$ in configuration space.

An analogous complementary behavior is found also for the corresponding LLL wave functions. They are exponentially localized in the $y$-variable, while in the other variable they behave like $1/x$. This is manifest on the asymptotic behavior of $\Psi_{00}(x, y)$ for $|x|, |y| \gg 1$. By the same method as before, one finds:

$$\Psi_{00}(x,y) \simeq \frac{e^{-ixy/2}e^{-y^2/2}}{2\pi^{5/4}} \left\{ -\frac{e^{2\pi(y+ix)}e^{-2\pi^2}}{|2\pi - y - ix|} + \frac{e^{\pi(y+ix)}e^{-\pi^2/2}}{|\pi - y - ix|} \right.$$
$$\left. - \frac{e^{-2\pi(y+ix)}e^{-2\pi^2}}{|2\pi + y + ix|} + \frac{e^{-\pi(y+ix)}e^{-\pi^2/2}}{|\pi + y + ix|} \right\},$$

which displays the exponential decay of $\left|\Psi_{00}(x, y)\right|$ in $y$ and the slow decay in $x$, as observed on Figure 3. Similar results may be obtained for the other basis functions $\Psi_{m0}(x, y)$ [**7**].

3.3.3. *Other bases.* Similar considerations can be made for the LLL bases obtained by (3.11) from other 1-D wavelet orthonormal bases. For instance [**7**]:

- The *Journé basis*, which does not come from MRA. This case is very similar to the Littlewood-Paley basis, since here too the mother wavelet has compact support in frequency space:

$$\widehat{\Psi}(\omega) = \begin{cases} (2\pi)^{-1/2}, & \text{if } 4\pi/7 \leqslant |\omega| \leqslant \pi \text{ and } 4\pi \leqslant |\omega| \leqslant 32\pi/7 \\ 0, & \text{otherwise.} \end{cases}$$

- *Spline bases*, for instance the order 1 splines, coming from the triangle or 'tent' function as scaling function.

In conclusion we may say that the orthonormal bases obtained so far are not yet sufficient for a good solution of the LLL problem, except at very low electron density. However, the construction sketched above suggests a general method for designing good bases, with localized functions and respecting the symmetry of a given lattice, for instance, a triangular lattice. This goes in the right direction, since Mani and von Klitzing have recently characterized the whole QHE as the transport of electrons in a (local) Wigner crystal [**19**]. The most promising point of such wavelet bases is the possibility of controlling the width of the (essential) supports, hence the overlap between basis functions at neighboring points, with help of the scaling parameter.

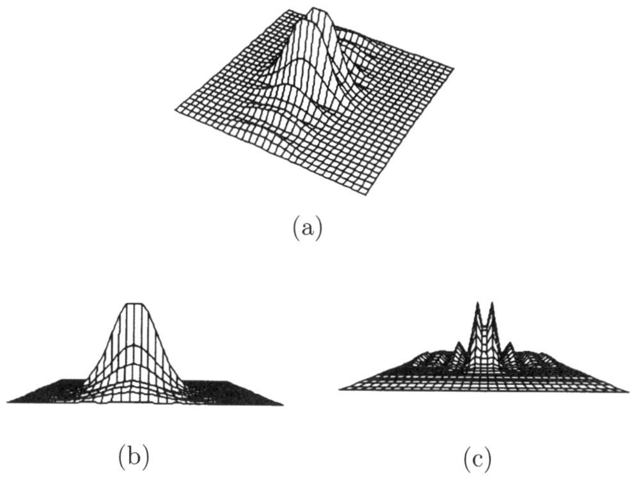

FIGURE 2. Modulus of the Haar LLL basis function $H_{00}(x, y)$: (a) In perspective; (b) The $x$-dependence; (c) The $y$-dependence.

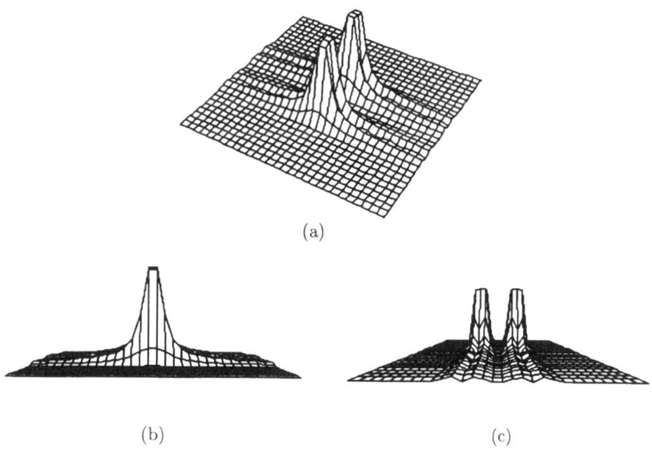

FIGURE 3. Modulus of the Littlewood-Paley LLL basis function $\Psi_{00}(x, y)$: (a) In perspective; (b) The $x$-dependence; (c) The $y$ dependence.

**3.4. Further variations on the same theme.** Recently [**8**], Bagarello has investigated another aspect of the trial wave function introduced in [**9**] for a two-dimensional system of electrons in Coulomb interaction. Namely he compares the ground state energy given by harmonic oscillator wave functions with that obtained with particular wavelets. It turns out that the latter always give better results, easily interpretable as localization properties of the wave function.

Consider an $N$-electron system in $\mathbb{R}^2$, with Hamiltonian

$$H = \sum_{i=1}^{N} H_o(i) + \frac{1}{2}\sum_{i\neq j} \frac{1}{|\vec{\mathbf{r}}_i - \vec{\mathbf{r}}_j|}, \quad H_o = \frac{1}{2}(p_x^2 + x^2) + \frac{1}{2}p_y^2 + p_x p_y.$$

Then, under the following canonical transformation:

$$Q = p_x + p_y, \quad P = -x, \quad Q' = p_y, \quad P' = x - y,$$

the free Hamiltonian turns into that of a harmonic oscillator, as before:

$$H_o = \frac{1}{2}(Q^2 + P^2),$$

and again one has an integral transform relating the $QQ'$ wave function to the original one:

$$\Psi(x,y) = \frac{1}{2}\int\!\!\int_{\mathbb{R}^2} e^{i[Q'(x-y)+Qx]}\Phi(Q,Q')\, dQ\, dQ'.$$

Here also the free Schrödinger equation admits factorized solutions

$$\Phi(Q,Q') = \pi^{-1/4}e^{-Q^2/2}\phi(Q'),$$

where $\phi(Q')$ is an arbitrary function. Choosing for the latter various basis functions, such as elements of the Littlewood-Paley, Haar or harmonic oscillator basis, one may compute the ground state energy $E_c$ of the $N$-electron system, with the wave function taken again as a Slater determinant, in the familiar Hartree-Fock manner. Actually $E_c$ is the sum of two terms, the direct one and the exchange term, but the latter is much smaller and may be neglected. As a preparation for the FQHE, one should also consider the electrons localized on the nodes of a lattice generated by magnetic translations, as explained in the preceding section.

Calculations of this type are presented in [**8**] for $N = 2$ (this case already displays the general features) and several configurations. For instance, two electrons localized at the origin (with different wave functions, otherwise the Slater determinant would vanish identically). Or two electrons on the $y$-axis, separated by a magnetic translation. In both cases, the value obtained for the energy of the trial ground state shows that the wavelet wave function is better localized than the harmonic oscillator wave function.

All these results strongly suggest that wavelets, preferably localized around the nodes of the triangular Wigner crystal, may be extremely useful for finding the ground state wave function of the FQHE.

## 4. Two-Scale Relations in 1-D Crystals

Obviously, any one-dimensional periodic system of period $a$ may be viewed as a $2a$-periodic system (this transformation is called 'inflation'). But how does the dynamics change? In particular, how do Bloch and Wannier functions (see below) transform under inflation? It turns out [**18**] that here too wavelets yield some information on the problem.

Consider an electron in a 1-D $a$-periodic lattice, with Hamiltonian

$$H(x;a) = -\frac{\hbar^2}{2m}\frac{d^2}{dx^2} + U(x), \quad U(x+a) = U(x). \tag{4.1}$$

The Schrödinger equation reads

$$H(x;a)\psi_{n,k}^{(a)} = \varepsilon_{n,k}^{(a)}\psi_{n,k}^{(a)}, \quad \pi n \leqslant k|a| \leqslant \pi(n+1), n \in \mathbb{Z}_+.$$

The eigenfunction $\psi_{n,k}^{(a)}$, called a Bloch function, factorized into a plane wave and a periodic function:

$$\psi_{n,k}^{(a)}(x) = e^{ikx}u_{n,k}^{(a)}(x), \quad u_{n,k}^{(a)}(x+a) = u_{n,k}^{(a)}(x)$$

($n$ indexes the energy band in the spectrum). The Bloch functions, in turn, may be expanded into the so-called Wannier functions:

$$\psi_{n,k}^{(a)}(x) = \sqrt{\frac{a}{2\pi}} \sum_{m\in\mathbb{Z}} \phi_n^{(a)}(x - am)e^{ikam}.$$

The Wannier functions $\{\phi_n^{(a)}(x - am), n \in \mathbb{Z}_+, m \in \mathbb{Z}\}$ form an orthonormal basis in $L^2(\mathbb{R})$.

In terms of the scaled variable $\bar{x}^{(1)} = x/a$, the scaled Hamiltonian reads

$$\overline{H}^{(1)}(\bar{x}^{(1)}) = 2ma^2\hbar^{-2}H(a\bar{x}^{(1)};a) = -\frac{d^2}{d\bar{x}^{(1)^2}} + \overline{U}^{(1)}(\bar{x}^{(1)}),$$

where $\overline{U}^{(1)}(\bar{x}^{(1)}) = 2ma^2\hbar^{-2}U(x)$ is 1-periodic. Now perform an inflation $a \mapsto 2a$. In terms of the scaled variable $\bar{x}^{(2)} = x/2a$, one gets

$$\overline{H}^{(2)}(\bar{x}^{(2)}) = -\frac{d^2}{d\bar{x}^{(2)^2}} + \bar{U}^{(2)}(\bar{x}^{(2)}),$$

where $\overline{U}^{(2)}(\bar{x}^{(2)}) = 2m(2a)^2\hbar^{-2}U(x)$ is also 1-periodic. Then one must distinguish between two cases:

- If $\overline{U}^{(1)} = \overline{U}^{(2)}$, one gets $\overline{H}^{(1)} = \overline{H}^{(2)}$ and inflation is a symmetry of the system. An example is a free electron. Then Bloch and Wannier functions obey two-scale relations (TSR) identical to those of MRA wavelets.
- If $\overline{U}^{(1)} \neq \overline{U}^{(2)}$, the general case, the connection with wavelets is not so direct any more, but one still has useful TSR.

We examine these two cases successively.

**4.1. Wannier functions for a free electron.** Take first a free electron, $U(x) = 0$, so that inflation is a symmetry. Then the Bloch function reduces to a plane wave, $u_{n,k}^{(a)}(x) = 1/\sqrt{2\pi}$, and the Wannier functions are obtained immediately:

$$\phi_n^{(a)}(x) = \frac{\sqrt{a}}{\pi x}\left[\sin\frac{\pi}{a}(n+1)x - \sin\frac{\pi}{a}nx\right],$$

or, in momentum space,

$$\widehat{\phi_n^{(a)}}(k) = \begin{cases} \sqrt{a}, & \text{if } n\pi \leqslant |k|a \leqslant (n+1)\pi \\ 0, & \text{otherwise.} \end{cases}$$

That is, the first Wannier function $\phi_1^{(a)}(x)$ coincides with the Littlewood-Paley mother wavelet (3.16) or (3.17)!

As a consequence, the Wannier functions obey the two-scale relations characteristic of MRA wavelets, for instance:

$$\phi_{4n}^{(a)}(x - am_2) = \sum_{m_1\in\mathbb{Z}} p_{m_1-2m_2}\sqrt{2}\phi_{4n}^{(2a)}(2x - am_1),$$

and one recovers the whole Littlewood-Paley multiresolution wavelet analysis.

**4.2. Wannier functions for a general 1-D crystal.** In the general case, inflation is not a symmetry of the crystal, but one may still derive useful TSR. Typically:

- for the Bloch functions: $\psi^{(a)}_{n,k} \simeq \alpha(k)\psi^{(2a)}_{2n,k} + \beta(k)\psi^{(2a)}_{2n+1,k}$
- for the Wannier functions: $\phi^{(2a)}_{4n}(x) = \sum_{m\in\mathbb{Z}} p_m \phi^{(a)}_{2n}(x - am)$, with coefficients $p_m$ independent of $a$.

The difficulty here is that, contrary to the free case and the MRA situation, the two sides of these TSR live on different lattices. For instance, $\phi^{(2a)}_0(x) \neq \phi^{(a)}_0(x)$.

In conclusion, there is a striking similarity between Wannier functions of a 1-D crystal and wavelets, and the two coincide for a free electron. It remains to be seen whether this is a mere curiosity or a physically useful information.

## 5. Outcome

As a general conclusion to [**1**] and the present paper, it seems clear that wavelets bring something new and useful, both in atomic and molecular physics, and in condensed-matter physics. On one hand, they yield a time-frequency representation that is crucial for unraveling fast transient phenomena. On the other, they provide bases, orthonormal or not, that are better adapted to a number of physical problems than traditional bases, mostly because of their good localization properties. As a result, wavelets often will substantially speed up calculations, a nonnegligible benefit.

Thus it seems that wavelets are here to stay, and that one has only to find new applications in various domains of quantum physics.

## Acknowledgements

It is a pleasure to thank T. A. Arias for communicating additional information about his work, and Z. Felfli, X. Gonze, R. Murenzi, C. J. Tymczak and X. Q. Wang for interesting discussions and comments about several aspects of this paper.

## References

1. J.-P. Antoine, Ph. Antoine, and B. Piraux, *Wavelets in atomic physics*, this volume.
2. J.-P. Antoine and F. Bagarello, *Wavelet-like orthonormal bases for the lowest Landau level*, J. Phys. A **27** (1994), 2471–2481.
3. J.-P. Antoine, P. Carrette, R. Murenzi, and B. Piette, *Image analysis with 2D continuous wavelet transform*, Signal Process. **31** (1993), 241–272.
4. J.-P. Antoine and R. Murenzi, *Two-dimensional directional wavelets and the scale-angle representation*, Signal Process. **52** (1996), 259–281.
5. T. A. Arias, data given on the WWW-site, http://web.mit.edu/muchomas/www/home.html.
6. ———, private communication.
7. F. Bagarello, *More wavelet-like orthonormal bases for the lowest Landau level: Some considerations*, J. Phys. A **27** (1994), 5583–5597.
8. ———, *Applications of wavelets in quantum mechanics: A pedagogical example*, J. Phys. A **29** (1996), 565–576.
9. F. Bagarello, G. Morchio, and F. Strocchi, *Quantum corrections to the Wigner crystal: A Hartree-Fock expansion*, Phys. Rev. B **48** (1993), 5306–5314.
10. K. Cho, T. A. Arias, J. D. Joannopoulos, and P. K. Lam, *Wavelets in electronic structure calculations*, Phys. Rev. Lett. **71** (1993), 1808–1811.
11. M. Y. Chou and S. Wei, *Wavelets in self-consistent electronic structure calculations*, Proc. March 1996 Meeting Amer. Phys. Soc., Session R19, Bull. Amer. Phys. Soc. **41** (1996), 746–748, and further papers in the same session.
12. C. K. Chui, *An introduction to wavelets*, Academic Press, San Diego, 1992.

13. I. Daubechies, *Ten lectures on wavelets*, SIAM, Philadelphia, PA, 1992.
14. R. Ferrari, *Two-dimensional electrons in a strong magnetic field: A basis for single-particle states*, Phys. Rev. B **42** (1990), 4598–4609.
15. X. Gonze, private communication.
16. M. Greiter and I. A. McDonald, *Hierarchy of quantized Hall states in double layer electron systems*, Nuclear Phys. B **410** (1993), 521–534.
17. F. D. M. Haldane, *Fractional quantization of the Hall effect: A hierarchy of incompressible quantum fluid states*, Phys. Rev. Lett. **51** (1983), 605–608.
18. K. Kaneda and T. Odagaki, *Two-scale relations in one-dimensional crystals and wavelets*, J. Phys. A **28** (1995), 4389–4406.
19. R. G. Mani and K. von Klitzing, *Fractional quantum Hall effect as an example of fractal geometry in nature*, Z. Phys. B **100** (1996), 635–642.
20. G. Morandi, *The role of topology in classical and quantum physics*, Lecture Notes in Phys. New Ser. m: Monogr., vol. 7, Springer-Verlag, Berlin, Heidelberg, 1992.
21. M. Moshinsky and C. Quesne, *Linear canonical transformations and their unitary representations*, J. Math. Phys. **12** (1971), 1772–1780.
22. C. J. Tymczak and X. Q. Wang, *Orthonormal wavelet bases for quantum molecular dynamics*, Phys. Rev. Lett. **78** (1997), 3654–3657, and private communication.
23. S. Wei and M.Y. Chou, *Wavelets in self-consistent electronic structure calculations*, Phys. Rev. Lett. **76** (1996), 2650–2653.
24. K. Yamaguchi and T. Mukoyama, *Wavelet representation for the solution of radial Schrödinger equation*, J. Phys. B **29** (1996), 4059–4071.

Institut de Physique Théorique, Université Catholique de Louvain, B–1348 Louvain-la-Neuve, Belgium

*E-mail address*: `Antoine@fyma.ucl.ac.be`

Centre de Recherches Mathématiques
CRM Proceedings and Lecture Notes
Volume **18**, 1999

# Wavelets in Atomic Physics

J.-P. Antoine, Ph. Antoine, and B. Piraux

## 1. Introduction

Atomic physics is a privileged domain of application for coherent states (CS). Laser-atom interactions, revival phenomena, Rydberg wave packets and various semi-classical situations are all instances in which a coherent state description is clearly well-adapted. The problem is, *which* coherent states? The canonical ones, associated to the harmonic oscillator, were invented by Schrödinger for modelling the quantum-to-classical transition and revived in the 60's for quantum optics (see [**1, 9, 14**] for a review). Various kinds of CS have been proposed for the Kepler-Coulomb problem, albeit with limited success. What about wavelets? After all, there are nothing but the CS associated to the affine groups in various space dimensions.

Now that would mean envisaging wavelets as *physical objects*, namely particular quantum states, in the same way as canonical CS are used in quantum optics (laser physics, or even QED). This is a tempting possibility, but so far very little has been achieved in this direction. Indeed there have been only a few applications of wavelets in atomic physics, and in all cases they were used as a *mathematical tool*, as in many other fields of physics.

In this paper we shall review the most significative of these applications, which fall into two main lines of research.

(i) An exciting phenomenon, both experimentally and theoretically, is the emission of light by an atom when hit by a very intense, ultrashort, laser pulse. The time profile of the emission spectrum reveals new information on the dynamics of this interaction process. Clearly, this time profile is out of reach of the traditional Fourier spectral methods. A time frequency analysis is required here, either with wavelets or with a Gabor approach (Short Time Fourier Transform—just another name for canonical CS!).

(ii) Both the extension of these phenomena to multielectron atoms and the electronic structure calculations (Hartree-Fock and generalizations) require the use of appropriate orthogonal bases for the description of the radial part of wave functions. Here (discrete) wavelet bases could adequately replace

1991 *Mathematics Subject Classification.* Primary: 81R30, Secondary: 81V45, 81V10.
This is the final version of the paper.

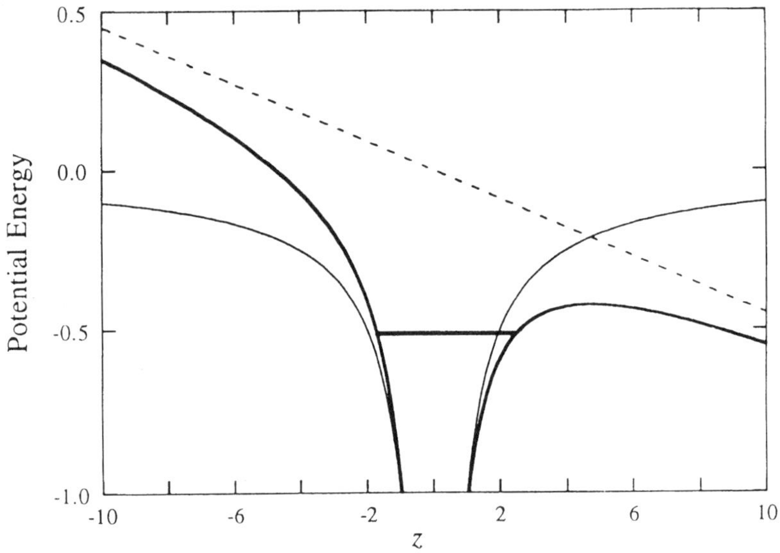


FIGURE 1. Instantaneous effective potential (bold solid line) for hydrogen in a field polarized along the $z$-axis, obtained by superposing the Coulomb potential (solid line) and the linear e.m. potential (dashed line). The ground state (indicated by a bold line inside of the well) can ionize by tunnelling through the suppressed barrier.

traditional plane waves on LCAO bases, and yield some hope of progress—although what we will describe below is more a program that explicit results.

In our opinion, what we have seen so far in these directions is only a beginning, and wavelets (including Gabor analysis) have a great potential in many different sectors of atomic (and molecular) physics.

## 2. Harmonic Generation in Atom-Laser Interaction

**2.1. Physical picture: a single atom in a strong laser pulse.** When an atom is exposed to a strong laser pulse, two competing processes may occur, the ionization of the atom and the emission of light. This emission process results from the oscillations of the atomic dipole at frequencies which are odd multiples (odd harmonics) of the driving field frequency (even harmonics are forbidden by parity conservation). It is now well known that harmonic generation provides an efficient source of XUV coherent radiation. The actual process of emission occurs in a very short time ($\leqslant$ 100 femtoseconds), which justifies the detailed study of its mechanism by means of a time frequency analysis.

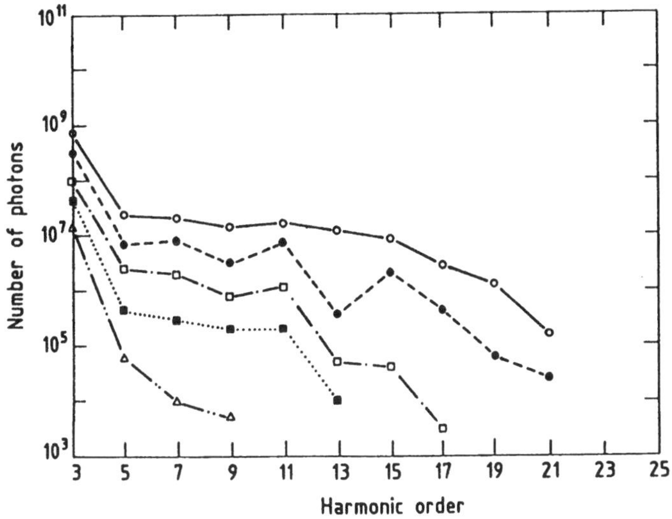


(a)

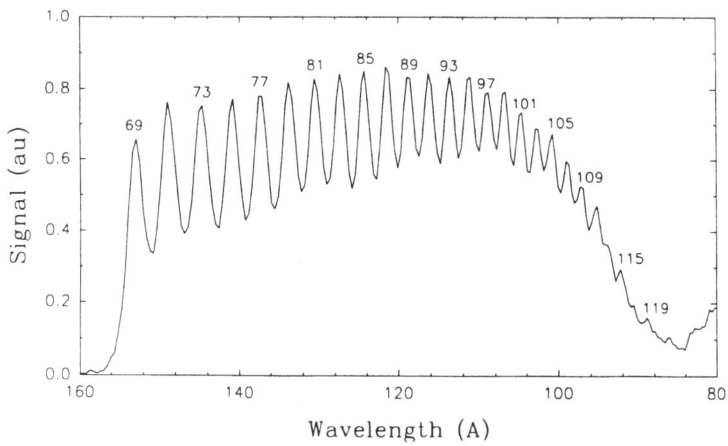


(b)

FIGURE 2. Harmonic spectrum: (a) Spectrum obtained in Xe for $\lambda = 1064$ nm and various laser intensities (from **[19]**); (b) Spectrum obtained in He for $\lambda = 1053$ nm and a laser intensity of $3 \times 10^{14} \mathrm{Wcm}^{-2}$ (from **[22]**); in this case, the first harmonics up to about the order 67 are filtered out. Note the sharp cut-off around the order 113.

Let us consider (for simplicity) the case of atomic hydrogen exposed to a strong laser pulse whose electric field is described classically as: $\mathbf{E} = \mathbf{E}_0(t) \cos \omega t$, where $\mathbf{E}_0$ is the pulse envelope. In these conditions, the electron "moves" both in the

Coulomb potential and in the electromagnetic (e.m.) potential, as shown in Figure 1. According to the semiclassical interpretation [**7, 16**], harmonic generation results from the following mechanism. The electron may tunnel through the potential barrier lowered by the e.m. field. When it is outside, it is accelerated by the laser e.m. field and may be driven back close to the nucleus. There, it may either be scattered or recombine with the nucleus into the ground state, emitting an harmonic photon. The emission spectrum exhibits characteristic features which depend on the laser intensity as shown in Figure 2. At low laser intensity, the spectrum of the emitted radiation decreases rapidly for increasing harmonic order, as expected (Figure 2a). At high laser intensity, the spectrum changes drastically: after a rapid decrease for the first orders, it exhibits a long plateau, followed by a sharp cut-off (Figure 2b). Experimentally, harmonic orders as high as the 135th order have been observed with a Nd-Glass laser, which corresponds to a wavelength of 7.8 nm (i.e. in the XUV regime) [**17**]. Clearly, the atomic response is highly nonlinear, since, as shown in Figure 2b, the 69th harmonic is as intense as the 109th. In contrast to the low intensity regime, where perturbative theory applies, nonperturbative methods have to be used to calculate the atomic response in the intense field regime, where the external field is of the same order or higher than the binding atomic field. A more realistic method consists in solving the time-dependent Schrödinger equation (as briefly described in the next section) [**15**]. Atomic dipoles calculated by this method have been analyzed with the standard Fourier spectral method, and the corresponding harmonic spectra exhibit the features described above. However, a deeper understanding of the mechanism requires a time-frequency analysis. In other words, it is important to trace *when* and *where* a given harmonic is emitted.

**2.2. Calculation for a one-electron atom.** As already mentioned in the introduction, the primary cause of harmonic emission is that, in the presence of the laser field, the electron of the atomic hydrogen oscillates back and forth, hence creates a dipole moment $d(t)$. Then, according to the Larmor formula, the energy radiated between frequencies $\omega$ and $\omega + d\omega$ is proportional to $|\widehat{a}(\omega)|^2$, where $\widehat{a}(\omega)$ is the Fourier transform of the dipole acceleration $a(t) = \ddot{d}(t)$. Therefore the problem consists in calculating numerically the acceleration $a(t)$ from the time-dependent Schrödinger equation:

$$i\frac{\partial}{\partial t}\psi(\mathbf{r},t) = \big(H_{at} + \mathbf{A}(t)\cdot\mathbf{p}\big)\psi(\mathbf{r},t). \tag{2.1}$$

In Equation (2.1), $H_{at}$ is the atomic Hamiltonian, the vector potential $\mathbf{A}$ is written in the dipole approximation, i.e. $\mathbf{A}(\mathbf{r},t) = \mathbf{A}(t)$ depends on time only, and is treated as a classical variable, and finally the quadratic term $\mathbf{A}^2$ has been gauged away. More precisely, the vector potential is taken along the $z$-axis:

$$\mathbf{A}(t) = A_o f(t) \sin\omega_L t\, \hat{\mathbf{e}}_z,$$

where $f(t)$ is the envelope of the pulse and $\omega_L$ the laser frequency. (Actually the shape of the pulse influences considerably the harmonic emission). In terms of the solution $\psi(\mathbf{r},t)$ of the Schrödinger equation (2.1), the atomic dipole (along the $z$-axis) reads

$$d(t) = \langle\psi(\mathbf{r},t)|z|\psi(\mathbf{r},t)\rangle. \tag{2.2}$$

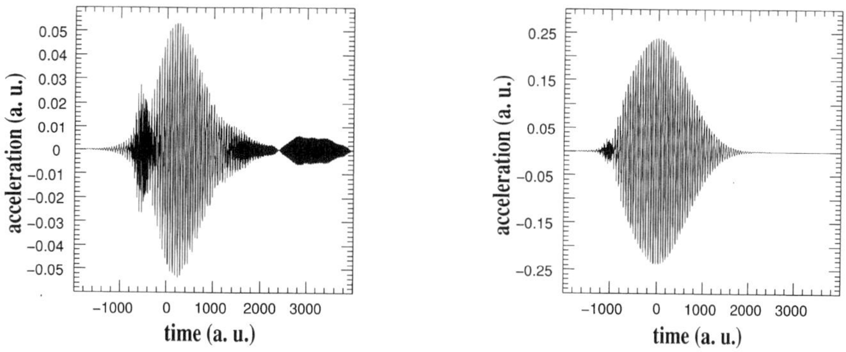


FIGURE 3. Acceleration of the atomic dipole as a function of time. This dipole is the result of the interaction of atomic hydrogen with a strong laser pulse of frequency $\omega_L = 0.118$ atomic units. The pulse is Gaussian and its full width at half maximum is 20 optical cycles. The peak intensity is (a) $2 \times 10^{14}\mathrm{Wcm}^{-2}$ and (b) $2 \times 10^{15}\mathrm{Wcm}^{-2}$.

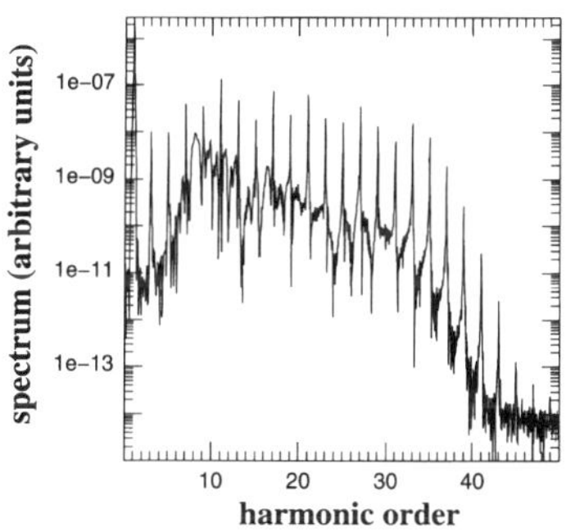


FIGURE 4. Harmonic spectrum in the case of the interaction of atomic hydrogen with a laser pulse of frequency $\omega_L = 0.047$ atomic units and peak intensity $10^{14}\mathrm{Wcm}^{-2}$. The pulse has a 4 optical cycle sinus square turn-on and -off and a flat top of 16 optical cycles.

Using Ehrenfest's theorem, the corresponding acceleration (along the $z$-axis) may be written as:

$$a(t) = \ddot{d}(t) = -\left\langle \psi(\mathbf{r},t) \left| \frac{z}{r^3} \right| \psi(\mathbf{r},t) \right\rangle + \frac{\partial A_z}{\partial t}. \tag{2.3}$$

Notice that, according to (2.3), we have to know the wave function $\psi(\mathbf{r},t)$ only for small $\mathbf{r}$. The next step is to expand $\psi(\mathbf{r},t)$ in an appropriate basis. A convenient choice is to take spherical harmonics for the angular part, and a Coulomb Sturmian basis $\{S_{nl}(r)\}$ for the radial part (the latter has good convergence properties)

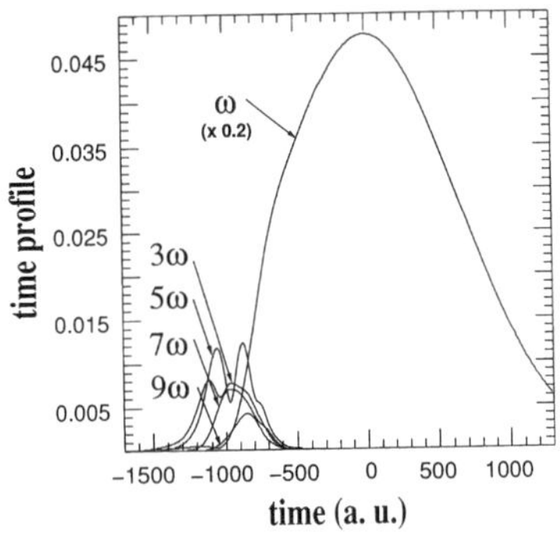


FIGURE 5. Time profile (from Gabor analysis) of the odd harmonics (1–9) produced as a result of the interaction of atomic hydrogen with the same laser pulse as in Figure 3b (from [**3**]).

[**13, 21**]. Thus we write:

$$\psi(\mathbf{r},t) = \sum_{nlm} a_{nl}(t)\frac{S_{nl}(r)}{r}Y_l^m(\theta,\varphi). \tag{2.4}$$

Inserting the expansion (2.4) into the Schrödinger equation (2.1), we obtain a set of first order differential equations for the coefficients $a_{nl}(t)$. We present in Figure 3 a typical result for the dipole acceleration. The curves show two regions of rapid oscillations. The first one starts well before the maximum of the pulse (taken as the origin of time) is reached and corresponds to the generation of harmonics. The second one is a residual effect: when the atom is not fully ionized after the passage of the pulse, it remains in a state which is a superposition of many atomic states, both bound and continuum states. As a result and in absence of dissipation, the dipole keeps oscillating, which creates the effect seen on Figure 3a, which has thus nothing to do with harmonic generation. This interpretation is confirmed by the fact that the effect disappears when the laser intensity is sufficiently high for the atom to be fully ionized, as shown in Figure 3b.

**2.3. Time-frequency analysis of the dipole acceleration: H(1s).** Taking now the Fourier transform $\widehat{a}(\omega)$ of the dipole acceleration $a(t)$, one obtains the power spectrum. A typical example is shown in Figure 4. This spectrum shows indeed the presence of a large number of odd harmonics of the laser frequency. But, of course, no time localization is provided. When is each harmonic emitted? What is its time profile? Answering those questions requires a time-frequency analysis of the acceleration:

$$a(t) \mapsto \tilde{a}(\alpha,\tau) = \int_{-\infty}^{\infty} \overline{g_{\alpha\tau}(t)}a(t)\,dt. \tag{2.5}$$

In this expression $\alpha \simeq \omega^{-1} > 0$ is a scale parameter and $\tau$ a time parameter (position in the spectrum). Two types of time-frequency analysis have been used in the present problem:

- a *Gabor transform* (Short Time FT)

$$g_{\alpha\tau}(t) = e^{it/\alpha}G(t-\tau),$$

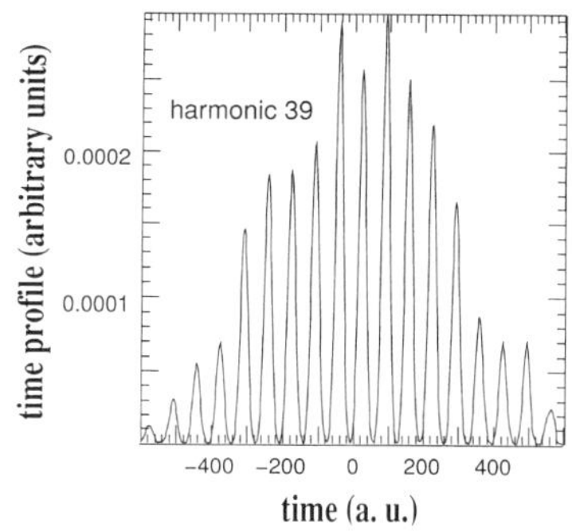


FIGURE 6. Time profile (from wavelet analysis) of the 39th harmonic emitted by atomic hydrogen exposed to the same pulse as in Figure 4. Note that the frequency bandwidth of the pulse exceeds in this case $2\omega_L$.

where $G$ is usually a Gaussian (and hence $g_{\alpha\tau}(t)$ is a canonical coherent state)

- a *wavelet transform*

$$g_{\alpha\tau}(t) = \alpha^{-1/2} w\left(\frac{t-\tau}{\alpha}\right),$$

where the analyzing wavelet $w(t)$ is often taken as a Morlet wavelet: $w(t) = e^{i\omega_o t}e^{-t^2/2}$ (strictly speaking, one should had a correction term to enforce the admissibility condition $\widehat{w}(0) = 0$, but this term is numerically negligible for $\omega_o \geqslant 6$).

This technique has been applied successfully in the case of an hydrogen atom in its ground state. First, we determine the emission strength as a function of time, for a given frequency, that is, we evaluate $\tilde{a}(\alpha, \tau)$ for fixed $\alpha$ [**3**]. One may also choose a different strategy and look at the full harmonic spectrum for a fixed time $\tau$ [**10**]. Finally, we present an example showing that time-frequency analysis also provides information on the instantaneous frequency of emission.

2.3.1. *Time profile of harmonics.*

EXAMPLE 1. Choosing for $\alpha$ in (2.5) the inverse of a fixed odd multiple of the laser frequency, one obtains the time profile of the corresponding harmonic [**3**]. A Gabor analysis yields the global shape of each harmonic (Figure 5). Two interesting points are visible on this picture. First, the emission of each harmonic starts at a given characteristic time, which depends on the laser intensity. Then, as their order increases, the harmonics are emitted during shorter time intervals. This implies that the linewidth of higher harmonics should broaden with their order. The fact that harmonic emission stops before the field has reached its maximum amplitude (in $t = 0$) is due to the rapid excitation and ionization of the atom. This effect has been observed experimentally [**23**].

The picture may be refined by using a better time resolution. This reveals a fine structure in the time profile, which is $2\omega_L$ periodic (Figure 6). One should note that, since the time resolution is better than half the optical period, the filter

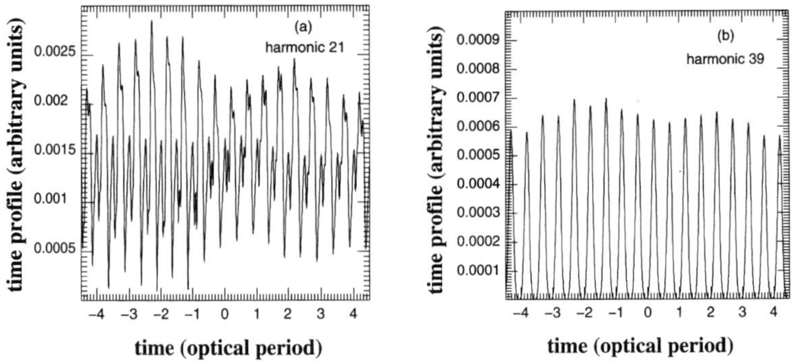


FIGURE 7. Time profile (from wavelet analysis) of (a) harmonic 21 (in the plateau) and (b) harmonic 39 (in the cut-off) for the same conditions as in Figure 4.

bandwidth is larger than $2\omega_L$. Therefore several harmonics are accepted by the the filter.

EXAMPLE 2. In lower frequency regime ($\omega_L = 0.047$ a.u.), the emission exhibits a longer plateau and a sharp cutoff as shown in Figure 4. The time profile of harmonics in the plateau (Figure 7a) and in the cutoff (Figure 7b) are totally different, which suggests two different dynamics of emission.

EXAMPLE 3. In the semiclassical description, the electron emits harmonics when it is close to the nucleus. Therefore, the harmonic emission may be controlled if one uses a laser beam with a polarization varying in time. When the polarization is linear, the electron oscillates back and forth, hence it comes periodically close to the nucleus, and harmonics are emitted, but, when the polarization is elliptic or circular, the electron stays far away and harmonic generation is suppressed. This polarization effect is demonstrated in Figures 8a and b, which show in parallel the time evolution of the harmonic 9 (8a) and the projection of the full wave function on the bare 1s state of atomic hydrogen (8b). Since the latter measures the probability of the electron being close to the nucleus, it oscillates when the field is linearly polarized, but remains constant in the circular or elliptic cases. As expected, the harmonic is totally suppressed when the polarization of the laser beam is circular and it reaches its maxima precisely when the polarization is linear.

2.3.2. *Harmonic spectrum at fixed time.* Alternatively, one may choose a fixed time $\tau = t_o$ and compute the full harmonic spectrum [**10**], either with a Gabor analysis, or with a wavelet transform (Figure 9). There is no significant difference between the two choices (except for hyper-Raman lines). These results have been obtained by using a two-state model for the atom.

2.3.3. *Instantaneous frequency of emission.* Actually, the frequencies of the harmonics are exact odd multiples of the frequency of the driving field only if the laser intensity remains constant. On the contrary, when the laser intensity increases, the frequencies are slighty above the exact multiples of $\omega_L$ (blueshift). To get a confirmation of that effect, we take a linear ramp as the pulse shape. Then all the harmonics have the same behavior: the Gabor coefficient grows rapidly in time

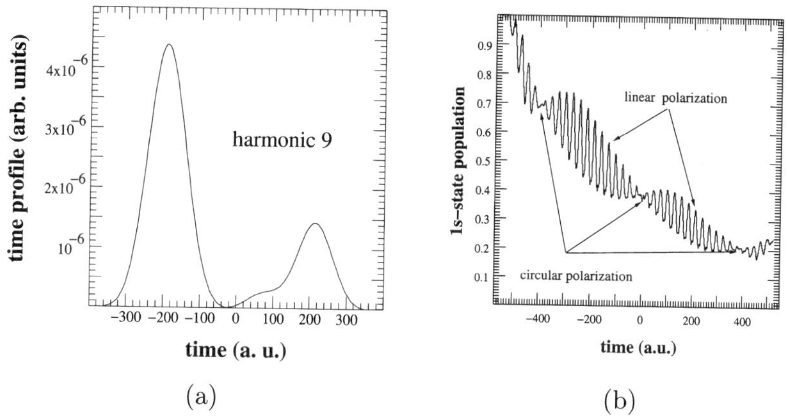


FIGURE 8. The polarization effect (from [**4**]): (a) Intensity of the 9th harmonic as a function of time (in atomic units) in the case where the atom is exposed to two perpendicularly polarized laser pulses of $10^{14}$ $\mathrm{Wcm}^{-2}$. The photon energies are 0.118 and 0.110 atomic units, respectively. Both pulses have a flat top and sine square turn-on and -off. The total duration of the pulse is 20 optical cycles. (b) Corresponding projection of the full wave function on the bare 1s state of atomic hydrogen.

($\sim$ intensity) and then reaches a plateau with oscillations. For a given harmonic, the *instantaneous frequency* of emission is obtained by means of another tool (familiar, for instance, in the determination of spectral lines [**8**]), namely, the time derivative $d/(d\tau)\phi(\alpha,\tau)$ of the phase $\phi(\alpha,\tau)$ of the corresponding Gabor coefficient. The computation fully confirms the effect (Figure 10): each of the harmonics is slighty blueshifted up to the moment (indicated by an arrow) when it reaches the saturation plateau.

2.3.4. *Wavelets or Gabor?* It is clear from this discussion that a time-frequency analysis is essential here. In particular, the temporal evolution of the harmonic emission, which is unaccessible to standard Fourier technique, sheds much light on the actual physical phenomenon. A natural question is the choice between wavelets and a Gabor analysis.

Globally, the two methods give similar results, provided the time resolution or the bandwidth of the analyzing functions are identical. However, wavelet analysis is more appropriate if we want to study *simultaneously* the time profile of several frequencies, keeping the same analyzing function. This results of course from the well-known property of wavelets, namely that $\Delta\omega/\omega$ is constant. In the case of harmonic generation, this remark is particular relevant in two cases:

(i) for studying very high order harmonics with a time resolution better than the optical cycle, and

(ii) when the atomic structure plays a more crucial role, for instance in the presence of hyper-Raman lines, resulting from bound-bound transitions during the interaction [**3**].

So the answer to the question, Wavelets or Gabor?, depends essentially on the physics of the problem at hand, so that *both* methods are needed in fact.

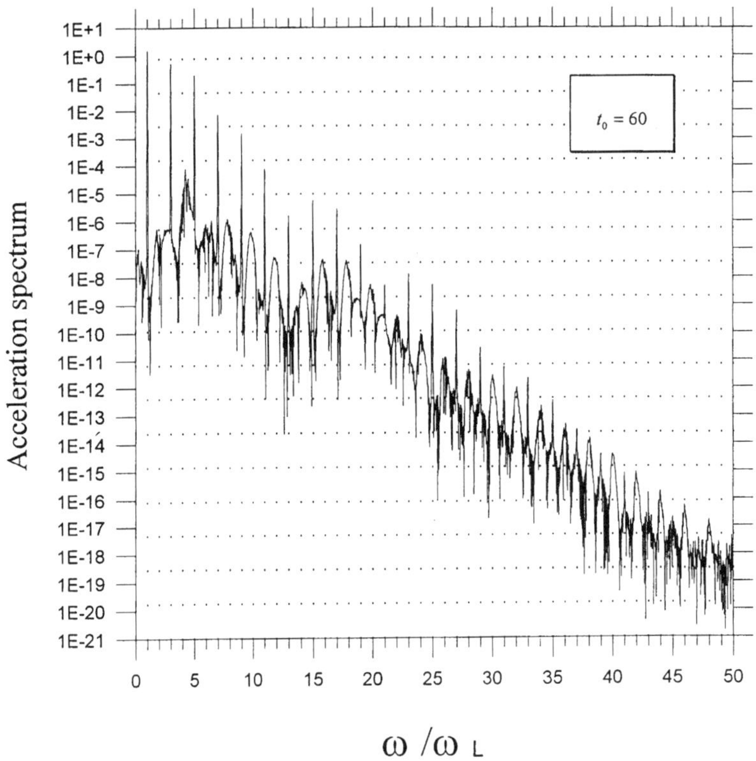


FIGURE 9. The full harmonic spectrum (Gabor transform of the dipole acceleration) emitted by a two-level atom excited by a laser pulse of intensity $8.8 \times 10^{13}$ Wcm$^{-2}$, at time $t_o \equiv \tau = 60$ optical cycles (from [**10**]).

## 3. Extension to Multielectron Atoms

**3.1. General problem.** The analysis of the interaction between a laser pulse and a one-electron atom, discussed in Section 2, rests on the possibility of solving numerically the time dependent Schrödinger equation. The resulting wave function may be used to describe various phenomena such as harmonic generation, multiphoton ionization, excitation of Rydberg wave packets, revivals, etc. The results thus predicted are generally in good agreement with experiment (whenever available). On the way, much physical insight has been obtained from a time-frequency analysis, wavelets or Gabor.

All that is even more true and interesting for two electron atoms, such as He or $H^-$. But now the situation is much more complicated, because of the correlation between the two electrons. It is hopeless to solve directly the time dependent Schrödinger equation and calculate a reasonable two electron wave packet $\Psi(\mathbf{r}_1, \mathbf{r}_2, t)$. On the other hand, the angular part of the latter is well under control, thanks to the familiar Racah algebra. Thus there remains to find a good radial

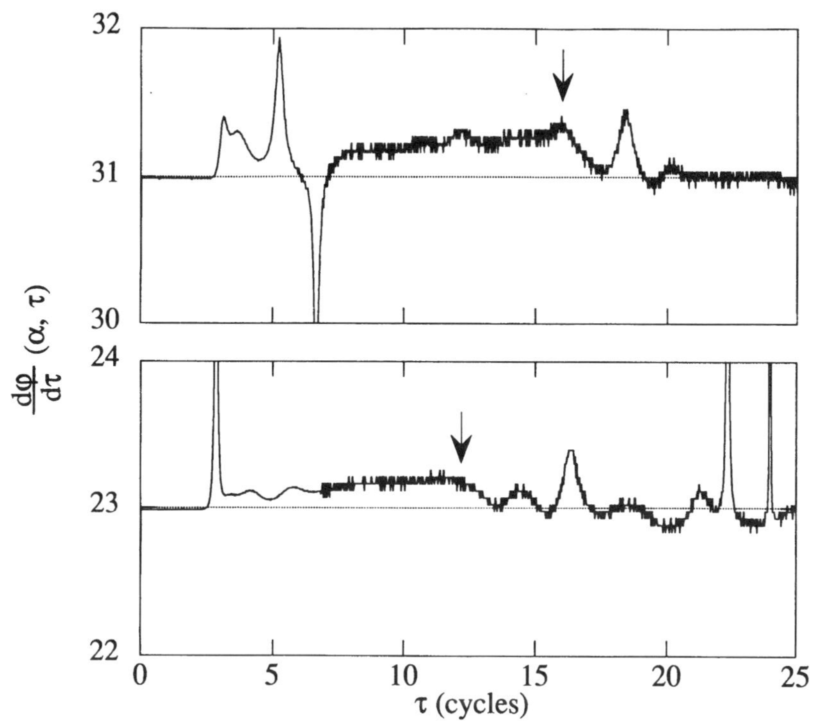


FIGURE 10. Time derivative of the phase of the Gabor coefficients for $\alpha^{-1} = 23\omega_L$ (lower part) and $31\omega_L$ (upper part). The pulse is a linear ramp of 20 optical cycles and a maximum intensity of $10^{14}$ $\mathrm{Wcm}^{-2}$. The arrows indicate the times when each harmonic reaches the saturation plateau (from [**24**]).

basis. Several candidates may be thought of, such as multiresolution-based orthogonal or bi-orthogonal wavelet bases, wavelet bases derived from affine hypergroups [**20**] or frames derived from Schrödinger coherent states.

Take, for instance, a single photon ionization

$$X + h\nu \rightarrow X^{+} + e^{-}.$$

The only measurable quantity is the cross-section, proportional to $\left|\langle f|H_{\mathrm{int}}|i\rangle\right|^2$, where $H_{\mathrm{int}}$ is the interaction hamiltonian. The initial state $|i\rangle$ is a bound state of the $X$ atom, the final state $|f\rangle \equiv |j,\, \epsilon\rangle$ describes the remaining ionic core $(j)$ and a photoelectron in a continuum state with energy $\epsilon$. As before, one may apply to the hamiltonian various approximations, such as the dipole approximation, the classical treatment of the e.m. field, etc. But the hard problem is to compute the initial and final wave functions. This leads us to a different use of wavelets, namely convenient bases for expansions of solutions of a partial differential equation. This situation has been encountered before, in the solution of nonlinear PDE's such as Burgers' equation [**18**]. But one has first to bring the problem to the stage of computation.

**3.2. Self-consistent methods.** One of the best answers to such questions is given by the self-consistent methods, namely the well-known Hartree-Fock method and its descendants. We shall sketch it very briefly here in order to give to a reader some feeling for the technique and the possible use of wavelets for improving it. Further details may be found in standard textbooks, e.g. [**5**].

3.2.1. *The Hartree-Fock method* (HF). The basic idea is to obtain the wave function $\Psi$ for an atom with $N$ electrons by a variational computation

$$\delta\langle\Psi|H_{\text{atom}}|\Psi\rangle = 0. \tag{3.1}$$

As input, one enforces the total antisymmetry of $\Psi$ by taking it as a Slater determinant

$$\Psi = \frac{1}{N!}\begin{vmatrix} \phi_1(1) & \dots & \phi_1(N) \\ \vdots & & \vdots \\ \phi_N(1) & \dots & \phi_N(N) \end{vmatrix},$$

where $\phi_i(j)$ is a one-electron wave function ('orbital') of the $j^{th}$ electron. It has the form

$$\phi_i(\mathbf{r}, \text{spin}) = \frac{1}{r} P_{nl}(r) Y_l^{m_l}(\theta, \varphi)\chi_{m_s}, \tag{3.2}$$

where $Y_l^{m_l}$ is a spherical harmonic (thus allowing the Racah algebra for angular momentum), $\chi_{m_s}$ is a spin orbital, $P_{nl}(r)$ is an unknown radial function and $i \equiv (n, l, m_l, m_s)$ is a collective quantum number ($m_s = \pm 1/2; m_l = -l, \dots l; l = 0, 1, \dots n-1; n = 1, 2, \dots$). One usually imposes orthonormality conditions

$$\langle\phi_i \mid \phi_j\rangle = \delta_{ij}$$

$$\int_0^\infty P_{nl}(r) P_{n'l'}(r)\, dr = \delta_{nn'}\delta_{ll'}$$

and thus rewrite the variational equation (3.1) with Lagrange multipliers $\{\lambda_{ij}\}$:

$$\delta\left\{\langle\Psi|H_{\text{atom}}|\Psi\rangle + \sum_{i<j}\lambda_{ij}\langle\phi_i \mid \phi_j\rangle\right\} = 0. \tag{3.3}$$

Inserting the ansatz (3.2), one obtains the Hartree-Fock equations that determine the radial functions $\{P_{nl}(r)\}$:

$$\left\{\frac{d^2}{dr^2} - \frac{l(l+1)}{r^2} + \frac{2}{r}[Z - Y(nl;r) - E_{nl}]\right\} P_{nl}(r) - \frac{2}{r}X(nl;r) = \sum_{n'}\lambda_{nln'l}P_{n'l}(r). \tag{3.4}$$

In this equation, $Y$ is called the direct term and represents a local (spherical) potential, whereas the so-called exchange term $X$ is nonlocal. These terms have the following form:

$$Y(nl;r) = \sum_{n'l'}\sum_k y_{nl,n'l'} Y^k(nl, n'l'; r)$$

$$X(nl;r) = \sum_{n'l'}\sum_k x_{nl,n'l'} Y^k(nl, n'l'; r) P_{n'l'}(r)$$

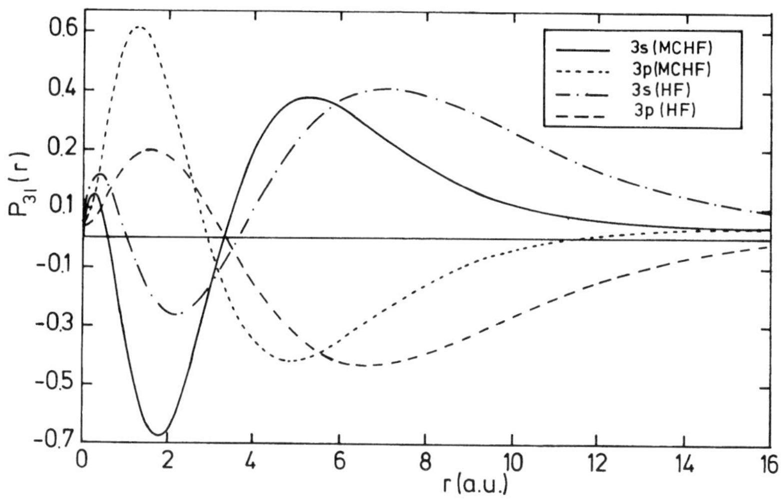


FIGURE 11. Comparison between 3s and 3p HF orbitals of helium with corresponding 5-configuration MCHF orbitals (courtesy of Z. Felfli).

where $y_{nl,n'l'}$, $x_{nl,n'l'}$ are coefficients obtained from the angular momentum algebra and

$$Y^k(nl, n'l'; r) = \int_0^\infty \frac{r_<^k}{r_>^{k+1}} P_{nl}(r') P_{n'l'}(r')\, dt', \quad r_< = \min(r, r'),\ r_> = \max(r, r').$$

Clearly (3.4) is a system of coupled, nonlinear equations in $P_{nl}$.

The usual technique consists in choosing for $P_{nl}$ a simple trial function, such as a Gaussian or a Slater function $r^{n-1}e^{-\alpha r}$, and solving the HF equation (3.4) by iteration. An alternative possibility (presently under consideration in several groups) is to take instead for $P_{nl}$ the elements of a suitable orthonormal wavelet basis. The idea is that the good localization properties of such bases may reduce considerably the number of terms needed.

Actually a similar program has been realized by a number of authors in the field of solid state physics, for calculating the electronic structure of various materials. Since treating a periodic crystal amounts to describe a single atom or molecule and then apply a Bloch transformation, this is really the same problem as discussed here. We refer to [**2**] for a review.

3.2.2. *The multiconfiguration Hartree-Fock method* (MCHF). Clearly the HF approximation is a drastic one, and it should be improved by taking electron correlations into account. Many methods have been developed to that effect, e.g. Configuration Interaction, many-body perturbation theory, R-matrix theory or the Multiconfiguration Hartree-Fock (MCHF) method [**5**]. In the latter case, for instance, the basic idea is to take for the total wave function a coherent superposition of several configuration wave functions, all with the same energy. The resulting wave function is in general more concentrated around the origin, thus oscillates faster, and is therefore more likely to be well represented by wavelets. (Figure 11 shows a typical example of MCHF vs. HF wave functions). The computation usually requires a large number of terms, but again, the good localization properties of wavelets should reduce significantly that number and accelerate the computation.

3.2.3. *A 1-D simplified* HF *equation.* In the same circle of ideas, a 1-dimensional HF equation has been recently studied with the help of wavelets [**11**].

One starts from the radial HF equation for the hydrogen atom, extended to $\mathbb{R}$ by antisymmetry, $x \mapsto -x$, and one obtains the differential equation

$$-\frac{1}{2}f''(x) - \frac{f(x)}{|x|} = Ef(x), \quad f \in L^2(\mathbb{R}, dx). \tag{3.5}$$

Choosing as analyzing wavelet the first derivative of a Gaussian, $w(x) \sim -xe^{-x^2/2}$, one takes the continuous wavelet transform of the two sides of (3.5), namely, with the usual notations ($a > 0, b \in \mathbb{R}$):

$$f(x) \mapsto \tilde{f}(a,b) \equiv \langle w_{ab} \mid f \rangle.$$

The result is a complicated integro-differential equation:

$$\begin{aligned} E\tilde{f}(a,b) &= \frac{3}{4a^2}\tilde{f}(a,b) - \frac{1}{2a}\frac{\partial \tilde{f}}{\partial a}(a,b) \\ &+ \frac{2}{\sqrt{\pi}}\frac{\partial}{\partial b}\int_0^\infty dp \int_b^\infty d\beta \frac{e^{-b^2p^2/1+2a^2p^2}}{(1+2a^2p^2)^{3/4}}\tilde{f}\left(\frac{a}{(1+2a^2p^2)^{1/2}}, \frac{\beta}{1+2a^2p^2}\right), \end{aligned}$$

that one solves by iteration. The result is a marked improvement over the standard Slater or Gaussian inputs (the comparison is easy, since the exact solution is known). One may of course object that the situation treated in [**11**] is too simplified and bears little resemblance to the real physical problem, but the idea looks nevertheless interesting.

**3.3. Combination with moment methods.** As a final item in the survey of possible applications of wavelets to atomic physics, we like to point out a new method for the calculation of energy levels in atoms, based on a clever combination of the WT with the well-known method of moments [**12**]. A account is given in the contribution of C. Handy and R. Murenzi to this volume.

## 4. Outcome

Looking in retrospect to the discussion above, one may conclude that the wavelet transform can be used profitably in various problem in atomic physics, under different aspects and for different purposes. For instance:

(i) The description of harmonic generation (and similar transient phenomena) requires a *time-frequency representation*, by wavelets or Gabor analysis.

(ii) The computation of $N$-electron atomic wave functions (HF,MCHF) demands a good *orthogonal basis* on the half line $\mathbb{R}^+$, and a wavelet basis seems well-adapted.

(iii) Finally, there is the possibility of using the continuous WT as an *integral transform*, just as Fourier, and thus obtain a new representation.

These three types of utilizations of wavelets have already given some promising results, in particular for (i). However all three consider the WT as a mathematical tool, whereas wavelets might also be used as genuine physical entities, namely as wave functions of a particular class of quantum states. This is the way canonical CS are used in quantum optics, for modelling the states of a laser, for example, or in QED, as a convenient description of incoming and outgoing states in the resolution of the infrared problem [**6**]. Many interesting phenomena could be described in that language (Rydberg atoms, semiclassical limit, revivals...). Finally there is also the possibility to use different wavelets, such as the CS associated to the Schrödinger group.

The only sensible conclusion is that much work remain to be done, but that hopes for progress are reasonably well-founded.

## Acknowledgements

It is a pleasure to thank Z. Felfli, C. Handy, R. Murenzi, and B. Torrésani for interesting discussions and comments about several aspects of this paper.

## References

1. S. T. Ali, J.-P. Antoine, J.-P. Gazeau, and U. A. Mueller, *Coherent states and their generalizations: A mathematical overview*, Rev. Math. Phys. **7** (1995), 1013–1104.
2. J.-P. Antoine, *Wavelets in molecular and condensed-matter physics*, in this volume.
3. Ph. Antoine, B. Piraux, and A. Maquet, *Time profile of harmonics generated by a single atom in a strong electromagnetic field*, Phys. Rev. A **51** (1995), R1750–R1753.
4. Ph. Antoine, B. Piraux, D. B. Milošević, and M. Gajda, *Generation of ultra-short pulses of harmonics*, Phys. Rev. A **54** (1996), R1761–R1764.
5. B. H. Bransden and C. J. Joachain, *Physics of atoms and molecules*, Longman, London and New York, 1983.
6. V. Chung, *Infrared divergence in quantum electrodynamics*, Phys. Rev. B **140** (1965), 1110–1122.
7. P. B. Corkum, *Plasma perspective on strong-field multiphoton ionization*, Phys. Rev. Lett. **71** (1993), 1994–1997.
8. N. Delprat, B. Escudié, Ph. Guillemain, R. Kronland-Martinet, Ph. Tchamitchian, and B. Torrésani, *Asymptotic wavelet and Gabor analysis: Extraction of instantaneous frequencies*, IEEE Trans. Inform. Theory **38** (1992), 644–664.
9. D. H. Feng, J. R. Klauder, and M. Strayer (eds.), Proc. Int. Sympos. *Coherent states: Past, present and future* (Oak Ridge 1993), World Sci. Publishing, Singapore, 1994.
10. E. Fiordilino and V. Miceli, *Temporal evolution of the spectrum emitted by a two-level atom in the presence of a laser field*, J. Modern Opt. **43** (1996), 735–751; S. De Luca and E. Fiordilino, *Wavelet temporal profile of high order harmonics emitted by a two-level atom in the presence of a laser pulse*, J. Phys. B **29** (1996), 3277–3292.
11. P. Fischer and M. Defranceschi, *Iterative process for solving Hartree-Fock equations by means of a wavelet transform*, Appl. Comput. Harmon. Anal. **1** (1994), 232–241.
12. C. Handy and R. Murenzi, *Continuous wavelet transform analysis of one dimensional quantum bound states from first principles*, Phys. Rev. A **54** (1996), 3754–3763; and in this volume.
13. E. Huens, B. Piraux, A. Bugacov, and M. Gajda, *Numerical studies of the dynamics of multiphoton processes with arbitrary field polarization: methodological considerations*, Phys. Rev. A **55** (1997), 2132–2143.
14. J. R. Klauder and B. S. Skagerstam, *Coherent states—applications in physics and mathematical physics*, World Sci. Publishing, Singapore, 1985.
15. K. C. Kulander, *Multiphoton ionization of hydrogen: A time-dependent theory*, Phys. Rev. A **35** (1987), 445–447.
16. K. C. Kulander, K. J. Schafer, and J. L. Krause, *Dynamics of short-pulse excitation, ionization and harmonic conversion*, Super-Intense Laser-Atom Physics; (B. Piraux, A. L'Huillier, and

K. Rzążewski, eds.), NATO Adv. Sci. Inst. Ser. B Phys. **316**, Plenum Press, New York, 1993, pp. 95–110.

17. A. L'Huillier and Ph. Balcou, *High-order harmonic generation in rare gases with a* 1-*ps* 1053-*nm laser*, Phys. Rev. Lett. **70** (1993), 774–777.
18. J. Liandrat, V. Perrier, and Ph. Tchamitchian, *Numerical resolution of the regularized Burgers equation using the wavelet transform*, Wavelets and Applications (Y. Meyer, ed.), Springer-Verlag, Berlin, and Masson, Paris, 1992, pp. 420–433.
19. L. A. Lompré, A. L'Huillier, M. Ferray, P. Monot, G. Mainfray, and C. Manus, *High-order harmonic generation in xenon: Intensity and propagation effects*, J. Opt. Soc. Amer. B **7** (1990), 754–761.
20. M. Rösler, *Radial wavelets and Bessel-Hingman hypergroups*, preprint, München (1996) (to appear).
21. M. Rotenberg, *Theory and application of Sturmian functions*, Adv. At. Mol. Phys. **6** (1970), 233–268.
22. R. A. Smith, J. W. G. Tisch, M. Ciarrocca, S. Augst, and M. H. R. Hutchinson, *Angularly resolved ultra high harmonic generation experiments with picosecond laser pulses*, Super-Intense Laser-Atom Physics (B. Piraux, A. L'Huillier, and K. Rzążewski, eds.), NATO Adv. Sci. Inst. Ser. B Phys. **316**, Plenum Press, New York, 1993, pp. 31–41.
23. T. Starczewski, J. Larsson, C.-G. Wahlström, J. W. G. Tisch, R. A. Smith, J. E. Muffet, and M. H. R. Hutchinson, *Time-resolved harmonic generation in an ionizing gas*, J. Phys. B **27** (1994), 3291–3301.
24. R. Taieb, A. Maquet, Ph. Antoine, and B. Piraux, *Time dependence of harmonic generation by a single atom*, Super Intense Laser Physics (H. G. Muller and M. V. Fedorov, eds.), NATO Adv. Sci. Inst. Ser. 3 High Tech. **13**, Kluwer, Dordrecht, 1996, pp. 445-454.

Institut de Physique Théorique, Université Catholique de Louvain, B–1348 Louvain-la-Neuve, Belgium

*E-mail address*: `Antoine@fyma.ucl.ac.be`

Laboratoire de Physique Atomique et Moléculaire, Université Catholique de Louvain, B–1348 Louvain-la-Neuve, Belgium

*E-mail address*: `Antoine@fyam.ucl.ac.be`

Laboratoire de Physique Atomique et Moléculaire, Université Catholique de Louvain, B–1348 Louvain-la-Neuve, Belgium

*E-mail address*: `Piraux@fyam.ucl.ac.be`

Centre de Recherches Mathématiques
CRM Proceedings and Lecture Notes
Volume **18**, 1999

# The Wavelet $\varepsilon$-Expansion and Hausdorff Dimension

G. Battle

## Introduction

The intimate connection between wavelets and the renormalization group is well-established [**2–4, 11**]. Indeed, Wilson [**19**] postulated the existence of wavelets for his renormalization group analysis more than ten years before the mathematics community began to construct them. His own motivation was to develop a hierarchical approximation for the non-perturbative application of the renormalization group transformation. On the other hand, the perturbative calculations were based on the standard ultraviolet cutoff for a scalar field theory, rather than on a wavelet cutoff or even a lattice cutoff. The $\varepsilon$-expansion was introduced in the UV formalism [**16**]. Feynman diagrams are easier to calculate because the kinetic, quadratic part of the Hamiltonian is a fixed point with respect to the renormalization group. The standard lattice cutoff does not have that advantage, while the wavelet cutoff was not considered because there was very little information about wavelets available at the time.

It has been shown in [**2, 3, 11, 12**] that the wavelet cutoff shares the advantage of the UV cutoff *and* has a lattice picture. The aim of this lecture is to advertise the merit of developing the $\varepsilon$-expansion in the wavelet formalism. First we review the scaling functions and averaging transformations. Section 2 introduces the renormalization group transformation associated with an averaging transformation and reviews the description of the Gaussian fixed point in the space of Gaussians. Third, we explain the generalized Ising model and the importance of renormalization group analysis. Section 4 revisits Wilson's introduction of $d = 4 - \varepsilon$ dimensions in the UV formalism. Next, we describe the wavelet cutoff with the associated lattice description and then introduce wavelet diagrams as analogs of Feynman diagrams. Finally, we suggest a way to introduce $d = 4 - \varepsilon$ dimensions in the wavelet context by systematic multi-scale omission of modes. In this formalism, $d = 4 - \varepsilon$ is actually the Hausdorff dimension [**18**] of a residual set of points in the continuum limit.

## 1. Averaging Transformations

A *configuration* is a real-valued function $\phi(\vec{n})$ on $\mathbb{Z}^d$, and an *averaging transformation* is an operator $T$ on the linear space of configurations which simply averages

1991 *Mathematics Subject Classification.* 82B05, 82B28.
This is the final version of the paper.

a function on $\mathbb{Z}^d$ to a function on $2\mathbb{Z}^d$ and then scales the latter back to a function on $\mathbb{Z}^d$. More precisely, $T$ is induced by a positive function on $\mathbb{Z}^d$—typically described by the sequence notation $(c_{\vec{n}})$—with the property

$$\sum_{\vec{n}} c_{\vec{n}} = 2^d. \tag{1.1}$$

The transformation is given by

$$(T\phi)(\vec{n}) = 2^{-d/2-s} \sum_{\vec{m}} c_{\vec{m}} \phi(2\vec{n} + \vec{m}), \tag{1.2}$$

where $s$ is the index of the Sobolev space involved in the continuum. The *Kadanoff-Wilson transformation* is given by the case $s = 1$ and

$$c_{\vec{n}} = \begin{cases} 1, & \vec{n} \in \{0,1\}^d, \\ 0, & \text{otherwise}. \end{cases} \tag{1.3}$$

This is the popular and most obvious choice of averaging, but in general, one may choose an averaging that is less localized. Let $\eta(\vec{x})$ be the *scaling function* associated with $(c_{\vec{n}})$—i.e.,

$$\eta(\vec{x}) = \sum_{\vec{m}} c_{\vec{m}} \eta(2\vec{x} - \vec{m}). \tag{1.4}$$

$\eta$ is the characteristic function of the unit cube in the Kadanoff-Wilson case. Let $\mathcal{T}$ be the continuum averaging transformation

$$(\mathcal{T}\xi)(\vec{n}) = \int \eta(\vec{x} - \vec{n})\xi(\vec{x})\, d\vec{x} \tag{1.5}$$

and $U_s(\lambda)$ the unitary group of scalings on the Sobolev space:

$$\big(U_s(\lambda)\xi\big)(\vec{x}) = e^{\lambda(d/2-s)}\xi(e^{\lambda}\vec{x}). \tag{1.6}$$

Then

$$T \circ \mathcal{T} = \mathcal{T} \circ U_s, \quad U_s \equiv U_s(\ln 2), \tag{1.7}$$

which is the relation between continuum averaging and lattice averaging.

## 2. The Renormalization Group Transformation

If $d\mu$ is a probability measure on the space of configurations and $T$ is an averaging transformation, then $\mathcal{R}_T(d\mu)$ is the probability measure defined by

$$\big\langle X(\phi)\big\rangle_{\mathcal{R}_T(d\mu)} = \big\langle X(T\phi)\big\rangle_{d\mu} \tag{2.1}$$

for random variables $X$. In statistical mechanics, $d\mu$ is realized as the $\Lambda = \mathbb{Z}^d$ limit of the probability measures

$$d\mu_\Lambda(\phi) = Z_\Lambda^{-1} e^{-H(\chi_\Lambda \phi)} \prod_{\vec{m} \in \Lambda} d\phi(\vec{m}), \tag{2.2}$$

$$Z_\Lambda = \int e^{-H(\chi_\Lambda \phi)} \prod_{\vec{m} \in \Lambda} d\phi(\vec{m}), \tag{2.3}$$

where $\Lambda$ ranges over finite subsets of $\mathbb{Z}^d$, $\chi_\Lambda$ is the characteristic function of $\Lambda$, and $H$ is a functional on the space of configurations with finite support. $H$ is the given Hamiltonian with the inverse temperature parameter absorbed as a factor, and $\mathcal{R}_T$ may be regarded as transforming $H$. Clearly, this (possibly non-local) expression in the variables $\phi(\vec{n}\,)$ must have the property that $Z_\Lambda < \infty$ for all finite $\Lambda \subset \mathbb{Z}^d$; in particular, it must be bounded below. The action of $\mathcal{R}_T$ on $H$ is given by

$$(2.4)\quad e^{-\mathcal{R}_T(H)(\phi')} = \lim_{\Lambda\to\mathbb{Z}^d}\Bigg[Z_\Lambda(T\phi=0)^{-1}\int e^{-H(\chi_\Lambda\phi)} \times \prod_{\vec{n}\,\in\Lambda\cap 2\mathbb{Z}^d}\delta\left(\phi'\left(\frac{1}{2}\vec{n}\right)-(T\phi)\left(\frac{1}{2}\vec{n}\right)\right)\prod_{\vec{n}\in\Lambda}d\phi(\vec{n}\,)\Bigg].$$

It takes an expansion method in statistical mechanics to rigorously control a single step, but only unit-scale fluctuations are integrated out.

The Hamiltonians of interest are of the form $H = H_0 + I$, where $H_0$ is a translation-invariant quadratic form. Although $H_0$ usually couples the lattice variables, such a Hamiltonian, by itself, is trivial in the deeper sense that the corresponding statistical mechanical model has no critical behavior. Nevertheless, the class of translation-invariant quadratic forms is the starting point in the RG (renormalization group) analysis. The probability measures are Gaussian and everything is computed in closed form, including the action of the renormalization group on translation-invariant quadratic forms, the space of which is preserved. Indeed, the formula (2.4) is unnecessary in this case; only (2.1) is needed for calculation. The Gaussian measure associated with $(H_0\phi,\phi)$ has mean zero and covariance

$$(2.5)\qquad \big\langle\phi(\vec{m})\phi(\vec{n}\,)\big\rangle_{H_0} = \frac{1}{(2\pi)^d}\left(\prod_{\iota=1}^{d}\int_{-\pi}^{\pi}dp_\iota\right)e^{-i\vec{p}\cdot(\vec{m}-\vec{n}\,)}H_0(\vec{p}\,)^{-1},$$

$$(2.6)\qquad H_0(\vec{p}\,) = \sum_{\vec{m}} H_0(\vec{m},0)e^{i\vec{m}\cdot\vec{p}},$$

$$(2.7)\qquad H_0(\phi) = \frac{1}{2}\sum_{\vec{m},\vec{n}} H_0(\vec{m},\vec{n}\,)\phi(\vec{n}\,)\phi(\vec{m}) \equiv \frac{1}{2}(H_0\phi,\phi),$$

while the effect of $\mathcal{R}_T$ on the covariance is obviously

$$(2.8)\qquad \big\langle\phi(\vec{m})\phi(\vec{n})\big\rangle_{\mathcal{R}_T(H_0)} = \big\langle(T\phi)(\vec{m})(T\phi)(\vec{n}\,)\big\rangle_{H_0}.$$

Therefore, if we set

$$(2.9)\qquad h(\vec{p}\,) = \sum_{\vec{n}} c_{\vec{n}}\, e^{i\vec{n}\,\cdot\vec{p}/2},$$

we may write

$$(2.10)\qquad \mathcal{R}_T(H_0)(\vec{p}\,) = \frac{2^{2d+2s}}{\sum_{\vec{\varepsilon}\in\{0,1\}^d}|h(\vec{p}+2\pi\vec{\varepsilon}\,)|^2 H_0(\vec{p}/2+\pi\vec{\varepsilon}\,)^{-1}}.$$

If $H_0$ has the property

$$(2.11)\qquad \lim_{\vec{p}\to 0}\frac{H_0(\vec{p}\,)}{|\vec{p}\,|^{2s}} = c \neq 0,$$

then iterations of the RG transformation result in:

$$(2.12)\qquad \lim_{N\to\infty}\mathcal{R}_T^N(H_0) = H_\infty,$$

where the fixed point $H_\infty$ is given by

$$(2.13)\qquad H_\infty(\vec{p}\,) = c\left(\sum_{\vec{\ell}}\frac{|\hat{\eta}(\vec{p}+2\pi\vec{\ell})|^2}{|\vec{p}+2\pi\vec{\ell}|^{2s}}\right)^{-1}.$$

This follows from the equation

$$\hat{\eta}(\vec{p}) = \hat{\eta}(0)\prod_{\rho=0}^{\infty}(2^{-d}h(2^{-\rho}\vec{p})) \tag{2.14}$$

generated by the momentum-space version of (1.4)—namely,

$$\hat{\eta}(\vec{p}) = 2^{-d}h(\vec{p})\hat{\eta}\left(\frac{1}{2}\vec{p}\right). \tag{2.15}$$

To understand the exponent $s$, suppose $H_0$ is given by the nearest-neighbor interaction. Then $s = 1$ is the appropriate choice for $T$ because

$$H_0(\vec{p}) = \sum_{\iota=1}^{d}(1 - \cos p_\iota) \tag{2.16}$$

in that case.

To apply RG analysis to a perturbation $H_0 + I$ of a translation-invariant quadratic form $H_0$, one takes (2.4) as the starting point and identifies fluctuations—i.e., amplitudes $(H_0\phi, \phi)$-orthogonal to the subspace of amplitudes that are not integrated out. Such integration variables factorize the Gaussian part of the measure in the partial integration formula. The fluctuation part of $\phi$ is given by

$$\tilde{\phi} = \phi - MT\phi, \tag{2.17}$$

where $M$ is the *minimizer*—i.e., $M\phi'$ minimizes $(H_0\phi, \phi)$ with respect to the constraint $T\phi = \phi'$. Thus $M$ is a right inverse of $T$, while $MT$ is an orthogonal projection with respect to the quadratic form. The partial integration is simplified by the decomposition

$$H_0(\phi) = H_0(\tilde{\phi}) + H_0(M\phi'), \tag{2.18}$$

but the perturbation $I(\phi)$ couples $M\phi'$ to $\tilde{\phi}$. One may expand the fluctuation part in "wavelets" that are mutually orthogonal with respect to the quadratic form:

$$\tilde{\phi}(\vec{n}\,) = \sum_{\kappa,\vec{m}} a_{\kappa\vec{m}}\psi_\kappa(\vec{n} - 2\vec{m}\,), \tag{2.19}$$

$$(H_0\tilde{\phi}, \tilde{\phi}) = \sum_{\kappa,\vec{m}} a^2_{\kappa\vec{m}}, \tag{2.20}$$

where $\kappa$ ranges over a finite set of exponentially localized expansion functions, all associated to the same block of lattice sites. The formula for the RG transformation can be realized as:

$$\begin{aligned} &e^{-\mathcal{R}_T(H)(\phi')} \\ &= e^{-H_0(M\phi')}\lim_{\Lambda\to\mathbb{Z}^d}\Bigg[\widetilde{Z}_\Lambda^{-1}\Big(\prod_\kappa\prod_{2\vec{m}\in\Lambda}\int da_{\kappa\vec{m}}\Big)\exp\Big(-\frac{1}{2}\sum_\kappa\sum_{2\vec{m}\in\Lambda}a^2_{\kappa\vec{m}}\Big) \\ &\qquad\times\exp\Big(-I\Big(M\phi' + \sum_\kappa\sum_{2\vec{m}\in\Lambda}a_{\kappa\vec{m}}\psi_\kappa(\cdot - 2\vec{m})\Big)\Big)\Bigg], \end{aligned} \tag{2.21}$$

$$\widetilde{Z}_\Lambda = \Big(\prod_\kappa \prod_{2\vec{m}\in\Lambda} \int da_{\kappa\vec{m}}\Big) \times \exp\Big(-\frac{1}{2}\sum_\kappa \sum_{2\vec{m}\in\Lambda} a^2_{\kappa\vec{m}} - I\Big(\sum_\kappa \sum_{2\vec{m}\in\Lambda} a_{\kappa\vec{m}}\psi_\kappa(\cdot - 2\vec{m})\Big)\Big), \tag{2.22}$$

where $\Lambda$ now indexes a cutoff on the set of well-localized expansion functions rather than a cutoff on the set of lattice sites.

## 3. The Model and Basic Concepts

The goal is to understand the critical behavior of the Ising model in $d$ dimensions, and the starting point is to approximate the "spin-up-or-spin-down" constraint with the Hamiltonian

$$H(\phi) = H_0(\phi) + \sum_{\vec{n}} \big(r\phi(\vec{n})^2 + u\phi(\vec{n})^4\big), \quad u > 0, \tag{3.1}$$

on the unconstrained configurations, where the quadratic form in this case is actually the nearest-neighbor interaction for that model:

$$H_0(\phi) = \frac{1}{2}\sum_{\langle \vec{m},\vec{n}\rangle} \big(\phi(\vec{m}) - \phi(\vec{n})\big)^2 \tag{3.2}$$

and therefore $H_0(\vec{p})$ is given by (2.16). The summand of the perturbation may be written in the form

$$V\big(\phi(\vec{n})\big) = u\Big(\phi(\vec{n})^2 + \frac{r}{2u}\Big)^2 \tag{3.3}$$

with the constant dropping out in the normalization of the measure, so for $r < 0$, $V$ is a double-well potential in the amplitude $\phi(\vec{n})$. This is a natural approximation to the Ising spin constraint.

The first application of the renormalization group transformation destroys the form of this Hamiltonian. The local perturbation becomes non-local, while $H_0(\phi)$ is no longer a nearest-neighbor interaction. There are also new terms involving non-local monomials in the amplitudes $\phi(\vec{n})$ of higher order, but they are *irrelevant* for $d > 3$—i.e., they are driven to zero exponentially fast with respect to iterations of the RG transformation. For $d = 3$, non-local monomials of order six have to be reckoned with. For $d < 4$, the RG flow is expected to be given by Figure 1. The *critical manifold* consists of Hamiltonians that are *critical* with respect to the fixed temperature, and this classification is defined by divergence of the *correlation length.* For Hamiltonians with respect to which the fixed temperature is sufficiently high, the two-point function has exponential decay:

$$\big\langle \phi(\vec{m})\phi(\vec{n})\big\rangle \sim ce^{-|\vec{m}-\vec{n}|/\xi}, \tag{3.4}$$

where $\xi$ is the correlation length. In the case where the long-distance decay is slower than any exponential decay, the correlation length is infinite and the Hamiltonian is said to be critical. The RG transformation itself has a simple effect on correlation length—namely,

$$\xi\big(\mathcal{R}_T(H)\big) = \frac{1}{2}\xi(H), \tag{3.5}$$

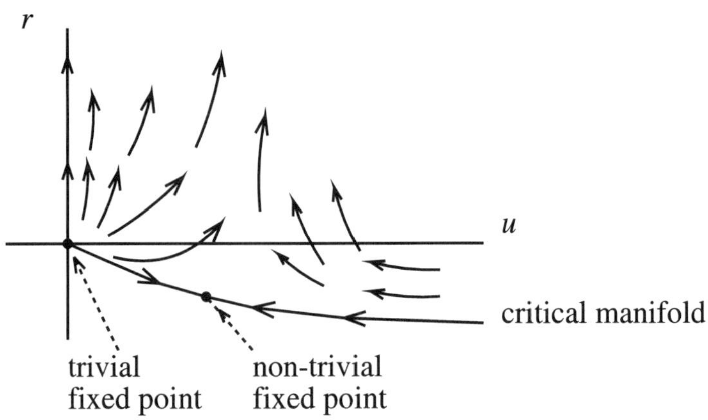


FIGURE 1. Renormalization group flow for $d < 4$.

so the critical manifold is preserved by the RG transformation and any Hamiltonian with finite correlation length is driven away from this unstable manifold by the RG flow. Obviously, a fixed point must either lie on the critical manifold or have zero correlation length. There is only one fixed point of the latter type, it is obviously stable, and it is called the *high-temperature fixed point.* A fixed point on the critical manifold is *trivial* if the probability measure is Gaussian—i.e., if the Hamiltonian is a quadratic form. For $d < 4$, the trivial fixed point is unstable in the $ur$-plane, while the non-trivial fixed point is semi-stable: attractive on the critical manifold and repulsive off it.

If $H$ is a critical Hamiltonian other than the trivial fixed point, then the two-point function is expected to have a power-law decay:

$$\langle \phi(\vec{m})\phi(\vec{n}) \rangle \sim c|\vec{m} - \vec{n}|^{2-d+\eta}. \tag{3.6}$$

$d - \eta$ is the *anomalous dimension*, as $\eta = 0$ yields the decay of the Green's function of the lattice Laplacian (= covariance of the Gaussian defined by $H_0(\phi)$). It is the attractive fixed point on the critical manifold that determines critical exponents such as $\eta$, and since that fixed point is non-trivial, it is possible that $\eta \neq 0$. Indeed, for $d = 2$, $\eta = \frac{1}{4}$, and for $d = 3$, approximations of $\eta$ range from .04 to .06. On the other hand, the $d = 4$ case has an entirely different RG flow given by Figure 2. There is no longer a non-trivial fixed point, but the trivial fixed point is now attractive on the critical manifold, so $\eta = 0$ in this case.

If an appropriate notion of fractional dimension is introduced, the non-trivial fixed point approaches the trivial fixed point as $d \nearrow 4$. Since the former type of fixed point is difficult to analyze, one would like to study it even in fractional dimension $d = 4 - \varepsilon$, and this motivates Wilson's $\varepsilon$-expansion. One simply realizes the non-trivial fixed point as an expansion about the trivial fixed point, where the small parameter is $\varepsilon$.

NOTE. Rigorous work on the renormalization group flow is concerned with neighborhoods of trivial fixed points for the most part. It has been shown [**12, 13**] that the attractive one in dimension $d = 4$ is indeed attractive and the repelling

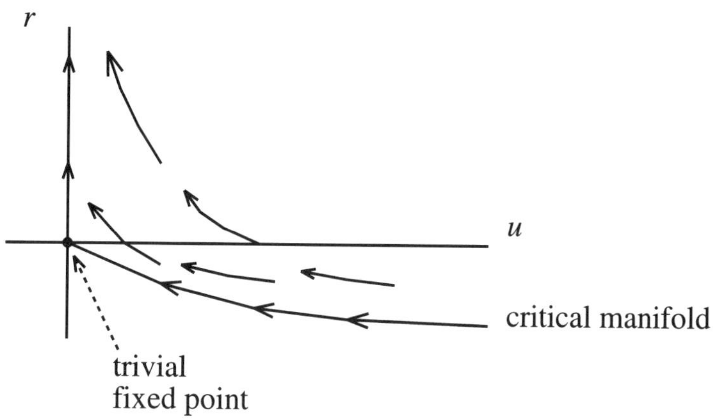


FIGURE 2. Renormalization group flow for $d = 4$.

one in dimension $d < 4$ is indeed repelling. The non-trivial fixed point is rigorously understood only for toy models such as the hierarchical model [**1,5–7,14,16,17,19**] or the planar model [**9, 18**].

## 4. The Wilson ε-Expansion

The Wilson approach [**16**] begins with replacing the lattice model with a continuum field model where an ultraviolet cutoff is imposed. In this case, $H_0(\vec{p}) = |\vec{p}\,|^2$, or equivalently,

$$H_0(\phi) = \frac{1}{2}\int (\nabla\phi)^2, \tag{4.1}$$

but more to the point, one imposes an ultraviolet cutoff on

$$H_0^{-1}(\vec{x}\,,\vec{y}\,) = \left(\prod_{\iota=1}^{d}\int dp_\iota\right) e^{i\vec{p}\cdot(\vec{x}\,-\vec{y}\,)}\frac{1}{|\vec{p}\,|^2}. \tag{4.2}$$

Define $d\mu_0$ as the Gaussian measure on $\mathcal{S}'(\mathbb{R}^d)$ with mean zero and covariance

$$\langle\phi(\vec{x}\,)\phi(\vec{y}\,)\rangle_{d\mu_0} = \left(\prod_{\iota=1}^{d}\int dp_\iota\right)_{|\vec{p}\,|\le 1} e^{i\vec{p}\cdot(\vec{x}-\vec{y}\,)}\frac{1}{|\vec{p}\,|^2} \equiv C(\vec{x}\, - \vec{y}\,). \tag{4.3}$$

The RG transformation is now defined a little differently. Let $d\tilde{\mu}_0$ be the Gaussian measure on $\mathcal{S}'(\mathbb{R}^d)$ with mean zero and covariance

$$\langle\phi(\vec{x}\,)\phi(\vec{y}\,)\rangle_{d\tilde{\mu}_0} = \left(\prod_{\iota=1}^{d}\int dp_\iota\right)_{1/2\le|\vec{p}\,|\le 1} e^{i\vec{p}\cdot(\vec{x}\,-\vec{y}\,)}\frac{1}{|\vec{p}\,|^2} \equiv \widetilde{C}(\vec{x}\, - \vec{y}\,). \tag{4.4}$$

Then the definition of $\mathcal{R}$ is:

$$e^{-\mathcal{R}(H)(\phi')} = \lim_{\Lambda\to\mathbb{R}^d}\left[\widetilde{Z}_\Lambda^{-1}\int e^{-u\int_\Lambda \phi^4 - r\int_\Lambda \phi^2 - H_0(\phi-\tilde{\phi})}\, d\tilde{\mu}_0(\phi)\right], \tag{4.5}$$

$$\widehat{\phi'}(\vec{p}) = \begin{cases} 2^{1-d/2}\hat{\phi}\left(\frac{1}{2}\vec{p}\right), & |\vec{p}| \le 1, \\ 0, & |\vec{p}| > 1, \end{cases} \tag{4.6}$$

$$\widetilde{Z}_\Lambda = \int e^{-u\int_\Lambda \tilde{\phi}^4 - r\int_\Lambda \tilde{\phi}^2}\, d\tilde{\mu}_0(\phi), \tag{4.7}$$

$$\widehat{\phi^\sim}(\vec{p}) = \begin{cases} 0, & |\vec{p}| < \frac{1}{2}, \\ \hat{\phi}(\vec{p}), & |\vec{p}| \ge \frac{1}{2}. \end{cases} \tag{4.8}$$

(Notice that the Sobolev index is $s = 1$.) One has

$$H_0(\phi) = H_0(\tilde{\phi}) + H_0(\phi - \tilde{\phi}) = H_0(\tilde{\phi}) + H_0(\phi') \tag{4.9}$$

because the operator $H_0$ is diagonal in momentum space and the quadratic form $(H_0\phi, \phi)$ is invariant with respect to the scaling used to define $\phi'$. This is the advantage offered by the ultraviolet cutoff, but the disadvantage is that it destroys whatever long-distance decay properties the lattice correlations may have. The usual solution to this technical difficulty is to replace this sharp ultraviolet cutoff with the Pauli-Villars regularization [**10, 12**] and introduce the free Euclidean measure as a tensor product of probability measures over momentum scales. This regularization is not only smooth in momentum space but real-analytic, while the tensor product avoids the resulting momentum overlap between scales when the RG transformation is applied to the free measure. We stick to the sharp ultraviolet cutoff for simplicity in our discussion and with this understanding.

Why is an RG analysis of this model supposed to reflect the long-distance behavior of the original lattice model? The reason is that, roughly speaking, there has only been a change in the *canonical manifold.* The canonical manifold is the parameterized class of initial Hamiltonians in an RG iteration, and the principle of *universality* states that the critical exponents of thermodynamic functions are the same for large classes of lattice systems. The reason is that the fixed-point analysis of an RG flow is independent of the choice of canonical manifold in the domain of a fixed point.

The perturbation theory for this partial functional integration reduces to Feynman diagrams, and we focus our discussion on a basic one. We have

$$\mathcal{R}(H)(\phi') = H_0(\phi') + \lim_{\Lambda\to\mathbb{R}^d}\left[\ln \widetilde{Z}_\Lambda - \ln\langle e^{-u\int_\Lambda \phi^4 - r\int_\Lambda \phi^2}\rangle_{d\tilde{\mu}_0}\right], \tag{4.10}$$

where the cumulant expansion yields

$$\begin{aligned} \ln\langle e^{-u\int_\Lambda \phi^4 - r\int_\Lambda \phi^2}\rangle_{d\tilde{\mu}_0} &= -u\int_\Lambda \langle\phi^4\rangle_{d\tilde{\mu}_0} - r\int_\Lambda \langle\phi^2\rangle_{d\tilde{\mu}_0} \\ &+ \frac{1}{2}\int_\Lambda d\vec{x}\int_\Lambda d\vec{y}\,\big\langle\big(u\phi(\vec{x})^4 + r\phi(\vec{x})^2\big)\big(u\phi(\vec{y})^4 + r\phi(\vec{y})^2\big)\big\rangle_{d\tilde{\mu}_0} \\ &- \frac{1}{2}\left(u\int_\Lambda \langle\phi^4\rangle_{d\tilde{\mu}_0} + r\int_\Lambda \langle\phi^2\rangle_{d\tilde{\mu}_0}\right)^2 + \text{higher order terms.} \end{aligned} \tag{4.11}$$

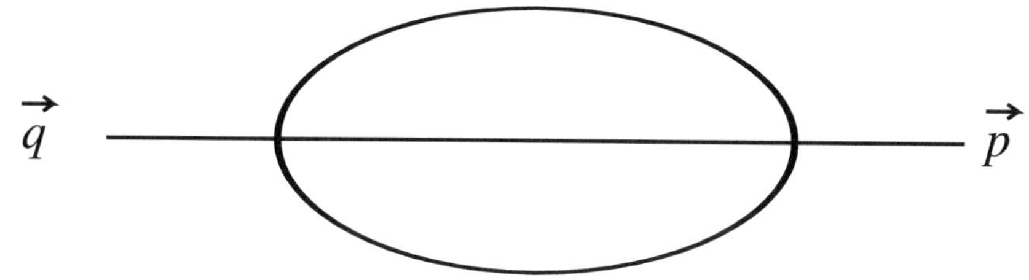


FIGURE 3. A Feynman diagram.

One contribution to $\mathcal{R}(H)(\phi')$ is 96 copies of

$$\frac{1}{2}u^2\int_\Lambda d\vec{x}\int_\Lambda d\vec{y}\,(\phi(\vec{x})-\tilde{\phi}(\vec{x}))(\phi(\vec{y})-\tilde{\phi}(\vec{y}))\widetilde{C}(\vec{x}-\vec{y})^3,$$

which may be written as

$$\frac{1}{2}u^2\int d\vec{p}\int d\vec{q}\,(\hat{\phi}(\vec{p})-\widehat{\phi^{\sim}}(\vec{p}))^*(\hat{\phi}(\vec{q})-\widehat{\phi^{\sim}}(\vec{q}))$$
$$\int d\vec{p}\,'\int d\vec{p}\,''\widetilde{C}(\vec{p}\,')\widetilde{C}(\vec{p}\,'')\widetilde{C}(\vec{p}-\vec{q}-\vec{p}\,'-\vec{p}\,'')$$

in the $\Lambda=\mathbb{R}^d$ limit. Scaling the integration, we get

$$2^{-4d-1}u^2\int d\vec{p}\int d\vec{q}\,\widehat{\phi'}(\vec{p})^*\widehat{\phi'}(\vec{q})\int d\vec{p}\,'\int d\vec{p}\,''\widetilde{C}\Big(\frac{1}{2}\vec{p}\,'\Big)\widetilde{C}\Big(\frac{1}{2}p''\Big)$$
$$\times\widetilde{C}\Big(\frac{1}{2}\vec{p}-\frac{1}{2}\vec{q}-\frac{1}{2}\vec{p}\,'-\frac{1}{2}\vec{p}\,''\Big),$$

which is represented by the Feynman diagram given by Figure 3, where the internal lines correspond to a smaller length scale. Another contribution to $\mathcal{R}(H)(\phi')$ is just

$$2^{4-d}u\int\phi'^4+4r\int\phi'^2,$$

so it follows that iterations of $\mathcal{R}$ will develop this diagram repeatedly. On the other hand, once such a diagram has appeared, the Hamiltonian generated by subsequent iterations will always include a transformation of that diagram where the internal lines correspond to still smaller length scales.

Wilson's notion of fractional dimension $d=4-\varepsilon$ simply extends the Jacobian expression for the $\int d\vec{p}\,'$ and $\int d\vec{p}\,''$ integrations to non-integer $d$ as follows:

$$\int d\vec{v}=\int_0^\infty dv\,v^{d-1}\int_{\mathbb{S}^{d-1}}d\Omega\longmapsto\int_0^\infty dv\,v^{d-1}\int_{\mathbb{S}^{\bar{d}-1}}d\Omega, \tag{4.12}$$

where $\bar{d}$ denotes the smallest integer $\geq d$ and $d\Omega$ is Lebesgue measure on the unit sphere $\mathbb{S}^{\bar{d}-1}$. Since the $\int d\vec{p}$ and $\int d\vec{q}$ integrations are not changed, this modification is strictly perturbative, but for fractional dimension, no one has been able to go beyond perturbation theory (except in the case of the hierarchical model) in the analysis of non-trivial fixed points. (In the case of the planar model, the perturbation series is convergent.) However, one may at least *define* fractional dimension non-pertubatively, by introducing the Jacobian scheme in the momentum integral for the covariance:

$$C(\vec{x}-\vec{y}) = \int_{|\vec{p}|\leq 1} d\vec{p}\, e^{i\vec{p}\cdot(\vec{x}-\vec{y})}\frac{1}{|\vec{p}|^2} \tag{4.13}$$
$$= \int_0^1 dp\, p^{d-3}\int_{\mathbb{S}^{d-1}} d\Omega(\vec{u})e^{ip\vec{u}\cdot(\vec{x}-\vec{y})}$$
$$\mapsto \int_0^1 dp\, p^{d-3}\int_{\mathbb{S}^{\bar{d}-1}} d\Omega(\vec{u})e^{ip\vec{u}\cdot(\vec{x}-\vec{y})}.$$

Once the extension to $d=4-\varepsilon$(with $\bar{d}=4$) has been chosen, the RG flow is studied by using expansions in powers of $\varepsilon$, which are asymptotic but not convergent.

## 5. The Wavelet Cutoff

The advantage in working with the standard ultraviolet cutoff on a continuum field is that the kinetic contribution $H_0(\phi)$ is a fixed point and the RG calculations reduce to the evaluation of Feynman diagrams. On the other hand, the cutoff is only qualitatively similar to the lattice cutoff, with the integration of high-momentum amplitudes only qualitatively similar to discrete averaging. The wavelet cutoff is technically more desirable because:

(a) it shares the advantage of the ultraviolet cutoff,
(b) it has a lattice picture with an averaging transformation,
(c) it admits the exponential decay of correlations as does the Pauli-Villars cutoff.

Section 2 already describes how the RG transformation induced by an averaging transformation can be written as an integration of unit-scale fluctuations, but the lattice quadratic form $H_0(\phi)$ is not a fixed point. Gawedzki and Kupiainen [**11,12**] replaced $H_0(\phi)$ by $H_\infty(\phi)$ but did not recognize the formalism as a wavelet cutoff in the continuum.

We return to the lattice with the averaging transformation $T$ defined as in (1.2) and with a translation-invariant quadratic form $(H_0\phi,\phi)$ satisfying (2.11). For arbitrary Sobolev index $s$, let $d\mu_s$ be the Gaussian measure on $\mathcal{S}'(\mathbb{R}^d)$ with mean zero and covariance operator $|\nabla|^{-2s}$. Since $H_\infty$ is given by (2.13),

$$\langle\phi(\vec{m})\phi(\vec{n})\rangle_{H_\infty} = \int d\vec{x}\int d\vec{y}\,\eta(\vec{x}-\vec{m})\eta(\vec{y}-\vec{n})\big\langle\phi(\vec{x})\phi(\vec{y})\big\rangle_{d\mu_s}, \tag{5.1}$$

where $\eta$ is the scaling function associated with $T$. This is the *lattice-continuum duality* of the Gaussian fixed point.

Where are the wavelets? Let $\mathcal{M}$ be the continuum minimizer—that is, for a lattice configuration $\phi$, define $\mathcal{M}\phi$ as the continuum configuration that minimizes $\big(|\nabla|^{2s}\varphi,\varphi\big)$ with respect to the constraint $\mathcal{T}\varphi=\phi$. Let $M_\infty\phi$ be the lattice configuration that minimizes $(H_\infty\psi,\psi)$ with respect to the constraint $T\psi=\phi$. Then

$$\mathcal{M}\circ M_\infty^n = U_s^{-n}\circ\mathcal{M}, \tag{5.2}$$

and if we consider the decomposition

$$\phi = M_\infty T\phi + \tilde{\phi}, \tag{5.3}$$

we can show that $\tilde{\phi}$ can be expanded in the form

$$\tilde{\phi}(\vec{n}) = \sum_{\kappa,\vec{m}}\alpha_{\kappa\vec{m}}\psi_\kappa^\infty(\vec{n}-2\vec{m}), \tag{5.4}$$

where the functions

$$\Psi_\kappa = \mathcal{M}\psi_\kappa^\infty \tag{5.5}$$

are wavelets—i.e.,

$$\Psi_\kappa(2^{-\rho+1}\vec{x} - 2\vec{m}) \perp \Psi_\kappa(\vec{x}), \quad \rho \geq 1, \vec{m} \neq 0, \tag{5.6}$$

with respect to the Sobolev norm. *The lattice field $\phi(\vec{n}\,)$ corresponds to a wavelet cutoff of the continuum field $\phi(\vec{x}\,)$.* The cutoff is very much like an ultraviolet cutoff, except the modes are discrete in scale and translation:

$$\phi = \sum_{\rho=1}^{\infty}\sum_{\kappa,\vec{m}} \alpha_{\kappa\vec{m}}^{(\rho-1)} U_s^{-\rho+1}\Psi_\kappa(\cdot - 2\vec{m}) \tag{5.7}$$

$$\langle \alpha_{\kappa\vec{m}}^{(\rho-1)} \alpha_{\kappa'\vec{m}'}^{(\rho'-1)} \rangle = \delta_{\kappa\kappa'}\delta_{\vec{m}\vec{m}'}\delta_{\rho\rho'}. \tag{5.8}$$

The RG transformation $\mathcal{R}_T$ based on the averaging is given by

$$\begin{aligned} &e^{-\mathcal{R}_T(H_\infty+I)(\phi')} \\ &= e^{-H_\infty(\phi')} \lim_{\Lambda\to\mathbb{Z}^d} \Bigg[\widehat{Z}_\Lambda^{-1}\Big(\prod_\kappa \prod_{2\vec{m}\in\Lambda}\int d\alpha_{\kappa\vec{m}}\Big)\exp\Big(-1/2\sum_\kappa\sum_{2\vec{m}\in\Lambda}\alpha_{\kappa\vec{m}}^2\Big) \\ &\qquad\times \exp\Bigg(-I\Big(M_\infty\phi' + \sum_\kappa\sum_{2\vec{m}\in\Lambda}\alpha_{\kappa\vec{m}}\psi_\kappa^\infty(\cdot - 2\vec{m})\Big)\Bigg)\Bigg], \end{aligned} \tag{5.9}$$

$$\begin{aligned} \widehat{Z}_\Lambda = \Big(\prod_\kappa \prod_{2\vec{m}\in\Lambda}\int d\alpha_{\kappa\vec{m}}\Big)&\exp\Big(-1/2\sum_\kappa\sum_{2\vec{m}\in\Lambda}\alpha_{\kappa\vec{m}}^2\Big) \\ &\times\exp\Bigg(-I\Big(\sum_\kappa\sum_{2\vec{m}\in\Lambda}\alpha_{\kappa\vec{m}}\psi_\kappa^\infty(\cdot - 2\vec{m})\Big)\Bigg), \end{aligned} \tag{5.10}$$

where $I(\phi)$ is the perturbation. One may certainly continue the RG analysis with the perturbation

$$\sum_{\vec{n}} V(\phi(\vec{n}\,)) = \sum_{\vec{n}} u\Big(\phi(\vec{n}\,)^2 + \frac{r}{2u}\Big)^2, \tag{5.11}$$

but we have already changed the canonical manifold, so it is appropriate to consider other choices that are qualitatively similar. In this case, it is technically desirable to choose a wavelet cutoff on the continuum version of the original perturbation. Thus

$$\begin{aligned} I(\phi) = r\int&\Big(\sum_{\rho=1}^{\infty}\sum_{\kappa,\vec{m}}\alpha_{\kappa\vec{m}}^{(\rho-1)}U_s^{-\rho+1}\Psi_\kappa(\cdot - 2\vec{m})(\vec{x})\Big)^2 d\vec{x} \\ &+ u\int\Big(\sum_{\rho=1}^{\infty}\sum_{\kappa,\vec{m}}\alpha_{\kappa\vec{m}}^{(\rho-1)}U_s^{-\rho+1}\Psi_\kappa(\cdot - 2\vec{m})(\vec{x})\Big)^4 d\vec{x}. \end{aligned} \tag{5.12}$$

One may write this in terms of $\phi$. Since iteration of (5.3) and (5.4) yields

$$\phi(\vec{n}\,) = \sum_{\rho=1}^{\infty}\sum_{\kappa,\vec{m}}\alpha_{\kappa\vec{m}}^{(\rho-1)}M_\infty^{\rho-1}\psi_\kappa^\infty(\cdot - 2\vec{m}), \tag{5.13}$$

it follows from (5.2) and (5.5) that

$$I(\phi) = r\int(\mathcal{M}\phi)^2 + u\int(\mathcal{M}\phi)^4. \tag{5.14}$$

Naturally, one uses (5.12) in the RG transformation, as the fluctuations are integrated out as wavelet amplitudes. For the quadratic form,

$$\begin{aligned} H_\infty(\phi) &= H_\infty(\phi') + H_\infty(\tilde{\phi}) \\ &= H_\infty(\phi') + \frac{1}{2}\sum_{k,\vec{m}} \alpha^2_{k\vec{m}} \end{aligned} \tag{5.15}$$

is the decomposition, and this is the property that the wavelet cutoff shares with the ultraviolet cutoff. Moreover, the wavelets we have introduced actually have exponential decay in the case $s = 1$, so the wavelet cutoff satisfies condition (c) above.

## 6. Wavelet Diagrams

The Feynman diagram in Figure 3 makes many different contributions to the Hamiltonians resulting from iterations of the RG transformation applied to the standard ultraviolet cutoff—all except one appearing as higher-order diagrams containing this diagram. We focus on the contribution which does not. After $N$ iterations of $\mathcal{R}$, this contribution is the momentum integral

$$\begin{aligned} c_N u^2 \int d\vec{p}\int d\vec{q}\,\widehat{\phi^{(N)}}(\vec{p})^*\widehat{\phi^{(N)}}(\vec{q})\int d\vec{p}\,'\int d\vec{p}\,''\widetilde{C}(2^{-N}\vec{p}\,')\widetilde{C}(2^{-N}p'') \\ \times \widetilde{C}(2^{-N}\vec{p} - 2^{-N}\vec{q} - 2^{-N}\vec{p}\,' - 2^{-N}\vec{p}\,''), \end{aligned}$$

where $c_N$ is a scaling constant and $\phi^{(N)}(\vec{x})$ is the scaling back of the non-fluctuating part of the field—i.e.,

$$\widehat{\phi^{(N)}}(\vec{p}) = \begin{cases} 2^{(1-d/2)N}\hat{\phi}(2^{-N}\vec{p}), & |\vec{p}| \le 1, \\ 0, & |\vec{p}| > 1. \end{cases} \tag{6.1}$$

On the other hand, if we apply $N$ iterations of $\mathcal{R}_T$ to the wavelet cutoff, the corresponding quantity is:

$$u^2 \sum_{\substack{k,\ell \\ L_k, L_\ell \ge 1}} \alpha_k\alpha_\ell \sum_{\substack{k_1,k_2,k_3 \\ L_{k_\iota} = 2^{-N}}} \int \Psi_k\Psi_{k_1}\Psi_{k_2}\Psi_{k_3}\int \Psi_\ell\Psi_{k_1}\Psi_{k_2}\Psi_{k_3},$$

where $L_\ell = 2^{\rho-1-N}$ for any wavelet index $\ell = (\rho-1, \vec{m}, \kappa)$. On one hand,

$$\Psi_\ell(\vec{x}) = L_\ell^{-d/2+1}\Psi_\kappa(L_\ell^{-1}\vec{x} - 2\vec{m}), \tag{6.2}$$

while on the other hand, $\alpha_\ell = \alpha^{(\rho-1)}_{\kappa\vec{m}}$. A convenient wavelet analog of Feynman diagrams has been used heavily by Federbush [**8**, private communication]. For this particular quantity, the Federbush diagram is given by Figure 4. The long dashes are *modes*, where a mode is a wavelet index. A solid vertical line is a $\int d\vec{x}$-integral of the product of those wavelets whose dashes lie on that vertical line, and a dashed horizontal line is an identification of the modes it connects. The height of a long dash illustrates the scale of the wavelet. The identification lines correspond to the internal lines of the Feynman diagram, and those identified modes have no $\alpha$-amplitudes in the quantity—they appear only numerically.

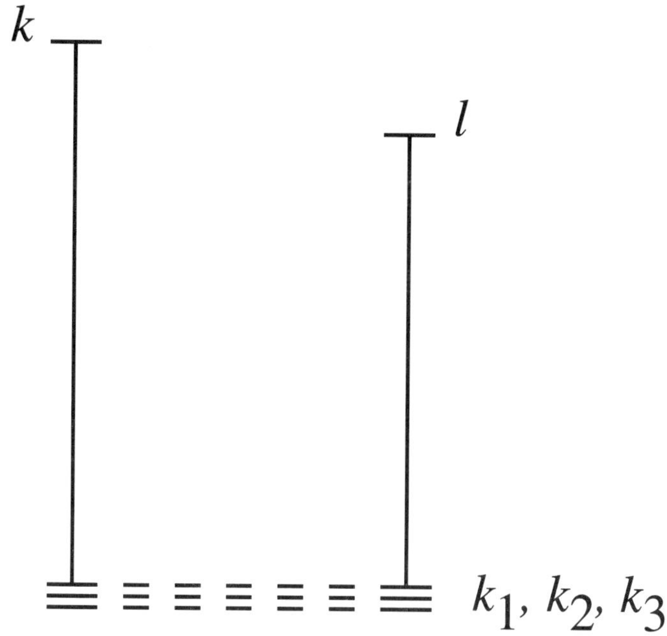


FIGURE 4. A wavelet diagram.

In the iteration of $\mathcal{R}_T$, this diagram appears again with each step that develops the previous occurrences. For clarity, fix the modes $k$ and $\ell$ and note that the corresponding quantity for the $n$th occurrence is initially

$$u^2\alpha_k\alpha_\ell \sum_{\substack{k_1,k_2,k' \\ 2^{-n}\le L_{k_\iota}\le 1/2, \\ L_{k'}=1/2}} \int \Psi_k\Psi_{k_1}\Psi_{k_2}\Psi_{k'} \int \Psi_\ell\Psi_{k_1}\Psi_{k_2}\Psi_{k'}$$

and develops into

$$u^2\alpha_k\alpha_\ell \sum_{\substack{k_1,k_2,k' \\ 2^{-N}\le L_{k_\iota}\le 2^{-N+n-1}, \\ L_{k'}=2^{-N+n-1}}} \int \Psi_k\Psi_{k_1}\Psi_{k_2}\Psi_{k'} \int \Psi_\ell\Psi_{k_1}\Psi_{k_2}\Psi_{k'}$$

after all $N$ iterations have occurred. In summary, the $N$ iterations generate the sum

$$u^2\alpha_k\alpha_\ell \sum_{\substack{k_1,k_2,k_3 \\ 2^{-N}\le L_{k_\iota}\le 1/2}} \int \Psi_k\Psi_{k_1}\Psi_{k_2}\Psi_{k_3} \int \Psi_\ell\Psi_{k_1}\Psi_{k_2}\Psi_{k_3},$$

so the modes that have no $\alpha$-amplitudes appearing are summed over all smaller (previous) scales independently. *The notion of fractional dimension in this wavelet picture can be defined in terms of a multi-scale restriction on the sum.*

Consider the momentum-space version of fractional dimension introduced in Section 4. A spherical shell of radius $v$ and thickness $\Delta v$ is assigned an integration volume roughly proportional to $v^{3-\varepsilon}\Delta v$—a factor $v^{-\varepsilon}$ times the true volume. This obviously becomes a small fraction of the volume for large momenta and a large multiple for small momenta. On the other hand, integrating over large momenta corresponds to summing over small-scale modes in the wavelet picture, so a wavelet analog is to restrict the sum over each $k_\iota$ in the wavelet diagram such that only a fraction $L^\varepsilon$ of all $L_{k_\iota} = L$ modes are summed over for each length scale $L$.

## 7. Wavelet $\varepsilon$-Expansion

We turn to the non-perturbative notion of fractional dimension—i.e., dimensional modification of the covariance. For the wavelet cutoff, the $d = 4$ covariance has the expansion

$$(7.1)\quad \langle \Phi(\vec{x})\Phi(\vec{y})\rangle_{d\mu_1} = \sum_{\rho=1}^{\infty}\sum_{\kappa,\vec{m}} (U_1^{-\rho+1}\Psi_\kappa(\cdot - 2\vec{m}))(\vec{x})(U_1^{-\rho+1}\Psi_\kappa(\cdot - 2\vec{m}))(\vec{y}),$$

so—following (4.13)—we may take the $d = 4-\varepsilon$ covariance to be

$$(7.2)\quad C(\vec{x},\vec{y}) = \sum_{\rho=1}^{\infty}\sum_{\kappa,\vec{m}} 2^{(\rho-1)\varepsilon}(U_1^{-\rho+1}\Psi_\kappa(\cdot - 2\vec{m}))(\vec{x})(U_1^{-\rho+1}\Psi_\kappa(\cdot - 2\vec{m}))(\vec{y}).$$

This is equivalent to making the replacement $\alpha_\ell \mapsto L_\ell^{\varepsilon/2}\alpha_\ell$ in all of the Gaussian expectations. In particular, the interaction is modified as follows:

$$(7.3)\quad \begin{aligned} I(\phi) &= r\int\Big(\sum_{L_k\geq 1}\alpha_k\Psi_k(\vec{x})\Big)^2 d\vec{x} + u\int\Big(\sum_{L_k\geq 1}\alpha_k\Psi_k(\vec{x})\Big)^4 d\vec{x} \\ &\mapsto r\int\Big(\sum_{L_k\geq 1}L_k^{\varepsilon/2}\alpha_k\Psi_k(\vec{x})\Big)^2 d\vec{x} + u\int\Big(\sum_{L_k\geq 1}L_k^{\varepsilon/2}\alpha_k\Psi_k(\vec{x})\Big)^4 d\vec{x}. \end{aligned}$$

However, if we impose a large-scale cutoff $L_k \leq 2^N$ on the interaction, the alternative suggested in the previous section is to make the replacement

$$I(\phi) \mapsto 2^{\varepsilon N} r\int\Big(\sum_{k\in W_{\varepsilon,N}}\alpha_k\Psi_k(\vec{x})\Big)^2 d\vec{x} + 2^{2\varepsilon N}u\int\Big(\sum_{k\in W_{\varepsilon,N}}\alpha_k\Psi_k(\vec{x})\Big)^4 d\vec{x}$$

with the definition of $W_{\varepsilon,N}$ guided by the rule to include only a fraction $(2^{-N}L)^\varepsilon$ of all $L_k = L$ variables $\alpha_k$ in the functional integration.

While motivated by the perturbation theory, this is certainly a non-perturbative modification. Consider the 16 blocks of scale-$2^{N-1}$ wavelets in the blocks of a scale-$2^N$ wavelet. Pick one of those blocks and throw away *all* wavelets—of scale $2^{N-1}$ or smaller—whose blocks lie in that block. Repeat the process for the 16 block of scale-$2^{N-2}$ wavelets lying in each remaining block of scale-$2^{N-1}$ wavelets. Iteration yields:

(1) $15/16$ of scale-$2^{N-1}$ wavelets remain,
(2) $(15/16)^2$ of scale-$2^{N-2}$ wavelets remain,
⋮
(n) $(15/16)^n$ of scale-$2^{N-n}$ wavelets remain.

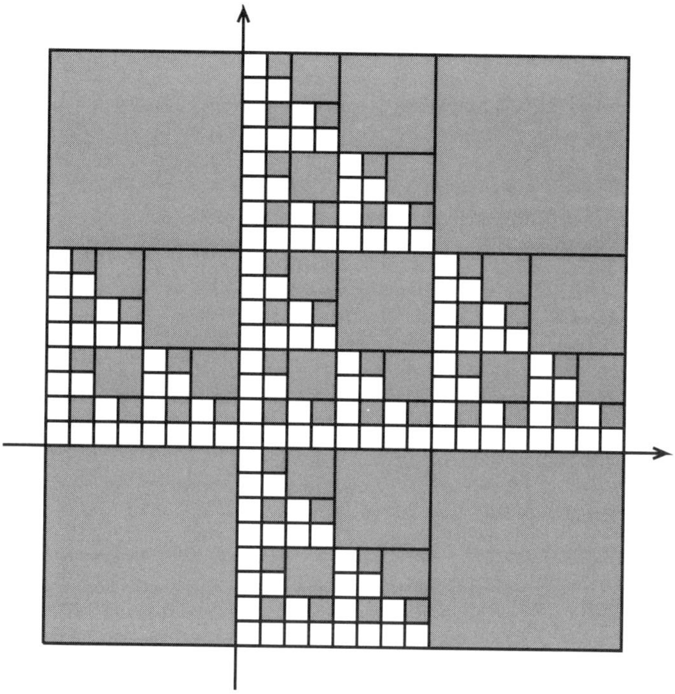

FIGURE 5. Blocks of omitted wavelets.

$W_{\varepsilon,N}$ is the total set of remaining wavelets. On the other hand, for $L = 2^{N-n}$,

$$\left(\frac{15}{16}\right)^n = 2^{(\log_2 15-4)n} = (2^{-n})^{4-\log_2 15} = (2^{-N}L)^{4-\log_2 15}, \tag{7.4}$$

so we have the desired restriction of the functional integration with $\varepsilon = 4 - \log_2 15$, or dimension $d = \log_2 15$. If we now consider the residual set of *points* in the continuum limit of this deletion of blocks, $d = \log_2 15$ is precisely the *Hausdorff dimension* of that set! Obviously, this scheme can be modified to yield a great range of small values for $\varepsilon$ (by allowing omissions to skip scales, etc.). In addition, this omission procedure extrapolates back to larger scales, so the choices are independent of $N$. We have the replacement

$$L_k^{\varepsilon/2}\alpha_k = 2^{N\varepsilon/2}(2^{-N}L_k)^{\varepsilon/2}\alpha_k \mapsto 2^{N\varepsilon/2}\alpha_k \tag{7.5}$$

with the omission of modes given explicitly. Naturally, this large power of $2^{\varepsilon/2}$ disappears after $N$ iterations of $\mathcal{R}_T$, since the averaging transformation is now given by:

$$(T\phi)(\vec{n}) = 2^{-3+\varepsilon/2}\sum_{\overrightarrow{m}} c_{\overrightarrow{m}}\phi(2\overrightarrow{m}+\vec{n}). \tag{7.6}$$

If we illustrate with *two*-dimensional configurations, where we pick $d = \log_2 3$, we have the picture given by Figure 5, where the shaded areas are the blocks of omitted wavelets.

## References


1. G. Baker, *Ising model with a scaling interaction*, Phys. Rev. B **5** (1972), 2622–2633.
2. G. Battle, *Wavelets: A renormalization group point of view*, Wavelets and Their Application (M. Ruskai *et al.*, eds.), Jones and Bartlett, Boston, MA, 1992, pp. 323–349.
3. ______, *Wavelet refinement of the Wilson recursion formula*, Recent Advances in Wavelet Analysis (L. Schumaker and G. Webb, eds.), Wavelet Anal. Appl., vol. 3, Academic Press, Boston, MA, 1994, pp. 87–118.
4. G. Battle and P. Federbush, *Ondelettes and phase cell cluster expansions, a vindication*, Comm. Math. Phys. **109** (1987), no. 3, 417–419.
5. P. Bleher and Y. Sinai, *Investigation of the critical point in models of the type of Dyson's hierarchical models*, Comm. Math. Phys. **33** (1973), 23–42.
6. P. Collet and J. Eckmann, *A renormalization group analysis of the hierarchical model in statistical mechanics*, Springer-Verlag, New York, 1978.
7. F. Dyson, *Existence of a phase-transition in a one-dimensional Ising ferromagnet*, Commun. Math. Phys. **12** (1969), no. 2, 91–107.
8. P. Federbush, *A mass zero cluster expansion.* II. *Convergence*, Comm. Math. Phys. **81** (1981), no. 4, 341–360.
9. G. Felder, *Construction of a nontrivial planar field theory with ultraviolet stable fixed point*, Comm. Math. Phys. **102** (1985), no. 1, 139–155.
10. G. Gallavotti, *The structure of renormalization theory*, Critical Phenomena, Random Systems, Gauge Theory (K. Osterwalder and R. Stora, eds.), North-Holland Publ., Amsterdam, 1986, pp. 467–492.
11. K. Gawedzki and A. Kupiainen, *A rigorous block spin approach to massless lattice theories*, Comm. Math. Phys. **77** (1980), 31–64.
12. ______, *Asymptotic freedom beyond perturbation theory*, Phénomènes Critiques, Systèmes Aléatoires, Théories de Jauge, Part I., II. (Les Houches, 1984) (K. Osterwalder and R. Stora, eds.), North-Holland Publ., Amsterdam, 1986, pp. 185–292.
13. ______, *Renormalization group study of a critical lattice model.* II. *The correlation functions*, Comm. Math. Phys. **83** (1982), no. 4, 469–492.
14. ______, *Non-Gaussian fixed points of the block spin transformation. Hierarchical model approximation*, Comm. Math. Phys. **89** (1983), no. 2, 191–220.
15. G. t'Hooft, *On the convergence of planar diagram expansions*, Comm. Math. Phys. **86** (1982), no. 4, 449–464.
16. G. Kogut and K. Wilson, *The renormalization group and the $\varepsilon$-expansion* , Phys. Rep. **12** (1974), 75–199.
17. H. Koch and P. Wittwer, *A non-Gaussian renormalization group fixed point for hierarchical scalar lattice field theories*, Comm. Math. Phys. **106** (1986), no. 3, 495–532.
18. R. Strichartz, *The way of analysis*, Bartlett and Jones, Boston, 1995.
19. K. Wilson, *Renormalization group and critical phenomena*, Phys. Rev. B **4** (1971), 3174–3205.



Mathematics Department, Texas A&M University, College Station, Texas 77843, USA

*E-mail address*: `battle@math.tamu.edu`

Centre de Recherches Mathématiques
CRM Proceedings and Lecture Notes
Volume **18**, 1999

# Modelling the Coupling Between Small and Large Scales in the Kuramoto-Sivashinsky Equation

J. Elezgaray, G. Berkooz, and P. Holmes

ABSTRACT. We show how wavelet decompositions can be used to approximate the coupling between small and large scales in the Kuramoto-Sivashinsky equation. Starting from the Galerkine equations obtained from the projection onto some orthogonal wavelet basis, we simplify the equations of motion of the small scales in such a way that the two and three point correlation statistics are preserved to a good approximation. The resulting small scale equations are those of the Kuramoto-Sivashinsky equation in a small domain.

## 1. Introduction

Renormalization group (RNG) procedures were first introduced in the context of critical phenomena [**22, 23**] and later extended to nonequilibrium systems such as turbulent flows [**8, 9, 25, 26**]. The common goal of any RNG procedure is the reduction of the number of relevant degrees of freedom. Namely, starting from some partial differential equation (for nonequilibrium processes) or hamiltonian (describing the interactions in an equilibrium system), any RNG method generates a family of 'renormalized' differential equations (or hamiltonians) by successive elimination of small scale modes. This set of renormalized equations is supposed to converge (in some sense) to a fixed point, which yields either a 'simpler' equation or the hamiltonian at the critical point. Invariably, all these procedures introduce higher-order nonlinearities into the coarse-grained equations to take into account the interactions with the small scale degrees of freedom eliminated at each step of the recursion. Not all the nonlinearities generated in this process behave in the same way. For instance, in the case of the Ising model in thermal equilibrium [**22, 23**], interactions involving four spins become negligible as the RNG iteration converges. Consequently, one can drop these terms, and only keep 'relevant' nonlinearities.

In the application of RNG to fluid turbulence, several approaches have been considered. In the work of Forster, Nelson and Stephen [**8**], Fournier and Frisch [**9**], and Yakhot and Orszag [**25, 26**], the RNG procedure relies strongly on the introduction of a white-noise, zero-mean forcing term into the Navier-Stokes momentum equation. The correlation function of this forcing term scales as $|k|^{\nu(\varepsilon)}$, where $\varepsilon \ll 1$ is a small parameter, and $k$ is the wavenumber. One then performs a perturbation

1991 *Mathematics Subject Classification.* Primary: 76M25; Secondary: 35Q35.
This is the final version of the paper.

expansion in powers of $\varepsilon$, and it can be shown [**25, 26**] that, in the limit $\varepsilon \to 0$, all the higher-order nonlinearities not present in the 'unrenormalized' Navier-Stokes equations are irrelevant interactions and can be neglected. In this limit, the effect of small scales upon large scales amounts simply to a 'renormalization' of the coefficients in front of the linear and nonlinear term. Unfortunately, the Kolmogorov inertial range corresponds to $\varepsilon = 4$, and it seems doubtful to still ignore the nonlinear interactions induced by the RNG procedure. Besides this, it is also puzzling to remark that the behaviour of small scales at the zeroth order of approximation is actually given by the random forcing term itself. The choice of the later is then crucial, and constitutes an essential and rather 'ad hoc' ingredient of this method.

The recursive-RNG method [**19, 28**], originally introduced by Rose [**19**] in the linear problem of turbulent advection of a passive scalar, uses a perturbative expansion in terms of the strength $\lambda$ of the nonlinear term $(u \cdot \nabla u)$ of the Navier-Stokes equation, which is eventually set to unity. A priori, no forcing term is needed to start the perturbative series, but in the case of free-decaying turbulence [**29**], additional hypothesis are required concerning the behaviour of the small scale velocity statistics.

In a very schematic way, the RNG method uses a spectral decomposition of the unknown $u(x,t)$, such as

$$u(x,t) = \sum_{\alpha} a_\alpha(t)\Psi_\alpha(x),$$

where $\{\Psi_\alpha(x)\}$ is some orthonormal basis, such as Fourier basis. If the evolution equation (e.g. Navier-Stokes equation) is written as

$$\dot{a}_\alpha = l_{\alpha\beta} a_\beta + n_{\alpha\beta\gamma} a_\beta a_\gamma, \tag{1.1}$$

where summation over repeated indices is assumed, and $a_\alpha$ variables are split into two sets $\{a_\alpha^<\}$ and $\{a_\alpha^>\}$ corresponding respectively to large and small scales, then the large scale equation reads:

$$\begin{aligned}\dot{a}_\alpha^< = l_{\alpha\beta}^{\ll} a_\beta^< + l_{\alpha\beta}^{<>} a_\beta^> + n_{\alpha\beta\gamma}^{\lll} a_\beta^< a_\gamma^< + n_{\alpha\beta\gamma}^{<><} a_\beta^> a_\gamma^< \\ + n_{\alpha\beta\gamma}^{\ll>} a_\beta^< a_\gamma^> + n_{\alpha\beta\gamma}^{<\gg} a_\beta^> a_\gamma^> .\end{aligned} \tag{1.2}$$

All the RNG methods perform some kind of short-time average (long compared to the time scale of the small scale variables, but short compared to the time scale of the large scales). This implicitly assumes a gap between the time scale of the long and short wavelengths. Using this assumption, the large scales equation is replaced by

$$\begin{aligned}\dot{a}_\alpha^< = l_{\alpha\beta}^{\ll} a_\beta^< + l_{\alpha\beta}^{<>} \langle a_\beta^> \rangle (a^<) + n_{\alpha\beta\gamma}^{\lll} a_\beta^< a_\gamma^< + n_{\alpha\beta\gamma}^{<><} \langle a_\beta^> \rangle (a^<) a_\gamma^< \\ + n_{\alpha\beta\gamma}^{\ll>} a_\beta^< \langle a_\gamma^> \rangle (a^<) + n_{\alpha\beta\gamma}^{<\gg} \langle a_\beta^> a_\gamma^> \rangle (a^<).\end{aligned} \tag{1.3}$$

The expressions $\langle a_\beta^> \rangle (a^<)$ and $\langle a_\beta^> a_\gamma^> \rangle (a^<)$ denote this short time average of $a_\beta$ and $a_\beta a_\gamma$ when the $a^<$ variables are considered as 'frozen'. The major difficulty in the RNG method is how to compute these time averages. As we mentioned, the addition of some random forcing [**8, 9, 25, 26**], supposed to be predominant at small scales, allows to analytically compute $\langle a_\beta^> \rangle (a^<)$ and $\langle a_\beta^> a_\gamma^> \langle (a^<)$ as a function of the $a^<$'s. Typically, this will yield linear and quadratic terms in the $a^<$'s, which will renormalize the coefficients of the corresponding terms in Eq. (1.3). Eventually, this procedure can be iterated, by selecting another set of small scales among the former large scales.

The purpose of this paper is to show how wavelet decompositions [**17**] can be used to obtain coarse-grained equations such as Eq. (1.3) in fluids dominated by coherent structures without using strong assumptions about the behaviour of the small scales. We have been initially motivated by the approach advocated in Ref. [**1**] to study the local dynamics of the coherent structures in the wall region of boundary layer turbulence. In this open flow, large coherent structures contain a major fraction of the energy. Furthermore, the violent bursting events in which they suddenly break-up also correspond to peaks in the turbulent energy production. In [**1**], N. Aubry *et al.* decomposed the velocity field into the Karhunen-Loeve basis, then used a Galerkin projection as well as eddy viscosity models to obtain a fairly low dimensional dynamical system modeling the interaction of a small number of coherent structures. One of the major difficulties raised by this procedure is due to the fact that the Karhunen-Loeve decomposition is identical to the Fourier decomposition in the directions in which the domain is translationally invariant (if any). In order to obtain a truly local model, an appropriate length scale has to be chosen containing enough structures to afford a reasonable interaction. Then, it is postulated that the influence of missing neighbouring structures at the boundary of the modelled subdomain can be supplied by periodizing the subdomain in the homogeneous direction.

In this paper, we show that the use of wavelet projections justifies the above procedure in the particular case of the Kuramoto-Sivashinsky equation [**12, 14, 20**]. More precisely, we show using numerical simulations how wavelet decompositions yield naturally a set of effective equations for the large scales of the system which preserve the statistics of the original equations, and in particular the energy transfer between small and large scales. Our main point is that wavelet decompositions are a well adapted tool to model the behavior of the small scales in terms of dynamics of the original equation in a small domain. Thus, no ad-hoc modelling is needed to parametrize the net effect of small scales upon large scales. We want to stress that the present approach (as well as that of the RNG methods) is not concerned about the approximation of particular solutions but rather aims to understand their statistical properties. This is to be compared with the approximation methods based on the notion of inertial manifold, that actually preserve the asymptotic behavior of (almost) any particular solution.

In Section 2 we present our approximation procedure. In Section 3 we discuss the numerical implementation and limitations of our method.

## 2. Wavelet Projections of the KS Equation

We will consider in this work the so called Kuramoto-Sivashinsky (KS) equation [**12, 14, 20**]:

$$\partial_t u + L^{-2} u_{xx} + L^{-4} u_{xxxx} + L^{-1} u u_x = 0, \tag{2.1}$$

where $u(x,t)$ is the value at time $t$ and position $x$ of the solution $u$, $L$ is a positive parameter hereafter called the 'length of the domain', and $x \in [0,1]$ is the spatial variable, which will be considered to be periodic. The formulation (2.1) is simply obtained using the rescaling $x' = Lx$, where $x' \in [0, L]$ describes a periodic domain of length $L$.

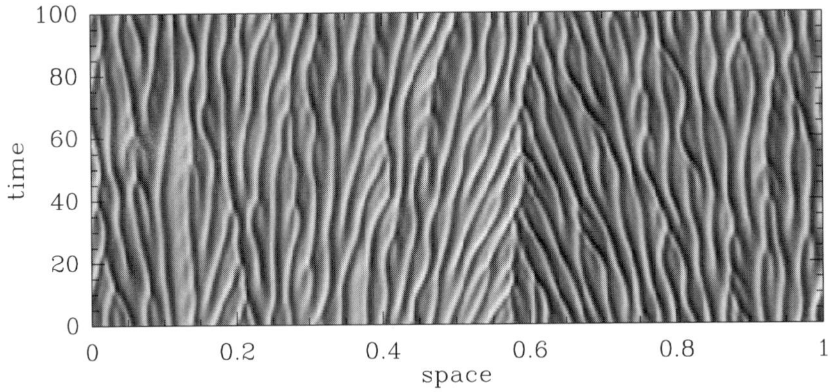


FIGURE 1. Space-time representation of a typical solution of Eq. (2.1) for the parameter $L = 400$. The maxima are represented in black, the minima in white.

The KS equation has of course no direct relation with the Navier-Stokes equation although it has been shown [**4**] that, for the large values of $L$, the solutions of the KS equation display local dynamical events which are reminiscent of heteroclinic cycles, as those observed in the modelling of the bursting phenomenon in the near wall region of a boundary layer [**1**]. This equation has also been thoroughly used in the literature as a test equation in order to validate several numerical methods, particularly those related to inertial manifolds and early versions of the RNG method [**10, 24**]. Recently, a lot of effort [**3, 11, 13, 21, 27**] has been devoted to the description of coarse grained equations describing the large scales of this equation. Moreover, the KS equation appears to be rather ubiquitous in the description of front fluctuations [**16**].

The Fourier analysis of the linear term of Eq. (2.1) shows that the second derivative $u_{xx}$ is a destabilizing term and acts as an energy source term while $u_{xxxx}$ is a stabilizing (dissipative) term whose effect increases at large wavenumbers. The competition between these two terms results in a characteristic wavelength $2\pi/\sqrt{2}L$, which corresponds to the most unstable wavenumber. The nonlinear term makes the transfert of energy between large and small scales, causing the continual appearance of peaks and troughs in $u(x,t)$. These coalescence and creation events are the coherent structures of our model problem. Notice also that this nonlinearity preserves the mean of the solution, which is then customarily set to zero. When the KS equation is integrated on a sufficiently long domain a characteristic pattern emerges, which we illustrate in Figure 1 for the specific case $L = 400$.

We will decompose the solutions $u(x,t)$ of Eq. (2.1) into some orthogonal basis $\{\Psi_{j,k}(x)\}$ of periodic wavelets [**17**], where the scale index $j = 0, 1, \ldots$ takes any non negative integer value, the position index $k = 0, 1, \ldots, 2^j - 1$ for each value of $j$, and there exists some function $\Psi(x)$ defined in the real line such that

$$\Psi_{j,k}(x) = 2^{j/2} \sum_{n=-\infty}^{\infty} \Psi\big(2^j(x+n) - k\big).$$

Typically, $\Psi(x)$ is a rapidly (exponentially) decreasing function which can effectively be considered as compactly supported for numerical purposes. In our numerical simulations, we will use a fifth order spline [**15, 18**] for $\Psi(x)$, although this specific detail will not enter the reasoning below. Using the rapid decay of $\Psi(x)$, one can easily obtain the estimate

$$\Psi_{j,k}(x) = 2^{j/2}\Psi(2^j x - k) + O\big(\exp(-2^j/S)\big), \tag{2.2}$$

provided $\Psi(x) \sim \exp\big(-|x|/S\big)$ as $x \to \infty$. Then, one recovers the usual picture of wavelet functions as being (approximately) rescaled and translated versions of the $\Psi(x)$ function. The greater the value of $j$, the better this approximation holds.

As usual, we represent the solution of the KS equation in this basis via

$$u(x,t) = \sum_\alpha a_\alpha(t)\Psi_\alpha(x),$$

where $\alpha$ denotes the multi-index $(j,k)$. We will also introduce an arbitrary scale $j_0$ such that the set $\{a_\alpha^< = a_{j,k}, j \le j_0\}$ represents the large scale component of the solution and $\{a_\alpha^> = a_{j,k}, j > j_0\}$ represents the small scale component. Specific values of $j_0$ will be fixed later. Let us write Eq. (2.1) in terms of the $a_\alpha$ variables:

$$\dot a_\alpha = l_{\alpha\beta}a_\beta + n_{\alpha\beta\gamma}a_\beta a_\gamma, \tag{2.3}$$

where

$$l_{\alpha\beta} = \int_0^1 \Psi_\alpha(x)(-L^{-2}\partial_{xx} - L^{-4}\partial_{xxxx})\Psi_\beta(x)\,dx$$

and

$$n_{\alpha\beta\gamma} = -L^{-1}\int_0^1 \Psi_\alpha(x)\Psi_\beta(x)\partial_x\Psi_\gamma(x)\,dx$$

are respectively the linear and nonlinear coupling coefficients, and summation over repeated indices is implied.

Our method to model small scales relies on a modified version of the equations of evolution of these scales:

$$\dot a_\alpha^> = \big[l_{\alpha\beta}^{\gg} + \delta l_{\alpha\beta}^{\gg}(t)\big]a_\beta^> + n_{\alpha\beta\gamma}^{\ggg>}a_\beta^> a_\gamma^> + f_\alpha^>(t), \tag{2.4}$$

where

$$\delta l_{\alpha\beta}^{\gg}(t) = n_{\alpha\beta\gamma}^{\gg<}a_\gamma^< + n_{\alpha\gamma\beta}^{><>}a_\gamma^< \tag{2.5}$$

is a perturbation of the linear term, and

$$f_\alpha^>(t) = n_{\alpha\beta\gamma}^{>\ll}a_\beta^< a_\gamma^< + l_{\alpha\beta}^{><}a_\beta^<$$

is an additive forcing term.

We will assume momentarily that the forcing terms $\delta l$ and $\delta f$ are 'small'. The key observation is that the equations governing the statistical properties of the $a^>$ variables, provided $\delta l$ and $\delta f$ are neglected, are those of the KS equation with parameter $\tilde L = L2^{-(j_0+1)}$, up to rescaling and small corrections.

More precisely, we define the sets of indices (or 'boxes') $B_k$ by

$$B_k = \big\{(j,k'), j \ge j_0+1, k2^{j-j_0-1} \le k' \le (k+1)2^{j-j_0} - 1\big\}, \quad k = 0,1,\ldots,2^{j_0+1}-1. \tag{2.6}$$

Provided $\delta l$ and $\delta f$ are neglected, we can decouple the $a^>$ equations into $2^{j_0+1}$ independent sets of equations, each one including only variables $a^>$ belonging to the same $B_k$ box, in a manner that preserves the statistics but not necessarily the dynamics.

The reason for this is easily seen from inspection of the moment equations associated with the unforced equation (2.4) ($f_\alpha^> = \delta_{\alpha\beta}^{\gg} = 0$). For example, consider the equations of motion for the second moment:

$$\text{(2.7)}\quad \frac{d}{d\tau}\langle a_\alpha^>(t)a_\beta^>(t+\tau)\rangle = \langle a_\alpha^>(t)l_{\beta\gamma}^{\gg}a_\gamma^>(t+\tau)\rangle + \langle a_\alpha^>(t)n_{\beta\gamma\delta}^{\ggg}a_\gamma^>(t+\tau)a_\delta^>(t+\tau)\rangle,$$

where $\langle\cdot\rangle$ is the time average. Let us denote by $|\alpha-\beta|$ the distance $|k2^{-j} - k'2^{-j'}|$ (modulo 1), where $\alpha = (j,k)$ and $\beta = (j',k')$. Using the estimate $\Psi(x) \sim \exp(-|x|/S)$ as $x \to \infty$, it is straightforward to obtain the following inequalities:

$$|l_{\alpha\beta}| \le C_1 2^{|j-j'|/2} d(|\alpha-\beta|)$$

and

$$|n_{\alpha\beta\gamma}| \le C_2 2^{|j-j'|/2} d(|\alpha-\beta|) 2^{|j-j''|/2} d(|\alpha-\gamma|),$$

where $C_1$ and $C_2$ are two constants, and $d(x) = \exp(x/S) + \exp((1-x)/S)$. We conclude that, for an appropriate value of $j_0$ and for $|\alpha-\beta| \ge 2^{-j_0}$, $l_{\alpha\beta}$ can be neglected, and that we may make analogous approximations for $n_{\alpha\beta\gamma}$. Using the fact that KS is translationally invariant, we can replace each of the moments in Eq. (2.7) by the analogous quantity for which all indices belong to the same $B_k$ box. For instance,

$$\text{(2.8)}\quad \langle a_\alpha(t)a_\beta(t+\tau)\rangle = \langle a_{\alpha-\Delta}(t)a_{\beta-\Delta}(t+\tau)\rangle,$$

where $(\alpha-\Delta), (\beta-\Delta) \in B_k$ and $\Delta$ is the translation parameter. Finally, using the approximate relation (2.2), it is easy to obtain the scaling relations between the coupling coefficients $l_{\alpha\beta}$ and $n_{\alpha\beta\gamma}$ for Eq. (1) with parameter $L$ and $\tilde{L} = L2^{-(j_0+1)/2}$:

$$\text{(2.9)}\quad \begin{aligned} l_{\alpha\beta}(L) &= l_{\alpha_1\beta_1}(\tilde{L})\big(1 + O(\exp^{-2^{j_{\min}}/S})\big) \\ n_{\alpha\beta\gamma}(L) &= 2^{-(j_0+1)/2} n_{\alpha_1\beta_1\gamma_1}(\tilde{L})\big(1 + O(\exp^{-2^{j_{\min}}/S})\big), \end{aligned}$$

where $\alpha = (j,k)$, $\alpha_1 = (j_1,k_1)$, $j_1 = j-(j_0+1)$, $k_1 = 2^{-(j_0+1)}k$, etc. ..., and $j_{\min} = \min(j,j',j'')$. Using these relations, one can easily see that the rescaled variables $\tilde{a}_{\alpha_1} = 2^{(j_0+1)/2}a_\alpha$ approximately satisfy the moment equations (2.7) arising from equation (1) for length parameter $\tilde{L}$. Assembling all these observations, one gets the relation claimed above, between the equation governing the statistical properties of solutions of (1) with parameter $\tilde{L}$ and those of (2.4) with no forcing.

The above considerations suggest to consider the following set of equations for the evolution of small scales:

$$\text{(2.10)}\quad \text{if } a_\alpha^> \in B_k, \dot{a}_\alpha^> = \sum_{a_\beta^> \in B_k} l_{\alpha\beta}^{\gg}(\tilde{L})a_\beta^> + 2^{-(j_0+1)/2} \sum_{a_\beta^>, a_\gamma^> \in B_k} n_{\alpha\beta\gamma}^{\ggg}(\tilde{L})a_\beta^> a_\gamma^> + \sum_{a_\beta^> \in B_k} \delta l_{\alpha\beta}^{\gg} a_\beta^> + f_\alpha^>,$$

where

$$l_{\alpha\beta}(\tilde{L}) = \int_0^1 \Psi_\alpha(x)(-\tilde{L}^{-2}\partial_{xx} - \tilde{L}^{-4}\partial_{xxxx})\Psi_\beta(x)\,dx$$

and

$$n_{\alpha\beta\gamma}(\tilde{L}) = -\tilde{L}^{-1} \int_0^1 \Psi_\alpha(x)\Psi_\beta(x)\partial_x\Psi_\gamma(x)\,dx$$

are respectively the linear and nonlinear coupling coefficients computed with parameter $\tilde{L}$. It is worth emphasizing the fact that, in Eq. (2.10), small scale variables in two different boxes $B_k$ and $B_{k'}$ are coupled only through the coupling with large scales. In other words, we drop direct couplings between different boxes, of the type $n^{\ggg}_{\alpha\beta\gamma}a^{>}_{\beta}a^{>}_{\gamma}$ where $a^{>}_{\beta} \in B_k$ and $a^{>}_{\gamma} \in B_{k'}$, while preserving all the couplings with large scales included in the forcing terms $\delta l_{\alpha\beta}$ and $f_\alpha$.

Therefore, the system of equations we will consider in the following is given by the set of equations (2.10) supplemented with the original large scale equations

(2.11)

$$\dot{a}^{<}_{\alpha} = l^{\ll}_{\alpha\beta}a^{<}_{\beta} + l^{<>}_{\alpha\beta}a^{>}_{\beta} + n^{\lll}_{\alpha\beta\gamma}a^{<}_{\beta}a^{<}_{\gamma} + n^{<><}_{\alpha\beta\gamma}a^{>}_{\beta}a^{<}_{\gamma} + n^{\ll>}_{\alpha\beta\gamma}a^{<}_{\beta}a^{>}_{\gamma} + n^{<\gg}_{\alpha\beta\gamma}a^{>}_{\beta}a^{>}_{\gamma}.$$

## 3. Numerical Results

In this section, we investigate by means of numerical simulations how the approximations involved in the first two terms of the equation (2.10) survive when the forcing terms $\delta l$ and $\delta f$ coming from the large scales are included. Our intuition is that these approximations can hold provided the amplitude of the forcing is small and its evolution well separated in time from the small scales. The amplitude of the forcing is essentially determined by the value of the cutoff scale $j_0$. It is well known that the energy of a typical solution of the KS equation is mainly concentrated around the wavelength $L/2\pi\sqrt{2}$, corresponding to the most unstable Fourier mode. In wavelet terms, this corresponds to the value of $j_M$ such that

$$2^{-(j_M+1)} < L/2\pi\sqrt{2} < 2^{-j_M}.$$

Thus, a necessary condition for our model equations (2.10) to behave satisfactorily is the inequality $j_0 < j_M$. For $L = 400$, $j_M = 6$.

On the other hand, the approximations stated in Eq. (2.9) do not hold for the 'top' coefficients (for which $j \sim j_0 + 1$) of each box $B_k$, where the approximate relation (2.2) is clearly wrong. Thus, we will introduce the parameter $\delta j$ in such a way that the equations (2.10) only hold for those coefficients such that $j \geq j_0+1+\delta j$. The wavelet coefficients for which $0 \geq j \geq j_0 + \delta j$ will then be follow the exact evolution equations (2.11).

In order to numerically integrate equations (2.11) and (2.10), we first perform a truncation in scale, and consider only values of $j \leq J_{\max}$, the value of $J_{\max}$ being proportional to the parameter $L$. We mostly considered the value $L = 400$, for which $J_{\max} = 8$ was used (this corresponds roughly speaking to 256 complex Fourier modes).

From a purely computational point of view, integrating equations (2.11) and (2.10) is a rather formidable task, mainly due to the large number of nonlinear couplings involved. In the simulations reported in this paper, we used a pseudo-spectral method to evaluate these terms. The r.h.s. of the large scales equations is easily evaluated by performing an inverse fast wavelet transform on the function $\partial_t u(x,t)$, which is computed in physical space using a collocation method. Evaluating the

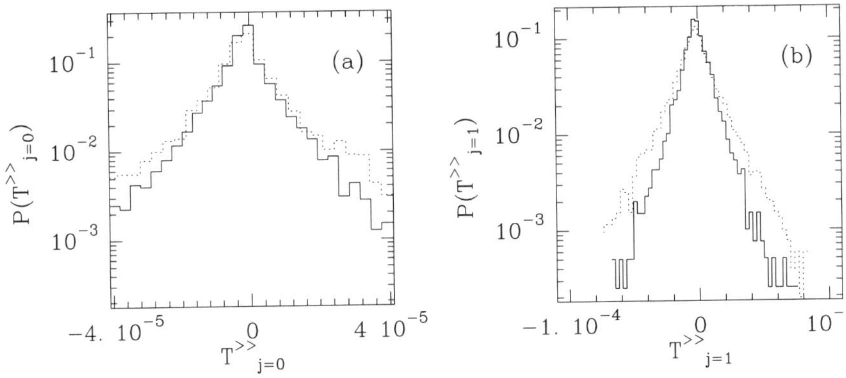

FIGURE 2. a) Probability distribution function of the transfer $T_\alpha^{\gg}$ term for $j = 0$, corresponding to the integration of the full KS equation (solid line) and to the model equations (2.11) and (2.10) (short-dashed line), for $j_0 = 2$. b) Same as in a) for the $j = 1$ transfer $T_\alpha^{\gg}$.

r.h.s. of Eq. (2.10) requires evaluating the expression $\partial_t \tilde{u}(x,t)$, where

$$\tilde{u}(x,t) = \sum a_\alpha^<(t)\Psi_\alpha(x) + \sum_{B_k} \sum_{a_{\tilde{\alpha}}^> \in B_k} 2^{(j_0+1)/2} a_{\tilde{\alpha}}^>(t)\Psi_{\tilde{\alpha}}(2^{-(j_0+1)}x). \tag{3.1}$$

We used the notation $\tilde{\alpha} = (j - j_0 - 1, k' - k2^{j-j_0-1})$ if $\alpha = (j, k')$. Then, it can be easily checked that the r.h.s. of Eq. (2.10) is obtained by wavelet transforming $\partial_t \tilde{u}(x,t)$.

In order to compare the statistics of the solutions of the exact with those of our model equations (2.11) and (2.10), we considered the probability distribution function (p.d.f.) of the two $\big(a_\alpha(t)a_\beta(t-\tau)\big)$ and three $\big(a_\alpha(t)a_\beta(t)a_\gamma(t-\tau)\big)$ point products. These enter for instance in the energy budget equation ($\tau = 0$)

$$\begin{aligned} \frac{da_\alpha^{<2}}{dt} &= l_{\alpha\beta}^{\ll} a_\alpha^< a_\beta^< + n_{\alpha\beta\gamma}^{\lll} a_\alpha^< a_\beta^< a_\gamma^< + \delta l_{\alpha\beta}^{\ll} a_\alpha^< a_\beta^< + f_\alpha^< a_\alpha^< \\ &\equiv l_{\alpha\beta}^{\ll} a_\alpha^< a_\beta^< + T_\alpha^{\ll} + T_\alpha^{<>} + T_\alpha^{\gg}, \end{aligned}$$

and have been studied in the RNG approach to the Navier-Stokes equations [**5, 28**] (we refer the reader to Ref. [**6**] for a comparison of the transfer statistics with $\tau \neq 0$). The time average of the transfer $T_\alpha^{\gg}$ is shown [**5, 28**] to correspond to a negative linear correction to the $l_{\alpha\alpha}\langle a_\alpha^2\rangle$ term. In Figure 2, we represent the p.d.f. of the transfer terms $T_{j=0,k=0}^{\gg}$ and $T_{j=1,k=0}^{\gg}$, as computed from the exact KS equation (continuous line), and using our model equations with parameters $j_0 = 2$ and $\delta j = 1$ (as expected, the simulation with $\delta j = 0$ gave unsatisfactory results). These two distributions agree fairly well, although our model seems to slightly enhance the transfer to and from small scales, as evidenced by the differences in the tails of the distributions. As can be seen in Figure 2, the average $\langle T_\alpha^{\gg}\rangle$ is actually negative, although very frequently the "instantaneous" transfer is actually positive. This is the mathematical translation of the common view that turbulent motion is equivalent, at large scales, to some sort of viscosity.

The agreement, however, becomes very poor for the transfer $T_{j=2,k=0}^{\gg}$ (not represented in Figure 2), that corresponds to the value of $j$ exactly at the cut-off between small and large scales. We don't know at the moment how to correct this. Let us note however that simulations at larger values of $L$ show that the disagreement is clearly confined to scales at or near the cut-off.

One can also play with the parameter $\delta$. Increasing the value of the later accordingly increases the number of small scale wavelet coefficients evolving with the exact KS equation. Unexpectedly, our simulations show that the agreement between the transfer statistics does not improve until rather large values of $\delta$ ($\delta = 4$ for $j_0 = 2$) are reached. This clearly amounts to include in the set of large scales *all* the energy containing modes. In this limit however, our model becomes rather trivial.

## 4. Conclusion

In this paper, we have used orthogonal wavelet projections to study models for the transfer of energy between small and large scales in the 1D periodic Kuramoto-Sivashinsky equation. We first showed that the structure of the small scale equations is such that, when neglecting couplings with the large scales, they can be approximated in a natural way by a set of $2^{j_0+1}$ replicas of the wavelet projection of Eq. (1) for length parameter $\tilde{L} = 2^{-(j_0+1)}L$. This approximation is intended only to preserve the statistics of the solutions, and not the "instantaneous" dynamics.

By means of numerical simulations, we then checked that coupling this set of $2^{j_0+1}$ independent subsystems with the large scales computed from the full simulation of Eq. (1) does actually reasonably reproduce the energy transfer between large and small scales, provided the scales involved in the transfer are well separated from the cut-off scale $j_0$.

The statistical picture suggested by our study is that of an assembly of identical short length subsystems, slowly driven via interactions with the large scales. A more detailed study [**4**] of the dynamics of the KS equation for values of $L \sim 400$ actually suggests the need for at least two kind of solutions in these subsystems, corresponding respectively to traveling waves and homoclinic cycles, in order to obtain reasonable tracking of the solutions of the full equation. Spatio-temporal representations of the solutions of the $L = 50$ KS equation strongly suggest that the latter does indeed exhibit these two types of dynamics.

## References

1. N. Aubry, P. Holmes, J. L. Lumley, and E. Stone, *The dynamics of coherent structures in the wall region of a turbulent boundary layer*, J. Fluid Mech. **192** (1988), 115–173.
2. M. Cannone, *Ondelettes, paraproduits et Navier-Stokes*, Diderot, Paris, 1995.
3. C. C. Chow and T. Hwa, *Defect-mediated stability: An effective hydrodynamic theory of spatiotemporal chaos*, Phys. D **84** (1995), 494–512.
4. H. Dankowicz, P. Holmes, J. Elezgaray, and G. Berkooz, *Local models of spatio-temporally complex fields*, Phys. D **90** (1996), 387–407.
5. J. A. Domaradzki, R. W. Metcalfe, R. S. Rogallo, and J. Riley, *Analysis of subgrid-scale eddy viscosity with use of results from direct numerical simulations*, Phys. Rev. Lett. **58** (1987), 547–550.
6. J. Elezgaray, G. Berkooz, and P. Holmes, *Large-scale statistics of the Kuramoto-Sivashinsky equation: A wavelet-based approach*, Phys. Rev. E **54** (1996), 224–230.
7. P. Federbush, *Navier and Stokes meet the wavelet*, Comm. Math. Phys. **155**, (1993), 219–248.
8. D. Forster, D. Nelson, and M. Stephen, *Large-distance and long-time properties of a randomly stirred fluid*, Phys. Rev. A **16** (1977), 732–749.
9. J. D. Fournier and U. Frisch, *Remarks on the renormalization group in statistical fluid dynamics*, Phys. Rev. A **28** (1982), 1000–1009.

10. H. Fujisaka and T. Yamada, *Theoretical study of a chemical turbulence*, Progr. Theoret. Phys. **57** (1977), 734–745 (related work).
11. F. Hayot, C. Jayaprakash, and Ch. Josserand, *Long-wavelength properties of the Kuramoto-Sivashinsky equation*, Phys. Rev. E **47** (1993), 911–915.
12. J. M. Hyman, B. Nicolaenko, and S. Zaleski, *Order and complexity in the Kuramoto-Sivashinsky model of weakly turbulent interfaces*, Phys. D **23** (1986), 265–292.
13. C. Jayaprakash, F. Hayot, and R. Pandit, *Universal properties of the two-dimensional Kuramoto-Sivashinsky equation*, Phys. Rev. Lett. **71** (1993), 12–15.
14. Y. Kuramoto, *Diffusion-induced chaos in reaction systems*, Progr. Theoret. Phys. **64** (1978), 346–367.
15. P. G. Lemarié, *Une nouvelle base d'ondelettes de $L^2(R^n)$*, J. Math. Pures Appl. (9) **67** (1988), 227–236.
16. P. Maneville, *Dissipative structures and weak turbulence*, Academic, Boston, 1990.
17. Y. Meyer, *Ondelettes et opérateurs*, Hermann, Paris, 1990.
18. V. Perrier and C. Basdevant, *Periodical wavelet analysis, a tool for inhomogeneous field investigations. Theory and algorithms*, Rech. Aérospat. **3** (1989), 53–67.
19. H. A. Rose, *Eddy diffusivity, eddy noise and subgrid-scale modelling*, J. Fluid Mech. **81** (1977), 719–734.
20. G. I. Sivashinsky, *Nonlinear analysis of hydrodynamic instability in laminar flames. I. Derivation of the basic equations*, Acta Astronautica **4** (1977), 1176–1206.
21. K. Sneppen, J. Krug, M. H. Jensen, C. Jayaprakash, and T. Bohr, *Dynamic scaling and crossover analysis for the Kuramoto-Sivashinsky equation*, Phys. Rev. A **46** (1992), 7351–7354.
22. K. G. Wilson, *The renormalization group: Critical phenomena and the Kondo problem*, Rev. Modern Phys. **47** (1975), 773–840.
23. K. G. Wilson and J. Kogut, *The renormalization group and the $\varepsilon$ expansion*, Phys. Rep. **12** (1974), 75–199.
24. V. Yakhot, *Large-scale properties of unstable systems governed by the Kuramoto-Sivashinsky equation*, Phys. Rev. A **24** (1981), 642–644.
25. V. Yakhot and S. A. Orszag, *Renormalization-group analysis of turbulence*, Proc. Internat. Congress Mathematicians (Berkeley, 1986), vols. 1, 2, Amer. Math. Soc., Providence, RI, 1987, pp. 1395–1399.
26. ———, *Renormalization group analysis of turbulence. I. Basic theory*, J. Sci. Comput. **1** (1986), 3–51.
27. S. Zaleski, *A stochastic model for the large scale dynamics of some fluctuating interfaces*, Phys. D **34** (1989), 427–449.
28. Y. Zhou and G. Vahala, *Reformulation of recursive-renormalization-group-based subgrid modeling of turbulence*, Phys. Rev. E **47** (1993), 2503–2519.
29. Y. Zhou, G. Vahala, and M. Hossain, *Renormalization-group theory for the eddy viscosity in subgrid modeling*, Phys. Rev. A **37** (1988), 2590–2598.

Centre de Recherche Paul Pascal, Av. Schweitzer, 33600 Pessac, France
*E-mail address*: elez@crpp.u-bordeaux.fr

BEAM Technologies Inc. and Sibley School of Mechanical and Aerospace Engineering, Cornell University, Ithaca, NY 14853, USA
*E-mail address*: gal@beamtech.com

Program in Applied and Computational Mathematics and Department of Mechanical and Aerospace Engineering, Princeton University, NJ 08544, USA
*E-mail address*: pholmes@rimbaud.Princeton.edu

Centre de Recherches Mathématiques
CRM Proceedings and Lecture Notes
Volume **18**, 1999

# Continuous Wavelet Transform Analysis of One Dimensional Quantum Ground States

C. R. Handy and R. Murenzi

ABSTRACT. The use of dilated and translated (generalized) moments, $\mu_{b,\gamma}(p)$, $\gamma \equiv 1/(2a^2)$, leads to a theoretically exact formalism by which to calculate the continuous wavelet transform, $\omega_t(b,a)$, of Sturm-Liouville Schrödinger problems in one space dimension, from first principles. Wavelet and non-wavelet based reconstruction methods can then be used to recover the desired solution. Utilizing the Eigenvalue Moment Method of Handy and Bessis, we are able to solve the infinite scale ($a = \infty$) problem to arbitrary precision, and then use standard integration methods to calculate $\mu_{b,\gamma}(p)$, for arbitrary $b, \gamma$ values. Since many important wavelet transforms are linear superpositions of $\mu_{b,\gamma}(p)$, the $\omega_t(b,a)$ transform is trivially obtained. We outline the general formalism and apply it to two problems, the quartic anharmonic oscillator, and the rational fraction potential $V(x) = gx^6/1 + \lambda x^2$.

## Overview

Over the last ten years, wavelet transform analysis has become a powerful tool in the analysis and synthesis of signals and images [**2**]. Its main contribution is the definition of a systematic process for identifying, extracting, and reconstructing the multiscale features of a signal through simultaneous time and frequency localization. Despite these successes, recent attempts at applying such methods to the study of linear differential and partial differential equations (ODEs and PDEs), such as the Schrödinger equation with a rational fraction potential, have not proven too successful partly because many of these equations do not admit a tractable wavelet transform representation [**19, 22, 23**]. Nevertheless, we can circumvent these difficulties by first examining a suitable *moment equation* representation for the given ODE or PDE. We limit the analysis only to the ground state, for simplicity. The extension to excited states is given elsewhere [**16**]. Our formalism applies to a broad class of wavelets; however, we shall limit the ensuing discussion to the *Mexican hat* case only, in order to facilitate the presentation.

Given a Sturm-Liouville Schrödinger problem, $-\partial_x^2\Psi(x) + V(x)\Psi(x) = E\Psi(x)$, with rational fraction potential, $V(x)$, its moment equation representation corresponds to a linear finite difference equation of (effective) order $1 + m_s$ involving the

1991 *Mathematics Subject Classification.* Primary: 34L10; Secondary: 44A60.
This is the final version of the paper.

*dilated* and *translated* moments

$$\mu_{b,\gamma}(p) \equiv \int_{-\infty}^{+\infty} x^p \Psi(b+x)\exp(-\gamma x^2)\,dx, \quad p \geq 0, \tag{0.1}$$

for the *Mexican hat* wavelet case studied in this work. Other wavelet choices are possible, with corresponding modifications in the definition of the dilated and translated moments.

All of the moments satisfying $p \geq 1+m_s$ are linearly dependent on the first $1+m_s$ moments, $\{\mu_{b,\gamma}(p \leq m_s)\}$, designated as the *initialization* or *missing moments.*

The wavelet transform will usually correspond to a linear combination of the lower order moments (refer to Eqs. (2.1-2). If one can generate the aforementioned moments, for all dilation and translation parameter values ($\gamma$ and $b$, respectively), one readily obtains the wavelet transform.

In this work we show how to generate a closed, coupled set of linear differential equations in $\gamma$, for the missing moments:

$$\partial_\gamma \mu_{b,\gamma}(i) = -\sum_{j=0}^{m_s} \mathcal{M}_{i,j}[\gamma, b, E]\mu_{b,\gamma}(j), \tag{0.2}$$

$0 \leq i,j \leq m_s$. These equations can be integrated, provided one has the starting (infinite scale, $a=\infty$, $\gamma = 1/(2a^2)$) values $\{\mu_{b,\gamma=0}(p)\}$, as well as the physical ground state energy value, $E$. The Eigenvalue Moment Method (EMM) developed by Handy and Bessis [**12, 13, 18**], provides us with this initial information.

Although the EMM approach is not the only method available for generating the required initial values (see references [**1, 8, 11, 20**]), it and the Rayleigh-Ritz *missing moment* variational approach in Ref. [**11**] are the only ones that emphasize the importance of the underlying *missing moments*, thereby facilitating the application of moment-based quantization methods to the continuous wavelet transform analysis of quantum systems.

Moment-quantization, in general, solves the Schrödinger problem at large scales. In order to obtain information at smaller scales (i.e. $\Psi(x)$), one can use the methods advocated here. Two approaches will be defined for reconstructing the wavefunction. The first uses the wavelet transform, calculated through the preceding procedure, combined with available wavelet-signal reconstruction formulas. The second reconstruction procedure is manifestly wavelet independent and makes use of certain asymptotic relations introduced in Section 2.

As noted, we shall combine EMM quantization and wavelet analysis in solving 1-D Schrödinger problems where the potential function, $V(x)$, can be any rational polynomial. More general potentials are allowed, provided the quantum problem can be transformed $(x \to y$ and $\Psi(x) \to \Phi(y))$ into a linear differential equation, $\sum_{i=0}^{p} C_i(y)\partial_y^i \Phi(y) = 0$, of arbitrary order, with rational fraction coefficients in the new variable, $C_i(y)$.

## 1. Continuous Wavelet Analysis

A continuous wavelet transform requires the selection of a wavelet function, $\omega(x)$, satisfying $\int_0^\infty (|\widehat{\omega}(k)|^2)/|k|\,dk < +\infty$, where $\widehat{\omega}(k)$ is the Fourier transform. In addition, it should have unit normalization, $\int |\omega(x)|^2 = 1$ [**10, 21**].

The wavelet transform of a one dimensional signal (wavefunction), $\Psi(x)$, is given by

$$\omega_t(a,b) \equiv \sqrt{a^{-1}} \int \omega\left(\frac{(x-b)}{a}\right) \Psi(x)\, dx, \tag{1.1a}$$

where $a > 0$ and $b$ define the scale and translation parameters, respectively [**10,21**]. The recovery of the wavefunction is possible through the formula

$$\Psi(x) = \sum_{m,n} \langle \omega_{m,n}(x) \mid \Psi(x) \rangle \tilde{\omega}_{m,n}(x), \tag{1.1b}$$

or

$$\Psi(x) = \sum_{m,n} \langle \tilde{\omega}_{m,n}(x) \mid \Psi(x) \rangle \omega_{m,n}(x), \tag{1.1c}$$

where $\omega_{m,n}(x) = a_0^{-m/2} \omega\big((x - nb_0 a_0^m)/(a_0^m)\big)$ define a frame (for arbitrary integers $m$ and $n$), $\widetilde{\omega}_{m,n}(x)$ is its dual frame [**7**].

For the case of "tight" frames we have $\widetilde{\omega}_{m,n}(x) = 2/(A+B)\omega_{m,n}(x)$, where $A/B = 1$. If the frame is not tight, $A/B \neq 1$, then there are two possibilities [**5, 7**]. The first is to compute the dual frame. For this case, if the frame consists of wavelets (dilations and translations of one mother wavelet, $\omega(x)$), then the dual must be computed for each translation, $\widetilde{\omega}_{0,n}$. Although, in principle, this entails the calculation of an infinite number of functions, in practice only a finite number are used. Despite this, there are some special wavelets, $\omega(x)$, and $b_0$ parameter values, for which even though the $\omega_{m,n}(x)$ are not close to defining a tight frame, nevertheless all the $\widetilde{\omega}_{m,n}(x)$ are dilated versions of one function [**5, 9**].

The second possibility applies to non-tight *snug* frames for which $A/B \approx 1$. In this case, one may take the first term in the defining series expansion for $\widetilde{\omega}_{m,n}(x)$ [**7**]. The resulting approximation yields the reconstruction formula (for the case $a_0 = 2$, $b_0 = 1$) [**6**]

$$\Psi(x) \approx \frac{2}{A+B} \sum_{m,n} \omega_{m,n} \frac{1}{\sqrt{2^m}} \omega\left(\frac{(x - n2^m)}{2^m}\right), \tag{1.2}$$

where $\omega_{m,n} \equiv \omega_t(2^m, n2^m)$, and $A + B = 6.819$.

One of the more popular wavelet functions is the *Mexican hat*, corresponding to $\omega(x) = -\mathcal{N}\partial^2 \exp(-1/2x^2) = \mathcal{N}(1-x^2)\exp(-1/2x^2)$, where $\mathcal{N} = 2/(\sqrt{3})\pi^{-1/4}$. We shall use this wavelet throughout this work.

The selection of the *Mexican hat* wavelet is a matter of convenience. In general, the methods developed here apply to any wavelet of the form $\partial_x^p e^{Q(x)}$, $p \geq 1$, provided $Q(x)$ is a polynomial, and $\mathrm{Lim}_{|x|\to\infty}\left[\partial_x^p e^{Q(x)}\right] = 0$.

The reconstruction formula in Eq. (1.2) is approximating the wavefunction through a superposition of localized oscillating structures. The wavefunction is being approximated in the $L^2$ norm. Through the Eigenvalue Moment Method discussed in the following section, we will be able to generate a finite number of the $\omega_{m,n}$ coefficients; thereby enabling the approximate reconstruction of the wavefunction.

It will be noted that an apparent paradox is created by Eq. (1.2) because the wavefunction, generally satisfying $\int \Psi(x)\, dx \neq 0$, is being represented as a superposition of wavelets each with integral zero, $\int \omega\big((x - n2^m)/(2^m)\big)\, dx = 0$. The resolution is that the wavelet-function reconstruction formulas apply in the $L^2$ norm, not in $L^1$. Thus, simply integrating both sides of Eq. (1.2) is incorrect. We are still guaranteed point-wise convergence. This 'paradox' is fully discussed by Daubechies in Ref. [**4**]. Our numerical results are in keeping with these theoretical observations.

We stress once more that our principal objective is to show how EMM theory can be used to generate the $\omega_{m,n}$ coefficients. There are many possible wavelet-reconstruction formulas, some involving Gaussians [**3**] as opposed to the Mexican hat wavelets in Eq. (1.2). We only utilize Eq. (1.2) as an example of how to utilize the generated wavelet coefficients in the reconstruction of the wavefunction. Also, the method can be extended to excited states [**16**].

## 2. Generating the $\omega_{m,n}$ Through the Eigenvalue Moment Method

From the preceding discussion, the Mexican hat wavelet transform for the wavefunction is given by (after a translation-change of variables):

$$\omega_t(a,b) = \mathcal{N}\sqrt{a^{-1}} \int_{-\infty}^{+\infty} \Psi(b+x)\big[1-(x/a)^2\big] \exp\Big(-\frac{1}{2}(x/a)^2\Big)\, dx, \tag{2.1}$$

or (defining $\gamma \equiv 1/(2a^2)$)

$$\omega_t(a,b) = \mathcal{N}(2\gamma)^{1/4}\big[\mu_{b,\gamma}(0) - 2\gamma\mu_{b,\gamma}(2)\big], \tag{2.2}$$

where

$$\mu_{b,\gamma}(p) \equiv \int_{-\infty}^{+\infty} x^p \Psi(b+x) \exp(-\gamma x^2)\, dx, \quad p \geq 0 \tag{2.3}$$

are the Hamburger moments of the measure $\Phi_{b,\gamma}(x) \equiv \Psi(b+x)\exp(-\gamma x^2)$.

The Hamburger moments satisfy the following differential equation with respects to the $\gamma > 0$ variable:

$$\partial_\gamma \mu_{b,\gamma}(p) = -\mu_{b,\gamma}(p+2). \tag{2.4}$$

For all one dimensional Schrödinger Hamiltonians with rational fraction potentials (as well as many other types of potentials which can be converted into rational fraction form after a suitable change of variables), all of the Hamburger moments become linearly dependent on the first $1+m_s$ order *missing moments* [**12, 13, 18**]:

$$\mu_{b,\gamma}(p) = \sum_{j=0}^{m_s} M_{E,b,\gamma}(p,j)\mu_{b,\gamma}(j), \tag{2.5}$$

where the energy, $E$, dependent coefficients $M_{E,b,\gamma}(p,j)$ are numerically or algebraically determinable, and must satisfy $M_{E,b,\gamma}(i,j) = \delta_{i,j}$, for $0 \leq i,j \leq m_s$. Clearly, inserting Eq. (2.5) into Eq. (2.4) yields a closed set of $1+m_s$ coupled, first order, linear differential equations

$$\partial_\gamma \mu_{b,\gamma}(i) = -\sum_{j=0}^{m_s} M_{E,b,\gamma}(i+2,j)\mu_{b,\gamma}(j), \quad \text{for } 0 \leq i \leq m_s. \tag{2.6}$$

For arbitrary $b$, given the physical energy and starting missing moment values $\{\mu_{b,0}(i) \mid 0 \leq i \leq m_s\}$ one can numerically integrate Eq. (2.6) and proceed to determine the wavelet transform Eq. (2.2). Facilitating this is the relation:

$$\mu_{b,0}(p) = \int_{-\infty}^{+\infty} x^p \Psi(b+x)\, dx = \int_{-\infty}^{+\infty} (x-b)^p \Psi(x)\, dx, \tag{2.7}$$

or (expanding)

$$\mu_{b,0}(p) = \sum_{q=0}^{p} \binom{p}{q} (-b)^{p-q} \mu_{0,0}(q). \tag{2.8}$$

The physical values for the energy, $E$, and $\mu_{0,0}(q) \equiv \int_{-\infty}^{+\infty} x^q \Psi(x)$ Hamburger moments, for $0 \leq q \leq m_s$, are determined by the Eigenvalue Moment Method (discussed in the following section). In practice, for each pair of values ($a = 2^m$, $b = n2^m$) at each point $b = n2^m$ we integrate Eq. (2.6) up to $\gamma = 1/2a^{-2} = 1/2(2^{-2m})$, which then allows us to generate the wavelet transform coefficients $\omega_{m,n}$.

An important observation is that the asymptotic behavior of the $\mu_{b,\gamma}(p)$ moments, with respects to $\gamma \to \infty$, determines the wavefunction:

$$\mathrm{Lim}_{\gamma \to +\infty}\, \mu_{b,\gamma}(p) = \left(\frac{1}{\sqrt{\gamma}}\right)^{(p+1)} \theta(p/2)\Psi(b), \quad p = \text{even}, \tag{2.9}$$

where $\theta(\rho) = \int_{-\infty}^{+\infty} y^{2\rho} \exp(-y^2)\, dy$. For $p = 0, 2$, we have $\theta(0) = \sqrt{\pi}$ and $\theta(1) = \sqrt{\pi}/2$. As will be clarified in the following section, an implicit normalization is assumed in the implementation of the EMM quantization. Our numerical comparisons between the wavelet-reconstructed wavefunction and the actual wavefunction (obtained from direct integration of the Schrödinger equation) will assume $\Psi(0)$ to have the value generated through Eq. (2.9) (approximated at a sufficiently large $\gamma$ value).

## 3. The Quartic Anharmonic Oscillator

We will apply the preceding formalism, in some detail, to two problems. The first corresponds to the quartic anharmonic oscillator Hamiltonian. The second, discussed in the next section, corresponds to a rational fraction potential problem. The details of the required EMM analysis may be found in the cited references.

The quartic anharmonic potential problem is defined by:

$$-\frac{d^2\Psi}{dx^2} + (mx^2 + gx^4)\Psi(x) = E\Psi(x). \tag{3.1}$$

Bearing in mind the formalism in Section 2, we need to define a moment equation for the configuration $\Phi_{b,\gamma}(x) \equiv \Psi(b + x)\exp(-\gamma x^2)$. One first obtains the differential equation:

$$\left[-\left[\frac{d^2}{dx^2} + 4\gamma x\frac{d}{dx} + 2\gamma + 4\gamma^2 x^2\right] + \left[m(x+b)^2 + g(x+b)^4\right]\right] \Phi_{b,\gamma}(x) = E\Phi_{b,\gamma}(x). \tag{3.2}$$

The Hamburger moments, $\mu_{b,\gamma}(p) = \int x^p \Psi(b+x)\exp(-\gamma x^2)\, dx$, satisfy the moment equation (for $p \geq 0$):

$$\mu_{b,\gamma}(p+4) = \sum_{i=1}^{3} C_i[\gamma]\mu_{b,\gamma}(p+i) + \left\{-4\frac{\gamma p}{g} + C_0[\gamma]\right\}\mu_{b,\gamma}(p) + \frac{p(p-1)}{g}\mu_{b,\gamma}(p-2), \tag{3.3}$$

where the coefficients are $C_3 = -4b$, $C_2[\gamma] = g^{-1}[4\gamma^2 - m - 6gb^2]$, $C_1 = -g^{-1}[2bm + 4gb^3]$, and $C_0[\gamma] = -g^{-1}[2\gamma + mb^2 + gb^4 - E]$.

The moment equation corresponds to a linear, homogeneous, fourth order finite difference equation (for nonnegative $p$ values). All of the moments are linearly dependent on the missing moments $\{\mu_{b,\gamma}(i) \mid 0 \le i \le 3\}$. This is expressed through the relation

$$\mu_{b,\gamma}(p) = \sum_{j=0}^{3} M_{E,b,\gamma}(p,j)\mu_{b,\gamma}(j), \tag{3.4a}$$

where

$$M_{E,b,\gamma}(i,j) = \delta_{i,j}, \quad \text{for } 0 \le i,j \le 3, \tag{3.4b}$$

and the remaining $M_{E,b,\gamma}(p > 3, j)$ coefficients satisfy the same moment equation as in Eq. (3.3).

The relation (from Eq. (2.6))

$$\frac{\partial}{\partial\gamma}\begin{pmatrix}\mu_{b,\gamma}(0)\\ \mu_{b,\gamma}(1)\\ \mu_{b,\gamma}(2)\\ \mu_{b,\gamma}(3)\end{pmatrix} = -\begin{pmatrix}\mu_{b,\gamma}(2)\\ \mu_{b,\gamma}(3)\\ \mu_{b,\gamma}(4)\\ \mu_{b,\gamma}(5)\end{pmatrix}, \tag{3.5}$$

now becomes (upon substituting Eq. (3.4a))

$$\frac{\partial}{\partial\gamma}\begin{pmatrix}\mu_{b,\gamma}(0)\\ \mu_{b,\gamma}(1)\\ \mu_{b,\gamma}(2)\\ \mu_{b,\gamma}(3)\end{pmatrix} = \begin{pmatrix} 0 & 0 & -1 & 0\\ 0 & 0 & 0 & -1\\ \mathcal{M}_{2,0}[\gamma] & \mathcal{M}_{2,1}[\gamma] & \mathcal{M}_{2,2}[\gamma] & \mathcal{M}_{2,3}[\gamma]\\ \mathcal{M}_{3,0}[\gamma] & \mathcal{M}_{3,1}[\gamma] & \mathcal{M}_{3,2}[\gamma] & \mathcal{M}_{3,3}[\gamma]\end{pmatrix}\begin{pmatrix}\mu_{b,\gamma}(0)\\ \mu_{b,\gamma}(1)\\ \mu_{b,\gamma}(2)\\ \mu_{b,\gamma}(3)\end{pmatrix} \tag{3.6}$$

(note, $\mathcal{M}_{2,0\le j\le 3} = -M_{E,b,\gamma}(4,j)$ and $\mathcal{M}_{3,0\le j\le 3} = -M_{E,b,\gamma}(5,j)$) where $\mathcal{M}_{2,0}[\gamma] = -C_0[\gamma]$, $\mathcal{M}_{2,1}[\gamma] = -C_1$, $\mathcal{M}_{2,2}[\gamma] = -C_2[\gamma]$, $\mathcal{M}_{2,3}[\gamma] = -C_3$, $\mathcal{M}_{3,0}[\gamma] = -C_0[\gamma]C_3$, $\mathcal{M}_{3,1}[\gamma] = -\big[C_1C_3 + C_0[\gamma] - 4\gamma/g\big]$, $\mathcal{M}_{3,2}[\gamma] = -\big[C_1 + C_3C_2[\gamma]\big]$, and $\mathcal{M}_{3,3}[\gamma] = -\big[C_2[\gamma] + C_3^2\big]$.

As indicated in the previous sections, given the starting values for the missing moments, $\{\mu_{b,\gamma=0}(i), (0 \le i \le 3)\}$, at any $b$, one can integrate the above equations. Facilitating the determination of these starting moment values is the relation in Eq. (2.8). Accordingly, one must determine the physical energy, $E$, and the physical missing moment values $\{\mu_{b=0,\gamma=0}(i), (0 \le i \le 3)\}$.

The $\mu_{0,0}(p)$ moments satisfy the moment equation

$$\mu_{0,0}(p+4) = g^{-1}\big[-m\mu_{0,0}(p+2) + E\mu_{0,0}(p) + p(p-1)\mu_{0,0}(p-2)\big], \tag{3.7}$$

for $p \ge 0$. Symmetric configurations, such as the ground state for the two problems being considered, will have all their odd order moments equal to zero, $\mu_{0,0}(\text{odd}) = 0$. For such configurations, one can streamline the formalism and effectively work with the even order Stieltjes moments $u(\rho) \equiv \mu_{0,0}(2\rho)$. Accordingly, only $u(0)$ and $u(1)$ become the nonzero missing moments in this case. An analogous formulation is possible for antisymmetric configurations; however, for simplicity, these are not discussed in this work.

The homogeneous nature of the moment equation implies the necessity for some normalization condition. One convenient choice is

$$u(0) + u(1) = 1. \tag{3.8}$$

Eliminating $u(0)$, all of the moments become linearly dependent on the unconstrained missing moment $u(1)$. This may be expressed through the relation $u(\rho) =$

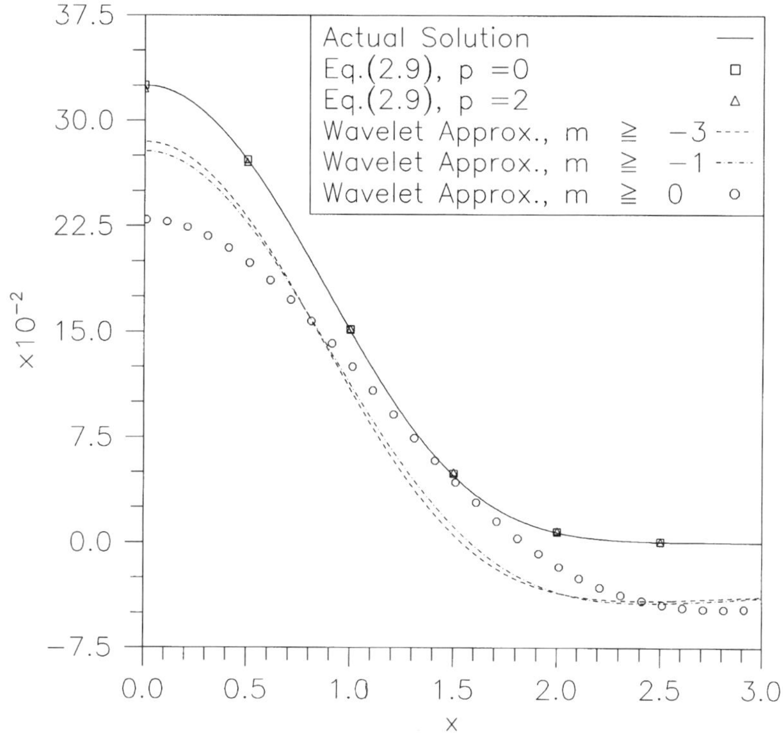


FIGURE 1. $\Psi_{\text{ground}}(x), V(x) = x^2 + x^4$. Reprinted from [**16**] with permission. ©1996 by the American Physical Society.

$\widehat{M}_E(\rho, 0) + \sum_{j=1}^{1} \widehat{M}_E(\rho, j)u(j)$, where the $\widehat{M}_E$ coefficients are readily determinable from Eq. (3.4) and the above normalization.

One now applies EMM quantization to determine the physical ground state values ($m = g = 1$) for the energy, $E$, and $u(1) = \mu_{0,0}(2)$ missing moment. One obtains $E = 1.39235164153$, $u(0) = .642670622325$, and $u(1) = .357329377675$. This plus $\mu_{0,0}(1) = \mu_{0,0}(3) = 0$, provide the starting information ($\mu_{b,\gamma=0}(0-3)$, from Eq. (2.8)) for integrating Eq. (3.6) through a fourth order Runge-Kutta ansatz.

In general, for $|b| < 5$, we could integrate Eq. (3.6) for $\gamma < O(10^2)$. Beyond this $\gamma$ value, one (or both) of the two even order moments, $\mu_{b,\gamma}(2,4)$, no longer remained positive (which they must be). Due to these limitations, we could only generate the wavelet coefficients $\omega_{m,n}$ corresponding to $m, n$ integer values ($b = n2^m, a = 2^m$) $|n2^m| < 5$ and $\gamma = 1/2a^{-2} < O(10^2)$. Specifically, we took $-3 \leq m \leq 3$.

One can compare the numerically observed limiting values for the asymptotic limits $\text{Lim}_{\gamma\to\infty} \sqrt{\gamma/\pi}\mu_{b,\gamma}(0)$ and $\text{Lim}_{\gamma\to\infty}(2\gamma^{3/2})/(\sqrt{\pi})\mu_{b,\gamma}(2)$. According to Eq. (2.9), they both must approach $\Psi(b)$. This was easily manifested by our integration results for $|b| \leq 2.5$. Accordingly, we took the value generated at $b = 0$ ($\Psi(0) = .325$) as the normalization condition in our (second) Runge-Kutta integration of the Schrödinger equation, in order to obtain the "exact" solution, for purposes of comparison.

Three results are illustrated in Figure 1:

(a) the estimated ground state $\Psi(b)$ wavefunction, given by Eq. (2.9) for the $p = 0, 2$ cases;
(b) the direct Runge-Kutta integration of the ground state wavefunction from the Schrödinger equation (utilizing the $\Psi(0) = .325$ value obtained by integrating Eq. (3.6) at $b = 0$ and $\gamma = O(10^2)$;
(c) and the Mexican hat wavelet—wavefunction reconstruction utilizing the formula in Eq. (1.2).

With respects to the latter, we show the effect of the wavelet reconstruction when successive smaller scales are included. Thus, $m \geq m_n$ means that only the corresponding terms in Eq. (1.2) were used in generating the wavelet approximant. In Figure 1, the cases $m_n = -3$ and $m_n = -2$ lie too close for graphing purposes. As such, the latter is not shown. In general, all of the results in Figure 1 are very good. The asymptotic estimates from Eq. (2.9) are exceptionally good.

## 4. A Rational Fraction Potential

We next consider the case of the rational fraction potential problem

$$-\partial_x^2\Psi(x) + \frac{gx^6}{1+\lambda x^2}\Psi(x) = E\Psi(x). \tag{4.1}$$

Let $\Phi_{b,\gamma}(x) \equiv \exp(-\gamma^2 x)\Psi(b+x)$, or $\Psi_b(x) = \exp(\gamma^2 x)\Phi_{b,\gamma}(x)$, where $\Psi_b(x) \equiv \Psi(b+x)$. The ensuing equation for $\Phi_{b,\gamma}$ is

$$\left[-\left[\frac{d^2}{dx^2} + 4\gamma x\frac{d}{dx} + 2\gamma + 4\gamma^2 x^2\right] + \left[\frac{g(x+b)^6}{1+\lambda(x+b)^2}\right]\right] \Phi_{b,\gamma}(x) = E\Phi_{b,\gamma}(x). \tag{4.2}$$

Multiplying both sides by $x^p\left(1+\lambda(x+b)^2\right)$ and performing the necessary algebraic expansions leads to a similar moment equation as before, although of higher order (6), with six missing moments $\{\mu_{b,\gamma}(i) \colon 0 \leq i \leq 5\}$:

$$\begin{aligned}\mu_{b,\gamma}(p+6) = &- 6b\mu_{b,\gamma}(p+5)\\ &+ \left[4\gamma^2\frac{\lambda}{g} - 15b^2\right]\mu_{b,\gamma}(p+4) + \left[8\frac{\lambda}{g}b\gamma^2 - 20b^3\right]\mu_{b,\gamma}(p+3)\\ &+ g^{-1}\left[4\gamma^2(1+\lambda b^2) - 4\gamma\lambda\left(p+\frac{5}{2}\right)\right.\\ &\qquad\left. - 15gb^4 + E\lambda\right]\mu_{b,\gamma}(p+2)\\ &+ g^{-1}\left[-2\lambda\gamma b(6+4p) - 6gb^5 + 2\lambda bE\right]\mu_{b,\gamma}(p+1)\\ &+ g^{-1}\left[\gamma(1+\lambda b^2)(E-2-4p)\right.\\ &\qquad\left. + \lambda(p+2)(p+1) - gb^6\right]\mu_{b,\gamma}(p)\\ &+ 2\lambda b g^{-1}p(p+1)\mu_{b,\gamma}(p-1)\\ &+ g^{-1}(1+\lambda b^2)p(p-1)\mu_{b,\gamma}(p-2).\end{aligned} \tag{4.3}$$

The ensuing analysis proceeds as in the previous case, except that the corresponding Stieltjes problem (for $b = \gamma = 0$) involves the missing moments $u(0)$, $u(1)$,

$u(2)$ with the normalization condition $u(0) + u(1) + u(2) = 1$. The EMM analysis yields the required starting information ($g = 1, \lambda = .1$): $E = 1.10500549453562$,

$$\mu_{0,0}(0) = .446117055534407,$$
$$\mu_{0,0}(2) = .238142672555330,$$
$$\mu_{0,0}(4) = .315740271910263.$$

The odd order moments must be zero, $\mu_{0,0}(1,3,5) = 0$.

Proceeding with a fourth order Runge-Kutta numerical integration, we could only integrate out to an order $\gamma \approx 7$, before encountering numerical instability problems. However, even within this relatively small $\gamma$ domain, the inverse $\gamma$ expansion behavior of $\mu_{b,\gamma}(p)$ was empirically satisfied for $\gamma \geq 4$. Specifically, from the integral representation in Eq. (2.3) one has

$$\mu_{b,\gamma}(p) = \gamma^{-(p+1)/2} \sum_{\substack{p+\eta=\text{even},\\ \eta\geq 0}}^{\infty} \frac{1}{\eta!}(\partial_b^\eta \Psi(b))\theta\left(\frac{p+\eta}{2}\right)\gamma^{-\eta/2}. \tag{4.4}$$

Consider fitting the Runge-Kutta iterates, $\mu_{b,\gamma_i}(p = 0, 2)$, for $I \leq i \leq I + N$, to a truncated analogue of the inverse $\gamma$ expansion ($\eta$ = even, and $\eta \leq 2N$), for $4 \leq \gamma_i < 7$:

$$\mu_{b,\gamma_i}(p = \text{even}) = \gamma^{-(p+1)/2} \sum_{\substack{p+\eta=\text{even},\\ \eta\geq 0}}^{2N} \frac{1}{\eta!}C_I^{(p)}(b,\eta)\theta\left(\frac{p+\eta}{2}\right)\gamma_i^{-\eta/2}. \tag{4.5}$$

According to Eq. (2.9), the leading term of this expansion should approximate the wavefunction

$$C_I^{(p)}(b,0) \approx \Psi_{\text{exact}}(b). \tag{4.6}$$

Our numerical results confirm this. That is, a fourth order Runge-Kutta integration directly on the Schrödinger equation, in order to obtain the exact ground state wavefunction, normalized by $\Psi_{\text{exact}}(0) \equiv C_I^{(p=0,2)}(0,0)$, concurred with the truncated inverse $\gamma$ expansion analysis $\Psi_{\text{exact}}(b) \approx C_I^{(p=0,2)}(b,0)$, for $|b| \leq 2.2$. The $C_I^{(p=0,2)}$ coefficient values given in Figure 2 were generated at $\gamma = 4$. Also, although the Runge-Kutta $\delta\gamma$ step size was $10^{-5}$, we generated the $C_I^{(p=0,2)}$ coefficients by sampling the Runge-Kutta iterates at intervals of $\gamma_{i+1} - \gamma_i = 10^{-1}$ (for $\gamma_i \approx 4$), to minimize round-off error fluctuations in inverting Eq. (4.5) to obtain the $C_I^{(p=0,2)}$s.

Unlike the quartic anharmonic oscillator case, we used the inverse $\gamma$ expansion to approximate the $\mu_{b,\gamma}(p)$ moments for $|b| \leq 2.2$ and $4 \leq \gamma < \infty$. Accordingly, the wavelet expansion coefficients $\omega_{m,n}$ are computed so as to satisfy $|b = n2^m| \leq 2$ and $\gamma = 1/(2(a = 2^m)^2)$ = arbitrary.

In Figure 2 we illustrate the exact ground state solution, the asymptotic estimates achieved through the truncated inverse $\gamma$ expansion ($N = 4$), and the wavelet reconstructed approximants. The results are consistent with the quality of the results for the quartic anharmonic oscillator problem previously discussed. A more comprehensive numerical discussion of this research will be presented elsewhere [**17**].

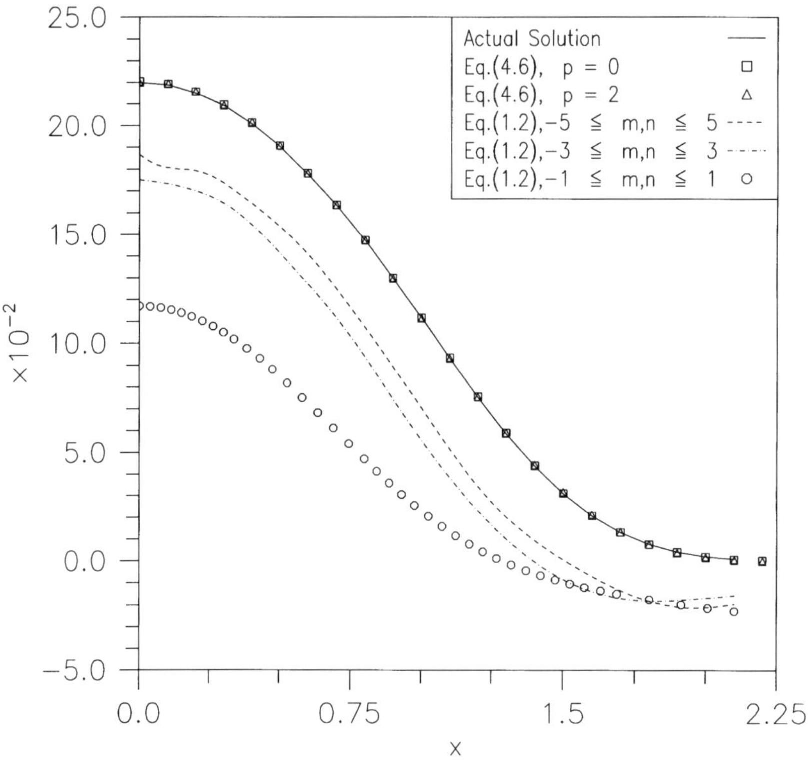


FIGURE 2. $\Psi_{\text{ground}}(x), V(x) = (x^6)/(1 + x^2/10)$.

## Conclusion

We have shown that moment quantization methods, when combined with a moment equation reformulation of the one dimensional Schrödinger equation, can be used to generate the wavelet transform, as well as reconstruct the solution. Our methods apply to any rational polynomial potential problem, and to any mother wavelet function $W(x) = -\mathcal{N}\partial_x^i e^{-Q(x)}$, provided $i \geq 1$ and $Q(x)$ is a polynomial satisfying $\text{Lim}_{x\to\mp\infty} e^{-Q(x)} = 0$ ($Q(x) = x^2/2$, for the Mexican hat wavelet). The multidimensional extension follows similarly [**15**]; however, the counterpart to the $\mathcal{M}$ matrix, on the right hand side of Eq. (0.2), will involve partial derivatives with respects to $b$.

The fact that the moment asymptotic limits in Eq. (2.9) approximate the solution much better than the dyadic formula in Eq. (1.2) might lead one to suspect that wavelet reconstruction is ineffective; however, in reality, Eq. (2.9) is equivalent to integrating over all scale and translation parameter values for an appropriate continuous mother wavelet function. To see this, consider the expression $U_S[a,b] \equiv 1/a\nu \int dx\, S\big((x-b)/a\big)\Psi(x)$, where the function $S(z)$ is sufficiently well behaved and goes to zero asymptotically; also, $\nu \equiv \int dz S(z) \neq 0$. One can then show [**14**] that the generalization of Eq. (2.9), $\text{Lim}_{a\to 0} U_S[a,b] = \Psi(b)$, is exactly equivalent to

$\Psi(b) = \int_0^\infty da 1/a^3 \int d\xi\, \omega_t(a,\xi)\widetilde{\omega}\big((\xi-b)/a\big)$, for mother wavelet and dual functions satisfying (through their Fourier transforms) $-k\partial_k \widehat{S}(k) = \sqrt{2\pi}\widehat{\omega}(k)\widehat{\widetilde{\omega}}(k)$ (note, contrary to Eq. (1.1a), we have $\omega_t(a,b) \equiv \int dx\, \omega((x-b)/a)\Psi(x)$. If both mother wavelet and dual correspond to the Mexican hat, $\omega(x) = \widetilde{\omega}(x) = -\partial_x^2 e^{-x^2/2}$, then $S(x) = (\sqrt{\pi})/2(3/2 - 1/4x^2)e^{-x^2/2}$. In this case ($\nu = \pi$), $U_S[a,b] = \sqrt{\gamma/2\pi}\big(3/2\mu_{\gamma/2,b}(0) - \gamma/2\mu_{\gamma/2,b}(2)\big)$. This is a linear superposition of the asymptotic moment results depicted in the Figures (i.e. $\sqrt{\gamma/2}\mu_{\gamma,b}(0) \to \Psi(b)$ and $(2_{\gamma^{3/2}})/(\sqrt{\pi})\mu_{\gamma,b}(2) \to \Psi(b)$). For the case $\omega(x) = -\partial_x^2 e^{-x^2/2}$ and $\widetilde{\omega}(x) = e^{-x^2/2}$, one obtains that $S(x) = (\sqrt{\pi})/2e^{-x^2/4}$. In this case, $U_S[a,b] = \sqrt{\gamma/(2\pi)}\mu_{\gamma/2,b}(0)$.

From the preceding one can see that the excellence of the asymptotic moment limit results depicted in Figures 1 and 2 (obtained over a relatively small domain of integration, $\gamma \in [0,\gamma_*]$, where $\gamma_* = O(10^2)$ and $O(7)$, for the quartic anharmonic oscillator and rational polynomial potentials, respectively) is due to the fact that they correspond to the integration over all translations and scales: $-\infty < b < +\infty$, $a_* \le a < \infty$, where $a_* = 1/(\sqrt{2\gamma_*})$.

## Acknowledgments

This work was supported, in part, by the National Science Foundation through the Center for Theoretical Studies of Physical Systems grant HRD 9450380. Additional support was provided through the Army Research Laboratory Advanced Sensors (Fed-Lab) Consortium, Cooperative Agreement DAAL01-96-2-0001. We would like to thank Professors Jean-Pierre Antoine, Ingrid Daubechies, Giorgio Mantica, Alfred Msezane, Xiao Qian Wang, and Guido Weiss for useful discussions.

## References

1. R. Blankenbecler, T. De Grand, and R. L. Sugar, *Moment method for eigenvalues and expectation values*, Phys. Rev. D **21** (1980), 1055–1061.
2. C. K. Chui, *An introduction to wavelets*, Academic Press Inc., New York, 1992.
3. K. Chuo, T. A. Arias, J. D. Joannopoulos, and P. K. Lam, *Wavelets in electronic structure calculations*, Phys. Rev. Lett. **71** (1993), 1808–1811.
4. I. Daubechies, *Different perspectives on wavelets* (I. Daubechies, ed.), Amer. Math. Soc., vol. 47, Providence, RI, 1993, pp. 5–6.
5. ———, *Ten lectures on wavelets*, SIAM, Philadelphia, PA, 1992.
6. ———, *Advances in spectrum analysis and array processing.* I. (S. Haykin, ed.), Prentice Hall, Englewood Cliffs, N.J., 1991.
7. ———, *The wavelet transform, time-frequency localization and signal analysis*, IEEE Trans. Inform. Theory **36** (1990), 961–1005.
8. F. M. Fernandez and J. F. Ogilvie, *Perturbation theory by the moment method applied to coupled anharmonic oscillators*, Phys. Lett. A **178** (1993), nos. 1-2, 11–16.
9. M. Frazier, B. Jawerth and G. Weiss, *The $\phi$-transform and applications to distribution spaces*, Function Spaces and Application (M. Cwikel, ed.), Lecture Note in Math., vol. 1302, Springer-Verlag, Berlin,1988, pp. 233–246.
10. A. Grossmann and J. Morlet, *Decomposition of Hardy functions into square integrable wavelets of constant shape*, SIAM J. Math. Anal. **15** (1984), no. 4, 723–736.
11. C. R. Handy, *A moment equation reformulation of Rayleigh-Ritz theory*, J. Phys. A **29** (1996), 4093–4104.
12. C. R. Handy and D. Bessis, *Rapidly convergent lower bounds for the Schrödinger-equation ground-state energy*, Phys. Rev. Lett. **55** (1985), 931–934.
13. C. R. Handy, D. Bessis, and T. D. Morley, *Generating quantum energy bounds by the moment method: A linear-programming approach*, Phys. Rev. A **37** (1988), 4557–4569.

14. C. R. Handy and R. Murenzi, *The interdependence of moment quantization and continuous wavelet transform analysis*, CAU preprint, 1997.
15. ———, *Exact generation of continuous wavelet transforms, and reconstruction, for Sturm-Liouville ODEs and PDEs*, Proceedings of 1997 SPIE Conference, no. 3078-64.
16. ———, *Continuous wavelet transform analysis of one-dimensional quantum bound states from first principles*, Phys. Rev. A (3) **54** (1996), no. 5, 3754–3763.
17. ———, *Continuous wavelet transform analysis of quantum systems with rational potentials*, J. Phys. A **30** (1997), no. 13, 4709–4729.
18. C. R. Handy, D. Bessis, G. Sigismondi, and T. D. Morley, *Rapidly converging bounds for the ground state energy of hydrogenic atoms in superstrong magnetic fields*, Phys. Rev. Lett. **60** (1988), 253–256.
19. S. Jaffard and Ph. Laurençot, *Wavelets: A tutorial in theory and application* (C. K. Chui, ed.), Academic Press Inc., San Diego, 1992.
20. J. P. Killingbeck, M. N. Jones, and M. J. Thompson, *Inner product methods for eigenvalue calculations*, J. Phys. A **18** (1985), no. 5, 793–801.
21. S. Mallat, *Multiresolution approximations and wavelet orthonormal bases of* $L^2(\mathbf{R})$, Trans. Amer. Math. Soc. **315** (1989), no. 1, 69–87.
22. T. Paul, *Functions analytic on the half-plane as quantum mechanical states*, J. Math. Phys. **25** (1984), no. 11, 3252–3263.
23. F. Plantevin, *Une application de la transformée en ondelettes continue à la mécanique quantique et analyse multi-résolution adaptive*, Ph.D. thesis, Univ. de Provence, France, 1992.



Department of Physics and Center for Theoretical Studies of Physical Systems, Clark Atlanta University, Atlanta, Georgia 30314, USA

*E-mail address*: `handy@pioneer.cau.edu`

Department of Physics and Center for Theoretical Studies of Physical Systems, Clark Atlanta University, Atlanta, Georgia 30314, USA

*E-mail address*: `murenzi@hubble.cau.edu`

Centre de Recherches Mathématiques
CRM Proceedings and Lecture Notes
Volume **18**, 1999

# Oscillating Singularities and Fractal Functions

A. Arneodo, E. Bacry, S. Jaffard, and J. F. Muzy

ABSTRACT. We use wavelet analysis to distinguish between cusps and oscillating singularities. We show that the latter are characterized by two exponents, the Hölder exponent $h$ and the oscillation exponent $\beta$ that accounts for the singular behavior of the instantaneous frequency of the function. For a large class of fractal functions whose singularities lie on Cantor sets, we present an extended multifractal formalism that allows us to compute the singularity spectrum $D(h, \beta)$ describing the fluctuations of both the Hölder and the oscillation exponents.

## 1. Introduction

This paper summarizes some rigorous results reported in Refs. [**1, 2**] about oscillating singularities and fractal functions. We first introduce the mathematical notion of *oscillating singularity*. A well known example of such singularity is the behavior around $x = x_0$ of the chirp function [**3, 20**]

$$f(x) = |x - x_0|^h \sin\left(\frac{1}{|x - x_0|^\beta}\right), \quad h > 0,\ \beta > 0, \tag{1.1}$$

which is of Hölder exponent $h(x_0) = h$. In signal analysis, this notion must cover functions whose *instantaneous frequency* increases fast at some time [**12**]. If one considers the primitive of the above chirp, it can be shown that fast accumulating oscillations cause cancellations so that the Hölder exponent becomes [**3, 20**] $h + 1 + \beta$. This behavior with respect to integrations is different from what happens for *cusp* singularities like $f(x) = |x - x_0|^h$, the Hölder exponent of its primitive being simply $h + 1$. In that respect, a cusp can be seen as an oscillating singularity with $\beta = 0$. Thus, in order to characterize fully a singular behavior one needs two exponents: the Hölder exponent $h$ and the *oscillation exponent* $\beta$. In this paper we propose a rigorous definition of the latter exponent based on computation of pointwise regularity. This will enable us to associate to any function $f$ two *oscillating singularity exponents* $(h, \beta)$ at every point $x_0$.

Our main purpose is to consider a class of functions that are multifractal in the sense that the dimension $D(h, \beta)$ of the set of points $x_0$ where the oscillating

---

1991 *Mathematics Subject Classification.* Primary: 26A15, 26A30, 28A80; Secondary: 26A16, 28A78, 33A65.

This is the final version of the paper.

singularity exponents are $(h, \beta)$ is a non trivial spectrum. This singularity spectrum generalizes the classical $D(h)$ singularity spectrum [**4, 5, 9, 14, 16, 18, 26–28**] to oscillating singularity exponents. We consider irregular functions explicitly constructed with the help of an orthonormal wavelet basis for which we show that $D(h, \beta)$ can be computed using an extended multifractal formalism based on wavelet analysis. Within the thermodynamic analogy, this new formalism turns out to be the equivalent of a "grand canonical" description that accounts for the statistics over "microstates" for which both the "energy" and the "number of particles" are fluctuating. Our purpose will be illustrated by some examples.

## 2. Cusps and Oscillating Singularities

**2.1. Pointwise regularity from wavelet analysis.** The wavelet transform of a real valued function $f$ according to the *analyzing wavelet* $\psi$ is defined as [**8, 10, 11, 15, 21, 23–25, 29**]

$$T_\psi[f](b,a) = \frac{1}{a}\int_{-\infty}^{\infty} \psi\left(\frac{x-b}{a}\right) f(x)\,dx, \tag{2.1}$$

where $a \in \mathbb{R}^{+*}$ and $b \in \mathbb{R}$. Generally $\psi$ is chosen to be well localized in both direct and Fourier spaces so that $T_\psi$ can be seen as an accurate space-scale analysis ($b$ is the space parameter whereas $1/a$ is the frequency parameter). The wavelet transform has been advocated as a powerful tool for characterizing the local regularity of functions [**17, 19, 22**]. In the following we will suppose that the analyzing wavelet is orthogonal to any polynomial, i.e., $\psi$ satisfies:

$$\int \psi(x) x^k\,dx = 0, \quad \forall k. \tag{2.2}$$

Let us introduce some useful notations that are fitted to our study. The first one is a weak form of the $\mathcal{O}$ notation of Landau, and the second one expresses the fact that two functions are of the same order of magnitude, disregarding "logarithmic corrections".

DEFINITION 1. If $F$ and $G$ are two functions, $F = \overline{\mathcal{O}}(G)$ if

$$\liminf \frac{\log|F|}{\log|G|} \geq 1, \tag{2.3}$$

and $F \sim G$ if

$$\lim \frac{\log|F|}{\log|G|} = 1. \tag{2.4}$$

Let $h = h(x_0)$ be the Hölder exponent of a function $f$ at $x_0$. The following proposition [**1, 2**] is a direct consequence of the main theorem proved is Refs. [**17**] and [**19**] concerning the characterization of pointwise regularity using the wavelet transform.

PROPOSITION 1. *Suppose that $f \in C^\varepsilon(\mathbb{R})$ for some $\varepsilon > 0$. The Hölder exponent of $f$ at $x_0$ is $h$ if and only if the two following conditions hold:*

- *In the neighborhood of $(b, a) = (x_0, 0)$*

$$\left|T_\psi[f](b,a)\right| = \overline{\mathcal{O}}\left(|b-x_0|^h + a^h\right). \tag{2.5}$$

- *There exists a sequence $(b_n, a_n) \to (x_0, 0)$ satisfying*

$$\left|T_\psi[f](b_n,a_n)\right| \sim |b_n - x_0|^h + a_n^h. \tag{2.6}$$

We will call such a sequence $(b_n, a_n)$ a *minimizing sequence* for $f$ at $x_0$. Let us suppose that there exists a minimizing sequence for $f$ such that $|b_n - x_0| = \overline{\mathcal{O}}(a_n)$ (since $|b - x_0| \leq Ca$ defines a cone centered at $x_0$, with a slight abuse, we will say that such a sequence is "in a cone"). Then, from the above proposition, if $f$ has an Hölder exponent $h(x_0)$ at $x_0$ then $\left|T_\psi(b, a)\right| = \overline{\mathcal{O}}(a^h)$. This formula, often written as $\left|T_\psi[f](x_0, a)\right| \sim a^{h(x_0)}$ in the literature, corresponds to the "classical" estimation of the singularity exponent from the scale behavior of the wavelet transform [**4, 5, 18, 26, 27**]. However, if we denote by $F$ the primitive of $f$, a simple integration by parts of the above equation leads to $\left|T_{\psi'}[F](x_0, a)\right| \sim a^{h(x_0)+1}$ and thus to the conclusion that $F$ is of Hölder exponent $h(x_0) + 1$ at $x_0$. As explained in the introduction, this is wrong for the chirp function defined in Eq. (1.1). Obviously for such a function a minimizing sequence cannot be in a cone because the strongest wavelet coefficients are localized on a "ridge" of equation $a = C|x - x_0|^{\beta+1}$ (where $C$ is a constant) that describes the variation of the instantaneous period [**12**]. Since the size $a_n$ of an oscillation is much smaller than its distance from $x_0$ ($a_n = C|x_n - x_0|^{\beta+1}$), chirps correspond to the case where the strongest coefficients are outside any cone. On the other side, one can easily check that singularities of the type $f(x) = |x - x_0|^h$ correspond to the case where the strongest coefficients are inside a cone, i.e., $|x_n - x_0| = \overline{\mathcal{O}}(a_n)$. This leads us to define cusps and oscillating singularities as follows [**1, 2**]:

DEFINITION 2. A function $f(x)$ is said to have a *cusp singularity* at the point $x_0$ if and only if there exists a minimizing sequence $(x_n, a_n)$ such that

$$|x_n - x_0| = \overline{\mathcal{O}}(a_n). \tag{2.7}$$

Conversely, $f$ is said to have an *oscillating singularity* at the point $x_0$ if and only if the singularity is not a cusp.

Let us characterize more precisely the location of minimizing sequences in the $(b, a)$-half-plane and introduce the notion of *oscillation exponent.*

**2.2. The oscillation exponent $\beta$.** As explained in the introduction, oscillating singularities are not fully characterized by their Hölder exponent. Indeed, in the case of the chirp defined in Eq. (1.1), the Hölder exponent $h$ at $x_0$ does not tell us how the instantaneous frequency goes to $\infty$ when $x$ goes to $x_0$. Ideally, we would like to have access to both exponents $h$ and $\beta$. Actually, as shown above, $\beta$ plays a very important role in the regularity of $f$ when this function is integrated [**20**]. Indeed, we have seen that the primitive of $f$ corresponds to a $(h + 1 + \beta, \beta)$ chirp. Thus, whereas the Hölder exponent increases by 1 for a cusp singularity, it increases by $1 + \beta$ for a chirp. More generally, if we fractionally $\varepsilon$-integrate $f$ around $x_0$ (see Ref. [**1**]), it will increase respectively by $\varepsilon$ or by $\varepsilon(1 + \beta)$. In that sense, a cusp can be seen as a chirp with $\beta = 0$. These remarks can be used for defining, in a general case, an exponent $\beta$ that will characterize the oscillations of a given singular behavior.

DEFINITION 3. The function $f$ is singular at $x_0$ with the exponents $(h, \beta)$ if and only if, $h$ is the Hölder exponent at $x_0$ and $\beta = (dh_\varepsilon)/(d\varepsilon)|_{\varepsilon=0^+} - 1$, where $h_\varepsilon$ is the Hölder exponent of the $\varepsilon$-primitive $f_\varepsilon$ of $f$ at $x_0$. $\beta$ is called the *oscillation exponent.*

This definition uses the fact that $h_\varepsilon(x_0)$ is right differentiable at $\varepsilon = 0$. Indeed it is proved in Ref. [**1**] that $h_\varepsilon$ is a concave, non decreasing differentiable function. The

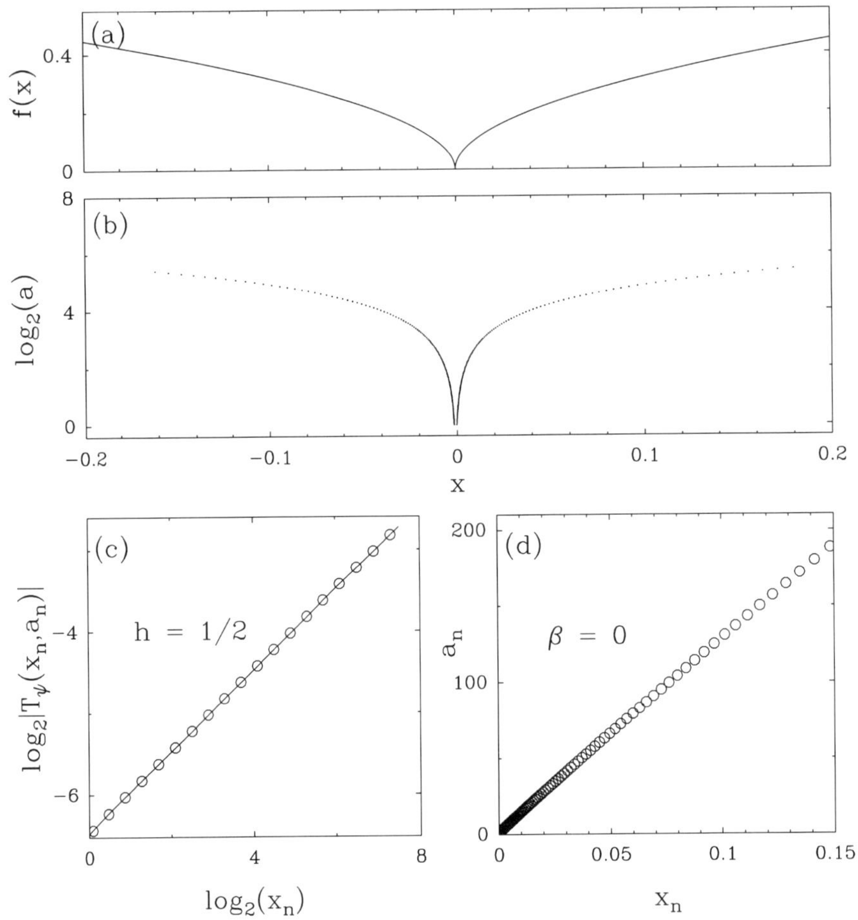


FIGURE 1. Hölder $\big(h(x_0)\big)$ and oscillation $\big(\beta(x_0)\big)$ exponents associated with a cusp singularity at $x = 0$. (a) Graph of the function $f(x) = |x|^{1/2}$. (b) Wavelet transform skeleton showing the positions of the wavelet transform modulus maxima for the signal in (a). The analyzing wavelet $\psi$ is the first derivative of the Gaussian function. These maxima fall on two maxima lines lying inside a cone $|x - x_0| = \overline{O}(a)$. (c) $\log_2\big|T_\psi[f](x_n, a_n)\big|$ vs $\log_2(x_n)$; the slope provides an estimate of $h(x_0) = 1/2$. (d) $a_n$ vs $x_n$; the fact that the points fall on a straight line indicates that $\beta(x_0) = 0$. In (c) and (d), the set of points $(x_n, a_n)$ defining a minimizing sequence corresponds to either one of the two maxima lines illustrated in (b).

following proposition relates the oscillation exponent $\beta$ to the position of minimizing sequences:

PROPOSITION 2. *Let $f(x)$ be a function that is singular at $x = x_0$ with the singularity exponents $\big(h(x_0), \beta(x_0)\big)$ (where $\beta(x_0)$ is defined as in Definition (3)). Then*

(a) *$x_0$ corresponds to a cusp singularity* $\Longleftrightarrow$ $\beta(x_0) = 0$,

(b) $x_0$ *corresponds to an oscillating singularity* $\Longleftrightarrow$ $\beta(x_0) > 0$,
(c) $\beta(x_0) = \max(0, \liminf \log(a_n)/\log|x_n - x_0| - 1)$ *where the* lim inf *is taken over all minimizing sequences.*

This proposition is proved in Ref. [**1**]. It states that the exponent $\beta$ that characterizes the variation of the regularity of $f$ when it is fractionally integrated can be recovered from minimizing sequences by studying the power-law behavior of the scale $a_n$ versus the distance to the singularity $|x_n - x_0|$.

In Figs. 1 and 2 are shown respectively a cusp and an oscillating singularity together with the numerical estimation of the corresponding exponents $h$ and $\beta$ using the wavelet transform. The "classical" multifractal formalism accounts for the distribution of the Hölder exponents $h$ only [**4, 5, 9, 14, 16, 18, 26–28**]. But unfortunately it leads to wrong results if oscillating singularities ($\beta \neq 0$) are involved in the considered fractal distribution (see below). In that case, $h$ gives a poor characterization of the singular behavior and the exponent $\beta$ appears to be essential. Our goal in the next section is to build a new statistical formalism that accounts for the fluctuations of both exponents $h$ and $\beta$.

## 3. Multifractal Formalism for Functions with Oscillating Singularities on Cantor Sets

Fractal functions are irregular functions for which the pointwise regularity can change from one point to another. They are generally characterized by the so-called singularity spectrum that associates to each set where the Hölder exponent is $h$, its fractal dimension $D(h)$. The paradigm of such functions are self-similar functions [**4, 5, 18, 26**] (hereafter denoted SS-functions). In Refs. [**5**] and [**18**], it has been proved that the multifractal properties of SS-functions can be described by a wavelet based multifractal formalism. In this section, we use the wavelet transform to generalize the notion of self-similarity and to construct a class of fractal functions whose oscillation exponent fluctuates from point to point. We then extend the notion of singularity spectrum to account for the statistics of these exponents and we show that this quantity can be computed within a generalized multifractal description.

### 3.1. Self-similarity in the wavelet half-plane.

We now construct a family of functions whose wavelet transform has self-similarity properties. These functions are not necessarily self-similar and we will show that they have oscillating singularities on fractal sets.

This construction is performed using an orthonormal wavelet basis and is the analog of the classical construction of Bernoulli measures on Cantor sets [**1, 2**]. We define

$$\psi_{j,k}(x) = \psi(2^j x - k), \quad j, k \in \mathbb{Z} \tag{3.1}$$

where the wavelet $\psi$ is $C^\infty$ and in the Schwartz class. Note that we use an $L^\infty$ normalization for wavelets, which is convenient to study Hölder regularity.

We denote by $B(= B_{j,k})$ the dyadic interval $[k2^{-j}, (k+1)2^{-j})$ where the wavelet is centered. We will use these intervals to index the wavelets, so that $\psi_B$ is the wavelet "centered " on the interval $B$. Because of the normalization we chose

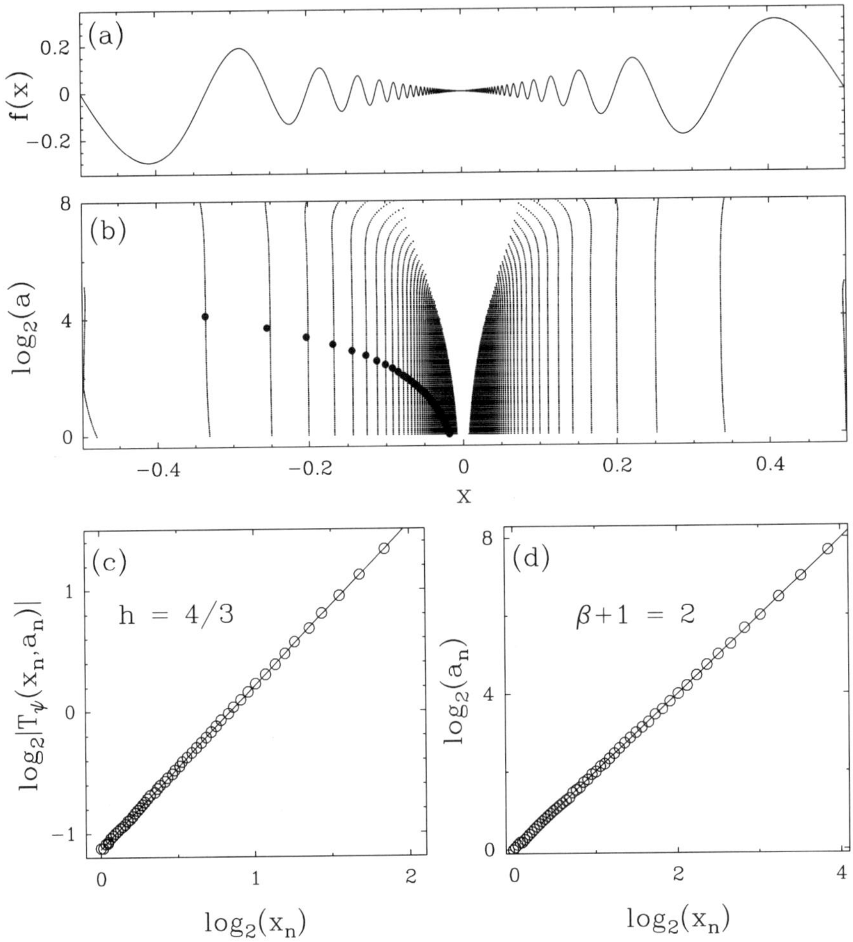


FIGURE 2. Hölder $\big(h(x_0)\big)$ and oscillation $\big(\beta(x_0)\big)$ exponents associated with an oscillating singularity at $x_0 = 0$. (a) Graph of the function $f(x) = |x|^{4/3}\sin(2\pi/x)$. (b) Corresponding wavelet transform skeleton. The analyzing wavelet $\psi$ is the first derivative of the Gaussian function. Along each maxima line $l_n$, the symbol ($\bullet$) marks the point $(x_n, a_n)$ where $\big|T_\psi[f]\big|$ is the greatest. The set of such points defines a minimizing sequence lying outside any cone. (c) $\log_2\big|T_\psi[f](x_n, a_n)\big|$ vs $\log_2(x_n)$; the slopes provides an estimate of $h(x_0) = 4/3$. (d) $\log_2(a_n)$ vs $\log_2(x_n)$; the slopes gives an estimate of $\beta(x_0) + 1 = 2$, i.e., $\beta(x_0) = 1$.

for wavelets, the wavelet coefficients of $f$ are given by

$$T(B) = 2^j \int f(x)\psi_B(x)\,dx. \tag{3.2}$$

Consider now the usual construction of a Cantor set:

- At level 0, we start with the interval $A_\emptyset = [0, 1]$

- At level 1, we construct $s$ disjoint intervals $A_1, \ldots, A_s$ included in $[0,1]$, where $A_k = [x_k, y_k)$ has length $|A_k| = y_k - x_k = \nu_k$, and $x_1 = 0$ and $y_s = 1$.
- At level $n$, we have $s^n$ subintervals $A_i$ indexed as usual by an $n$-tuple $i = (i_1, \ldots, i_n) \in \{1, \ldots s\}^n$; $A_i = [x_i, y_i)$ and the length of $A_i$ is

$$|A_i| = \nu_i = \nu_{i_1} \ldots \nu_{i_n}. \tag{3.3}$$

We denote by $K$ the Cantor set

$$K = \bigcap_{n \in \mathbb{N}} \Big( \bigcup_{i \in \{1,\ldots,s\}^n} A_i \Big). \tag{3.4}$$

We pick $s$ ratios for the scales $\lambda_1, \ldots, \lambda_s$ of the intervals indexing the wavelets such that

$$\forall k = 1, \ldots, s, \quad \lambda_k \leq \nu_k; \tag{3.5}$$

then, we construct on each interval $A_i$ a dyadic subinterval $B_i$ of length

$$|B_i| = \lambda_i = C(B_i)\lambda_{i_1} \ldots \lambda_{i_n} \tag{3.6}$$

where $C(B_i) \in [1/2, 1)$ is chosen such that $\lambda_i$ is a power of 2.

We still have to define the exact position of the wavelets and the size of the wavelet coefficients. For that we pick $\eta \in [0,1]$ and the interval $B_i$ is the unique dyadic interval of length $\lambda_i$ satisfying the condition

$$x_i + \eta\nu_i \in B_i \tag{3.7}$$

(it is thus a subinterval of $A_i$ for $n$ large enough). In order to define the wavelet coefficients, we pick $s$ ratios $\mu_1, \ldots, \mu_s$ in $[0,1)$, and the corresponding wavelet coefficient of the wavelet indexed by $B_i$ is

$$T(B_i)[f] = \mu_i = \mu_{i_1} \ldots \mu_{i_n}. \tag{3.8}$$

The function $f$ is completely determined by setting to zero all the other wavelet coefficients. An example of such construction is illustrated in Fig. 3 where $K$ is the triadic Cantor set. Fig. 4 shows other examples of such functions. Let us call $\mathcal{N}$ the class of functions constructed as above. Note that functions in $\mathcal{N}$ can be also constructed from continuous wavelet transform (see Ref. [**2**]).

Let $\mathcal{M}$ be the class of (linearly) SS-functions (see Refs. [**5**] and [**18**] for a definition). It can be easily shown that $\mathcal{M} \subset \mathcal{N}$ and corresponds to the special case where $\lambda_i = \nu_i$ $\forall i$ in the above construction [**3, 5, 18**]. Functions in $\mathcal{N}$ thus generalize SS-functions in the sense that they are characterized by a self-similar wavelet transform with a priori *different ratio in space and scale.* We will refer to such functions as WTSS-functions.

**3.2. Pointwise regularity and oscillation of WTSS functions.** In this section we show how the local Hölder exponent $h$ and the oscillation exponent $\beta$ can be computed for functions in $\mathcal{N}$.

The determination of the uniform Hölder regularity of $f \in \mathcal{N}$ is immediate: since all the wavelet coefficients $\mu_i$ satisfy

$$\frac{\log \mu_i}{\log \lambda_i} \geq \inf_{l=1,\ldots,s} \frac{\log \mu_l}{\log \lambda_l}, \tag{3.9}$$

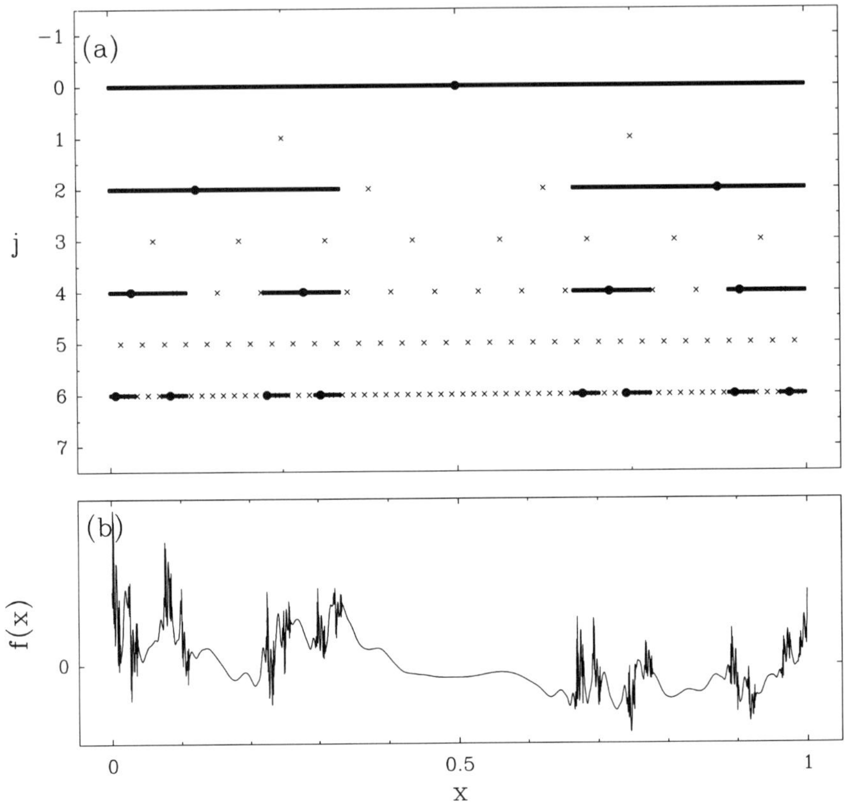


FIGURE 3. (a) Construction of the function $f$ on the dyadic grid in the case $s = 2$, $\nu_1 = \nu_2 = 1/3$, $\lambda_1 = \lambda_2 = 1/4$, $\mu_1 = 0.7$ and $\mu_2 = 0.8$ (symbols ($\bullet$) are for non zero coefficients). In this case $K$ is the triadic Cantor set. (b) Synthesis of the above decomposition using the Daubechies 9 wavelet.

then $f$ is $C^\gamma(\mathbb{R})$ where $\gamma$ is the best possible uniform regularity Hölder exponent:

$$\gamma = \inf_{l=1,\ldots,s} \frac{\log \mu_l}{\log \lambda_l}. \tag{3.10}$$

Let us note that if $x_0 \notin K$, there exists a neighborhood of $x_0$ which includes no dyadic interval indexing a nonvanishing wavelet coefficient of scale smaller than a certain $j_0$; $f$ is thus $C^\infty$ in a neighborhood of $x_0$, and we only have to consider the points of the Cantor set $K$.

If $x_0$ belongs to $K$ it is indexed by an infinite sequence $i \in \{1, \ldots, s\}^{\mathbb{N}}$. In Refs. [**1**] and [**2**], it is shown that the Hölder exponent at the point $x_0$ of the fractional $\varepsilon$-primitive of $f$ is given by

$$h_\varepsilon = \liminf \frac{\log \mu_i + \varepsilon \log \lambda_i}{\log \nu_i} \tag{3.11}$$

provided $\eta \notin K$. For the estimation of the oscillation exponent $\beta(x_0)$, let us introduce the two following notations:

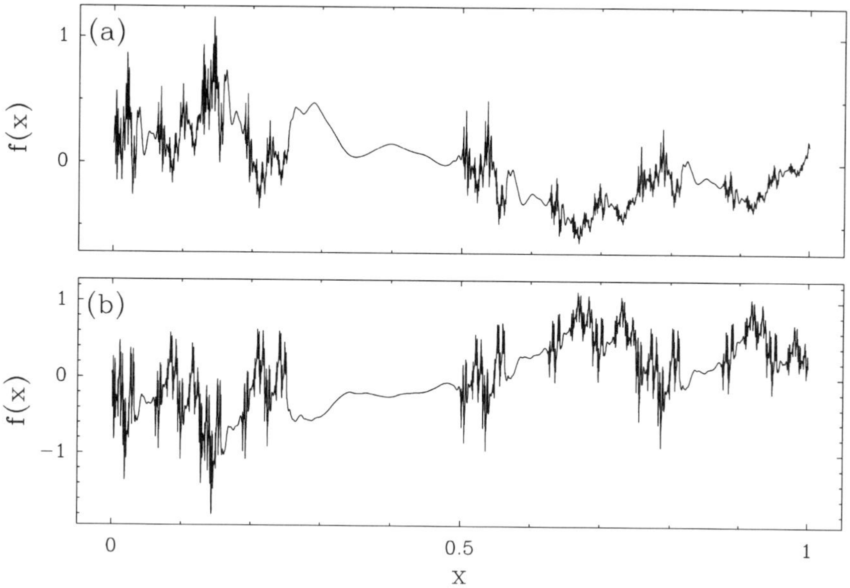


FIGURE 4. Examples of WTSS-functions that have oscillating singularities. These functions were iteratively constructed on the Daubechies 9 wavelet basis. (a) $\nu_1 = \nu_2 = \nu_3 = \nu_4 = 1/4; \lambda_1 = 1/8, \lambda_2 = \lambda_3 = \lambda_4 = 1/4; \mu_1 = 0.5, \mu_2 = 0, \mu_3 = 0.45, \mu_4 = 0.3$. (b) Same parameters as in (a) but $\mu_i = (\nu_i)^{(1/2)} = 1/2$ for $i = 1$ to 4.

DEFINITION 4. Let $(a_n)$ be a sequence of real numbers. A minimizing subsequence for $(a_n)$ is a subsequence $n_k$ such that

$$\liminf_{n\to\infty} a_n = \lim_{k\to\infty} a_{n_k}. \tag{3.12}$$

We denote by $S(a_n)$ the set of all minimizing subsequences for $(a_n)$; and if $(a_n)$ and $(b_n)$ are sequences, we define

$$\liminf_{(a_n)} b_n = \inf_{\{(n_k)\in S(a_n)\}} \liminf b_{n_k}, \tag{3.13}$$

where the infimum is taken over all minimizing subsequences $(n_k)$ for $(a_n)$.

For instance if $a_n = (-1)^n$, $\liminf_{(a_n)} b_n = \liminf b_{2n+1}$. Then, one can prove the following proposition [**1, 2**]:

PROPOSITION 3. *Let $f \in \mathcal{N}$ with $\eta \notin K$; let $x_0 \in K$. Then $f$ has at $x_0$ the oscillating singularity exponents*

$$(h, \beta) = \left(\liminf \left[\frac{\log \mu_i}{\log \nu_i}\right], \liminf_{(\log \mu_i / \log \nu_i)} \left[\frac{\log \lambda_i}{\log \nu_i}\right] - 1\right) \tag{3.14}$$

*where $i$ denotes the infinite sequence of symbols that indexes $x_0$.*

A direct consequence of this result is that WTSS-functions in $\mathcal{N}$ generally have oscillating singularities. However, if $f$ is self-similar, $\forall i, \lambda_i = \nu_i$ and thus $\forall x_0$, $\beta(x_0) = 0$; this shows that *singularities of self-similar functions are cusps.* This indicates that the usual singularity spectrum $D(h)$ suffices for characterizing the singularities of SS-functions. Moreover, since $\beta = 0$, this spectrum can be computed from the scaling behavior of partition functions defined from the wavelet coefficients along the line of the classical "canonical multifractal formalism" [**4, 5, 18, 26, 27**].

**3.3. Multifractal formalism for WTSS functions.** We define the $D(h,\beta)$ spectrum as the natural extension of the classical $D(h)$ spectrum for both Hölder and oscillation exponents:

$$D(h,\beta) = \mathrm{Dim}_H\, \mathcal{D}_{h,\beta}, \tag{3.15}$$

where $\mathcal{D}_{h,\beta}$ is the set defined by

$$\mathcal{D}_{h,\beta} = \{x, h(x) = h \text{ and } \beta(x) = \beta\} \tag{3.16}$$

and $\mathrm{Dim}_H$ is the Hausdorff dimension. Our goal in this section is to show that for WTSS-functions, $D(h,\beta)$ can be computed using a generalized multifractal formalism. Let $f \in \mathcal{N}$ and let us define the partition function associated to $f$:

$$\mathcal{Z}_n(q,p,\tau) = \sum_{i=(k_1\ldots k_n)} \mu_i^q \nu_i^{-\tau} \lambda_i^p. \tag{3.17}$$

This formula can be factorized in the following way:

$$\mathcal{Z}_n(q,p,\tau) = \left(\sum_{k=1}^{s} \mu_k^q \nu_k^{-\tau} \lambda_k^p\right)^n, \tag{3.18}$$

from which it is easy to prove that the function $P(q,p,\tau)$ defined by,

$$P(q,p,\tau) = \lim_{n\to+\infty} n^{-1} \log \mathcal{Z}_n(q,p,\tau) = \log\left(\sum_{k=1}^{s} \mu_k^q \lambda_k^p \nu_k^{-\tau}\right), \tag{3.19}$$

is real analytic and convex in each of its arguments and that there exists a real, concave analytic function $\tau(q,p)$ defined by

$$P\big(q,p,\tau(q,p)\big) = 0, \tag{3.20}$$

i.e. by,

$$\mathcal{Z}_n\big(q,p,\tau(q,p)\big) = 1. \tag{3.21}$$

In Refs. [**1**] and [**2**], we have proved the following theorem:

THEOREM 1. *Let $f \in \mathcal{N}$, $\mathcal{Z}_n(q,p,\tau)$ be its corresponding partition function defined in Eq.* (3.17) *and $\tau(q,p)$ the transition exponent defined in Eq.* (3.20). *Then the singularity spectrum $D(h,\beta)$ of $f$ is the* Legendre transform *of $\tau(q,p)$*:

$$D(h,\beta) = \min_{q,p}\big(qh + p(\beta+1) - \tau(q,p)\big). \tag{3.22}$$

This theorem is the analog of the classical result of the multifractal formalism that gives the singularity spectrum as the Legendre transform of the logarithm of some partition function. As an illustration, we show in Fig. 5, the $\tau(q,p)$ and $D(h,\beta)$ spectra of the multifractal function ($\in \mathcal{N}$) described in Fig. 4a.

From Theorem 1, one gets that the $D(h,\beta)$ singularity spectrum is a concave function. The support $\mathcal{S}$ of $D(h,\beta)$ is thus a convex set. If $s=2$, there always exists a linear combination of $h$ and $\beta$ which is constant, so that $D(h,\beta)$ is supported by a line. More generally, this set corresponds to the points $(x,y) \in \mathbb{R}^2$ such that $\exists (r_1,\ldots,r_s) \in [0,1]^s, \sum_i r_i = 1$ with $x = \sum_{k=1}^s r_k \log\mu_k / \sum_{k=1}^s r_k \log\nu_k$ and $y = \sum_{k=1}^s r_k \log\lambda_k / \sum_{k=1}^s r_k \log\nu_k$. In fact it is expected to correspond to the convex hull of the $\{(\log\mu_k/\log\nu_k, \log\lambda_k/\nu_k)\}_{1\le k\le s}$.

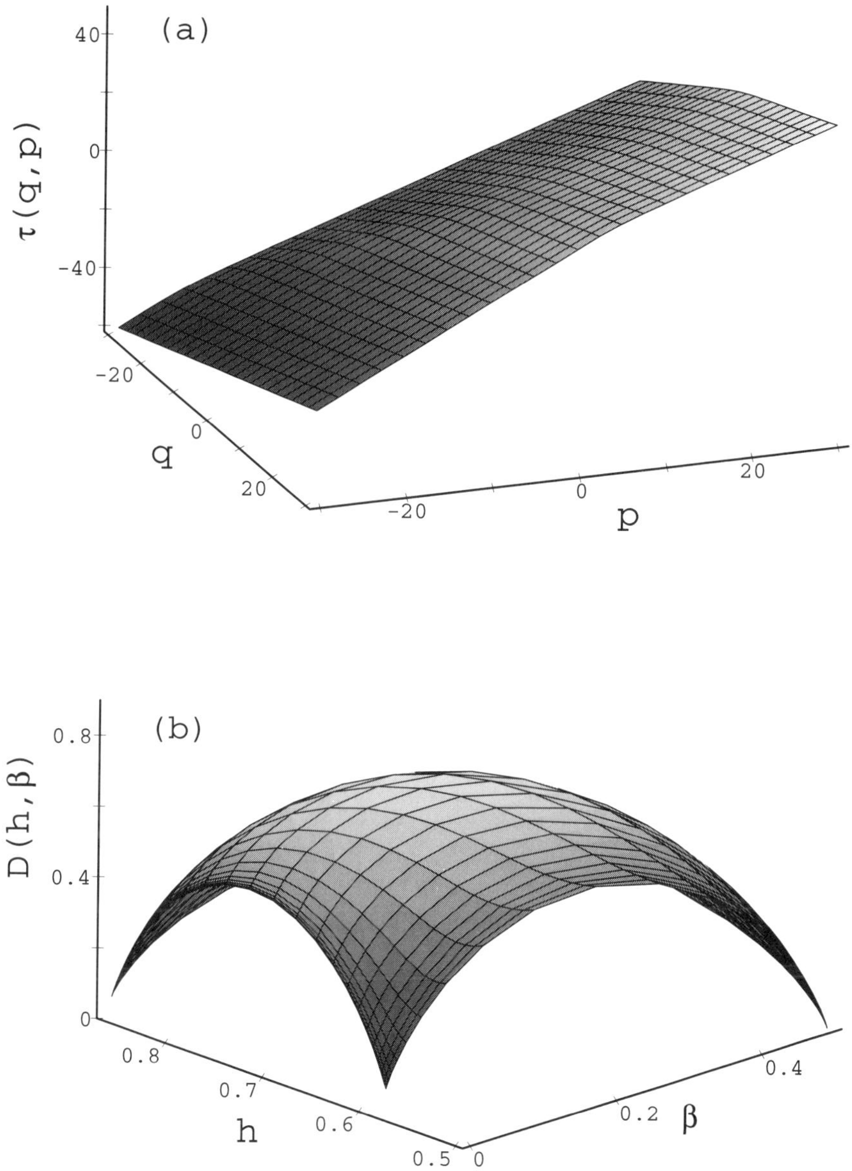


FIGURE 5. Multifractal spectra of the singular distribution ($\in \mathcal{N}$) described in Fig. 4a. (a) $\tau(q,p)$ spectrum. (b) $D(h,\beta)$ spectrum.

Since $\beta$ is linked to the derivative of $h_\varepsilon$ at $\varepsilon = 0$, one can prove (see Ref. [**1**]) that the singularity spectrum $D_\varepsilon(h,\beta)$ of the $\varepsilon$-primitive of $f$ is given by

$$D_\varepsilon(h,\beta) = D\big(h - \varepsilon(\beta+1), \beta\big). \tag{3.23}$$

Let us note that the singularity spectrum $D(h)$ of $f \in \mathcal{N}$ is simply obtained by projecting the graph of $D(h,\beta)$ on the plane $\beta = 0$:

$$D(h) = \sup_\beta D(h,\beta). \tag{3.24}$$

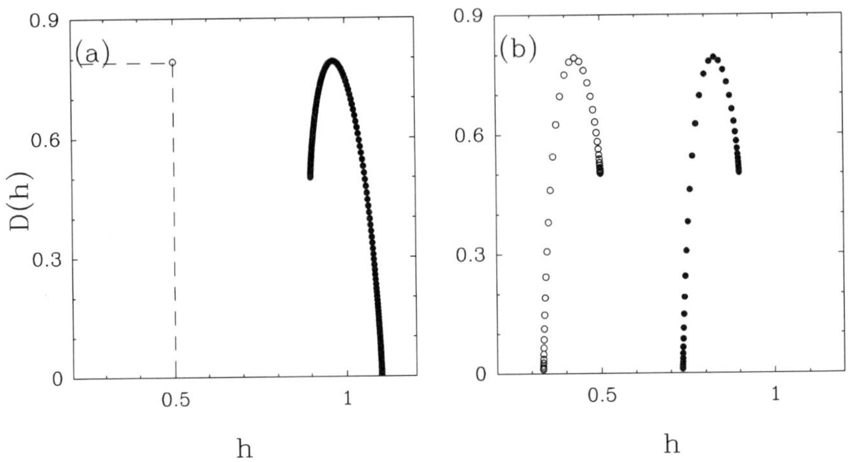


FIGURE 6. (a) $D(h)$ singularity spectrum of the monofractal function described in Fig. 4b ($\circ$) and of its $\varepsilon$-primitive ($\varepsilon = 0.4$, $\bullet$) which is clearly multifractal because of the presence of oscillating singularities (Eq. (3.23)). (b) The same $D(h)$ singularity spectra computed with the "classical" multifractal formalism. When oscillating singularities are present, usual methods lead to a wrong estimate of the $D(h)$ singularity spectrum and an $\varepsilon$-integration amounts to a simple shift of the spectrum.

Thus, as shown in Fig. 6a, one can build some monofractal distributions (in the sense that $D(h)$ is supported by a single point) whose primitives are multifractal. Indeed, for instance we suppose that for all $k$, we have $\log \mu_k = h_0 \log \nu_k$, and that $\lambda_k < \nu_k$. Then it is easy to check that all the singularities correspond to the same Hölder exponent $h(x) = h_0$ but with different values for $\beta(x)$. Thus, they are monofractal in terms of $h(x)$ but, because of the relation (3.23), they become multifractal when integrated.

Finally, let us mention that one of our motivations for considering oscillating singularities is that they can make the usual (one parameter) multifractal formalism (as described in Refs. [**4,5,9,14,16,18,26–28**] fail. Let us recall that in the classical formalism, the $D(h)$ singularity spectrum is obtained as the Legendre transform

$$D(h) = \min_q \big(qh - \tau(q)\big), \tag{3.25}$$

where $\tau(q)$ is defined e.g. as:

$$\tau(q) = -\liminf_{j \to +\infty} \frac{\log_2(\sum_k |T_{j,k}|^q)}{j}. \tag{3.26}$$

The functions (3.25) and (3.26) depend on the parameters $(\lambda_l)$ and $(\mu_l)$ only (but not on the $(\nu_l)$ upon which, from Eq. (3.24), the actual spectrum should obviously depend.) Thus, the multifractal formalism, as stated above in its canonical version, fails for the functions in $\mathcal{N}$. It is straightforward to see that Eq. (3.25) yields the spectrum of the SS-function associated to the parameters $\lambda_i = \nu_i$ and $\mu_i$: $\tau(q)$ is given by

$$\sum_{l=1}^{s} \mu_l^q \lambda_l^{-\tau(q)} = 1, \tag{3.27}$$

and $D(h)$ is the Legendre transform of this function. The inability of the canonical multifractal formalism to account for the presence of oscillating singularities is illustrated in Fig. 6b.

**3.4. Analogy with thermodynamics: a grand-canonical multifractal formalism.** It is well known that deep links exist between the multifractal formalism for self-similar distributions and the thermodynamic formalism defined from statistical physics [**4, 6, 7, 9, 13, 28**]. In particular, one can observe several analogies between thermodynamic potentials and multifractal spectra: for example, if a given point $x_0$ is identified with a *microstate* of a given statistical system (e.g. a spin configuration of some Ising model), the Hölder exponent $h(x_0)$ is the equivalent of the *energy* of this microstate and the $D(h)$ singularity spectrum corresponds to the *entropy* within a *microcanonical* description ($h$ is fixed). When $h(x_0)$ fluctuates from one point to another, the $\tau(q)$ function defined above is the logarithm of some *partition function* and can thus be seen as the analog of the *canonical free energy.* Here, we point out that the description we gave for WTSS-functions in the previous section can be seen as a *grand-canonical* description of a physical system where not only the energy varies from one state to another, but also the *number of particles.* For the sake of simplicity, let us suppose that $\nu_k = \nu$ is constant in the construction of Section 3.2. In that case, from Eq. (3.17) and by setting $\Gamma_n(q,p) = \sum_i \mu_i^q \lambda_i^p$, we get

$$\mathcal{Z}_n(q,p,\tau) = \nu^{-n\tau}\Gamma_n(q,p), \tag{3.28}$$

and thus, from the definition of $\tau(q,p)$,

$$\tau(q,p) = \lim_{n\to+\infty} \frac{\log \Gamma_n(q,p)}{\log(\nu^n)}. \tag{3.29}$$

From Eq. (3.14), one can introduce the expressions $\mu_i^q = e^{qh_i n \ln \nu}$ and $\lambda_i^p = e^{p(1+\beta_i)n\ln\nu}$ into $\Gamma_n$; we can then establish the following "dictionary" between this formalism and a grand-canonical description:

- $i \equiv$ microstate
- $-n\ln\nu \equiv$ volume $V$
- $-n\ln\nu \to +\infty \equiv$ thermodynamic limit
- $e^{-(qh_i+p(1+\beta_i))n\ln\nu} \equiv$ Boltzmann weight
- $h_i \equiv$ energy per unit of volume of the microstate $i$
- $1+\beta_i \equiv$ number of particles per unit of volume of the microstate $i$
- $q \equiv 1/kT$ (inverse temperature)
- $p \equiv$ chemical potential
- $\Gamma_n(q,p) \equiv$ grand canonical partition function
- $\tau(q,p) = \ln\Gamma_n/n\ln\nu \equiv$ grand potential
- $D(h,\beta) =$ microcanonical entropy
- $D(h,\beta) = \min_{q,p}\big(qh + p(\beta+1) - \tau(q,p)\big) \equiv$ principle of equivalence of ensembles in the thermodynamic limit.

## 4. Conclusion

To summarize, we have shown in this paper that a singular behavior must be described by two exponents: the Hölder exponent $h$ (the "strength" of the singularity) and the oscillation exponent $\beta$ (that quantifies the divergence of the instantaneous frequency). These two quantities can be easily characterized using wavelet analysis. We have defined a new multifractal formalism that allows us to estimate the $D(h, \beta)$ singularity spectrum for a large class of singular distributions involving both cusp and oscillating singularities. This newly defined multifractal formalism is equivalent to a grand-canonical description of the fluctuations of the two exponents $h$ and $\beta$. In a forthcoming work, we hope to present a new algorithm based on this grand-canonical multifractal formalism and that will overcome the intrinsic inadequacies of the classical canonical methods with respect to the detection of oscillating singularities.

## References

1. A. Arneodo, E. Bacry, S. Jaffard, and J. F. Muzy, *Singularity spectrum of multifractal functions involving oscillating singularities*, J. Fourier Anal. Appl. (1997) (to appear).
2. ———, *Oscillating singularities on Cantor sets: A grand-canonical multifractal formalism*, J. Stat. Phys. **87** (1997), nos. 1-2, 179–209.
3. A. Arneodo, E. Bacry, and J. F. Muzy, *Oscillating singularities in locally self-similar functions*, Phys. Rev. Lett. **74** (1995), 4823–4826.
4. ———, *The thermodynamics of fractals revisited with wavelets*, Physica A **213** (1995), 232–275.
5. E. Bacry, J. F. Muzy, and A. Arneodo, *Singularity spectrum of fractal signals from wavelet analysis: Exact results*, J. Stat. Phys. **70** (1993), 635–674.
6. R. Badii, *Conservation laws and thermodynamic formalism for dissipative dynamical systems*, Ph.D thesis, Univ. of Zurich, 1987.
7. T. Bohr and T. Tél, *The thermodynamics of fractals*, Directions in Chaos (Hao-Bai-Lin, ed.), vol. 2, World Scientific, Singapore, 1988, pp. 194–237.
8. C. K. Chui, *An introduction to wavelets*, Academic Press, Boston, 1992.
9. P. Collet, J. Lebowitz, and A. Porzio, *The dimension spectrum of some dynamical systems*, J. Stat. Phys. **47** (1987), 609–644.
10. J. M. Combes, A. Grossmann, and P. Tchamitchian (eds.), *Wavelets*, Springer, Berlin, 1989.
11. I. Daubechies, *Ten lectures on wavelets*, SIAM, Philadelphia, 1992.
12. N. Delprat, B. Escudié, P. Guillemain, R. Kroland-Martinet, Ph. Tchamitchian, and B. Torrésani, *Asymptotic wavelet and Gabor analysis: Extraction of instantaneous frequencies*, IEEE Trans. Inform. Theory **38** (1992), 644–664.
13. M. J. Feigenbaum, *Some characterization of strange sets*, J. Stat. Phys. **46** (1987), 919–924; *Scaling spectra and return times of dynamical systems*, J. Stat. Phys. **46** (1987), 925–932.
14. U. Frisch and G. Parisi, *On the singularity structure of fully-developed turbulence*, Turbulence and Predictability in Geophysical Fluid Dynamics and Climate Dynamics (M. Ghil, R. Benzi, and G. Parisi, eds.), North-Holland, Amsterdam, 1985, pp. 84–87.
15. A. Grossmann and J. Morlet, *Decomposition of Hardy functions into square integrable wavelets of constant shape*, SIAM J. Math. Anal. **15** (1984), no. 4, 723–736.
16. T. C. Halsey, M. H. Jensen, L. P. Kadanoff, I. Procaccia, and B. I. Shraiman, *Fractal measures and their singularities: The characterization of strange sets*, Phys. Rev. A **33** (1986), 1141–1151.
17. M. Holschneider and P. Tchamitchian, *Régularité locale de la fonction "non-différentiable" de Riemann*, Les Ondelettes en 1989 (Orsay, 1989), 102–124, 209–210; Lecture Notes in Math., 1438, Springer, Berlin, 1990.
18. S. Jaffard, *Multifractal formalism for functions.* I. *Results valid for all functions*, SIAM J. Math. Anal. **28** (1997), no. 4, 944–970; II. *Self-similar functions*, SIAM J. Math. Anal. **28** (1997), no. 4, 971–998.
19. S. Jaffard, *Exposants de Hölder en des points donnés et coefficients d'ondelettes*, C. R. Acad. Sci. Paris Série I Math. **308** (1989), no. 4, 79–81.

20. S. Jaffard and Y. Meyer, *Wavelet methods for pointwise regularity and local oscillations of functions*, Mem. Amer. Math. Soc. **123** (1996), no. 587, Providence, R.I.
21. P. G. Lemarié (ed.), *Les ondelettes en* 1989, Harmonic Analysis (Orsay, 1989), Lecture Notes in Math.S, vol. 1438, Springer-Verlag, Berlin, 1990.
22. S. Mallat and W. L. Hwang, *Singularity detection and processing with wavelets*, IEEE Trans. Inform. Theory **38** (1992), 617–643.
23. Y. Meyer, *Ondelettes*, Herman, Paris, 1990.
24. ——— (ed.), *Wavelets and applications*, Proc. Second Intern. Conf. (Marseilles, 1989) RMA: Res. Notes Appl. Math., vol. 20, Masson, Paris; Springer-Verlag, Berlin, 1992
25. Y. Meyer and S. Roques (eds.), *Wavelets: their past and their future*, Progress in Wavelet Analysis and Applications (Toulouse, 1992), Éditions Frontières, Gif-sur-Yvette, 1993, pp. 9–18,
26. J. F. Muzy, E. Bacry, and A. Arneodo, *The multifractal formalism revisited with wavelets*, Internat. J. Bifur. Chaos Appl. Sci. Engrg. **4** (1994), 245–302.
27. ———, *Wavelets and multifractal formalism for singular signals: Application to turbulence data*, Phys. Rev. Lett. **67** (1991), 3515–3518.
28. D. A. Rand, *The singularity spectrum $f(\alpha)$ for cookie-cutters*, Ergodic Theory Dynam. Systems **9** (1989), no. 3, 527–541.
29. M. B. Ruskai, G. Beylkin, R. Coifman, I. Daubechies, S. Mallat, Y. Meyer, and L. Raphael (eds.), *Wavelets and their applications*, Jones and Bartlett, Boston, 1992.

Centre de Recherche Paul Pascal, Avenue Schweitzer, 33600 Pessac, France
*E-mail address*: `arneodo@crpp.u-bordeaux.fr`

CMAP, École Polytechnique, 91128 Palaiseau, France
*E-mail address*: `bacry@cmapx.polytechnique.fr`

Département de Mathématiques, Université Paris XII, Faculté des Sciences et Technologie, 61 Avenue du Général de Gaulle, 94010 Créteil Cedex, France and C.M.L.A., ENS Cachan, 61 Avenue du Président Wilson, 94235 Cachan Cedex, France
*E-mail address*: `jaffard@cmla.ens-cachan.fr`

Centre de Recherche Paul-Pascal, Avenue Schweitzer, 33600 Pessac, France
*E-mail address*: `muzy2@hpondla.crpp.u-bordeaux-fr`

# Part 4

# Splines and Wavelets in Statistics

# Introduction and Summary

A fascinating and important problem in statistics is the recovery of an unknown function from noisy, sampled data. The more traditional methods of curve fitting such as kernel estimates, spline smoothers or truncated Fourier estimates are linear in the data and perform very well when the function to be estimated belongs to a known Sobolev class and is contaminated by Gaussian white noise. However, when the function exhibits spatial inhomogeneity, minimax linear estimators, that is, those that minimize the risk or expected loss among all linear estimates, can be arbitrarily outperformed by nonlinear ones. Influenced both by approximation theory and by decision theoretic arguments, Donoho and Johnstone developed a nonlinear method using orthonormal bases of compactly supported wavelets for function estimation in Gaussian white noise. Their method consists of thresholding the coefficients in the wavelet domain and has been shown to be asymptotically minimax for function estimation in the Gaussian white noise model over a large range of Besov spaces.

The papers presented here give us a taste of the many interesting ways that wavelets can be used to solve important function estimation problems in statistics. Different wavelet estimators are proposed to address some of the following questions:

(i) What happens if the function to be estimated has discontinuities?
(ii) What happens if the errors are no longer Gaussian?
(iii) Can wavelet estimation be made more robust?
(iv) What happens if the error distribution is no longer stationary?

Of course, other methods such as adaptive smoothing splines can be used to overcome the problem of spatial inhomogeneity in function estimation. A valuable contribution to the literature on the characterization of algorithms for smoothing splines is given here by Ramsay and Heckman. It is an accessible elucidation of important results of Anselone and Laurent.

The other four papers concentrate on wavelet methods.

As noted by Antoniadis, in many applications of curve estimation, the function to be estimated is only piecewise smooth and may have a finite number of jumps. He shows that for such functions in a Sobolev-type setting with normal errors, the wavelet based non-parametric regression estimator enjoys the same asymptotic convergence properties as in the case of smooth functions without discontinuities.

Averkamp and Houdré explore the interesting world of function estimation where the noise contamination is not necessarily Gaussian. They obtain important asymptotic minimaxity results for an "ideal denoising" scheme based on thresholding wavelet coefficients for a large class of symmetric error distributions, including the important family of scale mixtures of normals.

By generalizing both the family of functionals between point values and the family of interpolating functions beyond polynomials in the original interpolation scheme of Deslauriers and Dubuc, Donoho and Yu obtain many interesting refinement schemes and transforms. Of particular interest to statisticians because of its robustness possibilities is the non-linear refinement scheme based on median (rather than mean)-interpolation and imputation.

In many real life problems where data are measured over time, the error process can no longer be assumed to be stationary. In order to address this problem, von Sachs, Nason and Kroisandt use the class of locally stationary wavelet processes and study the estimation of the evolutionary wavelet spectrum by means of non-linear wavelet shrinkage of the wavelet periodogram.

Brenda Macgibbon

Centre de Recherches Mathématiques
CRM Proceedings and Lecture Notes
Volume **18**, 1999

# Wavelet Estimators for Change-Point Regression Models

A. Antoniadis

ABSTRACT. The objective of this paper is to contribute to the methodology available for dealing with the detection of discontinuities and the estimation of one dimensional piecewise smooth regression functions observed in white Gaussian noise over an interval.

Our approach is nonparametric in nature because the unknown function will not be assumed to have any specific form. The nonparametric methods comprise indirect and direct approaches. In "indirect" methods one detects and locates the change points prior to fitting the curve, and then uses one's favorite function estimation technique on each segment. In the "direct" approach, on the other hand, one tries to recover the unknown regression function while at the same time detect and preserves discontinuities. Our method follows the first approach.

In this paper it is shown that, provided discontinuities can be detected and located with sufficient accuracy, detection followed by wavelet smoothing enjoys optimal rates of convergence similar to those obtained in the smooth case.

## 1. Introduction

Wavelet based nonparametric regression has become a well known and mathematically founded technique for estimating smooth functions adaptively. Optimal rates of convergence have been thoroughly examined for different observation schemes and $L_2$ losses by many authors. In many applications, however, the function to be estimated presents isolated discontinuities, and is only piecewise smooth. To cite only few examples, such change point problems arise in quality control, medical monitoring, econometrics, study of growth in biology, signal processing and edge detection in image processing. At first glance, the known results on the optimal rates of convergence may suggest that only slower rates can be achieved for such functions with discontinuities. The object of this paper is to examine how well nonparametric wavelet based regression methods perform in this situation and how we can extend present techniques in order to obtain comparable rates of convergence.

To estimate a piecewise smooth regression function $f$ observed with noise, a number of strategies can be applied including parametric methods which do not

---

1991 *Mathematics Subject Classification.* Primary: 62G05; Secondary: 62G07.
This is the final version of the paper.

be concern us here. Nonparametric methods comprise both indirect and direct approaches. In "indirect" methods, one detects and locates the change points prior to fitting the curve, and then uses one's favorite function estimation technique on each segment. In the "direct" approach, on the other hand, one tries to recover the unknown regression function while at the same time detecting and preserving discontinuities.

There already exists a considerable body of literature on discontinuity detectors with good location properties that may be employed in the indirect approach. These include semiparametric smoothing splines methods proposed by Laurent and Utreras [**14**] Girard [**12**]; kernel based methods by Cany [**4**], Hall and Titterington [**13**], Müller [**20**], Wu and Chu [**29**], Speckman [**26**], Eubank and Speckman [**11**], Bunt, Koch and Pope [**3**]; local polynomial methods by McDonald and Owen [**19**], Leclerc and Zucker [**15**], Loader [**16**], Qui and Yandell [**25**]; wavelet methods proposed by Mallat and Hwang [**18**], Vercken and Potier [**23**], Wang [**27, 28**]; Fourier expansions in eigenfunctions arising from particular differential boundary value problems with Neumann conditions by Oudshoorn [**21**]. We present here a refinement of Wang's procedure for detecting and locating the discontinuities with optimal mean-square optimal rates in the context a Sobolev-type setting with Normal errors. The resulting nonparametric regression estimator, regardless of the presence of a finite number of jumps in the unknown regression function is then shown to enjoy the same asymptotic convergence rate as in the case of smooth functions without discontinuities.

An illustrative example conclude this paper.

## 2. Estimation in a White Noise Model

One frequently-used formulation of the change point problem is the following. Consider observing a process $Y_\tau(t)$ satisfying the following stochastic differential equation

$$dY_\tau(t) = f(t)\,dt + \tau\,dW(t), \quad 0 < t \leq 1, \tag{2.1}$$

with some prescribed initial value $X(0)$, which is either a random variable independent of the noise $W$ or constant, where $W$ is a standard Wiener process, $\tau$ is a formal noise level parameter which we think of as small, and $f$ is an unknown piecewise smooth function. By piecewise smooth function on $[0,1]$ we mean that, given a positive integer $m$ and a constant $Q > 0$:

(M1) $f \in L_2([0,1]) \cap L_\infty([0,1])$,

(M2) there are $d$ (unknown) points ($d$ bounded by $D$), $0 < b_1 < \cdots < b_d < 1$ points at which $f$ has a discontinuity and $b_{j+1} - b_j > \kappa$, $j = 0, \ldots, d$, where $\kappa$ is an unknown positive constant,

(M3) at each discontinuity point $b_i$, the left and right limits $f(b_i-)$ and $f(b_i+)$ exist and are such that $|f(b_i+) - f(b_i-)| \geq L$, where $L$ is a positive constant not necessarily known.

(M4) $f$ belongs to the piecewise Sobolev-ellipsoid, i.e.

$$\sum_{j=0}^{d} \int_{b_j}^{b_{j+1}} \left[f^{(m)}(t)\right]^2 dt \leq Q, \quad b_0 = 0, b_{d+1} = 1.$$

For this setting we refer to Oudshoorn [**21**], where Fourier expansions in eigenfunctions arising from a particular differential boundary value problem with Neumann conditions were used to obtain results similar to the ones of the present paper. For convenience, we denote by $\mathcal{H}(m,Q)$ the class of functions $f$ that satisfy assumptions M1–M4.

The statistical inference problem then consists in:

- estimating the number and position(s) of change point(s).
- recovering the piecewise smooth regression function $f$ without degrading its discontinuities.

REMARK 2.1. The white noise model (1) is a generalization of the diffusion model used by Pollak and Sigmund [**22**] to study change-point problems. It is also closely related to the following discrete regularly spaced nonparametric regression model:

$$Z_i = f(t_i) + \sigma\varepsilon_i \quad i = 1,\dots,n, \tag{2.2}$$

with $t_i = 1/n$, the $\varepsilon$'s independent standard normal errors, $\sigma > 0$ and $f$ an unknown function. By defining as in Wang [**28**] or Donoho and Johnstone [**10**], the regression process $\{Y_n(t), t \in [0,1]\}$ via $t_0 = 0$, $Y_n(0) = 0$ and $Y_n(t_i) = \sum_{j=1}^{i} Z_i$ for $i = 1,\dots,n$, with interpolation by a Wiener process $W$ for $t_i \leq t < t_{i+1}$, $Y_n$ appears as a white noise process with the function $f_n(t) = f(t_i)$ for $t_i \leq t < t_{i+1}$ and $\tau = \sigma n^{-1/2}$.

Since the assumptions M1–M4 mean in effect that we observe $d+1$ nonparametric regression problems which have been joined end-to-end without marking the endpoints, a natural strategy for estimation of $f$ is to attempt to reverse the process, that is:

(S1) Determine estimates $\hat{b}_j$, $j = 1,\dots,\hat{d}$ of jump points of $f$; set $\hat{b}_0 = 0$ and $\hat{b}_{\hat{d}+1} = 1$.
(S2) Break the process $\{Y_\tau(t), t \in [0,1]\}$ into $\hat{d}+1$ segments $\{Y_\tau(t),\ t \in [\hat{b}_j, \hat{b}_{j+1}[, j = 0,\dots,\hat{d}\}$.
(S3) Use a nonparametric wavelet shrinkage estimator adapted on each segment separately to estimate $f$ on that part of its domain.
(S4) Join the estimates together in the obvious way.

In order to achieve this task two steps in the previous process need to be specified: the discontinuity detector/locator (Step S1) and the nonparametric regression estimator for (Step S3). We shall concentrate on how well it is possible to estimate $f$ by this strategy.

The quadratic risk ($L_2$-loss) for an estimator $\hat{f}$ of $f$ is defined as

$$R(\hat{f}, f) = \mathbb{E}\{\|\hat{f} - f\|_2^2\}. \tag{2.3}$$

The minimax quadratic risk in estimating $f$ with respect to the class $\mathcal{H}(m,Q)$ is then defined by

$$R_\tau\big(\mathcal{H}(m,Q)\big) = \inf_{\hat{f}} \sup_{f\in\mathcal{H}(m,Q)} R(\hat{f}, f), \tag{2.4}$$

where the infimum is taken over all estimators $\hat{f}$.

The exact asymptotic behavior of the minimax risk for the class $\mathcal{H}(m,Q)$ satisfies

$$\lim_{\tau\downarrow 0} \tau^{-4m/2m+1} R_\tau\big(\mathcal{H}(m,Q)\big) = \gamma(m,Q), \tag{2.5}$$

where $\gamma(m,Q) = \big(Q(2m+1)\big)^{1/2m+1}\big(m/\pi(m+1)\big)^{2m/2m+1}$. We refer to Oudshoorn [**21**] for the derivation of this result.

Our purpose is to show that it is possible to estimate $f$ by wavelet methods with a risk that is nearly equal to this optimal asymptotic minimax risk. To do so we first use adaptative wavelet shrinkage to the data, in the case the jump points are known. Results from Donoho and Johnstone [**9**], produce an estimator $\tilde{f}$ with the ideal risk of Theorem 1 above, within a logarithmic factor. Of course this estimator depends on the (unknown) jump points. We realize Step S1 by estimating these jump points in an optimal fashion. To finally produce our estimator we replace in $\tilde{f}$ the jump points $b_j$ by their estimates, resulting in a asymptotically minimax estimator $\hat{f}$.

**2.1. Quasi-optimal estimation with known jump-points.** We follow the ideas of Donoho and Johnstone in deriving an optimal estimator of $f$ when the location of the jump points are assumed to be known. We will be using an orthogonal multiresolution analysis that "lives" on the interval as the one developed by Cohen, Daubechies and Vial [**6**] or the one developed by Anderson, Hall, Jawerth and Peters [**2**] . The starting point for such analyses, is any one of the compactly supported orthogonal bases in Daubechies, with $N$ vanishing moments and support $\psi =$ support $\phi = [0, 2N-1]$. The construction given by these authors involves the construction of multiscale spaces $V_j[a,b]$ and detail spaces $W_j[a,b]$ with properties analogous to those associated to multiresolution analyses on the real line, as well as collections of basis functions $\{\phi_{j,k}\}_k$ and $\{\psi_{j,k}\}_k$ spanning these spaces. At sufficiently fine scales, the approximation spaces are generated by exactly $2^j$ orthonormal basis functions and the same is true for the complement spaces $W_j[a,b]$. The result of these constructions is an orthonormal family of wavelets on $[a,b]$, with $N$ vanishing moments; together with an orthonormal family of scaling functions on $[a,b]$ at the coarsest scale under consideration, these adapted wavelets constitute an orthonormal basis of $L_2\big([a,b]\big)$. In addition, their regularity and vanishing moment properties ensure that they are unconditional bases for the Hölder spaces $C^s\big([a,b]\big)$ $s<r$, where $r$ is the regularity of the original wavelet basis, $\psi \in C^r$. The above constructions provide also fast wavelet transform of sequences of arbitrary length (see the above cited papers for details). For simplicity of notation, given a multiresolution decomposition $V_\ell\big([b_j,b_{j+1}]\big) \oplus \oplus_{j\geq\ell} W_j\big([b_j,b_{j+1}]\big)$, we hereafter identify $\psi^j_k$ with the sequence

$$\phi_{\ell,0},\ldots,\phi_{\ell,2^\ell-1},\psi_{\ell,0},\ldots,\phi_{\ell,2^\ell-1},\ldots,\psi_{j,0},\ldots,\psi_{j,2^j-1},\ldots\ .$$

Rewrite the segmented observed process in Step S2 with known segments $[b_j, b_{j+1}]$ into an equivalent sequence of observations $(j = 0,\ldots,d)$

$$Y_{k,(j)} = \theta_{k,(j)} + \tau\varepsilon_{k,(j)}, \quad j = 0,\ldots,d \quad k \in \mathbb{N},$$

where

$$Y_{k,(j)} = \int_{b_j}^{b_{j+1}} \psi^j_k(t)\,dY_\tau(t), \quad \theta_{k,(j)} = \int_{b_j}^{b_{j+1}} f(t)\psi^j_k(t)\,dt,$$

and

$$\varepsilon_{k,(j)} = \int_{b_j}^{b_{j+1}} \psi^j_k(t)\,dW(t).$$

The orthonormality of the above wavelet transform has a fundamental statistical consequence: the random variables $\varepsilon_{k,(j)}$ are independent and $N(0,1)$ distributed.

Moreover, the wavelet transform transforms estimators in one domain in the other domain with isometry of risks. The problem of recovering $f$ can then be seen as the one recovering those few coefficients of $f$ in each segment that are significantly different than 0 against a Gaussian white noise background. Donoho and Johnstone's **[9]** wavelet hard thresholding selection rule is an instance of this principle. Let

$$\eta_\lambda(y) = y\mathbf{I}\{|y| > \lambda\}$$

denote the hard threshold, which sets to zero data $y$ below $\lambda > 0$ in absolute value. As an estimate of the unknown function $f$ we now consider Donoho and Johnstone's shrinkage estimator, defined as follows:

$$\tilde{f}(t) = \sum_{k=0}^{\infty} \tilde{\theta}_{k,(j)}\psi_k^j(t), \quad t \in [b_j, b_{j+1}[, \quad j = 0, \dots, d, \tag{2.6}$$

where

$$\tilde{\theta}_{k,(j)} = \begin{cases} Y_{k,(j)} & \text{if } k < 2^\ell - 1 \\ \eta_s(Y_{k,(j)}), & \text{for } k \geq 2^\ell, \end{cases}$$

with $s = 2\tau/2m + 1\sqrt{2|\log(\tau) + \log\log(\tau)|}$ and $\ell = m$. The threshold and the method of estimation are optimal in the sense that the maximal quadratic risk of the pseudo-estimator $\tilde{f}$ over $\mathcal{H}(m, Q)$ is asymptotically nearly equal to the bound given in (2.5). In order to explain this, observe first that the threshold $s$ defining our estimator is asymptotically equivalent to the one used by these authors for finite sample regression models. We refer to Wang's paper (1995) for its derivation. Now, by the Parseval relation for the wavelet basis, the additivity of coordinate risks and independence of the noise, and the fact that on each segment we are dealing with a Sobolev ellipsoid, Theorem 13 of Donoho and Johnstone **[10]** applies on each segment, provided that the wavelet basis we are using is of regularity $r$ larger than $m$. The last mentioned theorem (and its proof), states that the quadratic (nearly) minimal risk for the sequence space is attained by the thresholded estimator and that it is equivalent, up to some constants that depend only the wavelet that is used, to the (nearly) minimal quadratic risk in the function problem. Using the fact that the number of segments for the segmented problem is bounded by $D + 1$ and using again the additivity of coordinate quadratic risks, we conclude easily that the maximal quadratic risk for $\tilde{f}$ over $\mathcal{H}(m, Q)$ is bounded above by $\gamma(m, Q)\tau^{4m/2m+1}\mathcal{O}(|\log(\tau)|)$. It is therefore, to a logarithmic factor, nearly minimal.

The discussion that follows in the next section focuses on the way in which appropriate estimates of the discontinuity locations may be fitted into the definition of $\tilde{f}$, in order to obtain asymptotically the same quality of estimation as that of $\tilde{f}$.

**2.2. A discontinuity locator.** As already mentioned in the introduction, many examples of estimators which detect and locate discontinuities exist in the literature. However we need here discontinuities locators that converge fast enough (in the mean-square sense) to the true discontinuity location. Condition M2 prevents the discontinuity detection and location step from finding discontinuities less than the prescribed distance $\kappa$ apart or to close to the boundary of the data. Assuming that such a condition is adopted, one can segment the interval $[O, 1]$ into intervals of size $h = \tau^2(\log(\tau^{-2}))^{1+\eta}$ for some fixed $\eta > 0$ and such that $2h < \kappa$ for $\tau$ small enough. Oudshoorn has used a similar grid to compute local left and right kernel smoothers for estimating the left and right limits of $f$ at these grid points, in order to locate the jumps. The method adopted here relies upon the continuous wavelet transform evaluated at these grid points and is much easier to compute and analyze. We denote by $G = \{g_i = i\tau^2,\ i = \lfloor h\tau^{-2}\rfloor, \dots, \lceil (1-h)\tau^2\rceil\}$ the corresponding grid.

Let $\psi$ be a compactly supported Daubechies wavelet regular enough to be differentiable and define $\psi_h(x) = h^{-1}\psi(x/h)$. The continuous wavelet transform of $f$ at $(h,x)$ is then given by $Tf(h,x) = \int \psi_h(x-u)f(u)\,du$. It is known (see e.g. Mallat and Hwang [**18**] or Daubechies [**7**]) that, at small scales $h$, $Tf(h,x)$ characterizes the local regularity of $f$. We therefore expect that $|Tf(h,g_i)|$ will be "large" if there is a discontinuity location $b_j$ at the vicinity of $g_i$. Indeed this is the case and we have the following proposition:

PROPOSITION 2.2. *If $\psi$ is differentiable, then for every $\delta > 0$ and $0 < \alpha < \eta/2$, uniformly in $f \in \mathcal{H}(m,Q)$, one has*

(i) $\mathbb{P}\{|TY(h,g_i)| > (\log\tau^{-2})^{-\alpha}\} = o(\tau^\delta)$, *as $\tau \downarrow 0$, for every point $g_i \in G$ such that $[g_i - h, g_i + h]$ does not contain any discontinuity location $b_j$,*

(ii) $\mathbb{P}\{|TY(h,g_i)| < (\log\tau^{-2})^{-\alpha}\} = o(\tau^\delta)$, *as $\tau \downarrow 0$, for every point $g_i \in G$ such that for some $j = 1,\dots,d$, $|g_i - b_j| < h/2$.*

Consider first a grid point $g_i$ such that no jump location $b_j$ is covered by the interval $[g_i - h, g_i + h]$. Since $m \geq 1$, the function $f$ is at least of Lipschitz regularity 1 at $g_i$. Now, by Theorem 2 of Mallat and Hwang, we have

$$|Tf(h,g_i)| \leq \text{const}\cdot h \leq \frac{1}{2}(\log\tau^{-2})^{-\alpha}, \tag{2.7}$$

for $\tau$ sufficiently small. At the scale $h$, the wavelet transform of the white noise, is a stationary Gaussian process with zero mean and variance $1/h$. Therefore, $TY(h,g_i)$ is a Normal random variable with mean $Tf(h,g_i)$ and variance $\tau^2/h$. Using the fact that for the standard normal distribution function $\Phi$ we have $1-\Phi(c) \leq \phi(c)/c$ as soon as $c > 1$, where $\phi$ denotes the standard normal density function, we have

$$\begin{aligned}
&\mathbb{P}\{|TY(h,g_i)| > (\log\tau^{-2})^{-\alpha}\} \\
&\qquad \leq \mathbb{P}\{|TY(h,g_i) - Tf(h,g_i)| > (\log\tau^{-2})^{-\alpha} - |Tf(h,g_i)|\} \\
&\qquad \leq \mathbb{P}\{|TY(h,g_i) - Tf(h,g_i)| > 1/2(\log\tau^{-2})^{-\alpha}\} \\
&\qquad \leq \sqrt{\frac{2}{\pi}}(\log\tau^{-2})^{\alpha-\eta/2-1/2}\exp\left\{-\frac{1}{8}(\log\tau^{-2})^{1+\eta-2\alpha}\right\} = o(\tau^\delta),
\end{aligned}$$

provided that $\alpha < \eta/2$.

For a grid point $g_i$ such that $|g_i - b_j| < h/2$, expression (A4) of Wang [**28**] applies and we have

$$|Tf(h,g_i)| = 2^{-1}|f(b_j+) - f(b_j-)|\left\{\left|\int \psi(u-t)\,\text{sign}(u)\,du\right| + o(1)\right\}$$

which by M3, we bound below by $L/3$. It then follows that

$$\begin{aligned}
\mathbb{P}\{|TY(h,g_i)| < (\log\tau^{-2})^{-\alpha}\} &\leq \mathbb{P}\{|TY(h,g_i) - Tf(h,g_i)| > L/4\} \\
&\leq \frac{2\sqrt{2}}{L\pi}(\log\tau^{-2})^{-1+\eta/2}\exp\left\{-\frac{L^2}{32}(\log\tau^{-2})^{1+\eta}\right\} = o(\tau^\delta).
\end{aligned}$$

Since more than one grid point $g_i$ may be close to a true discontinuity $b_j$, we find it necessary to match up as possible the true discontinuity points of $f$ with the points of the grid that detect them. We describe now such a matching algorithm. Let $\alpha < \eta/2$. Pair of adjacent points $(g_i, g_{i+1})$ in the grid $G$ will be called *linked* if both $|TY(h, g_i)|$ and $|TY(h, g_{i+1})|$ exceed $(\log \tau^{-2})^{-\alpha}$. According to Proposition 2.2, linked pairs consists of potential representatives of a true discontinuity point and its associated estimated values. If there exist more than one link for a point, we choose the best link as follows. Let $U$ be the subset of the grid $G$ consisting of all elements linked to some other element. Define a chain of length $\ell$ in $U$ to be a sequence $g_r < g_{r+1} < \cdots < g_{r+\ell}$ of elements in $U$ such that $g_i$ is linked to $g_{i+1}$ and no further elements can be added to the chain without destroying this property. Consider now all chains of length greater than $\lceil (\log \tau^{-2})^{1+\eta} \rceil$ and denote them by $C_j$, $j = 1, \ldots, \hat{d}$. If such a chain does not exist we set $\hat{d} = 0$. Finally we can now define the following estimates for possible jump points:

$$\hat{b}_j = \operatorname{argmax}_{g_i \in C_j} \{ |TY(h, g_i)| \}, \quad j = 1, \ldots, \hat{d}, \tag{2.8}$$

with $\hat{b}_0 = 0$ and $\hat{b}_{\hat{d}+1} = 1$. Having defined these estimators we now wish to control the probability of introducing spurious discontinuities (false positives). Such points are grid-points $g_i$ for which there is no $b_j$ in $[g_i - h, g_i + h]$ and nevertheless $|TY(h, g_i)| > (\log \tau^{-2})^{-\alpha}$ and therefore introducing false positives is equivalent to the event $\hat{d} > d$. Using assertion (i) of Proposition 2.2 with $\delta = 2 + 4m/2m + 1$, we have for this probability:

$$\mathbb{P}\{\hat{d} > d\} \leq \sum_{\substack{g_i \in G; \\ \text{no } b_j \in [g_i - h, g_i + h]}} \mathbb{P}\{ |TY(h, g_i)| > (\log \tau^{-2})^{-\alpha} \} = o\big((\tau^2)^{2m/2m+1}\big). \tag{2.9}$$

For the case of false negatives (undetected discontinuities) there will be at least one location $b_j$ such that there exists a grid point $g_i$ within a distance $h/2$ of $b_j$ having a $|TY(h, g_i)|$-value smaller than $(\log \tau^{-2})^{-\alpha}$. Using assertion (ii) of Proposition 2.2 with $\delta = 4m/2m + 1$ gives

$$\mathbb{P}\{\hat{d} < d\} \leq \sum_{b_j} \mathbb{P}\{ |TY(h, g_i)| < (\log \tau^{-2})^{-\alpha} \} = o\big((\tau^2)^{2m/2m+1}\big). \tag{2.10}$$

It follows from inequalities (9) and (10), that

$$\mathbb{P}\{\hat{d} \neq d\} = o\big((\tau^2)^{2m/2m+1}\big) \quad \text{as } \tau \downarrow 0. \tag{2.11}$$

Moreover, using the fact that $\mathbb{E}(\hat{b}_j - b_j)^2 \leq (2h)^2 + \mathbb{P}\{|\hat{b}_j - b_j| > 2h\}$ and Proposition 2.2, it is easy to see that

$$\mathbb{E}(\hat{b}_j - b_j)^2 \leq \mathcal{O}\Big(\big(\tau^2 (\log \tau^{-2})^{1+\eta}\big)^2\Big) = o\big((\tau^2)^{2m/2m+1}\big). \tag{2.12}$$

**2.3. Estimating the regression function.** To define the estimator in the case of unknown jump points it suffices now to substitute in Subsection 2.1, the unknown values $b_j$ and $b_{j+1}$ by their estimates $\hat{b}_j$ and $\hat{b}_{j+1}$. Such a substitution leads to the segmented observation process in Step S2 with segments $[\hat{b}_j, \hat{b}_{j+1}]$ into the equivalent sequence of observations ($j = 0, \ldots, \hat{d}$)

$$\widehat{Y}_{k,(j)} = \hat{\theta}_{k,(j)} + \tau \hat{\varepsilon}_{k,(j)}, \quad j = 0, \ldots, \hat{d} \quad k \in \mathbb{N},$$

where

$$\widehat{Y}_{k,(j)} = \int_{\hat{b}_j}^{\hat{b}_{j+1}} \hat{\psi}_k^j(t)\, dY_\tau(t), \quad \hat{\theta}_{k,(j)} = \int_{\hat{b}_j}^{\hat{b}_{j+1}} f(t)\hat{\psi}_k^j(t)\, dt,$$

and

$$\hat{\varepsilon}_{k,(j)} = \int_{\hat{b}_j}^{\hat{b}_{j+1}} \hat{\psi}_k^j(t)\, dW(t).$$

As suggested in Subsection 2.1, we propose then the following estimator for $f$,

$$\hat{f}(t) = \sum_{k=0}^{\infty} \tilde{\hat{\theta}}_{k,(j)} \hat{\psi}_k^j(t), \quad t \in [\hat{b}_j, \hat{b}_{j+1}[, \quad j = 0, \ldots, \hat{d}, \tag{2.13}$$

where

$$\tilde{\hat{\theta}}_{k,(j)} = \begin{cases} \widehat{Y}_{k,(j)}, & \text{if } k < 2^\ell - 1 \\ \eta_s(\widehat{Y}_{k,(j)}), & \text{for } k \geq 2^\ell, \end{cases}$$

with $s = 2\tau/2m + 1\sqrt{2|\log(\tau) + \log\log(\tau)|}$ and $\ell = m$.

Combining inequalities (2.11) and (2.12) together, it is easy to see that the complement of the event $A = \{d = \hat{d}\} \cap \{|\hat{b}_j - b_j| < 2h, j = 1, \ldots, d\}$ has a probability $o\big((\tau^2)^{2m/2m+1}\big)$ as $\tau$ goes to 0. Provided that $\tau$ is small enough the risk of the estimator defined by (2.13), can then be written as

$$R(\hat{f}, f) = \mathbb{E}\{\|\hat{f} - f\|^2 \mathbf{I}_A\} + \mathbb{E}\{\|\hat{f} - f\|^2 \mathbf{I}_{A^c}\}.$$

Using now the fact that $f$ is bounded by assumption M1 and an appropriate truncated version of $\hat{f}$, we obtain

$$R(\hat{f}, f) \leq \mathbb{E}\{\|\hat{f} - f\|^2 \mathbf{I}_A\} + o\big((\tau^2)^{2m/2m+1}\big).$$

To estimate the risk it is therefore sufficient to restrict ourselves to the case that $\hat{d} = d$ and $|b_j - \hat{b}_j| < 2h$ for all $j = 1, \ldots, d$ (the event $A$). Conditionally on this event, and by using Parseval's equality, the risk of $\hat{f}$ may be written as

$$R(\hat{f}, f) = \sum_{j=0}^{d} \sum_{k=0}^{\infty} \mathbb{E}\{(\tilde{\hat{\theta}}_{k,(j)} - \hat{\theta}_{k,(j)})^2\}.$$

Now, by definition of the estimated coefficients the risk can be decomposed as follows:

$$\begin{aligned} R(\hat{f}, f) &= \sum_{j=0}^{d} \sum_{k=0}^{2^\ell - 1} \mathbb{E}\{\tau^2 \hat{\varepsilon}_{k,(j)}^2\} \\ &\quad + \sum_{j=0}^{d} \sum_{k=2^\ell}^{\infty} \mathbb{E}\Big\{\big(\tilde{\hat{\theta}}_{k,(j)} - \eta_s(\hat{\theta}_{k,(j)}) + \eta_s(\hat{\theta}_{k,(j)}) - \hat{\theta}_{k,(j)}\big)^2\Big\} = R_1 + R_2, \end{aligned}$$

where $\eta_s$ denotes the hard thresholding function. The last summand $R_2$ can be further decomposed as follows

$$R = \sum_{j=0}^{d} \mathbb{E}(\widehat{S}_{(j)}) + R_{1,(j)} + R_{2,(j)} + R_{3,(j)},$$

where

$$\widehat{S}_{(j)} = \sum_{k=2^\ell}^{\infty} \mathbf{I}_{\{|\hat\theta_{k,(j)}|\le s\}}\hat\theta^2_{k,(j)} + \tau^2 \mathbf{I}_{\{|\widehat{Y}_{k,(j)}|>s\}}$$

$$R_{1,(j)} = \tau^2 \sum_{k=2^\ell}^{\infty} \mathbb{E}\big\{(\hat\varepsilon_{k,(j)}-1)^2 \mathbf{I}_{\{|\widehat{Y}_{k,(j)}|>s\}}\big\}$$

$$R_{2,(j)} = \sum_{k=2^\ell}^{\infty} \big[\mathbf{I}_{\{|\widehat{Y}_{k,(j)}|>s\}} - \mathbf{I}_{\{|\hat\theta_{k,(j)}|>s\}}\big]^2 \hat\theta^2_{k,(j)}$$

$$R_{3,(j)} = 2\tau \sum_{k=2^\ell}^{\infty} \mathbb{E}\big\{\hat\theta_{k,(j)}(\hat\varepsilon_{k,(j)}-2)\mathbf{I}_{\{|\widehat{Y}_{k,(j)}|>s\}}\mathbf{I}_{\{|\hat\theta_{k,(j)}|\le s\}}\big\}.$$

Using the fact that the estimated locations $\hat b_j$ belong to the grid $G$, and using Jensen's inequality it is easy to show that

$$R_1 \le \mathcal{O}\big(\tau^2 \log^3(\tau^{-2})\big).$$

A rather lengthly argument, that we omit due to the lack of space, and which parallels closely the one given by Oudshoorn, allows us to show that each remainders $R_{n,(j)}$, $n = 1, 2, 3$, behave, as $\tau \downarrow 0$, like

$$R_{n,(j)} = o(1)\mathbb{E}\{\widehat{S}_{(j)}\} + \mathcal{O}\big(\tau^2 \log^3(\tau^{-2})\big).$$

This implies that

$$R(\hat f, f) = \big(1+o(1)\big)\sum_{j=0}^{d} \mathbb{E}\{\widehat{S}_{(j)}\} + \mathcal{O}\big(\tau^2 \log^3(\tau^{-2})\big).$$

Recalling now that two jump points are at least within a distance $\kappa$ apart, that every estimate $\hat b_j$ differs from $b_j$ at most $2h$ on the event $A$ and that we are using compactly supported wavelets on the interval, for the particular scale $h$ that we are using each singularity belongs to one and only one influence cone of the wavelet transform. From these facts and the properties of the thresholding function as given in the appendix 2 of Donoho and Johnstone's paper [**8**], we conclude that

$$\sum_{j=0}^{d} \mathbb{E}\{\widehat{S}_{(j)}\} = R(\tilde f, f).$$

The resulting estimator $\hat f$ is therefore nearly minimax, since it shares the same minimal quadratic risk as $\tilde f$. The proof is rather lengthly and is available upon request from the author.

REMARKS. (1) In applied problems, the process $Y(t)$ is usually not observed at every point $t$ but is characterized by a uniform sampling which approximates $Y$ at a resolution that depends upon the sampling rate (see Remark 2.1). Now, he have already seen that the discrete regularly spaced nonparametric regression model given in Eq. (2) is equivalent to the white noise model $Y_n(t)$ given in Eq. (1) with $\tau = \sigma/\sqrt{n}$, provided that we take $f_n(t) = \sum_{i=0}^{n-1} f(t_i)\mathbf{I}_{\{t_i\le t<t_{i+1}\}}(t)$. Hence, on a common probability space, the process $\big(Y_n, \mathcal{H}(m,Q)\big)$ differs from $\big(Y, \mathcal{H}(m,Q)\big)$ by the difference between $f_n$ and $f$. Now, since $\mathcal{H}(m,Q)$ is a piece-wise Sobolev ellipsoid, it follows by Theorems 15 and 16 in Donoho and Johnstone [**10**, Section 9],

that the risks of the discrete nonparametric regression problem and of the white noise model are equivalent. Our estimation strategy therefore also holds for the discrete nonparametric regression model.

(2) For locating the discontinuities in Proposition 1, we have used the continuous wavelet transform. For efficient numerical implementation, no fast computational algorithm exists, and we need to discretize the scaling parameter $h$ along a sparse sequence. When the scale is discretized along a dyadic sequence, the wavelet transform can be computed with the fast dyadic wavelet transform of Mallat (see Abry [**1**]). The characterization theorems for local Lipschitz regularities for a sampled signal via the discrete dyadic wavelet transform remain valid (see Section 6 of Mallat and Hwang's [**18**]). Such Lipschitz regularity characterization does not hold in general for the discrete wavelet transform (see Abry [**1**]). This is due to the lack of translation invariance of the discrete wavelet transform. It is therefore imperative, in a discrete and practical setting, to use the discrete dyadic wavelet transform for estimating the location of the jump points.

(3) Usually $\tau$ is only known up to a scale factor. For example, using the relation between the discrete model and the white noise model we have $\tau = \sigma/\sqrt{n}$, and $\sigma$ is rarely known in practice. One way to estimate the unknown scale is to following Donoho and Johnstone (1994), that is to estimate $\sigma$ by the median absolute deviation of the $n - D$ smallest wavelet coefficients at the finest level available, divided by 0.6745. We are using here the $n - D$ smallest coefficients because in the presence of discontinuities it are exactly the largest coefficients at the finest level that carry the singularities, thus inducing a very biased estimation of $\sigma^2$. An alternative approach for the noise regression problem is to use a device suggested by Lombard [**17**], that is to transform the original white noise model into a white noise model driven by rank statistics. Important to note is that the structure of the regression function transformed in this way remains the same as in the original problem, but the scale parameter $\sigma^2$ is now equal to 1.

## 3. An Illustrative Example

The example concerns the Nile data given in Cobb [**5**] and displayed in the left panel of Figure 1. The data set consists of measurements of the annual volume of discharge from the Nile river at Aswan for the years 1870 to 1930. The measurements are meteorogically significant as evidence of a possible abrupt change in rainfall regime near the turn of the last century. Superimposed on the plot in the left panel of Figure 1 is the associated interpolating white noise model that we used for the analysis.

The right panel of Figure 1 shows the fit from our method with a change-point for the year 1898. The strategy implemented for treating this example had two steps. First, the detector/locator of Section 2 with a dyadic wavelet transform based on a cubic $B$-spline wavelet (Mallat's Wavel $C$-code) was applied to determine how many change-points (if any) are present and to estimate the exact location of the change-points. The grid that we used was a 64-point grid of the interval [1870, 1930] and $\alpha$ in Proposition 1 was taken equal to 1/2. One point only was identified as a change point, occurring between 1898 and 1899. Independent evidence that a change point at 1898 is real can be found in the rainfall records of most tropical

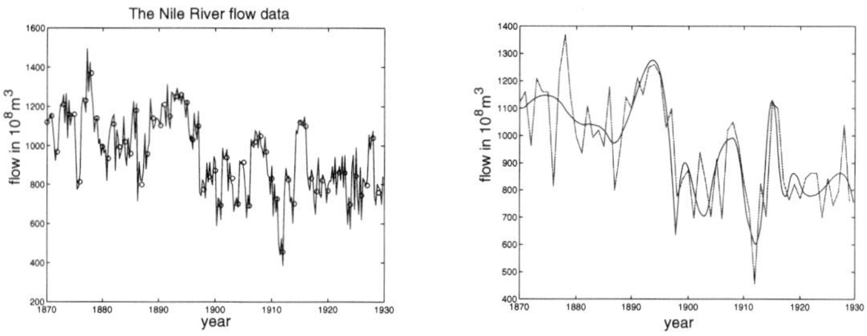


FIGURE 1. Left panel: The Nile river flow data (dots), with a white noise regression interpolation superimposed. Right panel: The wavelet estimator $\hat{f}$, adapted to the estimated change-point (solid curve) plotted together with the original data (grayed curve).

data (for a brief discussion see Cobb [**5**]. Finally the entire function was fitted with an estimated change point at 1898, using a Daubechies D8 adjusted to the interval wavelet and a hard thresholding rule with an estimated $\hat{\sigma} = 111$.

Despite the fact that this situation is no longer an asymptotic situation, we can see that the method performs quite well.

## Acknowledgment

This research was supported by the CNRS-INRIA project IDOPT 6 at IMAG-LMC of the University Joseph Fourier of Grenoble. I wish to thank K. Oudshoorn for providing me with a preprint of her 1995 paper prior to publication. S. Mallat and D. Donoho and I. Johnstone are gratefully acknowledged for providing respectively the Wavel and Wavelab statistical packages. I would like to thank the organizers of the workshop on splines and wavelets in statistics. I am also grateful to the anonymous referees for their suggestions.

## References

1. P. Abry, *Transformées en ondelettes—Analyses multirésolutions et signaux de pression en turbulance*, Ph. Thesis, Univ. Claude Bernard, Lyon, France, 1994.
2. L. Andersson, N. Hall, B. Jawerth, and G. Peters, *Wavelets on closed subsets of the real line*, Topics in the Theory and Applications of Wavelets (L. L. Schumacher and G. Webb, eds.) Academic Press, 1993 (to appear).
3. M. Bunt, I. Koch, and A. Pope, *Kernel-based nonparametric regression for discontinuous functions*, Statist. res. report no. SRR 012-95, The Australian National Univ., 1995.
4. J. Canny, *A computational approach to edge detection*, IEEE Trans. Pattern Analysis and Machine Intelligence **8** (1986), 679–698.
5. G. W. Cobb, *The problem of the Nile: Conditional solution to a changepoint problem*, Biometrika **65** (1978), 243–251.
6. A. Cohen, I. Daubechies, and P. Vial, *Wavelets on the interval and fast wavelet transforms*, Appl. Comput. Harmon. Anal. **1** (1993), 54–82.
7. I. Daubechies, *Ten lectures on wavelets*, CBMS-NSF Regional Conf. Ser. in Appl. Math., SIAM, Philadelphia, 1992.
8. D. L. Donoho and I. M. Johnstone, *Ideal spatial adaptation by wavelet shrinkage*, Biometrika **81** (1994), 425–455.
9. ———, *Adapting to unknown smoothness via wavelet shrinking*, J. Amer. Statist. Assoc. **90** (1995a), 1200–1224.
10. ———, *Minimax estimation via wavelet shrinkage*, Stanford Univ., 1995b (tech. report).

11. R. L. Eubank and P. L. Speckman, *Nonparametric estimation of functions with jump discontinuities*, Change-point Problems (E. Carlstein, H.-G. Müller, and D. Siegmund, eds.), IMS Lecture Notes Monograph Ser. **23** (1994), 130–143.
12. D. Girard, *Détection de discontinuités dans un signal par inf-convolution spline et validation croisée*, tech. report no. RR 702-I-M, Univ. Grenoble, 1990.
13. P. Hall and D. M. Titterington, *Edge preserving and peak-preserving smoothing*, Technometrics **34** (1992), 429–440.
14. P. J. Laurent and F. Utreras, *Optimal smoothing of noisy broken data with spline functions*, J. Approx. Theory and its Applications **2** (1986), 71–94.
15. Y. C. Leclerc and S. W. Zucker, *The local structure of image discontinuities in one dimension*, IEEE Trans. Pattern Analysis and Machine Intelligence **9** (1987), 341–355.
16. C. Loader, *Change point estimation using nonparametric regression*, Technical Report, AT&T Bell Laboratories, 1994.
17. F. Lombard, *Detecting change points by Fourier analysis*, Technometrics **30** (1988), 305–310.
18. S. Mallat and W. L. Hwang, *Singularity detection and processing with wavelets*, IEEE Trans. Inform. Theory **2** (1992), 617–643.
19. J. A. McDonald and A. B. Owen, *Smoothing with split linear fits*, Technometrics **28** (1986), 195–208.
20. H. G. Müller, *Change-points in nonparametric regression analysis*, Ann. Statist. **20** (1992), 737–761.
21. C. G. M. Oudshoorn, *Minimax estimation of a regression function with jumps: Attaining the optimal constant*, techn. report no. 934, Utrecht University, 1995.
22. M. Pollak and D. Siegmund, *A diffusion process and its applications to detecting a change in the drift of Brownian motion*, Biometrika **72** (1985), 267–280.
23. C. Potier and C. Vercken, *Spline fitting numerous noisy data with discontinuities*, Curves and Surfaces (Laurent *et al.*, eds.), Academic Press, 1994, 477–480.
24. P. Qui and B. Yandell, *A local polynomial jump detection algorithm in nonparametric regression*, Univ. of Wisconsin-Madison, 1994 (tech. report).
25. P. Qui, *Estimation of the number of jumps of jump regression functions*, Commun. Statist. Theory Methods **23** (1994), 2141–2155.
26. P. L. Speckman, *Detection of change-points in nonparametric regression*, Univ. Missouri-Columbia, 1994 (tech. report).
27. Y. Wang, *Jump and sharp cusp detection by wavelets with applications to estimation of functions with jumps*, Proc. of the 26th Symposium on the Interface (Fairfax, 1994), (John Sall and Ann Lehman, eds.), Comput. Sci. Statist., vol. 26, 1994.
28. ———, *Jump and sharp cusp detection by wavelets*, Biometrika **82** (1995), 385–397.
29. J. S. Wu and C. K. Chu, *Nonparametric function estimation and bandwidth selection for discontinuous functions*, Statist. Sinica **3** (1993), 557–576.

Laboratoire LMC–IMAG, Université Joseph Fourier, BP 53, 38042 Grenoble Cedex 09, France

*E-mail address*: antonia@imag.fr

Centre de Recherches Mathématiques
CRM Proceedings and Lecture Notes
Volume **18**, 1999

# Wavelet Thresholding for Non (Necessarily) Gaussian Noise: A Preliminary Report

R. Averkamp and C. Houdré

## 1. Introduction

We wish in these notes to present some preliminary results on denoising methods using the wavelet transform. We concentrate here on the wavelet thresholding method as applied to non (necessarily) Gaussian noise. More definitive results, new methods as well as the proofs of the statements given below will appear in the first author's doctoral thesis.

In the usual denoising setting by wavelet thresholding, one assumes that the target functions are members of a fixed subset, e.g. a ball, of a smoothness class such as a Besov space which can be characterized by the sequence of wavelet coefficients (Meyer [**M**]). For this target set of functions one then computes a suitable threshold and tries to prove that the risk for this threshold is close to the minimax risk in the class of all estimators [**CDLL, DL, D, DJ2, KP1, KP2, KP3, N, NS, NvS, W1, W2**]. Thresholding also provides an optimal (based on the MDL method) coding length of a signal in noise [**KP**]. It should also be mentioned that in this setting, linear shrinkers have been applied to the wavelet coefficients, see [**A, AV**], for such results.

Another approach introduced by Donoho and Johnstone ([**DJ1**] and [**DJKP**]) is the following method: Given noisy wavelet coefficients, i.e. the true wavelet coefficient plus a random term which represents the noise and assuming that one has knowledge of the true wavelet coefficients, an ideal estimator is to set a noisy coefficient to zero if the variance $\sigma^2$ of the noise is greater than the square of the true wavelet coefficient; otherwise the noisy coefficient is kept. Of course, the mean square error of this estimator is the minimum of $\sigma^2$ and of the square of the coefficient. Under the assumption of independent normal noise (see also [**JS**] for

1991 *Mathematics Subject Classification.* Primary 62G07, Secondary 60G70, 41A25.

The work of the first author was supported in part by the NSF Grant DMS-9632032, while this author was visiting the Southeast Applied Analysis Center, School of Mathematics, Georgia Institute of Technology, Atlanta, GA 30332, USA.

The work of the second author was supported in part by an NSF Mathematical Sciences Postdoctoral Fellowship, by the NIH Grant CA 59039-18 and by the NSF Grant DMS-9209130, while this author was visiting the Department of Statistics, Stanford University, Stanford, CA 94305, USA and by the NFS Grant DMS-9803239.

This is the final version of the paper.

normal correlated noise), these authors show that the soft thresholding estimator achieves a risk at most $O(\log n)$ times the risk of this ideal estimator. Moreover, no estimator is asymptotically better. In contrast to the smoothness class approach, this scheme does not try to minimize the maximal mean square error for a fixed class of functions, but tries to minimize the quotient of the mean square error and of a functional of the function itself. This functional depends on the wavelet coefficients and measures how easily the estimation can be performed. It is clear that it is more difficult to denoise a signal whose energy is uniformly distributed over the wavelet coefficients than to denoise a signal whose energy is concentrated in a few big wavelet coefficients. This ease of estimation is measured by the mean square error of the ideal estimator. The "ideal method" does not require any a-priori knowledge on the function to be denoised, but might not be optimal if a smoothness class information is available. For many "smooth" functions "most of" the wavelet coefficients are rather small and only a small part of the wavelet coefficients are big. This means that the risk of the ideal estimator is small and this, in turn, implies that the risk of soft thresholding is small for these functions.

We investigated here this "ideal denoising" scheme for non (necessarily) Gaussian noise, in which case the noise in the wavelet coefficients is no longer identically distributed nor independent. This creates new problems, but for some classes of noises, it is still possible to prove the asymptotic minimaxity of the soft thresholding estimator.

Let us now further describe the model and the method. Our data is discrete, and we have the observations

$$X_i = f_i + e_i, \quad i = 1, \ldots, n = 2^\ell,$$

where $f_i \in \mathbb{R}$ are parameters we want to estimate and where the $e_i$, the noise in the observations, are random variables with mean zero and constant variance $\sigma^2$.

We assume that the $f_i$ are samples of a function defined on $(0,1)$, i.e. $f_i = f(i/n)$, and so the $f_i$ actually depend on $n$.

To this data we apply a discrete wavelet transform $W_n : \mathbb{R}^n \to \mathbb{R}^n$, and for simplicity a periodic version is used in which case $W_n$ is an orthogonal transformation.

The heuristic for this scheme is that for a multiresolution analysis with scaling function $\phi$

$$2^\ell \int_{\mathbb{R}} \phi\big((t - i/2^\ell)2^\ell\big) f(t)\, dt \approx f(i/2^\ell),$$

if $\ell$ is large $\int_{\mathbb{R}} \phi(x)\, dx = 1$.

We thus think of the $X_i$ as noisy versions of the left side of the above formula.

Let $Y = W_n(X)$ be the empirical wavelet coefficients, let $\theta = W_n(f)$ and let $z = W_n(e)$. Since $W_n$ is linear $Y_i = \theta_i + z_i$, $i = 1, \ldots, n$, and since it is an orthogonal transformation, the respective mean square error in estimating $\theta$ and $f$ are equal. Therefore, instead of estimating the $f_i$ directly, we can estimate the $\theta_i$.

Assuming some knowledge of the true wavelet coefficients $\theta$, consider the following estimator for $\theta_i$: $\tilde{\theta}_i = Y_i$ if $\theta_i^2 \geq \sigma^2$ and $\tilde{\theta}_i = 0$ if $\theta_i^2 \leq \sigma^2$. In plain words, an empirical wavelet coefficient is kept if its contribution to the energy of the function is greater than the variance of the noise, otherwise it is killed.

The performance of other estimators (in particular of the soft thresholding operator $\delta_\lambda(x) = (|x| - \lambda)_+ \operatorname{sgn}(x)$ when applied to $Y$) will be compared to the benchmark

$$B_n(\theta,\sigma^2) = \sigma^2 + \sum_{i=1}^{n} \min(\theta_i^2,\sigma^2)$$

which is the mean square error of $\tilde{\theta}$ plus $\sigma^2$. Since the mean square error of $\tilde{\theta}$ is 0 if $\theta = 0$ we add $\sigma^2$. This is close to assuming that at least one $\theta_i^2$ is greater than the variance $\sigma^2$. Notice that $B_n(\theta,\sigma^2)$ is small in comparison to $n\sigma^2$ (the total variance in the signal $X$, i.e. the sum of the variances of each component) if $\theta$ (the wavelet transform) is sparse. In fact $B_n(\theta,\sigma^2)$ is itself a measure for the sparsity of $\theta$.

If the $e_i$ are i.i.d. normal random variables, then the $z_i$ are normal and independent since $W_n$ is an orthogonal transformation. This is the original case investigated by Donoho and Johnstone. The following result from multivariate decision theory is theirs and is crucial for their results.

THEOREM 1.1. *Let $Y_i = \theta_i + z_i$, $i = 1,\dots,n$, where the $\theta_i$ are parameters of interest and the $z_i$ are* i.i.d. *normal random variables with mean zero and variance $\sigma^2$. Let $\lambda_n = \sigma\sqrt{2\log n}$, then*

$$\sup_{\theta\in\mathbb{R}^n} \frac{E\|\delta_{\lambda_n}(Y)-\theta\|^2}{B_n(\theta,\sigma^2)} \leq (1+2\log n).$$

In his thesis, Gao [**G**] proves a similar result for i.i.d. random variables with exponential tails. In fact, a stronger result of this type holds for a wider class of distributions.

THEOREM 1.2. *Let $Y_i = \theta_i + z_i$, $i = 1,\dots,n \geq 18$, where the $\theta_i$ are parameters of interest and the $z_i$ are identically distributed symmetric random variables with distribution $\mu$ and $Ez_1^2 = \sigma^2$. Then, the equation*

$$\left(2\int_\lambda^\infty (x-\lambda)^2\mu(dx)\right)(n+1) = \lambda^2+\sigma^2,$$

*has a unique positive solution $\lambda_n$. Moreover,*

$$\Lambda_n = \sup_{\theta\in\mathbb{R}^n} \frac{E\|\delta_{\lambda_n}(Y)-\theta\|^2}{B_n(\theta,\sigma^2)} = \inf_\lambda \sup_{\theta\in\mathbb{R}^n} \frac{E\|\delta_{\lambda}(Y)-\theta\|^2}{B_n(\theta,\sigma^2)},$$

*where $\Lambda_n = (\lambda_n^2+\sigma^2)/(\sigma^2(1+1/n))$.*

The great advantage of the normal i.i.d. noise assumption in the data is the preservation of these properties under the orthogonal transformation $W_n$. If the noise in the samples is not normal, the noise in the wavelet coefficients is no longer independent nor identically distributed. So actually the above theorem cannot be applied to empirical wavelet coefficients in the non–normal case. However, we have the following:

THEOREM 1.3. *Let $Y_i = \theta_i + z_i$, $i = 1,\dots,n \geq 18$, where the $\theta_i$ are parameters of interest and the $z_i$ are symmetric random variables with distribution $\mu_i$ and $Ez_i^2 = \sigma^2$. Then, the equation*

$$\left(2\int_\lambda^\infty (x-\lambda)^2\mu_i(dx)\right)(n+1) = \lambda^2+\sigma^2,$$

*has a unique positive solution* $\lambda_{n,i}$. *Let* $\lambda_n \geq \sup_i \lambda_{n,i}$ *and let* $\Lambda_n = (\lambda_n^2 + \sigma^2)/(\sigma^2(1+1/n))$, *then*

$$\sup_{\theta\in\mathbb{R}^n} \frac{E\|\delta_{\lambda_n}(Y)-\theta\|^2}{B_n(\theta,\sigma^2)} \leq \Lambda_n.$$

We want to apply the above theorem to the case where $Y_i = \theta_i + z_i$ are empirical wavelet coefficients. Since the $z_i$ are linear combinations of the initial noise, $\lambda_{n,i}$ is quite complicated to compute. An alternative is to find an upper bound for the $\lambda_{n,i}$ which only depends on the initial noise. Since $W_n$ is orthogonal, we know that $z_k = \sum_{i=1}^n w_{k,i}^{(n)} e_i$ with $\sum_{i=1}^n w_{k,i}^{(n)\,2} = 1$. The tail behavior of $z_k$ is thus crucial for $\lambda_{n,k}$. If $e_1$ has bigger tails than the normal distribution, then the central limit theorem applies and the distribution of $\sum_{i=1}^n w_{k,i}^{(n)} e_i$ "tends" to the normal distribution, i.e. its tail is becoming smaller. This is the heuristic which leads to the following:

THEOREM 1.4. *Let* $\mu$ *be a distribution on* $\mathbb{R}$ *which is a variance mixture of normal distributions, i.e. it is absolutely continuous with density*

$$\int_{(0,\infty)} \phi_a(x)\nu(da),$$

*where* $\nu$ *is a probability measure on* $(0,\infty)$ *and where* $\phi_a$ *is the centered normal density of variance* $a$. *Let* $X_i$, $i = 1,\ldots,n$, *be* i.i.d. *random variables with distribution* $\mu$. *Then, for any* $a_1,\ldots,a_n \in \mathbb{R}$, $b_1,\ldots,b_n \in \mathbb{R}$, *whose respective absolute values are in non–increasing order with moreover* $\sum_{i=1}^k a_i^2 \leq \sum_{i=1}^k b_i^2$, $k = 1,\ldots,n$,

$$E\left(\left|\sum_i a_i X_i\right| - \lambda\right)_+^2 \leq E\left(\left|\sum_i b_i X_i\right| - \lambda\right)_+^2,$$

*for all* $\lambda > 0$.

By applying Theorem 1.3 and Theorem 1.4 to the situation where the observations are empirical wavelet coefficients, one easily obtains:

PROPOSITION 1.5. *Let* $X_i = f_i + e_i$, $i = 1,\ldots,n \geq 18$, *where the* $f_i$ *are parameters of interest (for example samples from a function) and the* $e_i$ *are* i.i.d. *random variables whose distribution* $\mu$ *is a variance mixture of normal distributions. Let* $Y = W_n(X)$ *and let* $\theta = W_n(f)$. *Then the equation*

$$\left(2\int_\lambda^\infty (x-\lambda)^2\mu(dx)\right)(n+1) = \lambda^2 + \sigma^2,$$

*has the unique positive solution* $\lambda_n$. *Moreover,*

$$\sup_{\theta\in\mathbb{R}^n} \frac{E\|\delta_{\lambda_n}(Y)-\theta\|^2}{B_n(\theta,\sigma^2)} \leq \Lambda_n,$$

*where* $\Lambda_n = (\lambda_n^2 + \sigma^2)/(\sigma^2(1+1/n))$.

PROOF. Let $i \in 1,\ldots,n$, then $z_i = W_n(e)_i = \sum_{j=1}^n w_{i,j}^{(n)} e_j$, for $w_{i,j}^{(n)} \in \mathbb{R}$ with $\sum_{j=1}^n w_{i,j}^{(n)\,2} = 1$, since $W_n$ is an orthogonal transformation. Without loss of generality assume that the $w_{i,j}^{(n)\,2}$ are in non-increasing order (for the index $j$). Set

$(b_1, b_2, \dots, b_n) = (1, 0, \dots, 0) \in \mathbb{R}^n$. Then it follows from Theorem 1.4 that

$$E\Big(\Big|\sum_j w_{i,j}^{(n)} e_j\Big| - \lambda\Big)_+^2 \le E\Big(\Big|\sum_j b_j e_j\Big| - \lambda\Big)_+^2$$

for all $\lambda > 0$. But the above are respectively equal to $2\int_\lambda^\infty (x-\lambda)^2 \mu_i(dx)$ and $2\int_\lambda^\infty (x-\lambda)^2 \mu(dx)$, where $\mu_i$ is the distributiuon of $z_i$. This implies that the solution $\lambda_n$ of $2\int_\lambda^\infty (x-\lambda)^2 \mu(dx)(n+1) = \lambda^2 + \sigma^2$ is larger than the solution $\lambda_{n,i}$ of $2\int_\lambda^\infty (x-\lambda)^2 \mu_i(dx)(n+1) = \lambda^2 + \sigma^2$. Applying Theorem 1.3 completes the proof. □

It is easy to see that a density $f$ is a variance mixture of normal densities if and only if $f$ is symmetric and $f(\sqrt{x})$, $x \in [0, \infty)$, is a Laplace transform (see, e.g. [**AM**]). In turn, it is well known [**F**] that a function $g$ is a Laplace transform if and only if it is completely monotone, i.e.

$$(-1)^n g^{(n)}(x) \ge 0, \quad \forall x \in [0, \infty), \quad \forall n \in \mathbb{N}.$$

Thus,

$$\frac{1}{(1+x^n)}, n \ge 1 \text{ and } e^{-x^d}, 0 < d \le 1,$$

are examples of completely monotone functions, and densities (up to multiplicative normalizing constants) like

$$\frac{1}{(1+x^2)^n}, n \ge 1 \text{ and } e^{-x^{2d}}, 0 < d \le 1,$$

are mixtures of normal densities.

If the $e_i$ are not normal mixtures, but if there exists a random variable $v$ whose distribution is a normal mixture with $P(|e_1| > x) \le KP(|v| > x)$, $K \ge 1$, and $\lambda_n$ is the positive solution of $E(|Kv| - \lambda)_+^2 (n+1) = \lambda^2 + \sigma^2$ then

$$\sup_{\theta \in \mathbb{R}^n} \frac{E\|\delta_{\lambda_n}(Y) - \theta\|^2}{B_n(\theta, \sigma^2)} \le \Lambda_n$$

where again $\Lambda_n = (\lambda_n^2 + \sigma^2)/((1 + 1/n)\sigma^2)$. This is a consequence of a version of the concentration principle ([LT, 4.2]).

If the noise is normal, thresholding achieves the minimax rate in the class of all estimators, i.e.

$$\lim_{n\to\infty} \inf_{\hat\theta} \sup_{\theta \in \mathbb{R}^n} \frac{E\|\hat\theta - \theta\|^2}{B_n(\theta, \sigma^2)} \Lambda_n^{-1} = 1,$$

where $\lambda_n$ and $\Lambda_n$ are now computed for the normal distribution [**DJ1**]. For a special class of distributions one can also show that soft thresholding is asymptotically near minimax, i.e. the 1 in the above right hand-side is replaced by a constant. This result is a natural consequence of the next two theorems. In the first one, an asymptotic rate for $\lambda_n$ is given for a class of distributions, while the rate of the minimax risk in the class of all estimators is the subject of the second theorem.

THEOREM 1.6. *Let $\mu$ be an absolutely continuous distribution on $\mathbb{R}$ with density* $\exp(-h)$, *where $h$ is symmetric about zero, continuous, increasing on $[0, \infty)$, and with variance $\sigma^2$. Further, let*

$$\liminf_{x\to\infty} \frac{h(cx)}{h(x)} > 1,$$

*for all* $c > 1$. *Let* $\lambda_n$ *be the solution of*

$$\left(2\int_{\lambda}^{\infty}(x-\lambda)^2\mu(dx)\right)(n+1) = \lambda^2+\sigma^2, \quad \lambda > 0,$$

*and let* $\Lambda_n = (\lambda_n^2+\sigma^2)/(\sigma^2(1+1/n))$. *Then,*

$$\lim_{n\to\infty}\frac{h^{-1}(\log n)}{\lambda_n} = 1, \text{ and } \lim_{n\to\infty}\frac{(h^{-1}(\log n))^2+\sigma^2}{\sigma^2\Lambda_n} = 1.$$

THEOREM 1.7. *Let* $X_i = f_i + e_i$, $i = 1,\dots,n = 2^\ell$, *where the* $f_i$ *are parameters of interest (for example samples from a function) and the* $e_i$ *are* i.i.d. *random variables whose distribution* $\mu$ *is absolutely continuous with density* $\exp(-h)$, *where* $h$ *is symmetric about zero, continuous and increasing on* $[0,\infty)$. *Further let*

$$\limsup_{x\to\infty}\frac{h(cx)}{h(x)} < \infty, \quad \forall c > 1.$$

*Applying a wavelet transform* $W_n$ *with finite filter length to the* $X_i$, *where for every* $n$ *the same type of wavelet is used, let* $Y = W_n(X)$ *and let* $\theta = W_n(f)$. *Then,*

$$\liminf_{n\to\infty}\inf_{\hat\theta}\sup_{\theta\in\mathbb{R}^n}\frac{E\|\hat\theta-\theta\|^2}{B_n(\theta,\sigma^2)}\left(h^{-1}(\log n)\right)^{-2} > 0,$$

*where the infimum is over all the estimators* $\hat\theta$ *of* $\theta$.

In Theorem 1.6 and 1.7, the conditions on $h$ are simultaneously fulfilled if

$$\liminf_{x\to\infty}\frac{h(cx)}{h(x)} > 1,$$

and

$$\limsup_{x\to\infty}\frac{h(cx)}{h(x)} < \infty,$$

for all $c > 1$. This is in particular the case if $h$ is a (fractional) polynomial in $|x|$. So, all in all, and when the noise is a normal mixture, the minimax result holds only if $h$ behaves like a fractional polynomial $|x|^d$, $d \in (0,2]$. The upper bound 2 for $d$ is needed, because if $d > 2$ then the distribution of the noise is no longer "close" to a variance mixture of normal distributions (see the examples after Theorem 1.6). Also, note that the conditions of Theorem 1.7 are not satisfied if $h(x) \sim d\log|x|$, $d > 0$ which is equivalent to requiring that the density of the noise behaves like $1/|x|^d$, as $|x| \to +\infty$.

It would be desirable to generalize Theorem 1.4 to a wider class of distributions, or to obtain a version of Theorem 1.4 which directly describes the behavior of the tails of the sums. There are results of this type for special classes of distributions, where the tail of the sums is bounded by the tail of the initial random variable [**B**]. This type of questions is presently under active investigation.

However, for other kind of distributions one cannot expect to get bounds similar to the bounds in Theorem 1.4. For an example let us consider the symmetric Bernoulli noise, i.e. noise taking the values $\pm 1$ with equal probability.

In this case, something similar to Theorem 1.4 is rather unlikely. Nevertheless, the central limit theorem applies and moreover, since the tail of the distribution of the noise is very thin one can make use the following type of well known inequalities:

Let $X_i$, $i = 1, \ldots, n$ be i.i.d. symmetric Bernoulli random variables, $a_i \in \mathbb{R}$, $i = 1, \ldots, n$, then

$$P\Big(\sum_{i=1}^{n} a_i X_i > t\|a\|_2\Big) \le e^{-t^2/2},$$

where $\|a\|_2$ is the $\ell^2$–norm of the $a_i$. Of course this inequality is not optimal, there are tighter ones, but anyway one can use this simple inequality to compute thresholds and bounds for $\Lambda_n$. Thus the values we get for $\lambda_n$ are $O(\sqrt{\log n})$, which implies $\Lambda_n = O(\log n)$, which is the same as in the gaussian case.

Thanks to the central limit theorem, this rate for $\Lambda_n$ is the best one can get, i.e. if one applies thresholding estimators to the wavelet transform $Y = \theta + z$ of a signal contaminated with Bernoulli noise, then

$$\inf_{(\lambda_{n,i})_{1\le i\le n}} \sup_{\theta\in\mathbb{R}^n} \frac{E\|\delta_{\lambda_{n,i}}(Y) - \theta\|^2}{B_n(\theta, \sigma^2)} = O(\log n).$$

Again, proofs and details will appear in Averkamp's thesis.

## References


[AV] U. Amato and D. T. Vuza, *Besov regularization, thresholding and wavelets for smoothing data*, Numer. Funct. Anal. Optim. **18** (1997), nos. 5–6, 461–493.

[AM] D. F. Andrews and Mallows, *Scale mixtures of normal distributions*, J. Royal Statist. Soc. Ser. **B36** (1974), 99–102.

[A] A. Antoniadis, *Smoothing noisy data with tapered coiflets series*, Scand. J. Statist. **23** (1996), 313–330.

[B] N. K. Bakirov, *Extrema of the distributions of quadratic forms of Gaussian variables*, Theory Probab. Appl. **34** (1989), 207–215.

[CDLL] A. Chambolle, R. A. DeVore, N. Y. Lee, and B. J. Lucier, *Nonlinear wavelet image processing: Variational problems, compression, and noise removal through wavelet shrinkage* (preprint).

[DL] R. A. DeVore and B. Lucier, *Fast wavelet techniques for near optimal image processing*, IEEE Military Comm. Conf. (1992), 2–12.

[D] D. L. Donoho, *De-noising by soft-thresholding*, IEEE Trans. Inform. Theory **41** (1994), no. 3, 613–627.

[DJ1] D. L. Donoho and I. M. Johnstone, *Ideal spatial adaptation by wavelet shrinkage*, Biometrika **81** (1994), no. 3, 425–455.

[DJ2] ———, *Neo-classical minimax problems, thresholding and adaptive function estimation*, Bernoulli **2** (1996), no. 1, 39–62.

[DJKP] D. L. Donoho, I. M. Johnstone, G. Kerkyacharian, and D. Picard, *Wavelet shrinkage: Asymptotia?*, J. Roy. Statist. Soc. Ser. B **57** (1995), no. 2, 301–369.

[F] W. Feller, *An introduction to probability theory and its applications*, vol II, 2nd ed., John Wiley & Sons, Inc. New York Wiley, 1971.

[G] H. Y. Gao, *Wavelet estimation of spectral densities in time series analysis*, Ph.D. Thesis, Univ. California, Berkeley, 1993.

[JS] I. M. Johnstone and B. W. Silverman, *Wavelet threshold estimators for data with correlated noise*, J. Roy. Statist. Soc. Ser. B **59** (1997), no. 2, 319–351.

[KP1] G. Kerkyacharian and D. Picard, *Estimation de densité par méthode de noyau et d'ondelettes: Les liens entre la géometrie du noyaux et les contraintes de régularité*, C. R. Acad. Sci. Paris Sér. I **315** (1992), no. 1, 79–84.

[KP2] ———, *Density estimation in Besov spaces*, Statist. Probab. Lett. **13** (1992), no. 1, 15–24.

[KP3] ———, *Density estimation by kernel and wavelet methods: Optimality of Besov spaces*, Statist. Probab. Lett. **18** (1993), no. 4, 327–336.

[KP] H. Krim and J.-C. Pesquet, *On the statistics of best bases criteria*, Wavelets and Statics (A. Antoniadis *et al.*, eds), Proc. of the 15th French-Belgian meeting of statisticians

(Villard de Lans, 1994), Lecture Notes in Statist. vol. 103, Springer-Verlag, New York, 1995, pp. 193–207.

[LT] M. Ledoux and M. Talagrand, *Probability in Banach spaces*, Springer-Verlag, Berlin, 1991.

[M] Y. Meyer, *Ondelettes et opérateurs.* I: *Ondelettes*, Hermann, Éditeurs des Sciences et des Arts, Paris, 1990.

[N] M. H. Neumann, *Spectral density estimation via nonlinear wavelet methods for stationary non-Gaussian time-series*, J. Time Ser. Anal. **17** (1996), no. 6, 601–633.

[NS] M. H. Neumann and V. G. Spokoiny, *On the efficiency of wavelet estimators under arbitrary error distributions*, Math. Methods Statist. **4** (1995), no. 2, 137–166.

[NvS] M. H. Neumann and R. von Sachs, *Wavelet thresholding: Beyond the Gaussian* i.i.d. *situation*, Wavelets and Statistics (Villard de Lans, 1994), 301–329 Lecture Notes in Statist., vol. 103, Springer, New York, 1995.

[W1] Y. Wang, *Function estimation via wavelet shrinkage for long-memory data*, Ann. Statist. **24** (1996), 466–484.

[W2] ______, *Minimax estimation via wavelets for indirect long memory data*, J. Statist. Plann. Inference **64** (1997), no. 1, 45–55.

Institut für mathematische Stochastik, Freiburg University, Eckerstrasse 1, 79104 Freiburg, Germany, and Southeast Applied Analysis Center, School of Mathematics, Georgia Institute of Technology, Atlanta, GA 30332, USA

*E-mail address*: `averkamp@pollux.mathematik.uni-freiburg.de`

Southeast Applied Analysis Center, School of Mathematics, Georgia Institute of Technology, Atlanta, GA 30332, USA

*E-mail address*: `houdre@math.gatech.edu`

Centre de Recherches Mathématiques
CRM Proceedings and Lecture Notes
Volume **18**, 1999

# Deslauriers-Dubuc: Ten Years After

D. L. Donoho and Thomas P. Y. Yu

ABSTRACT. Ten years ago, Deslauriers and Dubuc introduced a process for interpolating data observed at the integers, producing a smooth function defined on the real line. In this note we point out that their idea admits many fruitful generalizations including:

1. Interpolation of other linear functionals of $f$ (not just point values), yielding other refinement schemes and biorthogonal wavelet transforms;
2. Interpolation of vector-valued data, yielding vector refinement schemes and multiwavelets;
3. Interpolation of nonlinear functionals, yielding nonlinear refinement schemes and nonlinear wavelet transforms;
4. Interpolation by other families than polynomials, yielding e.g. segmented interpolation and segmented transforms.

We refer to a variety of recent work on these schemes.

## 1. Deslauriers-Dubuc Interpolation

Suppose we are given the point-values $(\nu_k = f(k)\colon k \in \mathbb{Z})$ of a function $f$ evaluated at the integers. We wish to interpolate, getting a function $\tilde{f}(x)$ defined for $x \in \mathbb{R}$, obeying $\tilde{f}(k) = \nu_k$.

Dubuc [**11**] and Deslauriers and Dubuc [**4, 5**] introduced a multiscale refinement technique for this problem. For integer $L \geq 0$, let $D = 2L+1$ be an odd integer $> 0$. They obtained such a function $\tilde{f}$ by interpolating the data at the integers to a function defined on the binary rationals by repeated application of a two-scale refinement transformation. If $\tilde{f}$ has already been defined at all binary rationals with denominator $2^j$, $j \geq 0$, their process extends it to all binary rationals with denominator $2^{j+1}$, i.e. all points halfway between previously defined points. Specifically, to define the function at $(k+1/2)/2^j$ when it is already defined at all $k/2^j$, we fit a polynomial $\pi_{j,k}$ to the data $\big(k'/2^j, \tilde{f}(k'/2^j)\big)$ for $k' \in \big\{(k-L)/2^j, \dots, (k+L+1)/2^j\big\}$—this polynomial is unique—and set

$$\tilde{f}\big((k+1/2)/2^j\big) \equiv \pi_{j,k}\big((k+1/2)/2^j\big).$$

1991 *Mathematics Subject Classification.* Primary: 39B12, 41A58; Secondary: 65D05, 26A18.

This is the final version of the paper.

For $D = 1$ this is just linear interpolation, pure and simple. For $D = 3, 5, 7, \ldots$, Deslauriers and Dubuc showed that this scheme defines a function which is uniformly continuous at the rationals and hence has a unique continuous extension to the reals. They showed that this extension is actually $C^R$ for $R = R(D)$ an increasing function of $D = 2L + 1$.

We illustrate the basic calculation of the local polynomial $\pi_{j,k}$ in Figure 1 using $L = 1$ and so $D = 3$.

We illustrate the sequence of refinements in Figure 2 by showing the sequence of refinements of a Kronecker sequence $(\nu_k^{(0)})$ defined by $\nu_k^{(0)} = 1_{\{k=0\}}$. The "$x$" markers display the given data and the "$o$" markers display the successive generations of interpolated data.

Let $\phi$ denote the result of refining the Kronecker sequence; this depends on the order $D$ defining the interpolation scheme and so can be denoted $\phi = \phi_{(D)}$. Figure 3 shows results for $D = 1, 3, 7, 11$.

The function $\phi_{(D)}$ is often called the *fundamental function* of the DD interpolation scheme. The terminology is justified by the fact that if DD refinement is applied to data $(\nu_k)_{k\in\mathbb{Z}}$, the result of refinement is a superposition of $\phi$ and its shifts:

$$\tilde{f}(x) = \sum_{k\in\mathbb{Z}} \nu_k \phi(x-k).$$

This of course follows from the linearity and shift invariance of the refinement process. As discussed above, the Deslauriers-Dubuc fundamental function is compactly supported and $C^R$ for $R = R(D)$.

## 2. Surprises

The DD refinement scheme is simple and beautiful; in fact the simplicity is deceptive. On the one hand, to understand the smoothness properties of $\phi_{(D)}$ requires entry into a considerable literature on properties of solutions to refinement schemes; we do not pursue this question in this paper, and instead merely note the existence of work such as [**3, 12, 1**]. On the other hand, the DD refinement scheme admits of certain very interesting generalizations which we would like to describe in this paper. Moreover, the scheme is connected to the rapidly growing literature on wavelet transforms, and by this connection makes it possible to consider various generalizations of wavelet transforms; we also will describe some of these generalizations in this paper.

In short, the point of this paper is to explain our belief that the DD refinement scheme is a natural point of departure for defining a number of new refinement schemes and mathematical transforms. It is pleasure to have the opportunity to describe this belief and to thank Gilles Deslauriers, Serge Dubuc and Brenda MacGibbon for a chance to visit Montreal and describe a part of this work.

**2.1. Connection with wavelets.** We begin by sketching the connection between DD-Interpolation and Wavelet transforms. As described above, the result of interpolating point samples $(f(k))_{k\in\mathbb{Z}}$ furnishes a function $\tilde{f}$ which depends linearly on $f$. Hence we can define a linear operator $P_0$ by the equation $\tilde{f} = P_0 f$. As described above, we have the explicit representation $(P_0 f)(x) = \sum_k f(k)\phi(x-k)$.

Define now $\phi_{j,k}(x) = \phi(2^j x - k)$, a version of the DD fundamental function that is dilated to width $\approx 2^{-j}$ and translated to location $\approx k/2^j$. Then define the operator $P_j$ by $(P_j f)(x) = \sum_k f(k/2^j)\phi_{j,k}(x)$. Equivalently, $P_j$ represents the

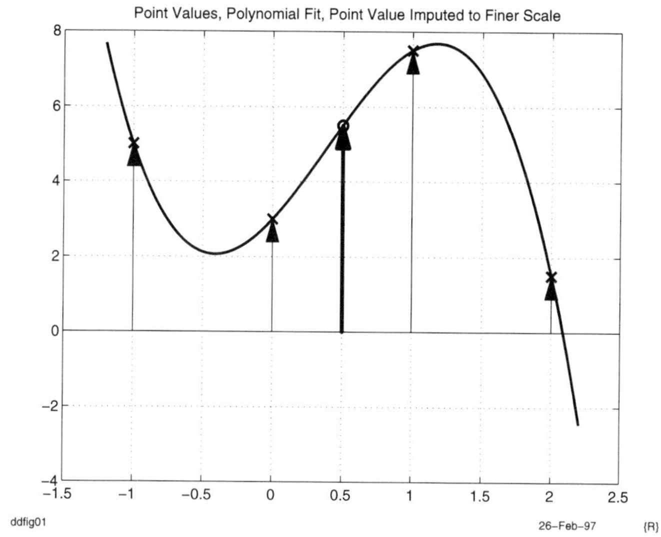


FIGURE 1. Imputation at $x = 1/2$ from local polynomial of degree 3 fit to value s at $x \in \{-1, 0, 1, 2\}$.

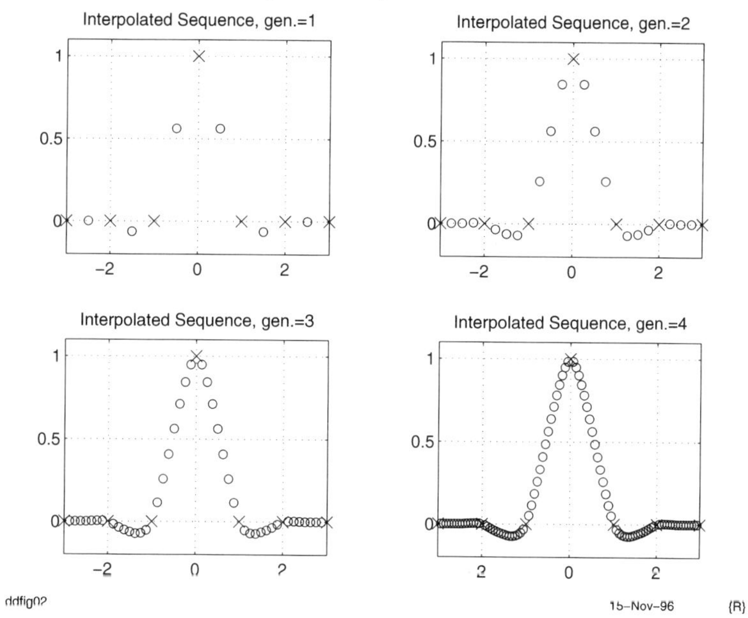


FIGURE 2. A sequence of refinements to successively finer and finer scales. $D = 3$.

result of DD interpolation with data given at sites in the dyadic grid $2^{-j}\mathbb{Z}$ rather than at integer sites. The key facts about the operator $P_j$ can be expressed by two relations. First, with $\phi$ the DD fundamental function,

$$P_j \phi = \phi \tag{2.1}$$

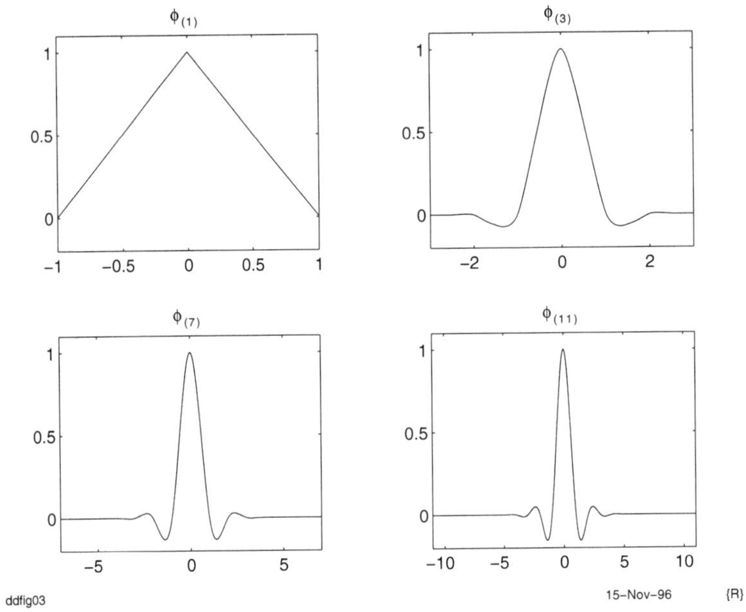


FIGURE 3. Fundamental solutions for $D = 1, 3, 7, 11$.

for all $j = 0, 1, 2, \ldots$. Second, if $\pi$ is a polynomial of degree $D$, $P_j\pi = \pi$ for all $j = 0, \pm1, \pm2, \ldots$. Let $V_j$ be the range of $P_j$, i.e. the $L^\infty$ closure of all sums $\sum_k \nu_k \phi_{j,k}$ with finitely many nonzero terms. Then from (2.1) we have $P_{j+h}\phi_{j,k} = \phi_{j,k}$ for all $h > 0$, and so $V_j \subset V_{j+h}$ for all $h > 0$.

Define now the difference operator $Q_j = P_{j+1} - P_j$ and the difference space $W_j = \text{span}(Q_j)$. This operator describes the difference between interpolating a function known at all sites in $2^{-j-1}\mathbb{Z}$ and only at sites in $2^{-j}\mathbb{Z}$. Now the range of $Q_j$ is in $V_{j+1}$, so it has a representation

$$Q_j f = \sum_k q_{j,k}(f)\phi_{j+1,k}$$

where $q_{j,k}(f) = \big(P_{j+1}(f)(k/2^{j+1} - P_j(f)(k/2^{j+1})\big)$.

Now obviously $Q_j f$ is composed of terms of scale $2^{-(j+1)}$. Because for nice functions, $f = \lim_{j\to\infty} P_j f$ and we have the telescoping sum $P_j = P_0 + \sum_{h=0}^{j-1}(P_{h+1} - P_h)$, we write, formally

$$f = P_0 f + \sum_{j\geq 0} Q_j f, \tag{2.2}$$

obtaining a decomposition of $f$ into a superposition of contributions from scales $2^{-0}, 2^{-1}, \ldots$. These contributions are parameterized by coefficients $(\nu_{0,k})_k$ (the samples of point values at the unit scale) and finer scale coefficients $\big((q_{j,k}(f)\big)_{j,k}$. The number of coefficients associated with a given interval grows with $j$ like $2^j$, and the collection of those coefficients associated with a given interval can be thought of as a kind of "Pyramid", with a few coefficients "near the top"—where $j$ is near 0—and many coefficients "lower down"—where $j$ large.

Since $(P_j f)(k/2^j) = (P_{j+1}f)(2k/2^{j+1}) = f(k/2^j)$, we have $Q_j f(k/2^j) = 0$, so $q_{j,2k}(f) = 0$, and so actually at most half the coefficients $q$ in the pyramid representation are nonzero. In short the pyramid representation is redundant. It is possible to obtain a nonredundant wavelet representation which is at least twice as sparse as the pyramid representation. Let $\psi_{j,k} = \phi\big(2^{j+1}x - (2k+1)\big)$ denote a version of the DD function dilated and translated to live at the sites $(2k+1)/2^{j+1}$ where the corresponding coefficients $q_{j,2k+1}$ are potentially nonzero; these are at $(k+1/2)/2^j$, halfway between the sites at which $P_j$ is interpolating. Then we can write

$$Q_j f(x) = \sum_k \alpha_{j,k}\psi_{j,k}(x) \tag{2.3}$$

with $\alpha_{j,k} = (Q_j f)\big((k+1/2)/2^j\big)$. For example, if $D = 1$ then $\alpha_{j,k} = f\big((k+1/2)/2^j\big) - \big(f(k/2^j) + f\big((k+1)/2^j\big)\big)/2$, while more generally

$$\alpha_{j,k}(f) = (f - \pi_{j,k})\big((k+1/2)/2^j\big). \tag{2.4}$$

Using (2.3) we can rewrite (2.2) as

$$f = \sum_k \beta_{0,k}\phi_{j,k} + \sum_{j\geq 0}\sum_k \alpha_{j,k}\psi_{j,k}. \tag{2.5}$$

This is a nonredundant sum of the same form as commonly used in the wavelet transform: a coarse-scale sum of "father wavelets" $\phi_{j,k}$ and a series of finer and finer scale dilates and translates $\psi_{j,k}$ of a "mother wavelet".

The association of a function $f$ with its coefficients $\big((\beta_{j,k})_k, (\alpha_{0,k})_k, (\alpha_{1,k})_k, \dots\big)$ is a linear operator, well-defined for every continuous $f$. As in [**6**] we call this the Interpolating Wavelet Transform (IWT), because it is built on wavelets which arise from translation and dilation of $\phi$, and so have various interpolating properties. The wavelet coefficient functionals $\alpha_{j,k}(f)$ have an interesting interpretation. By (2.4) they measure the extent to which the point values of $f$ at a given scale can be predicted by the point values at the next coarser scale. If $f$ is very smooth, of course, then $\alpha_{j,k}(f)$ will be small at large $j$ corresponding to fine scales $2^j$.

Unlike traditional wavelet transforms, this transform is defined by point values rather than integrals. It is only nicely defined on spaces of continuous functions; this places limitations on the extent to which it has properties analogous to the traditional wavelet transform. Thus, it is not defined for arbitrary square integrable functions (such functions do not have precisely defined point values) and so there is no hope of a "Parseval relation" relating "coefficient energy in the wavelet domain" to "energy in the original domain". Despite this limitation, the IWT has many of the same properties as traditional wavelet transforms. For example, it can characterize smoothness spaces [**6**]. Suppose $0 < \delta < R(D)$. Then a bounded function $f$ is Hölder-$\delta$ if and only if

$$|\alpha_{j,k}| \leq C \cdot 2^{-j\delta}, \quad j \geq 0,\ k \in \mathbb{Z}.$$

There are in fact a range of norm-equivalence results [**6**]. Let $(\sigma, p, q)$ denote a triple of indices for a Besov Space $B^\sigma_{p,q}$ or a Triebel Space $F^\sigma_{p,q}$; then we can define norms on the wavelet coefficients so that

$$\|\alpha\|_{b^\sigma_{p,q}} \asymp \|f\|_{B^\sigma_{p,q}}, \quad \|\alpha\|_{f^\sigma_{p,q}} \asymp \|f\|_{F^\sigma_{p,q}}$$

whenever $R(D) > \sigma > 1/p$. Informally, if we have a functional class in the Besov/Triebel scales (this includes Hölder and Sobolev classes), and if all members of the class are continuous (so $\sigma > 1/p$), and if the fundamental function $\phi$ is a member of the class (so $R(D) > \sigma$), then the decay of the interpolating wavelet coefficients tells whether $f$ is in the class.

In short, starting from the DD refinement scheme, we can define a wavelet-like transform, the coefficients of the transform have a nice interpretation in terms of agreement with a local polynomial fit, and the coefficients of this transform characterize smoothness properties.

**2.2. Generalization.** We can generalize the original DD refinement scheme in two ways. First, we can change the functionals being interpolated; second, we can change the interpolation family.

2.2.1. *Choice of functional.* Consider the following *Functional Interpolation Problem.* Let $I_{0,k}$ denote the interval $[k, k+1)$. Suppose we are given data $\nu_k = \lambda(f \mid I_{0,k})$ where $\lambda(f \mid I)$ is a functional of $f$ associated with the behavior of $f$ in the vicinity of the interval $I$. We wish to find a function $\tilde{f}$ which agrees with the prescribed values of the $\lambda$ functionals at scale 0.

Consider the following two-scale refinement procedure, for obtaining functional values associated to the finer collection of intervals $(I_{j+1,k})_k$ starting from values on the coarser collection of intervals $(I_{j,k})_k$.

[$\lambda$1] **($\lambda$-Interpolation).** For each interval $I_{j,k}$, find a polynomial $\pi_{j,k}$ of degree $D$ satisfying the $\lambda$-*interpolation condition*:

$$\lambda(\pi_{j,k} \mid I_{j,k+h}) = \nu_{j,k+h} \quad \text{for} \quad -L \leq h \leq L. \tag{2.6}$$

[$\lambda$2] **($\lambda$-Imputation).** Obtain imputed values at a finer scale by setting

$$\nu_{j+1,2k+h} = \lambda(\pi_{j,k} \mid I_{j+1,2k+h}) \quad \text{for} \quad h = 0, 1. \tag{2.7}$$

This is a generalization of the DD refinement procedure, where $\lambda(f \mid I_{j,k})$ plays the role which the point value $f(k/2^j)$ played in the original procedure. We intend here that $\lambda(f \mid I_{j,k})$ may be either a linear or nonlinear functional, that it may be scalar valued or vector valued. We will discuss all these possibilities in the section to follow.

We don't wish to make light of some serious questions that need to be resolved in attempting this generalization. Q1: for a given family of functionals $\lambda(\cdot \mid I_{j,k})$, is the interpolation problem solvable by polynomials: can one always find a polynomial that satisfies the constraints (2.6)? Q2: if so, does the process converge when iterated repeatedly, producing a limit function $\tilde{f}$? Q3: if so, does the limit function $\tilde{f}$ have any useful properties? Q4: does the procedure yield useful fast algorithms?

2.2.2. *Choice of interpolation family.* Another generalization is to go beyond use of polynomials for interpolation. Let $\mathcal{G}$ be a finite-dimensional parametric family of functions. We can generalize the polynomial interpolation problem to the $(\lambda, \mathcal{G})$-*interpolation problem*, defining a new two-scale refinement procedure by fitting and imputing locally from $\mathcal{G}$. This procedure aims to obtain functional values associated to the finer collection of intervals $(I_{j+1,k})_k$ starting from values on the coarser collection of intervals $(I_{j,k})_k$.

[G1] **($(\lambda, \mathcal{G})$-Interpolation)**. For each interval $I_{j,k}$, find an element $g_{j,k} \in \mathcal{G}$ satisfying

$$\lambda(g_{j,k} \mid I_{j,k+h}) = \nu_{j,k+h} \quad \text{for} \quad -L \leq h \leq L. \tag{2.8}$$

[G2] **($(\lambda, \mathcal{G})$-Imputation)**. Obtain imputed values at a finer scale by setting

$$\nu_{j+1,2k+h} = \lambda(g_{j,k} \mid I_{j+1,2k+h}) \quad \text{for} \quad h = 0, 1. \tag{2.9}$$

Here the family $\mathcal{G}$ can be composed of functions for which the interpolation problem is solvable. Below we will give an example with so-called "split polynomials". There are several questions that need to be resolved here to make sense of this so-called generalization. These are precisely analogous to Q1-Q4 mentioned in connection with $\lambda$-interpolation.

## 3. Average-Interpolation

Consider the case where $\lambda(f \mid I_{j,k}) = \text{Ave}(f \mid I_{j,k}) = |I_{j,k}|^{-1} \int_{I_{j,k}} f(x)\,dx$ and we use polynomial interpolation methods. Suppose we have an array $(a_{0,k})_{k=-\infty}^{\infty}$ which represents averages of a function $f$ on dyadic intervals $I_{0,k}$. We may synthesize mock-averages at scale $j+1$ from scale $j$ by the following *Average-Interpolating Refinement* procedure. For integer $L > 0$, let $D = 2L$ be an *even* integer greater than 0.

[A1] (Average-Interpolation). At each site $k$, find the polynomial $\pi_{j,k}$ of degree $D$ which generates the same averages in the neighborhood $(a_{j,k'}, k' = k - L, \ldots, k + L)$, i.e.

$$\text{Ave}_{j,k'}\, \pi_{j,k} = a_{j,k'}, \quad k' = k - L, \ldots, k + L.$$

[A2] (Average-Imputation). Define the mock-averages at the next finer scale as averages of the AI polynomial. On the two halves of the interval we get

$$a_{j+1,2k+h} = \text{Ave}_{j+1,2k+h}\, \pi_{j,k} \quad h = 0, 1.$$

Figure 4 illustrates the fitting of an Average-Interpolating polynomial. Actually, the average-interpolation procedure is well-defined. The desired polynomial has $D+1$ coefficients and is required to satisfy $D+1$ constraints; the Average-Interpolation problem can be written in terms of a nonsingular system of linear equations.

Figure 5 illustrates the results of several iterations of refinement. Figure 6 illustrates the fundamental solution $\varphi = \varphi_{(D)}$—refinement of a Kronecker sequence $a_{0,k}^{(0)} = 1_{\{k=0\}}$—for a range of values of $D = 2L$. It turns out that this fundamental solution is continuous and has a smoothness $\overline{R}(D)$ that increases with roughly linearly $D$. Indeed, there is a special relationship between the fundamental solution $\varphi$ of Average-Interpolation ($D = 2L$) and the fundamental solution $\phi$ of Deslauriers-Dubuc interpolation ($D = 2L + 1$). In fact [**7**]

$$\phi'(x) = \varphi(x + 1) - \varphi(x).$$

Evidently, with $R(2L + 1)$ denoting the smoothness possessed by DD interpolation for $D = 2L+1$, AI with $D = 2L$ yields one fewer derivative: $\overline{R}(2L) = R(2L+1) - 1$.

Average-Interpolation was introduced in Donoho (1993) and Harten (1994). It can be used to construct a wavelet transform, that is a decomposition

$$f = \sum_k \beta_{0,k} \varphi_{0,k} + \sum_{j \geq 0} \sum_k \alpha_{j,k} \overline{\psi}_{j,k}, \tag{3.1}$$

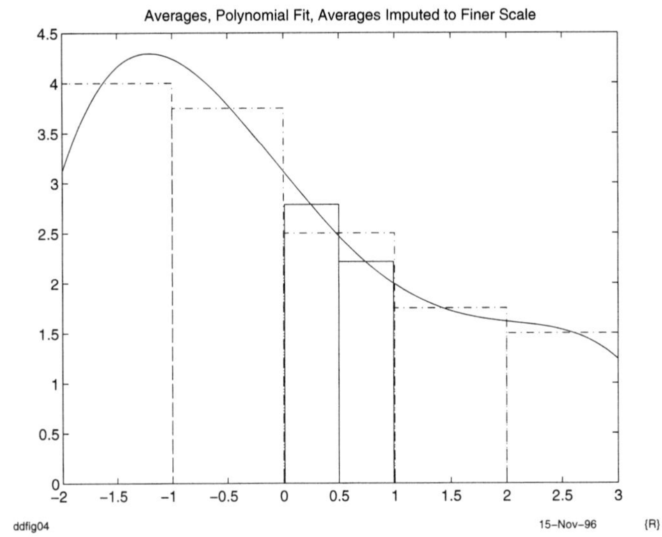


FIGURE 4. Imputation of averages over $[0, 1/2)$ and $[1/2, 1)$ from averages over $[k, k+1)$ for $k \in \{-2, -1, 0, 1, 2\}$.

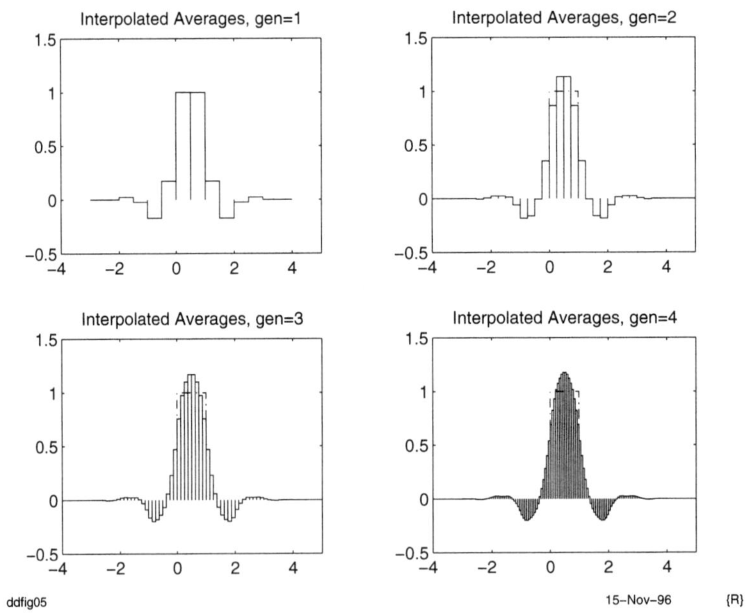


FIGURE 5. Sequence of refinements by imputing averages to finer and finer scales.

where $\varphi_{0,k}$ is a translated version of the AI fundamental function, and $\overline{\psi}_{j,k}(x) = 2^{j/2}\overline{\psi}(2^j x - k)$ is a scaled and translated version of the mother wavelet $\overline{\psi}(x) = \big(\varphi(x+1/2) - \varphi(x)\big) \,/\, \sqrt{2}$.

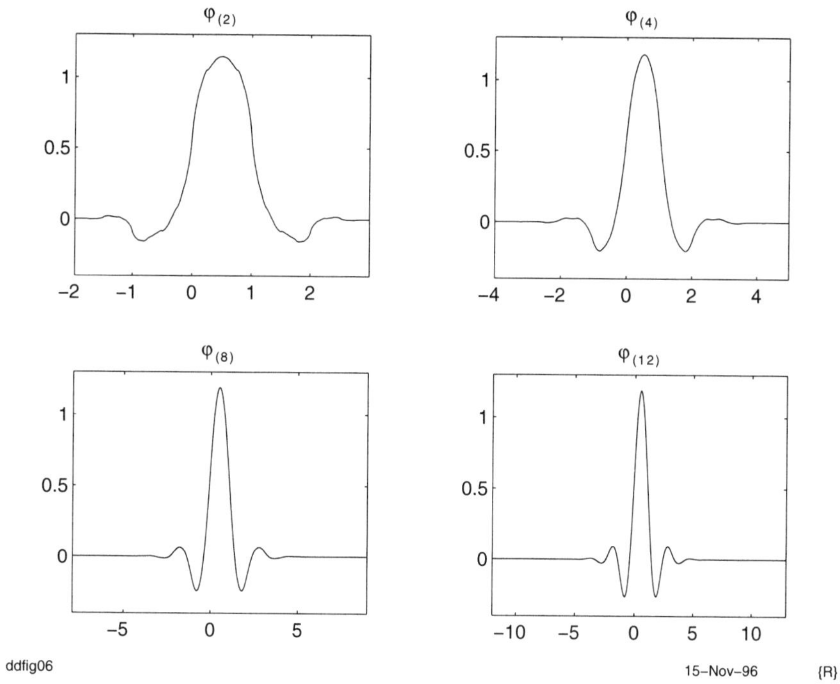


FIGURE 6. Fundamental solutions to Average-Interpolation, for degrees $D = 2, 4, 8, 12$.

The wavelet coefficient functionals $\alpha_{j,k}(f)$ for this transform obey the relation

$$\alpha_{j,k}(f) = 2^{-j/2}\big(\text{Ave}\{f \mid I_{j+1,2k}\} - \text{Ave}\{\pi_{j,k} \mid I_{j+1,2k}\}\big)$$

and so they measure the extent to which the averages of $f$ at a given scale cannot be predicted from the averages at the next coarser scale.

This transform is well-defined for all locally integrable functions and the coefficients obey various norm-equivalence results. Let $(\sigma, p, q)$ denote a triple of indices for a Besov Space $B^{\sigma}_{p,q}$ or a Triebel Space $F^{\sigma}_{p,q}$: then we can define norms on the wavelet coefficients so that [**7**]

$$\|\alpha\|_{b^{\sigma}_{p,q}} \asymp \|f\|_{B^{\sigma}_{p,q}} \quad \|\alpha\|_{f^{\sigma}_{p,q}} \asymp \|f\|_{F^{\sigma}_{p,q}},$$

whenever $\overline{R}(D) > \sigma > 1/p - 1$. Informally, the transform can characterise every space in the Besov/Triebel scales which consists of locally integrable functions and which contains $\varphi$. The transform itself is not orthogonal; as it turns out [**7**], it is identical to a family of biorthogonal wavelet transforms developed from another point of view by Cohen, Daubechies, and Feauveau (1991).

## 4. Hermite Interpolating Refinement

Now we consider an example of the general $\lambda$-interpolation picture where $\lambda$ is vector-valued. Suppose we are given the values of a function and its first $m - 1$ derivatives at the integers:

$$f^{(\ell)}(k) = \nu^{\ell}_{0,k}, \quad k \in \mathbb{Z},\ 0 \le \ell < m,$$

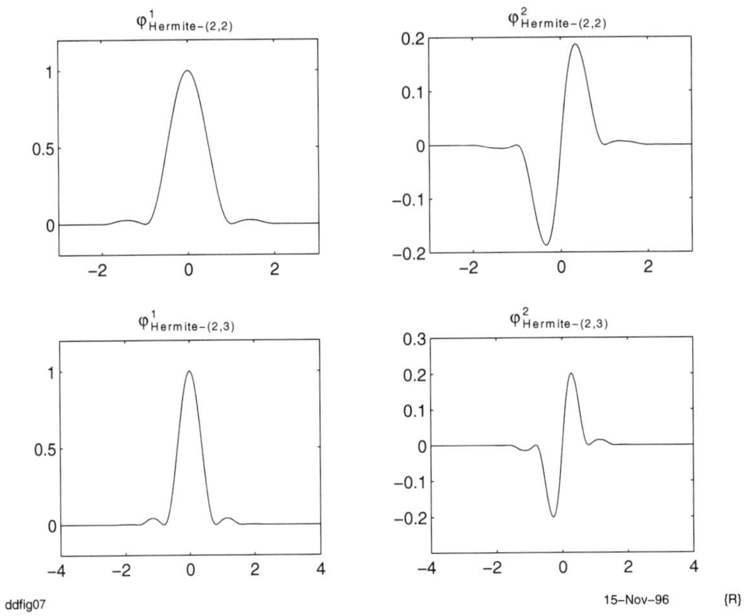


FIGURE 7. Fundamental solutions of Average-Interpolation, for $m = 2$, and $L = 2, 3$.

where $f^{(\ell)} = d^\ell/dx^\ell f$. Hermite interpolation is the problem of finding a smooth function $\tilde{f}(x)$ obeying the given conditions.

A two-scale refinement scheme can be developed, based on imputing the point-values and derivatives at the half integers $f^{(\ell)}(k/2) = \nu^\ell_{1,2k+1}$, using local polynomial interpolation from the values at the integers, and then iterating at finer scales. Fix $L > 0$ and set $\widetilde{D} = (2L+2)m-1$. The process works as follows, for each $k \in \mathbb{Z}$.

[H1] (Hermite Interpolation). Find a polynomial $\tilde{\pi}_{0,k}$ of degree $\widetilde{D}$ with the prescribed point values and derivatives $\nu^\ell_{0,k'}$ for $k'$ in a neighborhood of $k$:

$$\tilde{\pi}^{(\ell)}_{0,k}(k+h) = \nu^\ell_{0,k+h}, \quad -L \le h \le L+1. \tag{4.1}$$

[H2] (Hermite Imputation). Calculate the point values and derivatives of $\tilde{\pi}_{0,k}$ at the next finer scale:

$$\nu^\ell_{1,2k+1} = \tilde{\pi}^{(\ell)}_{0,k}(k+1/2), \quad k \in \mathbb{Z},\ 0 \le \ell < m. \tag{4.2}$$

In the case $m = 1$ this is just the scheme of Deslauriers-Dubuc Interpolation. This scheme was proposed by Merrien [**16**] and studied by Dyn and Levin [**13**]. It is known that the Hermite-Interpolation problem (4.1) is well-posed. Precise information about the smoothness of solutions has been obtained by Dyn, Levin, and Yu (1996).

There are $m$ fundamental solutions $\phi^\ell$ to this problem, corresponding to the refinement of the different Kronecker sequences $(\nu_k^{(l,\ell)})_{k,l}$ defined by $\nu_k^{l,\ell} = 1_{\{k=0\}} \cdot 1_{\{l=\ell\}}$ (i.e. refinement of sequences with all point values and derivatives zero except for one specific point value or derivative in one specific location). Figure 7 shows different fundamental solutions for $m = 2$ and two different values of $L$ Hervé (1994) has shown how this general type of interpolation can be used to construct a multi-interpolating multi-wavelet-type transform. As the reader should by now expect, the coefficients of such a transform measure the extent to which values and first $m-1$ derivatives at points in a grid of a certain scale can be predicted from values and derivatives at points in the next coarser scale grid. The transform is only well defined on functions which are $C^{m-1}$ or better; but within those limitations, the obvious generalization of [**6**] will show that the multi-wavelet coefficients have good norm characterization properties.

## 5. Moment-Interpolating Refinement

Here is another example where $\lambda$ is vector-valued. Let $m \geq 1$ be fixed and let $\chi_{0,0}^{\ell}$ denote the $\ell$-th Legendre polynomial associated to the interval $[0,1]$. For example, $\chi_{0,0}^{0}(x) = 1_{[0,1]}(x)$, $\chi_{0,0}^{1}(x) = (x-1/2)\cdot\sqrt{12}\cdot 1_{[0,1]}(x)$. These are monomials which are localized to the interval $[0,1]$ and orthogonalized to corresponding localized monomials of lower degree. Let $\chi_{j,k}^{\ell}(x) = 2^{j/2}\chi_{0,0}^{\ell}(2^j x - k)$ be the same functions transported to "live" on $I_{j,k}$.

*Moment interpolating refinement* is an answer to the following problem: Given the local moments of a function $f$,

$$\mu_{0,k}^{\ell} = \langle f, \chi_{0,k}^{\ell}\rangle, \quad k \in \mathbb{Z},\ 0 \leq \ell < m, \tag{5.1}$$

construct a "smooth" function $\tilde{f}$ matching those moments. We may define a two-scale refinement operator which, for a given sequence $(\mu_{j,k}^{\ell})_k$ of moment vectors $\mu_{j,k}^{\ell} = \langle f, \chi_{j,k}^{\ell}\rangle$ delivers a "predicte" sequence $(\mu_{j+1,k}^{\ell})_k$ of moment vectors at the next finer scale. Fix $L > 0$ and set $D = (2L+1)m - 1$. The process works as follows, for each $k \in \mathbb{Z}$.

[$\mu$1] (Moment Interpolation). Find a polynomial $\pi_{j,k}$ of degree $D$ with the prescribed moments $\mu_{j,k'}^{\ell}$ for $k'$ in a neighborhood of $k$:

$$\langle \pi_{j,k}, \chi_{j,k+h}^{\ell}\rangle = \mu_{j,k+h}^{\ell}, \quad -L \leq h \leq L. \tag{5.2}$$

[$\mu$2] (Moment Imputation). Calculate the moments of $\pi_{j,k}$ at the next finer scale:

$$\mu_{j+1,2k+h}^{\ell} = \langle \pi_{j,k}, \chi_{j+1,2k+h}^{\ell}\rangle, \quad h = 0, 1. \tag{5.3}$$

The Moment Interpolation Problem (5.2) has a solution. It turns out that the vector equations (5.2) impose exactly $D+1$ constraints on $\pi_{j,k}$ which is a polynomial of degree $D$. These constraints can be proven to be linearly independent over the space of polynomials of degree $D$.

The case $m = 1$ is really just Average-Interpolation. There are $m$ fundamental solutions $\varphi^{\ell}$ to this problem, corresponding to Moment-Interpolating Refinement of the different Kronecker sequences $(\nu_k^{(l,\ell)})_{k,l}$ defined by $\nu_k^{l,\ell} = 1_{\{k=0\}}\cdot 1_{\{l=\ell\}}$ (i.e. interpolation of sequences of prescribed moments which all vanish except for one prescribed moment at one prescribed site). Figure 8 shows different fundamental solutions for $m = 2$ and two different values of $L$. Donoho, Dyn, Levin, and Yu (1996) have shown how this general type of interpolation is closely connected with Hermite Interpolation, and that the smoothness of solutions to this problem can be derived easily from the smoothness of solutions to the corresponding Hermite Interpolation problem with the same value of $m$ and $L$. They also showed that the refinement scheme can be used to construct a multi-wavelet-type transform. It is defined for all locally integrable functions, and the multi-coefficients allow

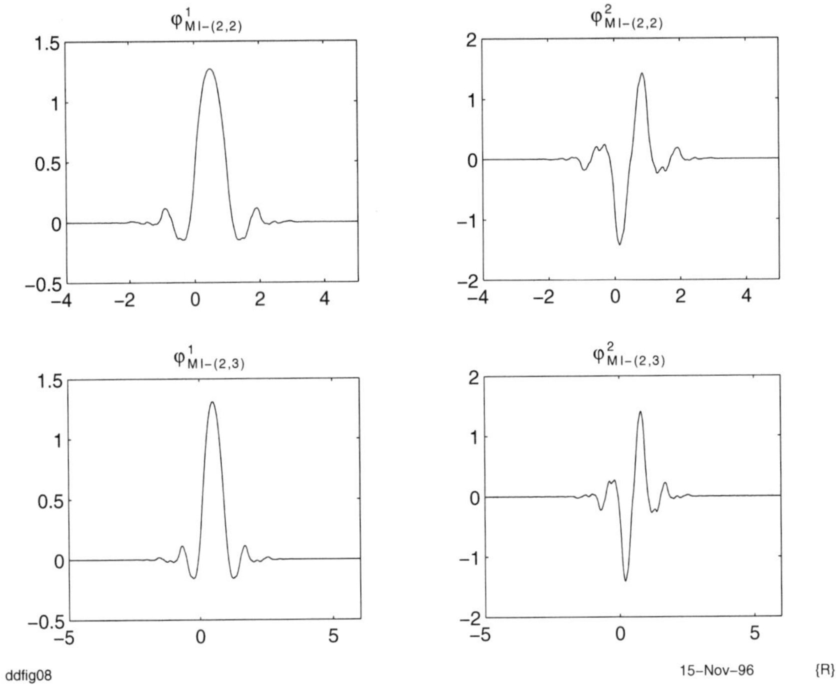


FIGURE 8. Fundamental solutions of Moment-Interpolation, for $m = 2$ and $L = 2, 3$.

characterization of various smoothness spaces. They indicate applications to the study of recursive partitioning schemes.

## 6. Nonlinear Refinement: Median-Interpolation

Given a function $f$ on an interval $I$, let $\operatorname{med}(f \mid I)$ denote a median of $f$ defined by

$$\operatorname{med}(f \mid I) = \inf\{\mu : m(t \in I : f(t) \geq \mu) \geq m(t \in I : f(t) \leq \mu)\}.$$

It turns out to be slightly nicer to work with triadic rather than dyadic refinement in this setting. Suppose we are given a triadic array $\{m_{j,k}\}_{k=0}^{3^j-1}$ of numbers representing the medians of $f$ on the triadic intervals $I_{j,k} = [k3^{-j}, (k+1)3^{-j})$:

$$m_{j,k} = \operatorname{med}(f \mid I_{j,k}), \quad 0 \leq k \leq 3^j.$$

Median-interpolating refinement can be defined starting from a fixed even integer $D$. The basic idea is, starting from data on medians over coarser-scale triadic blocks to impute medians over finer-scale blocks. There are two steps.

[M1] **(Median-Interpolation)**. For each interval $I_{j,k}$, find a polynomial $\pi_{j,k}$ of degree $D = 2L$ satisfy the *median-interpolation condition*:

$$\operatorname{med}(\pi_{j,k} \mid I_{j,k+h}) = m_{j,k+h} \quad \text{for} \quad -L \leq h \leq L. \tag{6.1}$$

[M2] **(Median-Imputation)**. Obtain (pseudo-) medians at the finer scale by setting

$$m_{j+1,3k+h} = \operatorname{med}(\pi_{j,k} \mid I_{j,3k+h}) \quad \text{for} \quad h = 0, 1, 2. \tag{6.2}$$

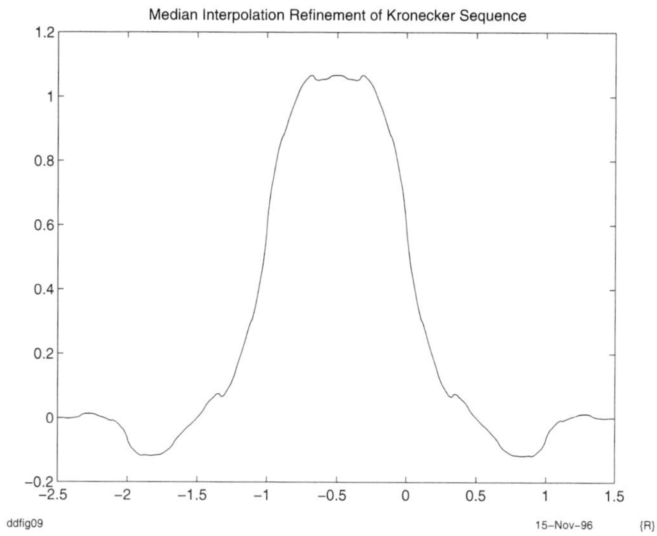


FIGURE 9. Refinement limit of a Kronecker sequence for Median-Interpolation for $D = 2$.

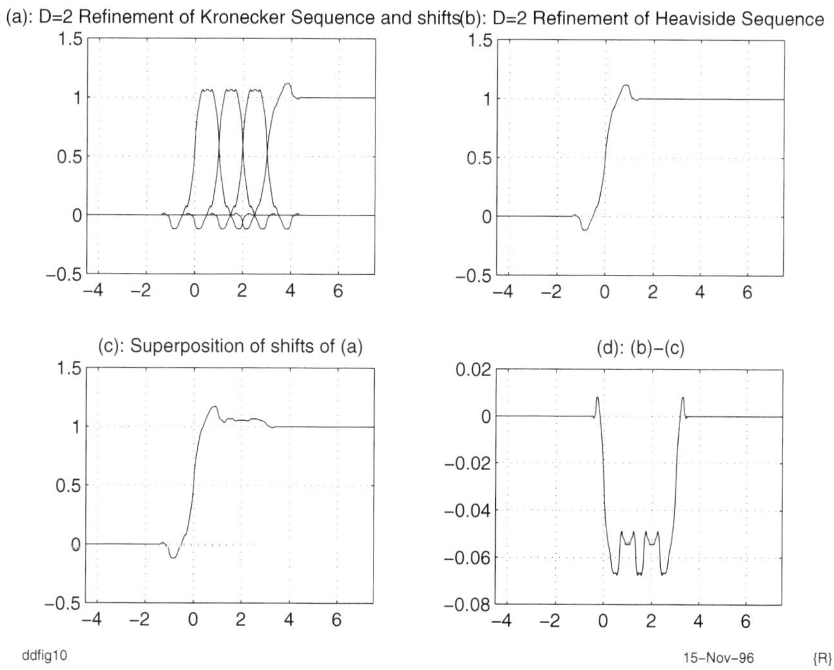


FIGURE 10. Nonlinearity of Median-Interpolating refinement.

*If* this can be done, it provides a systematic way to impute behavior at finer scales from behavior at coarse scales. The questions Q1-Q4 of Section 2.1 come up naturally. Such questions are studied in [**10**], where it is shown that for $D = 2$, they have affirmative answers. In that paper, the authors develop an effective median interpolation algorithm and show that it can be used to generate a nonlinear refinement scheme with continuous (in fact almost Lipschitz) solutions.

The solution from refining a Kronecker sequence is depicted in Figure 9.

As the refinement scheme is not linear, the result from refining an arbitrary sequence is not a linear superposition of shifts of this function. Indeed, Figure 10 depicts the result of refining a Heaviside sequence $m_{0,k} = 1_{\{k \geq 0\}}$, which can be viewed as a superposition of Kronecker sequences $\sum_{\ell \geq 0} \delta^{(\ell)}$, and then compares this with the superposition of separate components.

Donoho and Yu have constructed nonlinear pyramid transforms, but not wavelet transforms, starting from this refinement scheme, and used them to analyse heavily non-Gaussian data. In such settings the robustness of the median gives an important benefit over various non-robust linear wavelet procedures.

## 7. Non-Polynomial Refinement: Segmented Interpolation

We now give an example where the family $\mathcal{G}$ is not polynomial; more properly, it consists of *split-polynomials*. That is, there is a segmentation point $\tau$ and $\mathcal{G}$ consists of polynomials as well as pairs of polynomials of degree $D$, one of which is thought of as a right-polynomial and one of which is thought of as a left-polynomial. The functionals $\lambda(f \mid I_{j,k})$ are just averages $\text{Ave}(f \mid I_{j,k})$.

Consider, as in [**8**], the following segmented refinement procedure, with segmentation point $\tau \in [0,1]$. Given a sequence of averages $a_{j,k}$, $0 \leq k < 2^j$, we synthesise mock averages at finer scales by the following procedure:

[1] At each site $k$ which is more than $L$ sites away from the segmentation point $\tau$, use the average interpolating procedure to find the polynomial $\pi_{j,k}$ of degree $D = 2L$ which generates the same averages in the neighborhood $(a_{j,k'}, k' = k-L, \dots, k+L)$.

[2] At each site $k$ which is at most $L$ sites away from the segmentation point $\tau$, we distinguish two cases.

- [2a] If $\tau \notin I_{j,k}$, then find the polynomial $\pi_{j,k}$ of degree $D$ which generates the same averages in the neighborhood $(a_{j,k'})_{k' \in N(k)}$, where $N(k)$ consists of the $D+1$ nearest neighbors of $k$ which are all on the same side of the segmentation point as $k$.
- [2b] If $\tau \in I_{j,k}$, then fit, by constrained least squares, left and right polynomials $\pi^L_{j,k}$ $\pi^R_{j,k}$ of degree $D$ to the block averages in the neighborhoods on the right and left of the segmentation point, respectively; the constraint is that the piecewise polynomial $\pi^\tau_{j,k}$ which is $\pi^L$ on the left of $\tau$ and $\pi^R$ on the right of $\tau$ should have an average equal to the average $a_{j,k}$.

[3] In cases [1] and [2a], define the mock-averages at the next finer scale as averages of the polynomial. On the two halves of the interval we get

$$a^\tau_{j+1,2k+h} = \text{Ave}_{j+1,2k+h}\, \pi_{j,k}, \quad h = 0, 1.$$

In case [2b], the steps are the same, only using the piecewise polynomial $\pi^\tau_{j,k}$.

This segmented Average-Interpolating resolution has a variety of properties; a key one being that *if $\pi^\tau$ is a piecewise polynomial of degree $D$, with one knot, at $\tau$, then the refinement process recovers $\pi^\tau$ exactly.* Applications of this procedure to determining optimal segmentations are described in [**8**]. The basic idea is to measure the quality of a segmentation by an "entropy" of the coefficients $\alpha^{\tau}_{j,k}$; the "best" segmentation point $\tau$ then has coefficients with the smallest entropy $E\big((\alpha^{\tau}_{j,k})_{j,k}\big)$.

## 8. Conclusion

When Deslauriers and Dubuc developed their interpolation scheme, they also made available a general pattern of thought that can be applied in other settings. In this brief note we have shown that by generalizing the family of functionals beyond point values and generalizing the family of interpolating functions beyond polynomials, we can obtain a variety of interesting refinement schemes and transforms. We are confident there will be many other examples illustrating this theme.

This note was written following the principle of *Reproducible Research.* Code to generate the figures can be obtained by pointing a web browser at `http://www-stat.stanford.edu/~wavelab`.

## Acknowledgments

This research was partially supported by NSF DMS-92-09130 and DMS-95-05151, by an AFOSR Multi-University-Research Initiative, and by other sponsors. DLD would like to thank Nira Dyn and David Levin of the Mathematics Department at Tel Aviv University for plenty of conversations and interaction during his visit there in 1996.

## References

1. A. S. Cavaretta, W. Dahmen, and C. A. Micchelli, *Stationary subdivision*, Mem. Amer. Math. Soc. **453** (1991), 1–186.
2. A. Cohen, I. Daubechies, and J. C. Feauveau, *Biorthogonal bases of compactly supported wavelets*, Comm. Pure Appl. Math. **45** (1990), 485–560.
3. I. Daubechies and J. Lagarias, *Two-scale difference equations*, I. *Global regularity of solutions*, SIAM J. Math. Anal. **22** (1991), 1388–1410.
4. G. Deslauriers and S. Dubuc, *Interpolation dyadique*, Fractals, Dimensions Non-Entières et Applications, Masson, Paris, 1987.
5. ———, *Symmetric iterative interpolation processes*, Constr. Approx. **5** (1989), 49–68.
6. D. L. Donoho, *Interpolating wavelet transforms*, Stanford Univ., 1992, (tech. report), `http://www-stat.stanford.edu/pub/donoho/iwt.ps`.
7. ———, *Smooth wavelet decompositions with blocky coefficient kernels*, Recent Advances in Wavelet Analysis (L. Schumaker and F. Ward, eds.), Academic Press, 1993.
8. ———, *On minimum entropy segmentation*, Wavelets: Theory, Algorithms and Applications (C.K. Chui, L. Montefusco, and L. Puccio, eds.), San Diego, Academic Press, 1994.
9. D. L. Donoho, N. Dyn, D. Levin, and T. P. Y. Yu, *Smooth multiwavelet duals of Alpert bases by Moment-Interpolation, with applications to recursive partitioning*, Stanford Univ., 1996, (tech. report), `http://www-stat.stanford.edu/pub/donoho/moment.ps`.
10. D. L. Donoho and T. P. Y. Yu, *Robust nonlinear wavelet transforms by median interpolation*, Stanford Univ., 1996, (tech. report) `http://www-stat.stanford.edu/pub/donoho/median.ps`.
11. S. Dubuc, *Interpolation through an iterative scheme*, J. Math. Anal. Appl. **114** (1986), 185–204.
12. N. Dyn, J. A. Gregory, and D. Levin, *Analysis of uniform binary subdivision schemes for curve design*, Constr. Approx. **7** (1991), 127–147.
13. N. Dyn and D. Levin, *Analysis of Hermite-type subdivision schemes*, Approximation Theory VIII (Charles K. Chui and Larry L. Schumaker, eds.), World Scientific Publishing, 1995.
14. A. Harten, *Multiresolution representation of cell-averaged data*, UCLA Comput. Appl. Math. report CAM 94-21, 1994.

15. L. Hervé, *Multiresolution analysis of multiplicity d: Applications to dyadic interpolation*, ACHA **1** (1994), 299–315.
16. J. L. Merrien, *A family of Hermite interpolants by bisection algorithms*, Numer. Algorithms **2** (1992), 187–200.
17. T. P. Y. Yu, N. Dyn, and D. Levin, *Smoothness of solutions to Hermite refinement schemes*, manuscript, 1997.

Department of Statistics, Stanford University, Stanford, CA 94305, USA
*E-mail address*: `donoho@stat.stanford.edu`

Department of Mathematical Sciences, Rensselaer Polytechnic Institute, Troy, New York 12180-3590, USA
*E-mail address*: `pyu@rpi.edu`

Centre de Recherches Mathématiques
CRM Proceedings and Lecture Notes
Volume **18**, 1999

# Some Theory for $L$-Spline Smoothing

J. O. Ramsay and N. Heckman

Abstract. We consider the problem of estimating a smoothing spline where the penalty on the smoothing function $g$ is $\int(Lg)^2$. An algorithm requiring $O(n)$ operations where $n$ is the number of values being smoothed is developed. The technique relies on results obtained by Anselone and Laurent [**1**], and these are described in a form that may be more accessible to some readers.

## 1. Introduction

Data smoothing plays a very large and ever increasing role in modern statistical analysis. Because the number of observations $n$ to be smoothed is often large, a smoothing algorithm is only generally useful if the function $g$ smoothing the data can be computed in a number of operations that is proportional to $n$ or $n \log n$, that is, $O(n)$ or $O(n \log n)$. The path-breaking paper by Anselone and Laurent [**1**], drawing on earlier unpublished work by Atteia [**2**] and Greville [**4**], dealt with spline smoothing in a very general and abstract context, and pointed specifically to the possibility of an $O(n)$ algorithm. The papers by Reinsch [**6, 7**]) and subsequent development by Hutchison and de Hoog [**5**] enabled the development of $O(n)$ smoothing software in the cubic polynomial spline case, that is, when the penalty on $g$ is $\int(D^2g)^2$.

We want to put the results of Anselone and Laurent [**1**] to work in the more general context of $L$-spline smoothing, where the penalty $\int(D^2g)^2$ is replaced by a more general $\int(Lg)^2$, $L$ being an arbitrary linear differential operator with possibly variable coefficients.

Because the exposition in Anselone and Laurent [**1**] may not be easy to follow for those with limited exposure to the functional analytic literature on splines of that era, a good deal of this paper is given over to an orderly exposition of their results. This is followed in the final section with details on an $O(n)$ $L$-spline smoothing algorithm.

## 2. Vector Spaces

We will be using the four following vector spaces:

---

1991 *Mathematics Subject Classification.* Primary: 62G07; Secondary: 46N30.
This is the final version of the paper.

**$\mathcal{L}^2$:** The space of functions defined on a closed bounded real interval $T$ that have Lebesgue integrable squares. We will use $x$ and $y$ to indicate two such functions, and $x(t)$ and $y(t)$ to denote their values at an argument $t \in T$.

**$\mathcal{H}^m$:** This is the space of functions on $T$ possessing $m-1$ absolutely continuous derivatives and an $m$th derivative in $\mathcal{L}^2$. We will use $g$ and $h$ to denote functions in this space.

**$\mathcal{R}^m$ and $\mathcal{R}^n$:** The Euclidean vector spaces of dimension $m$ and $n$, respectively. Elements of either space will be denoted by $\mathbf{x}$ and $\mathbf{y}$, with the context determining which space is intended. Also, $\mathcal{R}^n$ will be associated with an *argument sequence*

$$\mathbf{t} = (t_1, \ldots, t_n)', \quad t_j \in T.$$

**$\mathcal{R}^n \times \mathcal{L}^2$:** Elements of this product space will be denoted by $\langle \mathbf{x}, x \rangle$ and $\langle \mathbf{y}, y \rangle$.

Each of these vector spaces has a "usual" inner product. In $\mathcal{L}^2$, the usual inner product of $x$ and $y$ is simply $\int x(t)y(t)\,dt$. In $\mathcal{H}^m$, the usual inner product between $g$ and $h$ is $\sum_{j=0}^{m-1} \int_T g^{(j)}h^{(j)} + \int_T g^{(m)}h^{(m)}$. In $k$-dimensional Euclidean space, the inner product is simply the vector cross-product. In Section 3, we will define other inner products on these spaces. These inner products, while topologically equivalent to the usual inner products, will be more convenient for the smoothing spline problem.

## 3. Operators

We will be using the following four linear operators.

**Evaluation Operator $A$:**

$$A \colon \mathcal{H}^m \to \mathcal{R}^n \mid Ax = \big(x(t_1), \ldots, x(t_n)\big)'.$$

The notation $Ax = x(\mathbf{t})$ also refers to the value of this mapping. It is well known that this operator is continuous in the usual norms on $\mathcal{H}^m$ and $\mathcal{R}^n$.

**Differential Operator $L$:**

$$L \colon \mathcal{H}^m \to \mathcal{L}^2 \mid Lx = w_0 x + w_1 Dx + \cdots + w_{m-1} D^{m-1}x + D^m x$$

where $w_j, j = 0, \ldots, m-1$ are weight functions with suitable regularities. We will assume that $L$ is continuous with closed range and that $= A(\ker L)$ is an $m$-dimensional subspace of $\mathcal{R}^n$.

**Constraint Operator $B$:**

$$B \colon \mathcal{H}^m \to \mathcal{R}^m \mid Bx = (B_1 x, \ldots, B_m x)'$$

where $B_j, j = 1, \ldots, m$ are continuous linear functionals defined on $\mathcal{H}^m$. We assume that the range of $B$ is $\mathcal{R}^m$.

**Composite Operator $Z$:**

$$Z \colon \mathcal{H}^m \to \mathcal{R}^n \times \mathcal{L}^2 \mid Zg = (Ag, Lg).$$

Associated with each of these operators are their kernel and range spaces:

- range $A$ is of dimension $n$ and $\ker A$ is of codimension $n$
- $\ker L$ is of dimension $m$. We shall use $u$ to refer to a function in $\ker L$
- range $B$ is of dimension $m$ and $\ker B$ is of codimension $m$.

Here, the codimension of a subspace $\mathcal{G}$ is the dimension of its orthogonal complement $\mathcal{G}^\perp$, which is defined in Section 4.

**3.1. Two operator constraints.** We impose on these operators two complementarity constraints:

$$\ker B \cap \ker L = \{0\} \tag{3.1}$$

$$\ker A \cap \ker L = \{0\}. \tag{3.2}$$

The first constraint will be used in Section 4 to define a convenient inner product structure on $\mathcal{H}^m$. The second constraint is essential to ensure that a spline smooth is uniquely defined. In fact, this second constraint has a simple interpretation. Let $\mathbf{u}$ be a vector of $m$ linearly independent functions that span $\ker L$. Then $h \in \ker L$ is equivalent to $h = \mathbf{d}'\mathbf{u}$ for some vector $\mathbf{d} \in \mathcal{R}^m$. Now let $\mathbf{U} \in \mathcal{R}^{nm}$ be the $n \times m$ matrix containing the values of these functions at the argument values $\mathbf{t}$

$$\mathbf{U}_{ij} = u_j(t_i).$$

Then constraint (3.2) means simply that $\mathbf{Ud} \neq 0$ for all possible $\mathbf{d}$ or, in short, that $\mathbf{U}$ is of full column rank.

## 4. Inner Products and Hilbert Spaces

Each of the vector spaces above will be given an inner product structure which is topologically equivalent to the usual inner product. We will use the same notation for each vector space and its associated Hilbert space. Some of these inner products are defined using the above operators

$$\begin{aligned}
\langle x, y\rangle_{\mathcal{L}^2} &= \int_T x(t)y(t)\,dt \\
\langle \mathbf{x}, \mathbf{y}\rangle_{\mathcal{R}^m} &= \mathbf{x}'\mathbf{y} \\
\langle \mathbf{x}, \mathbf{y}\rangle_{\mathcal{R}^n} &= \mathbf{x}'\mathbf{W}\mathbf{y} \\
\langle g, h\rangle_{\mathcal{H}^m} &= (Bg)'(Bh) + \int_T Lg(t)Lh(t)\,dt \\
&= \langle Bg, Bh\rangle_{\mathcal{R}^m} + \langle Lg, Lh\rangle_{\mathcal{L}^2}
\end{aligned} \tag{4.1}$$

where $\mathbf{W}$ is a symmetric positive definite matrix of order $n$. Note that (3.1) guarantees that, if $\langle g, g\rangle_{\mathcal{H}^m} = 0$ and so $Bg = 0$ and $Lg = 0$, then $g = 0$.

In addition, we also want to define an inner product on the space $\mathcal{R}^n \times \mathcal{L}^2$ as follows:

$$\big\langle (\mathbf{x}, x), (\mathbf{y}, y)\big\rangle_{\mathcal{R}^n \times \mathcal{L}^2} = \mathbf{x}'\mathbf{W}\mathbf{y} + \lambda \int_T x(t)y(t)\,dt \tag{4.2}$$

for some $\lambda > 0$.

We shall rely on the symbols used in a specific inner product to indicate which inner product is intended in order to avoid tedious repetition of inner product subscripting.

Associated also with an inner product is the concept of an orthogonal complement to a subspace. For subspace $\mathcal{G}$ the orthogonal complement is denoted

$$\mathcal{G}^\perp = \{g \mid \langle h, g\rangle = 0 \ \forall h \in \mathcal{G}\}.$$

A useful orthogonal complement is $(\ker L)^\perp$. Since $\ker B$ is of codimension $m$ and, by the definition of $\langle\ ,\ \rangle_{\mathcal{H}^m}$, is a subspace of $(\ker L)^\perp$, also of codimension $m$, we see that

$$\ker B = (\ker L)^\perp.$$

## 5. The Smoothing Problem

We want to find $h \in \mathcal{H}^m$ that is simultaneously

- close in a least squares sense to an observed vector $\mathbf{y} \in \mathcal{R}^n$, and
- small in the sense that $Lh$ is close to 0.

To this end we use the composite inner product to define the criterion

$$\begin{aligned} F_\lambda(h,\mathbf{y}) &= (\mathbf{y} - Ah)'\mathbf{W}(\mathbf{y} - Ah) + \lambda \int_T [Lh(t)]^2\, dt \\ &= \|\mathbf{y} - Ah\|^2 + \lambda\|Lh\|^2 \\ &= \big\|(\mathbf{y},0) - Zh\big\|^2 \end{aligned} \tag{5.1}$$

and we seek

$$F_\lambda(h,\mathbf{y}) = \min_{g\in\mathcal{H}^m} F_\lambda(g,\mathbf{y}).$$

The minimizer $h$ is called the smoothing spline solution.

## 6. The Reproducing Kernel and the Evaluation Operator

The evaluation functional $\rho_j h = h(t_j)$ is continuous in $\mathcal{H}^m$ and consequently the Riesz Representation Theorem states that there exists a function $k_j \in \mathcal{H}^m$ such that

$$\langle k_j, h\rangle = \rho_j h = h(t_j) \text{ for all } h \in \mathcal{H}^m.$$

These functions $k_j$ are defined by the *reproducing kernel* $k$ with values $k(s,t)$ such that

$$k(t,\cdot) \in \mathcal{H}^m \text{ for all } t \in T$$
$$\big\langle k(t,\cdot), h\big\rangle = h(t) \text{ for all } h \in \mathcal{H}^m.$$

It will be convenient to write $k$ as $k = k_1 + k_2$, where $k_1$ is the reproducing kernel for functions in $\ker L$ and $k_2$ is the reproducing kernel for functions in $(\ker L)^\perp = \ker B$. One easily shows that, for all $h \in \ker B$, $\big\langle h, k_2(t,\cdot)\big\rangle = h(t)$.

We now discuss how to compute $k_2$ in terms of the *Green's function* associated with the operator pair $(B, L)$. The Green's function $G$ with values $G(t,w)$ satisfies

$$\int_T G(t,w)Lh(w)\,dw = h(t), \quad h \perp \ker L. \tag{6.1}$$

In many cases, the function $G$ is easily computed (see, for example, Schumaker [**8**], and Dalzell and Ramsay [**3**]).

Since, for all $h \perp \ker L$, $h(t) = \big\langle h, k_2(t,\cdot)\big\rangle = \int_T \big(Lk_2(t,\cdot)\big)(w)(Lh)(w)\,dw$ we see from (6.1) that

$$Lk_2(t,\cdot) = Lk_2(\cdot,t) = G(t,\cdot). \tag{6.2}$$

Here $Lk_2(t,\cdot)$ means applying $L$ to $k_2$ with $t$ considered fixed. To calculate $k_2$, use the reproducing kernel $k$ and (6.2):

$$k_2(s,t) = \big\langle k_2(s,\cdot), k(t,\cdot)\big\rangle = \big\langle k_2(s,\cdot), k_2(t,\cdot)\big\rangle = \int_T G(t,w)G(s,w)\,dw. \tag{6.3}$$

We shall need the vector defined by

$$\mathbf{g}(w) = \big(Lk(\mathbf{t},\cdot)\big)(w) = G(\mathbf{t},w) \in \mathcal{R}^n.$$

If we use the notation

$$\mathbf{k} = k(\mathbf{t}, \cdot) = \big(k(t_1, \cdot), \dots, k(t_n, \cdot)\big)'$$

to refer to the vector of the $n$ functions $k_1, \dots, k_n$, then we can abuse inner product notation slightly to say

$$Ah = \langle \mathbf{k}, h \rangle.$$

## 7. Adjoint Operators

If $M$ is a continuous linear operator mapping Hilbert space $H$ into Hilbert space $G$, the relation

$$\langle Mh, x \rangle_G = \langle h, M^* x \rangle_H$$

defines a mapping $M^*\colon G \to H$ called the *adjoint* of $M$. If $H$ has a reproducing kernel $k$, then adjoints are easily calculated. By the reproducing quality of $k$ and the definition of adjoint

$$(M^* x)(t) = \big\langle M^* x, k(t, \cdot) \big\rangle_H = \big\langle x, Mk(t, \cdot) \big\rangle_G. \tag{7.1}$$

We need the following adjoint operators to solve the spline smoothing problem.

**Evaluation adjoint.** Using (7.1), we easily calculate

$$A^*\colon \mathcal{R}^n \to \mathcal{H}^m \mid A^* \mathbf{x} = \mathbf{k}' \mathbf{W} \mathbf{x}. \tag{7.2}$$

**Differential operator adjoint.** First we write the reproducing kernel $k$ as $k = k_1 + k_2$ with $k_1(t, \cdot) \in \ker L$ and $k_2(t, \cdot) \in (\ker L)^\perp$ for all $t$, as in Section 6. By (7.1) and then (6.2)

$$(L^* x)(t) = \big\langle x, Lk(t, \cdot) \big\rangle = \big\langle x, Lk_2(t, \cdot) \big\rangle = \big\langle x, G(t, \cdot) \big\rangle.$$

Therefore

$$(L^* x)(t) = \int_T G(t, w) x(w)\, dw.$$

**Composite adjoint.** By the additive nature of $Z^*$ and our inner product on $\mathcal{R}^n \times \mathcal{L}^2$, we see that

$$Z^*(\mathbf{x}, x) = A^* \mathbf{x} + \lambda L^* x.$$

Finally, associated with these adjoint operators are their kernels, and of particular importance is the subspace $\mathcal{G} \subset \mathcal{R}^n \times \mathcal{L}^2$ defined by

$$\mathcal{G} = \ker Z^*.$$

An important theorem for our purposes is the Closed Range Theorem: range $M$ is closed if and only if range $M^*$ is closed. We also use the fact that the closure of the range of $M^*$ is always equal to $(\ker M)^\perp$. Thus the ranges of $L^*$ and $A^*$ are closed and range $L^* = (\ker L)^\perp = \ker B$.

## 8. Some Preliminary Results

The proof of the existence and uniqueness of a smoothing spline is quite simple once some initial results are established. The characterization of this spline as well as the design of an efficient algorithm for estimating it also depend heavily on these results.

**8.1. The range Z is closed.** The range of $Z$ is the cross product of the range of $A$ and the range of $L$. Both of these ranges are closed. Since our norm on $\mathcal{R}^n \times \mathcal{L}^2$ is the usual cross-product norm, the range of $Z$ is closed in this norm.

**8.2. The dimension of $\mathcal{G} = \ker \mathbf{Z}^*$ is $\mathbf{n} - \mathbf{m}$.** The relation $Z^*(\mathbf{x}, x) = 0$ implies that

$$\lambda L^* x = -A^* \mathbf{x} = h$$

for some $h \in \mathcal{H}^m$. This is equivalent to

$$h \in A^* \mathcal{R}^n \cap L^* \mathcal{L}^2. \tag{8.1}$$

We now show that

$$A^* \mathcal{R}^n \cap L^* \mathcal{L}^2 = A^* \left[ (A \ker L)^{\perp} \right].$$

Suppose that $g = A^* \mathbf{x}$ for some $\mathbf{x} \in \mathcal{R}^n$. It suffices to show that $g \in L^* \mathcal{L}^2$ if and only if $\mathbf{x}$ is perpendicular to $A \ker L$. But $\mathbf{x}$ is perpendicular to $A \ker L$ if and only if, for all $u \in \ker L$, $\langle \mathbf{x}, Au \rangle = 0$, that is $0 = \langle A^* \mathbf{x}, u \rangle$ which, by our assumption on $g$, is equal to $\langle g, u \rangle$. But $\langle g, u \rangle = 0$ for all $u \in \ker L$ if and only if $g \in (\ker L)^{\perp} = \text{range } L^*$.

By assumption, $\dim(A \ker L) = m$ and therefore $\dim(A \ker L)^{\perp} = n - m$. Since $A^*$ is injective, $\dim[A^*(A \ker L)^{\perp}]$ is also $n - m$.

## 9. The Main Existence and Uniqueness Result for the Smoothing Spline

THEOREM 9.1. *There exists a unique $s \in \mathcal{H}^m$ minimizing criterion* (5.1) *for any $\mathbf{y} \in \mathcal{R}^n$. This minimizing function $s$ satisfies*

$$(\mathbf{y}, 0) - Zs \perp \text{range } Z. \tag{9.1}$$

*Thus* $(\mathbf{y}, 0) - Zs \in (\text{range } Z)^{\perp} = \ker Z^*$ *or equivalently*

$$(\mathbf{y} - As, -Ls) \in \mathcal{G}. \tag{9.2}$$

PROOF. Since range $Z$ is closed, it is a fundamental property of Hilbert spaces that there exists a unique $s$ such that $Zs$ is the projection of $(\mathbf{y}, 0)$ on range $Z$. Moreover, it is also a property of a projection on to a subspace that the residual $(\mathbf{y}, 0) - Zs$ is orthogonal to this projected value.

This is the counterpart of the well-known fact that the least squares residual $\mathbf{y} - \hat{\mathbf{y}} = \mathbf{y} - \mathbf{X}\beta$ satisfies $(\mathbf{y} - \hat{\mathbf{y}})'\mathbf{X} = 0$. □

## 10. Characterizing the Smoothing Spline and Defining an Efficient Smoothing Algorithm

First we need to define a basis of dimension $n - m$ for $\mathcal{G}$. Then the result in (9.2) will allow us to compute the values $As$ easily.

**10.1. A basis for $\mathcal{G}$.** Recall that $\mathbf{U}$ is the $n$ by $m$ matrix containing the evaluation of a basis for $\ker L$ at $\mathbf{t}$. Let the matrix $\mathbf{Q} \in \mathcal{R}^{n, n-m}$ be chosen so as to have full column rank and satisfy $\mathbf{Q}'\mathbf{W}\mathbf{U} = 0$. Although there are many ways to choose $\mathbf{Q}$ via a QR-decomposition, we shall be particularly interested in it having a band structure and, in Section 10.3, will show how this may be achieved.

The matrix $\mathbf{Q}'$ can be thought of as a generalized divided difference matrix. For instance, suppose that $\mathbf{W}$ is the identity matrix and that the $t_i$'s are evenly spaced. If $m = 2$ and $L = D^2$, then $u_1(t) = 1$ and $u_2(t) = t$ span $\ker L$. The first column of $\mathbf{U}$ contains one's and the second column is equal to $\mathbf{t}$. We can take $\mathbf{Q}'_{ii} = \mathbf{Q}'_{i,i+2} = 1$, $\mathbf{Q}'_{i,i+1} = -2$, and $\mathbf{Q}'_{ij} = 0$ otherwise. Thus $\mathbf{Q}'$ is the usual second divided difference matrix. If $t_{i+1} - t_i \equiv \Delta$, $m = 2$ and $L = D^2 + \omega^2 I$, then $u_1(t) = \cos \omega t$ and $u_2(t) = \sin \omega t$ span $\ker L$. We can take $\mathbf{Q}'_{ii} = \mathbf{Q}'_{i,i+2} = 1$, $\mathbf{Q}'_{i,i+1} = -2\cos(\omega\Delta)$ and $\mathbf{Q}'_{ij} = 0$ otherwise.

We can use $\mathbf{Q}$ to define a set of basis functions for $\mathcal{G}$ as follows. Recall that $\mathbf{g}(w) = G(\mathbf{t}, w)$.

PROPOSITION.

$$\begin{bmatrix} \lambda\mathbf{Q} \\ -\mathbf{g}'\mathbf{W}\mathbf{Q} \end{bmatrix} \text{ is a basis for } \mathcal{G} = \ker Z^*$$

PROOF. Let any $x$ be of the form

$$x = (\lambda\mathbf{Q}\mathbf{c}, -\mathbf{g}'\mathbf{W}\mathbf{Q}\mathbf{c})$$

for some $\mathbf{c} \in \mathcal{R}^{n-m}$. Then

$$\begin{aligned} Z^*x &= A^*(\lambda\mathbf{Q}\mathbf{c}) - \lambda L^*\mathbf{g}'\mathbf{W}\mathbf{Q}\mathbf{c} \\ &= \lambda\mathbf{k}'\mathbf{W}\mathbf{Q}\mathbf{c} - \lambda\Big(\int G(\cdot, w)\mathbf{g}'(w)\,dw\Big)\mathbf{W}\mathbf{Q}\mathbf{c} \\ &= \lambda\mathbf{k}'\mathbf{W}\mathbf{Q}\mathbf{c} - \lambda\mathbf{k}_2'\mathbf{W}\mathbf{Q}\mathbf{c} \end{aligned} \tag{10.1}$$

where $\mathbf{k}_2 = k_2(\mathbf{t}, \cdot) = \big(k_2(t_1, \cdot), \dots, k_2(t_n, \cdot)\big)'$ and $k_2$ is the reproducing kernel for functions in $\ker B$. Recall that $k = k_1 + k_2$ where $k_1$ is the reproducing kernel for functions in $\ker L$. Since $\mathbf{Q}'\mathbf{W}\mathbf{U} = 0$, $\mathbf{k}'\mathbf{W}\mathbf{Q} = \mathbf{k}_2'\mathbf{W}\mathbf{Q}$. Consequently, $Z^*x = 0$ and so $x \in \mathcal{G}$.

Obviously the proposed basis is of dimension $n - m$, which is the dimension of $\mathcal{G}$. □

**10.2. The solution for As.** Since $(\mathbf{y}, 0) - Zs \in \mathcal{G}$, it follows that there is a vector $\mathbf{c} \in \mathcal{R}^{n-m}$ such that

$$\begin{aligned} \mathbf{y} - As &= \lambda\mathbf{Q}\mathbf{c} \\ -Ls &= -\mathbf{g}'\mathbf{W}\mathbf{Q}\mathbf{c} \end{aligned} \tag{10.2}$$

or, equivalently,

$$(\mathbf{y}, 0) - Zs = (\lambda\mathbf{Q}\mathbf{c}, -\mathbf{g}'\mathbf{W}\mathbf{Q}\mathbf{c}).$$

Because

$$Zs = (\mathbf{y}, 0) - (\lambda\mathbf{Q}\mathbf{c}, -\mathbf{g}'\mathbf{W}\mathbf{Q}\mathbf{c}) \in \operatorname{range} Z = \mathcal{G}^\perp$$

it follows that, for all $\mathbf{d} \in \mathcal{R}^{n-m}$

$$\begin{aligned} \big\langle Zs, (\lambda\mathbf{Q}\mathbf{d}, -\mathbf{g}'\mathbf{W}\mathbf{Q}\mathbf{d})\big\rangle &= 0 \\ &= (\mathbf{y} - \lambda\mathbf{Q}\mathbf{c})'\mathbf{W}(\lambda\mathbf{Q}\mathbf{d}) \\ &\quad + \lambda\int \big(\mathbf{g}(t)'\mathbf{W}\mathbf{Q}\mathbf{c}\big)\big(-\mathbf{g}(t)'\mathbf{W}\mathbf{Q}\mathbf{d}\big)\,dt \\ &= (\mathbf{y} - \lambda\mathbf{Q}\mathbf{c})'\mathbf{W}(\lambda\mathbf{Q}\mathbf{d}) \\ &\quad - \lambda\mathbf{c}'\mathbf{Q}'\mathbf{W}\Big(\int \mathbf{g}(t)\mathbf{g}(t)'\,dt\Big)\mathbf{W}\mathbf{Q}\mathbf{d}. \end{aligned} \tag{10.3}$$

But, from (6.3)

$$\int \mathbf{g}(t)\mathbf{g}(t)'\,dt = \mathbf{K}$$

where $\mathbf{K} \in \mathcal{R}^{nn}$ contains the values of the reproducing kernel for $\ker B$ defined by

$$k_{ij} = \big\langle G(t_i, \cdot), G(t_j, \cdot)\big\rangle.$$

Defining matrix $\mathbf{H} \in \mathcal{R}^{n-m,n-m}$ by

$$\mathbf{H} = \mathbf{Q}'\mathbf{WKWQ},$$

we have that

$$\mathbf{d}^t(\mathbf{H} + \lambda\mathbf{Q}'\mathbf{WQ})\mathbf{c} = \mathbf{d}^t\mathbf{Q}'\mathbf{Wy}$$

for all $\mathbf{d} \in \mathcal{R}^{n-m}$. Therefore

$$\begin{aligned} (\mathbf{H} + \lambda\mathbf{Q}'\mathbf{WQ})\mathbf{c} &= \mathbf{Q}'\mathbf{Wy} \\ \mathbf{y} - As &= \lambda\mathbf{Qc}. \end{aligned} \tag{10.4}$$

**10.3. A band structured matrix Q.** Evidently equations (10.4) can be solved in $O(n)$ operations if the coefficient matrices are band structured. One can easily find $\mathbf{Q}$ to produce banded matrices for the case

$$\begin{aligned} Bh &= \big(h(0), Dh(0), \ldots, D^{m-1}h(0)\big)' \\ T &= [0, a] \\ \mathbf{W} &\quad \text{diagonal.} \end{aligned} \tag{10.5}$$

The algorithm is explained here and summarized in Section 11.

In this case we have that the Green's function is

$$G(t, w) = \left\{ \begin{array}{ll} \mathbf{u}'(t)\mathbf{v}(w) & w \leq t \\ 0 & w > t \end{array} \right\}$$

where $\mathbf{u}$ is the $m$-vector of functions spanning $\ker L$ and $\mathbf{v}$ is a vector of functions $v_1, \ldots, v_m$ calculated from $\mathbf{u}$ (see Section 11). Therefore

$$\begin{aligned} k_{ij} &= \int_T G(t_i, w)G(t_j, w)\, dw \\ &= \sum_{r,s} u_r(t_i)u_s(t_j) \int_0^{\min\{t_i,t_j\}} v_r(w)v_s(w)\, dw \\ &\equiv \mathbf{u}'(t_i)\mathbf{F}(\min\{t_i, t_j\})\mathbf{u}(t_j). \end{aligned}$$

To construct banded matrices we seek $\mathbf{Q}$, $n$ by $n-m$ such that $\mathbf{Q}'\mathbf{WU} = 0$. We also require that $\mathbf{Q}$ has $ij$th element $q_{ij}$ equal to zero unless $i = j, j+1, \ldots, j+m$. (See Section 11 to calculate $\mathbf{Q}$.)

For $\mathbf{Q}$ of this form and $\mathbf{W}$ diagonal, $\mathbf{Q}'\mathbf{WQ}$ is banded.

We now show that $\mathbf{H}$ is band structured.

Write

$$\mathbf{H}_{st} = \sum_{i,j} q_{is}\mathbf{W}_{ii}\mathbf{u}'(t_i)\mathbf{F}\big(\min\{t_i, t_j\}\big)\mathbf{u}(t_j)\mathbf{W}_{jj}q_{jt} \equiv \Sigma_1 + \Sigma_2$$

where

$$\Sigma_1 = \sum_{i \leq j} q_{is}\mathbf{W}_{ii}\mathbf{u}'(t_i)\mathbf{F}(t_i)\mathbf{u}(t_j)\mathbf{W}_{jj}q_{jt}$$

and

$$\Sigma_2 = \sum_{j<i} q_{is}\mathbf{W}_{ii}\mathbf{u}'(t_i)\mathbf{F}(t_j)\mathbf{u}(t_j)\mathbf{W}_{jj}q_{jt}.$$

Write

$$\Sigma_1 = \sum_{i=1}^{n} q_{is}\mathbf{W}_{ii}\mathbf{u}'(t_i)\mathbf{F}(t_i) \sum_{j=i}^{n} \mathbf{u}(t_j)\mathbf{W}_{jj}q_{jt}.$$

But $\sum_{j=1}^{n} \mathbf{u}(t_j)\mathbf{W}_{jj}q_{jt}$ is the transpose of the $t$th row of $\mathbf{Q'WU}$, which is equal to 0 by construction of $\mathbf{Q}$. Thus, since $q_{jt} = 0$ unless $t \le j \le t+m$, $\sum_{j=i}^{n} \mathbf{u}(t_j)\mathbf{W}_{jj}q_{jt} = 0$ unless $t < i \le t+m$. But, in the outer summation in $\Sigma_1$, $q_{is} = 0$ unless $s \le i \le s+m$. Therefore, $\Sigma_1 = 0$ whenever the intervals $[s, s+m]$ and $[t, t+m]$ are non-overlapping. Similarly, $\Sigma_2 = 0$ whenever the intervals $[s, s+m]$ and $[t, t+m]$ are non-overlapping. Therefore $\mathbf{H}$ is banded.

Therefore with this choice of $\mathbf{Q}$, (10.4) can by solved in $O(n)$ operations.

## 11. The Recipe for O(n) L-Spline Smoothing

Suppose that (10.5) holds.

1. Compute the Green's function as follows (see, e.g. Schumaker, 1981):
   (a) Compute the order $m$ Wronskian matrix
$$\begin{bmatrix} u_1 & Du_1 & \dots & D^{m-1}u_1 \\ \vdots & \vdots & & \vdots \\ u_m & Du_m & \dots & D^{m-1}u_m \end{bmatrix}.$$
   (b) Define $\mathbf{v}$ as the last row of the inverse of this matrix.
   (c)
$$G(t,w) = \left\{ \begin{array}{ll} \mathbf{u}'(t)\mathbf{v}(w) & w \le t \\ 0 & w > t \end{array} \right\}.$$
2. Compute $n$ by $n-m$ matrix $\mathbf{Q}$ satisfying:
   - $\mathbf{Q}$ is band-structured with $m+1$ nonzero values in column $i$ starting with $q_{ii}$, $i = 1, \dots, n-m$.
   - $\sum_{\ell=0}^{m} q_{i+\ell,i} w_{i+\ell,i+\ell} u_j(t_{i+\ell}) = 0$, $j = 1, \dots, m$.

   For each $i$ the computation of the $m+1$ coefficients $q_{i+\ell,i}$, $\ell = 0, \dots, m$ can be done by the QR-decomposition of the $m+1$ by $m$ matrix $(\mathbf{WU})_i$, which contains the value $w_{i+\ell,i+\ell}u_j(t_{i+\ell})$ in row $\ell$ column $j$. The required coefficients are in the $(m+1)$st column of $\mathbf{Q}_i$ where
$$(\mathbf{WU})_i = \mathbf{Q}_i\mathbf{R}_i$$
   is the QR-decomposition.

   From Section 10.3, we know that $\mathbf{H} = \mathbf{Q'WKWQ}$ is band-structured (as well as symmetric non-negative definite) with $m$ nonzero bands above and below the diagonal.
3. Next
   (a) Solve the band-structured linear equation
$$(\mathbf{H} + \lambda\mathbf{Q'WQ})\mathbf{c} = \mathbf{Q'Wy}$$
   for $\mathbf{c} \in \mathcal{R}^{n-m}$.
   (b) Compute
$$\begin{aligned} \mathbf{s} &= \mathbf{y} - \lambda\mathbf{Qc} \\ &= \left[\mathbf{I} - \lambda\mathbf{Q}(\mathbf{H} + \lambda\mathbf{Q'WQ})^{-1}\mathbf{Q'W}\right]\mathbf{y}. \end{aligned}$$
   Vector $\mathbf{s}$ contains the values $s(t_1), \dots, s(t_n)$.

This algorithm can run into trouble in practice if any of the matrices $(\mathbf{WU})_i$ are of less than rank $m$ or badly conditioned. This will happen if adjacent argument values are either identical or too close. In this event these values may be replaced by a single value, $n$ reduced by 1, and the corresponding weight $w_i$ doubled. This is a standard procedure in spline smoothing.

## Acknowledgments

The authors wish to acknowledge the support of the Natural Sciences and Engineering Research Council of Canada through grants A-320 and A-7969 , respectively. Requests for this paper should be directed to the first author at Department of Psychology, 1205 Dr. Penfield Ave., Montréal, Québec, H3A 1B1, Canada.

## References

1. P. M. Anselone and P. J. Laurent, *A general method for the construction of interpolating or smoothing spline-functions*, Numer. Math. **12** (1968), 66–82.
2. M. Atteia, *Théorie et applications des fonctions-spline à plusieurs variables*, Ph.D. thesis, Univ. Grenoble, 1966.
3. C. Dalzell and J. O. Ramsay, *Computing reproducing kernels with arbitrary boundary constraints*, SIAM J. Sci. Comput. **14** (1993), 511–518.
4. T. N. E. Greville, *Numerical procedures for interpolation by spline functions*, SIAM J. Numer. Anal. **1** (1964), 53–68.
5. M. F. Hutchison and F. R. de Hoog, *Smoothing noisy data with spline functions*, Numer. Math. **47** (1985), 99–106.
6. C. Reinsch, *Smoothing by spline functions*, Numer. Math. **10** (1967), 177–183.
7. ———, *Smoothing by spline functions. II*, Numer. Math. **16** (1970), 451–454.
8. L. L. Schumaker, *Spline functions: Basic theory*, John Wiley & Sons, New York, 1981.

Department of Psychology, 1205 Dr. Penfield Ave., Montréal, Québec, H3A 1B1
*E-mail address*: ramsay@psych.mcgill.ca

Department of Statistics, Univ. of British Columbia, Vancouver, British Columbia, V6T 1Z2
*E-mail address*: nancy@stat.ubc.ca

Centre de Recherches Mathématiques
CRM Proceedings and Lecture Notes
Volume **18**, 1999

# Spectral Representation and Estimation for Locally Stationary Wavelet Processes

R. von Sachs, G. P. Nason, and G. Kroisandt

ABSTRACT. This article develops a wavelet decomposition of a stochastic process which parallels a time-localized Cramér (Fourier) spectral representation. We provide a time-scale instead of a time-frequency decomposition and, hence, instead of thinking as scale in terms of "inverse frequency" we start from genuine time-scale building blocks or "atoms".

Using this class of locally stationary wavelet (LSW) processes, a doubly-indexed array of processes $\{X_{t,T}\}_{t=1,\dots,T}, T \geq 1$, we develop a theory for the estimation of the "evolutionary wavelet spectrum". Our asymptotics are based on rescaling in time-location which allows us to perform rigorous estimation theory starting from a single stretch of observations of $\{X_{t,T}\}$. This evolutionary wavelet spectrum measures the local power in the variance-covariance decomposition of the process $\{X_{t,T}\}$ at a certain scale and a (rescaled) time location.

It is possible to estimate the evolutionary wavelet spectrum by means of a wavelet periodogram or scalogram, i.e. the squared coefficients from a discrete or stationary wavelet transform (SWT). However, in order to estimate the spectrum consistently we have to smooth the periodogram. In this article we choose to smooth by non-linear wavelet shrinkage of the wavelet periodogram treated as a function of rescaled time-location with respect to another orthogonal wavelet basis (this parallels recent work in curve estimation).

Finally, we suggest an inverse transformation of the smoothed wavelet periodogram that estimates a localized autocovariance of the original stochastic process. Some numerical simulations and an application to a medical time series, which shows a particular non-stationary behaviour, indicate the usefulness of the SWT in our approach.

## 1. Introduction

The classical Cramér spectral decomposition of a *stationary* stochastic process $X_t, t \in \mathbf{Z}$, may be written

$$X_t = \int_{-\pi}^{\pi} \exp(i\omega t)\, dZ(\omega) = \int_{-\pi}^{\pi} A(\omega) \exp(i\omega t)\, d\zeta(\omega), \tag{1}$$

1991 *Mathematics Subject Classification.* Primary: 62M10; Secondary: 62M15.

The work of the first author was partially supported by DFG Grant Sa 474/3-1.

This is the final version of the paper.

where $dZ(\omega)$ and $d\zeta(\omega)$ are orthogonal and orthonormal increment processes respectively. In the last decade there have been several generalizations away from the assumption of stationarity. We will focus on the description of second order structure of $\{X_t\}$ which is possibly *changing over time.* To motivate this description we start from the spectral decomposition of the autocovariance function, $r_X(\tau)$, of a stationary process $X_t$,

$$r_X(\tau) = \int_{-\pi}^{\pi} f(\omega)\exp(i\omega\tau)\,d\omega, \tag{2}$$

where the spectral density $f(\omega)$ is linked to $Z(\omega)$ and the spectral measure $F(\omega)$ by

$$E\left|dZ(\omega)\right|^2 = dF(\omega) = f(\omega)\,d\omega, \tag{3}$$

(if $dF(\omega)$ is absolutely continuous with respect to the Lebesgue measure). Imposition of a time-dependency on the spectrum $f(\omega)$ means that one is mainly looking for *localized* decompositions, which are necessary if the (second order) structure of the process is no longer time-shift invariant but changes over time $t$. One possibility (again purely in the sense of nonparametric modelling) is the introduction of a time dependency in the amplitude function $A(\omega)$, i.e. consideration of

$$X_t = \int_{-\pi}^{\pi} A_t(\omega)\exp(i\omega t)\,d\zeta(\omega), \tag{4}$$

where $A_t(\omega)$ denotes the square root of a time-varying spectrum $f_t(\omega) = \left|A_t(\omega)\right|^2$, and $d\zeta(\omega)$ is an orthonormal increment process. More specific models of this kind of *time-frequency* decomposition include *oscillatory processes* [**Priest**] and the recent modification to *locally stationary processes* ([**Da**] cf. Remark 2.8 in Section 2), which are the statistical generalizations of (1).

Recently, local atomic decompositions (wavelets, wavelet libraries) have become popular [**RiVe**] for the analysis of deterministic signals as an alternative to non-local Fourier representations [**Fla**]. The question immediately arises: is it possible and meaningful to apply such atomic decompositions to, not necessarily stationary, stochastic processes?

**1.1. An example.** Figure 1 shows a subsample of an electrocardiogram (heart rate) recording

of an 66 day old baby sampled at 0.25Hz from 21:17:59 to 06:27:18. The original series was sampled at 1Hz however the power at high frequencies was found to be insignificant and so a 0.25Hz subsampled version was derived. It is unlikely that this will be a stationary time series. For one thing, of interest to paediatricians, the "heart rate" varies considerably over time and changes significantly between periods of sleep and waking. However, we shall show that a wavelet periodogram smoothed using non-linear wavelet shrinkage can reveal important features of interest that cannot be elicited from the time series or standard periodogram techniques alone. Indeed, some time-frequency methods were also used and these did not reveal much of interest. This type of time series motivates us to develop the class of locally stationary wavelet processes (although of course, it is not the *only* plausible model). We return to this example in Section 5.

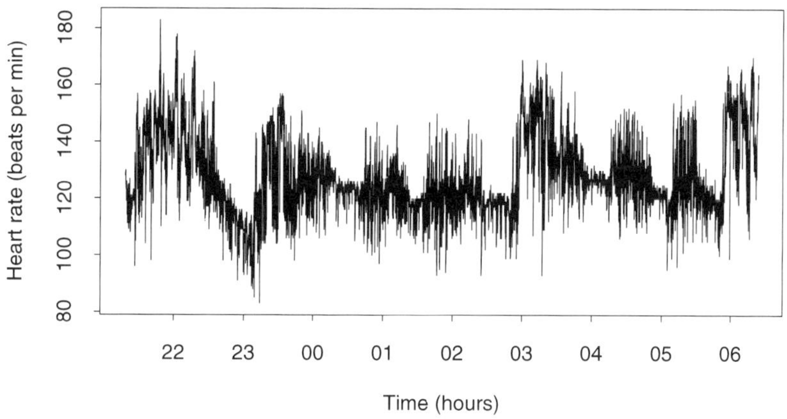


FIGURE 1. Electrocardiogram recording of 66 day old baby. Series is sampled at 0.25Hz and is recorded from 21:17:59 to 06:27:18.

**1.2. Locally stationary wavelet processes.** In this article, we show that for nonstationary situations, i.e. typically where $r_X(t,s) = \mathrm{Cov}(X_t, X_s)$ is a function of both time points $t$ and $s$, it is important to introduce a fully local concept.

Instead of using the set of harmonics $\{\exp(it\omega) \mid \omega \in [-\pi, \pi]\}$ we will introduce a decomposition where we switch to a family of locally supported wavelets $\{\psi_{j,k}(t) \mid j \in \mathbb{Z}^-; k \in \mathbb{Z}\}$, where $\psi_{j,k}(t) = 2^{j/2}\psi(2^j t - k)$, for a sufficiently well concentrated mother wavelet $\psi(t)$. With this we deliver a time-scale instead of a time-frequency decomposition and, hence, instead of thinking as scale in terms of "inverse frequency" we start from genuine time-scale building blocks or "atoms" $\psi_{j,k}(t)$. In other words we represent processes in terms of wavelets with random amplitudes, to be precise:

$$X_t = \sum_{j=-\infty}^{-1} \sum_{k=-\infty}^{\infty} \psi_{j,k}(t)\eta_{j,k}, \tag{5}$$

where $\eta_{j,k}$ is a measure in the time-scale half plane and the representation is understood to be in the mean squared sense. By some correlation property of the atoms to be specified later, we consider the process to be decomposed into atoms which match the typical local correlation length of $X_t$ in a local neighborhood of $t$ (i.e. the length of its memory or the interval in time around $X_t$ that is locally highly correlated). Assume $\psi(t)$ is localized at time 0 then typically the variance ("power") of a coefficient $\eta_{j,k}$ will be large if at time $t = 2^{-j}k$ the correlation length of $X_t$ is approximately of the order of magnitude of the "wavelength" of the atom $\psi_{j,k}$ (basically the length of the support of this wavelet), which is proportional to $2^{-j}$. The promising aspect of this model is that it allows local (time) variation, in a way that fast "oscillations" are modeled to change quickly and slow "oscillations" to change slowly. Here, the restriction to discretely indexed wavelets is mainly so that we can embed our new concept into a statistical (estimation) theory. One can certainly think about more general continuously-indexed models [**vS**] and their usefulness.

We emphasize that we do not *generalize* the concept of a traditional stationary process in the sense that we deliver a broader class which automatically includes the former one. Rather we *switch* to a different class of processes which we consider as a useful alternative for certain situations.

We introduce a *wavelet spectrum* which is the expected value of the squared modulus of the coefficients $\eta_{j,k}$. This quantity turns out to be a localized wavelet spectral measure in the time-scale plane. It measures the local power in the variance-covariance decomposition of the process $\{X_t\}$ at a certain scale $j$ and a time location $k$ and delivers a time-scale decomposition in as much as (4) represents a time-frequency decomposition.

Although decomposition (1) is obviously the most appropriate one for (classical) stationary processes our new approach can be used to test for stationarity. If in (5), $X_t$ is stationary, then the distribution of the coefficients $\eta_{j,k}$ will be independent of $k$. Indeed, for stationary processes the wavelet spectrum is constant over time location $k$.

However, in order to both properly identify and estimate a wavelet spectrum from a single finite-length stretch of the process $\{X_t\}$ we need to restrict the time-variation of the second order structure of $\{X_t\}$. This is exactly as in the situation of time-varying Fourier spectra $f_t(\omega)$, where no rigorous (asymptotic) estimation theory is possible without control of an otherwise arbitrary time variation of the spectrum (cf. [**Da, vSS, NvS**]).

This leads us to the definition of the class of "locally stationary" wavelet (LSW) processes. In this class of doubly-indexed processes $\{X_{t,T}\}_{t=1,\dots,T}, T \geq 1$, our asymptotics are based on rescaling in the time-location index $k$ of the random coefficients $\eta_{j,k}$. This rescaling allows us to do rigorous estimation theory starting from a single stretch of observations of $\{X_{t,T}\}$. Hence, the evolutionary wavelet spectrum turns out to be a localized density of a wavelet spectral measure in the time-scale plane, where we deal with a rescaled time on the unit interval.

Theoretical representations and their characteristic quantities turn out to be useful only when they are accompanied by some rigorous approach of how to estimate these quantities. Hence, in Section 3 following the development of our model in Section 2, we study an empirical wavelet spectrum and its asymptotics. This estimator of the theoretical wavelet spectrum is based on a scalogram or "wavelet periodogram", i.e. the squared coefficients from a discrete wavelet transform of the data. In addition to its statistical properties, we present some interesting links to the ordinary periodogram and the Wigner-Ville spectrum in the time-frequency plane (see also [**NvS**]). We also briefly discuss the possibility of using a translation-equivariant transform, the *Stationary Wavelet Transform* (SWT, see [**NaSi**]) instead of the discrete wavelet transform.

However, like the ordinary periodogram the wavelet periodogram is not consistent. Hence we have to modify the wavelet periodogram by some sort of local smoothing, in contrast to averaging over all wavelet periodograms of one scale (as, e.g., in [**AGF**]), where all local information in time is lost. In Section 4, we will introduce a particular local smoothing: non-linear thresholding of the wavelet periodogram with respect to another orthonormal wavelet basis of $L_2([0,1])$. By rescaling in time our wavelet periodograms can be considered as functions of local time in $[0,1]$. Also, in this section we briefly mention inverse transformation of the smoothed wavelet periodogram in order to estimate localized autocovariances (and local variances, also) of the original stochastic process.

In the last section we provide some numerical simulations which also indicate the usefulness of the SWT [**NaSi**] in our approach.

Most of the existing work on wavelets with stochastic processes (e.g. [**CaMa, CaHo, KaSh, CT**]) does not aim to give a decomposition with respect to an (orthogonal) increment process in the time-scale plane. The first three papers focus on probabilistic approximations and do not cover estimation. Indeed [**KaSh**] develop a wavelet process representation where they study conditions on the properties of a general measure to provide a weakly (Fourier) stationary process. Finally, the paper by [**MoCh**] uses the wavelet periodogram approach but only for stationary time series in the classical sense. We should also mention that the idea of using localized representations with truly local basis functions is certainly not completely new. For example Section 13.5 of [**Wal**] gives an example of a cyclostationary process that can be expanded into a series of box-car (Haar scaling) functions with uncorrelated coefficients. Another related field is the analysis and synthesis of $1/f$-processes with wavelets: In the work of [**AGF**], e.g., in the particular situation of a stationary increment process, the time marginal of the scalogram was used in order to estimate the spectral exponent of the fractional Brownian motion. One can certainly think about generalizations to more non-stationary situations which would call for the more localized wavelet approaches as being introduced in our work.

## 2. LSW Processes and Evolutionary Wavelet Spectra

In this section we define our special model used in order to be able to identify and estimate the wavelet spectrum of a certain class of processes $\{X_t\}$ *from a single stretch* of observations from these processes.

DEFINITION 2.1. *Define a class of LSW processes, a sequence of doubly-indexed stochastic processes $\{X_{t,T}\}_{t=1,\dots,T}, T \geq 1$ with the following representation in the mean-square sense with respect to a given wavelet basis $\{\psi_{jk}(t)\}_{jk}$ of $L_2(\mathbf{R})$, and an orthonormal random sequence of increments $\xi_{jk}$*

$$X_{t,T} = \sum_{j=-J}^{-1} \sum_{k} w^0_{j,k;T} \psi_{jk}(t) \xi_{jk}, \tag{6}$$

*where $2^J = T$, $\psi_{jk}(t) = 2^{j/2}\psi(2^j t - k)$, $j = -1, -2, \dots, -J(T) = -\log_2(T)$, $k \in \mathbf{Z}$, and possessing the following properties:*

(1) *$E\xi_{jk} = 0$ for all $j, k$. Hence $EX_{t,T} = 0$ for all $t$ and $T$.*
(2) $\operatorname{Cov}(\xi_{jk}, \xi_{\ell m}) = \delta_{j\ell}\delta_{km}$.
(3) *There exists for each $j \leq -1$ a Lipschitz-continuous function $W_j(z)$ on $(0,1)$ such that*

$$\sup_k \left| w^0_{j,k;T} - W_j\left(\frac{k\ 2^{-j}}{T}\right) \right| = O(T^{-1}) \quad \text{for} \quad T \to \infty,$$

*where for each $j = -1, \dots, -J(T) = -\log_2(T)$ the $\sup$ is over $k = 1, \dots, 2^j T$.*
(4) *The real-valued wavelet basis $\{\psi_{jk}\}_{jk}$ is orthonormal and the wavelets have compact support.*
(5) $\sum_{j=-\infty}^{-1} |W_j(z)|^2 < \infty$, *for all $z \in (0,1)$.*

REMARK 2.2. The second assumption of uncorrelated increments is restrictive. In practice, a broader class of processes which allows for correlation within or even between scales would probably be more useful. The present model merely serves as a starting point that enables us to develop rigorous estimation theory and clearly present our ideas. We believe that our methods can also be applied to more realistic models. For example, one that permits restricted dependencies, such as ARMA models, within each scale. We will discuss possible generalizations in more detail in [**vSNK**].

Assumption 2.1 is crucial. It allows us to asymptotically identify the model coefficients and leads to a unique representation (6) given the fixed wavelet basis. This is because the smoothness assumption on $W_j(z)$ controls the variation of each coefficient $w^0_{j,k}$ as a function of $k$ so that it can only vary slowly enough. Basically the smoothness says that only a neighbourhood of locations determines this variation and this neighbourhood becomes asymptotically arbitrarily small in the rescaled time $z$.

The orthogonality in Assumption 2.1 needs some extra justification if we want to compute empirical wavelet coefficients from discrete data $X_{1,T}, \ldots, X_{T,T}$. To use Mallat's [**Mal**] pyramid algorithm and to be able to properly derive asymptotic properties of the empirical coefficients we identify the (vectors of) wavelet functions $(\psi_{-1,k}, \psi_{-2,k}, \ldots, \psi_{-J,k})$ with corresponding (vectors of) quadrature mirror filter operators $(G, GH, GH^2, \ldots, GH^{J-1})$ where $H = \{h_k\}$ and $G = \{g_k\}$ are the low and bandpass filters of the Mallat pyramid algorithm respectively. In all that follows, the symbol $\psi_{jk}$ stands for $(GH^{-j-1})_k$.

Assumption 2.1 guarantees both the convergence of $\sum_j |w^0_{j,k;T}|^2$ in the limit as $T \to \infty$, due to Assumption 2.1, and the finiteness and convergence of the variance of the process itself. By Proposition 2.7 below, $\mathrm{Var}\{X_{[zT],T}\} \to c(z, 0) < \infty$, for all $z \in (0, 1)$.

REMARK 2.3. The reader may have noted that we have slightly departed from the usual wavelet scale numbering scheme. In this article scale zero is the scale upon which the "data" live. Scale $-1$ is the scale which contains the finest resolution wavelet detail and scale $-J$ the coarsest.

The advantage of this numbering scheme is that it enables us to keep the length of the support of the finest wavelets fixed and constant with respect to $T$. However, the interpretation of the coefficients at the coarsest scale change asymptotically. As a näive illustration of this subtle point consider $T = 16$ points sampled every 1/16 seconds. Then the $-1$ scale points "cover" every 1/8s, the scale $-2$ points "cover" every 1/4s and eventually scale $-J$ "covers" the whole series over 1s. Now suppose we double the number of samples $T = 32$ *but still sample every* 1/16*s*. Clearly the scale $-J$ now "covers" 2s, not 1s. So the interpretation of the coarsest scale $-J$ changes as do all scales above which are fixed relative to this scale. This makes intuitive sense as longer and longer cycles may be observed in the series with growing $T$.

However, recall that asymptotically more and more data points are going to live on the finest scale, which hence plays the same role as for usual wavelet estimation schemes and allows for asymptotic statistical inference for all scales $-j$ with $2^{-j} = o(T)$.

For a given LSW process as in Definition 2.1 we define the following localized density of a wavelet spectral measure in the (rescaled-) time-scale plane.

DEFINITION 2.4. *Let*

$$S_j(z) := \left|W_j(z)\right|^2, z \in (0,1). \tag{7}$$

$\{S_j(z)\}_{j\le -1}$ *is the "evolutionary wavelet spectrum" of the sequence* $\{X_{t,T}\}$ *with respect to* $\{\psi_{jk}\}$.

*Using Assumption* 2.1 *of Definition* 2.1,

$$S_j(z) = \lim_{T\to\infty} \left|w^0_{j,2^j[zT];T}\right|^2, \quad \forall z \in (0,1). \tag{8}$$

This quantity can be thought of as measuring the local power of the decomposition of the variance/covariance of the process $\{X_{t,T}\}$ at scale $j$ and an (arbitrary) location $z \in (0,1)$. Hence equation (6) delivers a time-scale decomposition which parallels a time-frequency decomposition (a localized Cramér spectral representation described in Remark 2.8 below). However, we start from genuine time-scale building blocks (atoms). Note that the evolutionary wavelet spectrum $S_j(z)$ is defined only for $z \in (0,1)$, as boundaries do not make sense in this model, completely analogously to the local time-frequency model of Remark 2.8.

REMARK 2.5. Representation (6) has an analogue for the decomposition of the "local autocovariance" $\mathrm{Cov}\{X_{[zT]-\tau,T}; X_{[zT]+\tau,T}\}$, $z \in (0,1)$ and $\tau$ an integer (see Section 4 for more discussion). The following definition and proposition demonstrate that asymptotically this covariance tends to a local autocovariance which is reminiscent of the inverse wavelet transform of the evolutionary wavelet spectrum.

DEFINITION 2.6. *Define the local autocovariance* $c(z,\tau)$ *by*

$$c(z,\tau) = \sum_{j=-\infty}^{-1} S_j(z) \sum_k \psi_{jk}(-\tau)\psi_{jk}(\tau). \tag{9}$$

PROPOSITION 2.7. *As* $T \to \infty$

$$\mathrm{Cov}\{X_{[zT]-\tau,T}; X_{[zT]+\tau,T}\} \to c(z,\tau).$$

PROOF. Using (6) and Assumption 2.1 of Definition 2.1 we have

$$\begin{aligned} \mathrm{Cov}\{X_{[zT]-\tau,T}; X_{[zT]+\tau,T}\} &= \sum_{jk} |w^0_{j,k;T}|^2 \psi_{jk}\big([zT]-\tau\big)\psi_{jk}\big([zT]+\tau\big) \\ &= \sum_{j\ell} \left|w^0_{j,2^j[zT]+\ell;T}\right|^2 \\ &\qquad \psi_{j,2^j[zT]+\ell}\big([zT]-\tau\big)\psi_{j,2^j[zT]+\ell,T}\big([zT]+\tau\big) \\ &= \sum_{j\ell} \left|w^0_{j,2^j[zT]+\ell;T}\right|^2 \psi_{j\ell}(-\tau)\psi_{j\ell}(\tau) \\ &= \sum_{jk} \left|w^0_{j,2^j[zT]+k;T}\right|^2 \psi_{jk}(-\tau)\psi_{jk}(\tau). \end{aligned} \tag{10}$$

Now we show that (10) converges to $c(z,\tau)$. By Definitions 2.1 and 2.4 we have

$$|w^0_{j,2^j[zT]+k;T}|^2 = S_j(z+k/T) + O(T^{-1})$$

and since $S_j$ is Lipschitz

$$S_j(z+k/T) = S_j(z) + O(k/T).$$

Therefore (10) becomes

$$\sum_{jk} |w^0_{j,2^j[zT]+k;T}|^2 \psi_{jk}(-\tau)\psi_{jk}(\tau) = c(z,\tau) + \sum_{jk} O(T^{-1})(k+1)\psi_{jk}(-\tau)\psi_{jk}(\tau). \tag{11}$$

The second term converges to zero at a rate $O(T^{-1})$ because for fixed $\tau$ and $j$ the number of non-zero summands is finite (due to the compactness of the wavelets). Actually, for fixed $\tau$ and $j$ tending to $-\infty$ the number of terms in the sum over $k$ goes to zero with $2^j\tau$. Since the wavelets are bounded, i.e., $|\psi_{jk}(-\tau)\psi_{jk}(\tau)| \leq M$,

$$\sum_{jk} (k+1)\psi_{jk}(-\tau)\psi_{jk}(\tau) = O(1). \quad \square \tag{12}$$

Note also that Assumption 2.1 of Definition 2.1 on the summability of $W_j(z)$ guarantees that the sum over all $j$ in equation (9) is well defined if $T \to \infty$.

REMARK 2.8. Assumption 2.1 in Definition 2.1 is motivated by R. Dahlhaus' definition [**Da**] of "(Fourier) locally stationary processes" with the following time-frequency or generalised (time-varying) Cramér spectral representation:

$$X_{t,T} = \int_{-\pi}^{\pi} A^0_{t,T}(\lambda) \exp i\lambda t \, d\xi(\lambda), \tag{13}$$

where:

(a) $\{d\xi(\lambda)\}_\lambda$ is an orthonormal increment process on $[-\pi,\pi]$;
(b) $\sup_{t,\lambda} |A^0_{t,T}(\lambda) - A(t/T,\lambda)| = O(T^{-1})$,

where $A(u,\lambda), u \in (0,1)$ is a smooth function, $\lambda \in [-\pi,\pi]$ and the "evolutionary spectrum" is $f(u,\lambda) = |A(u,\lambda)|^2$. Note that this class of processes is a direct generalization of the class of stationary processes, i.e. the broader contains the smaller class. Our class of LSW processes, however, focuses on a different time-varying spectral behaviour, which allows to include, e.g., "chirp-like" phenomena, with a smoothness in their spectra which can hardly be modeled by "Fourier locally stationary processes". Those, on the other hand allow for best, i.e., sparsest representations of signals being highly concentrated in frequency, as, e.g., modulated pure sinusoids. These differences, quite naturally, will show up in the attempt to estimate the corresponding spectral structure for which either time- or frequency-localized methods will prove more useful. However, there is certainly an overlap (i.e. a "transition") between the two classes which allows to study interesting relationships between the two "evolutionary spectra". This will be given in the next section.

## 3. Estimation

We only consider the situation of a fixed underlying wavelet basis which is *a priori* given for the representation (6). We do not consider the problem of choice of best basis fitting to a given set of observations $X_1, \ldots, X_T$ nor the question of what happens if a different wavelet basis is chosen. Both are interesting questions for future work.

Recall that in the following definition of the empirical wavelet coefficients the symbol $\psi_{jk}$ stands for $(GH^{-j-1})_k$, see Remark 2.2.

DEFINITION 3.1. *Let*

$$d_{j,k;T} := \sum_{t=1}^{T} X_{t,T}\psi_{jk}(t)$$

*be the empirical wavelet coefficients, where $X_{t,T}$ is a LSW process as given by* (6).

In fact, in this definition the sum over $t$ is finite. For fixed $j$ the actual number of summands does not change with $T$; it is always $M_j = M \cdot 2^{-j}, j = -1, \ldots, -J$, where $M$ denotes the length of the support of $\psi(t)$ on the finest scale 0.

Now we define our key statistic which plays the same role in wavelet spectral analysis as the classical periodogram does in traditional (Fourier) spectral analysis.

DEFINITION 3.2. *Let*

$$I^j_{k,T} := |d_{j,k;T}|^2 \tag{14}$$

*be the wavelet periodogram of* $\{X_{t,T}\}$.

See also Nason and Silverman [**NaSi**] whose definition coincides with ours. It may be useful at this point to examine Figure 2 which shows the wavelet periodogram (Figure 2c) of a simulated signal (Figure 2b). In fact, for this example the wavelet periodogram is constructed using a stationary wavelet transform (SWT) which produces wavelet coefficients with respect to all possible shifted wavelet bases. This builds some robustness into the estimation and smoothing procedures since exact alignment with the classical dyadic wavelet structure is no longer necessary for good estimation.

We now further assume that the $\xi_{jk}$ are normally distributed. Therefore $\{X_{t,T}\}$ is also Gaussian. The wavelet periodogram satisfies properties analogous to those satisfied by the classical Fourier periodogram as follows.

PROPOSITION 3.3.

**(expectation):**

$$EI^j_{2^j[zT],T} = S_j(z) + O(T^{-1}) \;\forall z \in (0,1).$$

**(variance):**

$$\operatorname{Var} I^j_{2^j[zT],T} = 2S_j^2(z) + O(T^{-1}) \;\forall z \in (0,1).$$

**(covariance):**

$$\operatorname{Cov}\{I^j_{k,T}; I^l_{m,T}\} = 2\delta_{jl}\delta_{km}\left\{S_j\left(\frac{k}{2^j\,T}\right)^2 + O(T^{-1})\right\}.$$

PROOF. First observe that the following holds in mean square sense:

$$\begin{aligned}
d_{j,k;T} &= \sum_{t=1}^{T} X_{t,T}\psi_{jk}(t) \\
&= \sum_{t=1}^{T}\left\{\sum_{l,m} w^0_{l,m;T}\psi_{lm}(t)\xi_{lm}\right\}\psi_{jk}(t) \\
&= \sum_{l,m} w^0_{l,m;T}\xi_{lm}\left\{\sum_{t=1}^{T}\psi_{lm}(t)\psi_{jk}(t)\right\} = w^0_{j,k;T}\xi_{jk},
\end{aligned}$$

by the orthonormality of the wavelets. Hence, since the $\xi_{lm}$ are uncorrelated

$$\mathrm{Cov}\{d_{j,k;T}; d_{l,m;T}\} = |w^0_{j,k;T}|^2 \delta_{jl}\delta_{km}.$$

This implies that as $T \to \infty$

$$EI^j_{2^j[zT],T} = |w^0_{j,2^j[zT];T}|^2 = S_j(z) + O(T^{-1}),$$

by Assumption 2.1 of Definition 2.1. Further,

$$\begin{aligned}\mathrm{Var}\, I^j_{2^j[zT],T} &= \mathrm{Var}\, |d_{j,2^j[zT];T}|^2 \\ &= |w^0_{j,2^j[zT];T}|^4\, \mathrm{Var}(\xi^2_{j,2^j[zT]}) \\ &= 2|w^0_{j,2^j[zT];T}|^4 = 2S_j(z)^2 + O(T^{-1}),\end{aligned}$$

where we use that $\xi^2_{j,2^j[zT]}$ is $\chi^2_1$-distributed because $\xi_{jk}$ is Gaussian.

Finally, due to the Theorem of Isserlis for Gaussian variables

$$\mathrm{Cov}\{I^j_{k,T}; I^l_{m,T}\} = 2|w^0_{j,k;T}|^4 \delta_{jl}\delta_{km} = 2\delta_{jl}\delta_{km}\left\{S_j\left(\frac{k}{2^jT}\right)^2 + O(T^{-1})\right\}. \quad \square$$

The following lemma gives some useful connection between time-scale and time-frequency analysis (it is of no originality, and can be found in more generality in, for example, [**Fla**], Equation (3-70)).

LEMMA 3.4.

$$I^j_{k,T} = \int \sum_{t=1}^{T} I_{t,T}(\omega) W^{\psi}_{jk}(t,\omega)\, d\omega, \tag{15}$$

*where*

$$I_{t,T}(\omega) = \sum_{\{s:1\le t-s/2,t+s/2\le T\}} X_{t-s/2,T}X_{t+s/2,T}\exp(-i\omega s),$$

*is the localised "pre-periodogram" (defined in* [**NvS**]*), which is a Wigner-Ville local spectral estimate, and*

$$W^{\psi}_{jk}(t,\omega) = \sum_{\tau} \psi_{jk}\left(t - \frac{\tau}{2}\right)\psi_{jk}\left(t + \frac{\tau}{2}\right)\exp(-i\omega\tau)$$

*is the Wigner-Ville distribution of the wavelet* $\psi_{jk}$.

This lemma will be useful for the following corollary which collects some relations between the wavelet periodogram and the classical Fourier spectral density for a stationary process, and the evolutionary Fourier spectrum for a locally stationary process (13) defined by Dahlhaus [**Da**], cf. Remark 2.8.

COROLLARY 3.5. (1) *If* $\{X_{t,T}\}$ *is a classical second-order stationary process* $\{X_t\}$ *with spectral density* $f(\omega)$*, then*

$$\lim_{T\to\infty} EI^j_{k,T} = \int_{-\pi}^{\pi} f(\omega)|\widehat{\psi_{jk}}(\omega)|^2\, d\omega \tag{16}$$

*which is independent of* $k$ *(as* $|\widehat{\psi_{jk}}(\omega)|^2$ *is) and where*

$$\widehat{\psi_{jk}}(\omega) := \sum_{s=-\infty}^{\infty} \psi_{jk}(s)\exp(i\omega s).$$

*The quantity $\widehat{\psi_{jk}}$ is not independent of $k$ but is a phased-shifted version of $\widehat{\psi_{j0}}$. However, taking absolute values cancels the phase shift. For notational simplicity we define $\psi_j = \psi_{j0}$.*

*With equation (16), for stationary processes, $S_j(z) = \int f(\omega)\left|\widehat{\psi_j}(\omega)\right|^2 d\omega$, turns out to be independent of $z$ which is expected from a meaningful definition and estimator.*

(2) *If $\{X_{t,T}\}$ is locally stationary in the sense of (13) with evolutionary Fourier spectrum $f(u,\omega)$, then*

$$\lim_{T\to\infty} EI^j_{2^j[uT],T} = \int f(u,\omega)\left|\widehat{\psi_j}(\omega)\right|^2 d\omega. \tag{17}$$

*In this case, the evolutionary wavelet spectrum $S_j(z)$ is related to the evolutionary Fourier spectrum by the following equation:*

$$S_j(z) = \int f(z,\omega)\left|\widehat{\psi_j}(\omega)\right|^2 d\omega. \tag{18}$$

PROOF OF 1.

$$\begin{aligned}
\lim_{T\to\infty} EI^j_{k,T} &= \lim_{T\to\infty} E\left|\sum_{t=1}^{T} X_{t,T}\psi_{jk}(t)\right|^2 \\
&= \lim_{T\to\infty} E\left|\sum_{t=1}^{T}\int \exp(i\omega t)\, dZ(\omega)\ \psi_{jk}(t)\right|^2 \\
&= \lim_{T\to\infty} E\left|\int\left(\sum_{t=1}^{T}\psi_{jk}(t)\exp(i\omega t)\right) dZ(\omega)\right|^2 \\
&= \lim_{T\to\infty}\int\left|\sum_{t=1}^{T}\psi_{jk}(t)\exp(i\omega t)\right|^2 f(\omega)\, d\omega \\
&= \int\left|\widehat{\psi_{jk}}(\omega)\right|^2 f(\omega)\, d\omega.
\end{aligned}$$

PROOF OF 2. This proof is based on two approximations: First,

$$EI_{t,T}(\omega) = \sum_s c_T(t/T,s)\exp(-i\omega s),$$

where, with Remark 2.8 and equation (13),

$$\begin{aligned}
c_T(u,s) &:= \operatorname{Cov}\left\{X_{[uT-s/2],T}; X_{[uT+s/2],T}\right\} \\
&= \int_{-\pi}^{\pi} A\big([uT-s/2]\,/\,T,\lambda\big)\overline{A\big([uT+s/2]/T,\lambda\big)}\exp(i\lambda s)\, d\lambda + O(T^{-1}).
\end{aligned}$$

Here the idea is, that as $T\to\infty$, due to the smoothness of $A$ in the first argument,

$$c_T(u,s)\to c(u,s) := \int_{-\pi}^{\pi}\left|A(u,\lambda)\right|^2\exp(i\lambda s)\, d\lambda.$$

The second approximation leads to the remainder of

$$\sum_t \int d\omega E I_{t,T}(\omega) W^{\psi}_{jk}(t,\omega) = \sum_t \int d\omega E I_{t+k2^{-j},T}(\omega) W^{\psi}_j(t,\omega)$$
$$= \int d\omega f(u,\omega) \left|\widehat{\psi_j}(\omega)\right|^2 + O\left(\frac{M_j}{T}\right).$$

For the remainder, the idea is to use $f(t + k2^{-j}/T, \omega) = f(k2^{-j}/T, \omega) + O(t/T)$, and

$$\int d\omega \sum_t^{M_j} t W^{\psi}_j(t,\omega) = O(M_j),$$

which—for less regular spectra—needs techniques as in [**NvS**], proof of Lemma 3.2. □

## 4. Smoothing and Inverting the Wavelet Periodogram

Proposition 3.3 demonstrates that the wavelet periodogram is *not* a consistent estimator of the wavelet spectrum and needs to be smoothed. Nason and Silverman [**NaSi**] developed a heuristic method based on linear Fourier smoothing. However, alternative methods could be used. For example, kernel smoothing or a moving average. Of course, the question of how to choose the smoothing parameter then arises.

Here, we will shortly discuss non-linear thresholding of $I^j_{k,T}$ as a function of $z = 2^{-j}k/T$ with respect to either a traditional discrete (DWT) or a stationary wavelet transformation (SWT) with an orthonormal wavelet basis $\{\tilde{\varphi}_{j_0 m}(z), \widetilde{\psi}_{\ell m}(z)\}$ of $L_2([0,1])$. Here $\widetilde{\psi}_{\ell m}(z) = 2^{\ell/2}\widetilde{\psi}(2^{\ell}z - m)$, where $j_0$ is the coarsest scale in the scheme and $\ell \geq j_0$, $m = 0, \ldots, 2^{\ell} - 1$. This completely parallels existing wavelet curve estimation techniques, as in [**Do, NvS, JoSi**].

Empirical wavelet coefficients of a DWT are provided by projecting, for fixed $j$ the wavelet periodogram $I^j_{[zT]}$ onto the wavelets $\widetilde{\psi}_{\ell m}(z)$.

The SWT is described in [**NaSi**] and it has the advantage that it provides a translation-invariant denoising technique. This is achieved by obtaining DWTs of all possible shifts of the original data in a computationally efficient way. In this way, the actual number of wavelet coefficients per scale $j$ is considerably increased from order $T \cdot 2^j$ for the DWT (if applied to $T$ data points) to order $T$ instead. Hence, for statistical estimation purposes, this number serves as kind of effective sample size on scale $j$.

With our asymptotics, as described in Remark 2.3, both for DWT and SWT there are a growing number of coefficients $I^j_{k,T}$ if we are on a fixed scale $j$, fixed relatively to the finest scale. Then, as on this finest scale the number of coefficients grows with order $O(T)$, it does so on all the fixed scales below, which are bounded away from the coarsest scale. Hence, $S_j$ can be estimated with increasing accuracy as $T$ increases.

In order to do denoising we apply non-linear thresholding to the empirical wavelet coefficients. For now, we only consider the DWT case. The resulting reconstruction (denoised estimator) is obtained by inverting the DWT using only those coefficients which remain after soft or hard thresholding. In order to appropriately determine the threshold $\lambda = \lambda(j, \ell, m, T)$ the asymptotic properties of the empirical coefficients have to be explored. It turns out that, under the assumption of normality and with a wavelet $\tilde{\psi}$ of bounded variation, bias and variance exactly behave like for traditional nonparametric curve estimation with wavelets. Their exact form will be reported elsewhere [**vSNK**] but basically the variance is of order $O\big((T \cdot 2^j)\big)$, i.e. of order of the effective sample size on scale $j$. Also, the squared bias is of at most this order (depending on the smoothness assumption on the evolutionary wavelet spectrum). Note, that as always in wavelet curve estimation, one has to keep away from the finest scale $\ell$, i.e. $2^\ell = o(T \cdot 2^j)$. Also, asymptotically one has to keep away from the coarsest scale $-J = O(\log T)$, i.e. $2^{-j} = o(T)$, as on this scale there is no asymptotically increasing number of "data" $I^j_{k,T}$.

Assuming process normality, we can proceed as in [**vSS**] to derive an appropriate threshold by exploiting the asymptotic form of the variance of the empirical coefficients, see again [**vSNK**]. Including an additional factor of order $\log(T \cdot 2^j)$, a threshold estimator $\tilde{I}^j_{k,T}$ of the evolutionary wavelet spectrum which is based on this threshold can be shown to achieve the asymptotically near-optimal minimax rate of the $L_2$-risk between estimator and spectrum. This is based on existing results on quadratic forms of independent Gaussian variables, which are $\chi^2$-distributed. For soft-thresholding it is possible to straightforwardly derive results due to the equivalence to the situation of optimal recovery (see [**DJKP**]). For non-normality, techniques as in [**NvS**] should be applied.

We mention that in practice some modification might be appropriate as for this approach the thresholds depend on the unknown wavelet spectrum. This is exactly as it is for stationary Fourier spectrum estimation with wavelet thresholding of periodograms and also for the time-dependent version of this, see [**vSS**]. In both latter cases we experienced quite good results with thresholding of the logarithm of the periodogram, as this transformation stabilizes the variance of the empirical wavelet coefficients. Hence application of a universal threshold (see, e.g., [**Do**]) might be appropriate.

We also consider to apply denoising to the wavelet periodogram based on the SWT. The SWT actually increases the effective sample size of the wavelet periodogram to be $T$ on *each* scale, i.e. independent of $j$, and was used to produce the denoised estimator of Figure 4. Due to extra correlation between resulting empirical wavelet coefficients closer investigations are necessary to show that basically the same asymptotic results hold as for the case of DWT denoising. For those we like to refer to [**vSNK**].

Finally, in a preliminary, rather exploratory, investigation we found that it might be useful, in addition to look at the actual tableau of smoothed squared (stationary) wavelet coefficients $\tilde{I}^j_k$, also to plot the sum over $j$ of these. This quantity can be considered as a *local variance* estimate as it measures the power of the signal content over local time $k$.

More generally we define the following localized autocovariances and their respective empirical counterparts (estimators based on inverting smoothed wavelet periodograms $\tilde{I}^j_k$). Let

$$c(z, \tau) = \sum_j S_j(z) \sum_k \psi_{jk}(-\tau)\psi_{jk}(\tau) \tag{19}$$

be the *local autocovariance* of the LSW process, as being derived in Remark 2.5 and Definition 2.6. Then, replacing the wavelet spectrum $S_j(z)$ by its empirical analog, the smoothed wavelet periodogram $\tilde{I}^j_{k,T}$, should allow us to consistently

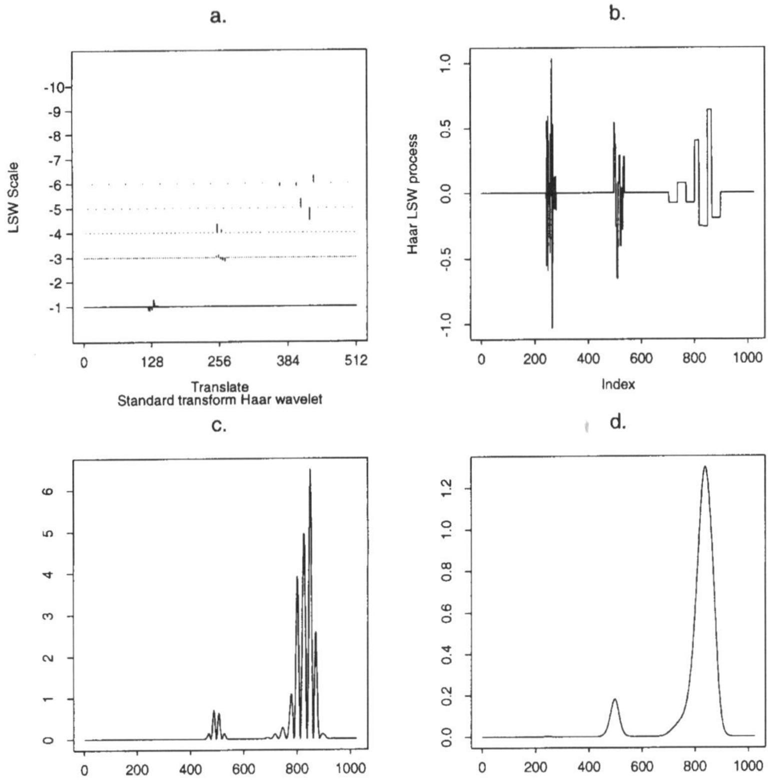


FIGURE 2. a. Simulated Haar LSW process coefficients for a discrete process of 1024 points; b. Simulated Haar LSW process (DWT inverse of a.); c. Scale $-5$ coefficients from squared SWT of b. using Haar wavelets; d. Non-linear wavelet smooth of logarithm of c. using universal thresholding and Daubechies' Least asymmetric wavelets $N = 10$.

estimate the local autocovariance $c(z, \tau)$. We conjecture that this is so, at least in an $L_2([0, 1])$ sense in $z$. More exact results on this will be reported soon elsewhere [**vSNK**].

## 5. Simulation and Application to Real Data Examples

We now illustrate these concepts using a simulated and real data example.

**5.1. Simulated example.** Figure 2 shows simulated LSW Haar coefficients (a) and process (b) produced by placing Gaussian Haar coefficients into a wavelet tableau (produced by `wavegrow()` which is part of the WaveThresh3 package, see [**Na**]). It is useful to think of Figure 2b as the "original" process for which we just happen to know the truth (Figure 2a). We then performed the SWT using Haar

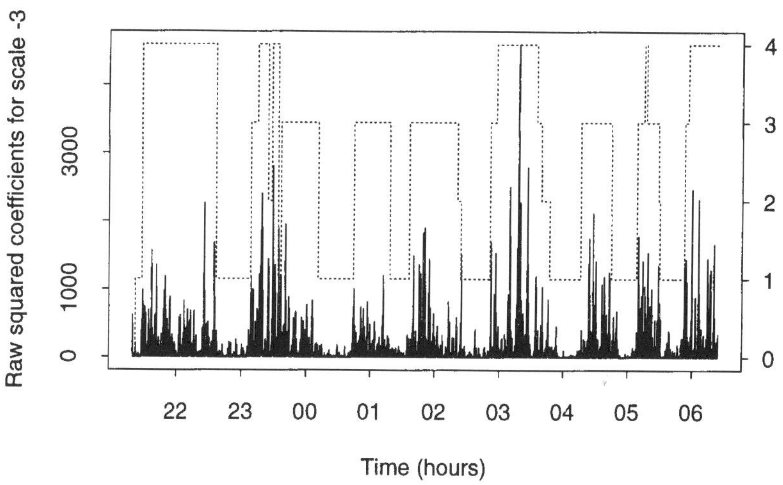


FIGURE 3. The (stationary) wavelet periodogram $I^{-3}_{k,8192}$ for baby heart rate data (solid line) for the particular scale $j = -3$. The dotted line indicates the sleep state for the baby and its values are indicated by the right-hand axis (0=unknown, 1=asleep, 2=hard to tell, 3=active sleep, 4=awake).

wavelets and squared the coefficients (function `LocalSpec()` in WaveThresh3). Figure 2c shows the scale $-5$ coefficients, $I^{-5}_{k,1024}$, from this squared SWT. This level of coefficients should contain information about the medium scale effects within the process—the activity towards the right of the plot (note in this case we know that the activity is exactly at scales $-6$ and $-5$ because of plot a, in practice we would look at all scales. This simulated example is just to verify that the activity turns up at the correct level). Figure 2d shows the non-linear wavelet smoothed version of c and is our estimate for $S_{-5}(z)$. This was obtained by thresholding the DWT of logarithm of c. using Daubechies' [**Dau**] least-asymmetric wavelets $\widetilde{\psi}$, $N = 10$, as described in Section 4. The inverse DWT was applied to the thresholded coefficients and then exponentiated. Clearly large "power" occurs where it should, although there is some bleed-over of power from scale $-4$ near the middle of the plot.

**5.2. Baby heart beat example.** This example continues from the discussion given in Section 1.1. Figure 3 shows the raw wavelet periodogram coefficients for scale $-3$ for the baby ECG data series shown in Figure 1. Figure 4 shows the results of a non-linear wavelet smoothing of the wavelet periodogram. Here Daubechies' [**Dau**] least-asymmetric wavelets $N = 10$ were used to form the SWT. The logarithms of the squared coefficients $I^{-3}_{k,8192}$ were then directly subjected to non-linear stationary wavelet smoothing using universal soft thresholding on all scales with a MAD-based estimate of the noise level using Daubechies' least-asymmetric $N = 10$ wavelets. This resulted in our estimate of $S_{-3}(z)$ given in Figure 4 (solid line). The overlying dotted line in that figure refers to the baby's sleep state as judged by a trained human observer. The observer classifies the baby's state of sleep as

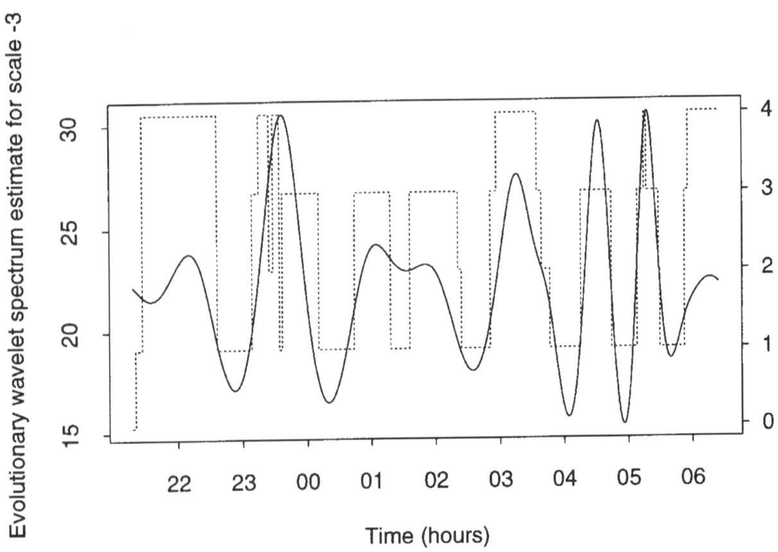


FIGURE 4. Estimate of evolutionary wavelet spectrum $S_{-3}(z)$ for baby heart rate data (solid line). The dotted line indicates the sleep state for the baby and its values are indicated by the right-hand axis (0=unknown, 1=asleep, 2=hard to tell, 3=active sleep, 4=awake).

unknown, quiet sleep, hard to tell, active sleep and awake as indicated in the caption to the figure. It is clear that there is some, if somewhat imperfect, relationship between the estimate of $S_{-3}(z)$ and the sleep state. In particular, periods of active and wakefulness occur whilst the estimate of $S_{-3}(z)$ is large and periods of quiet sleep when it is small. In effect, the $S_{-3}$ quantity is measuring the local stochastic variability of the signal at that scale and it correlates fairly well with the sleep state variable. Objective measures of heart-rate variability are of great interest to paediatricians especially when linked to external observables such as sleep state. More on this example can be found in [**vSNK**].

## 6. Acknowledgments

The authors would like to thank Prof. Peter Fleming, Dr. Andrew Sawczenko and Jeanine Young of the Institute of Child Health, Royal Hospital for Sick Children, Bristol for supplying the ECG data. Guy Nason was supported in part by the U.K. Engineering and Physical Sciences Research Council Grant GR/K70236.

This paper was written while Rainer von Sachs visited the Department of Statistics, Stanford University. It is also based on intensive discussions during his visit to the Department of Mathematics, University of Bristol. Rainer von Sachs, who was supported by a grant of the Deutsche Forschungsgemeinschaft, would like to thank a lot of people at both institutions for providing a warm and fruitful atmosphere of hospitality and work. Also, many thanks to Kai Schneider, who was the first to discuss related ideas with him long before this work got started.

## References


[AGF] P. Abry, P. Goncalves, and P. Flandrin, *Wavelets, spectrum analysis and $1/f$-processes*, Lecture Notes in Statist. **103** (1995), Springer-Verlag, New York, 15–29.

[CaHo] S. Cambanis and C. Houdré, *On the continuous wavelet transform of second-order random processes*, IEEE Trans. Inform. Theory **41** (1995), 628–642.

[CaMa] S. Cambanis and E. Masry, *Wavelet approximation of deterministic and random signals: Convergence properties and rates*, IEEE Trans. Inform. Theory **40** (1994), 1013–1029.

[CT] B. Cheng and H. Tong, *A theory of wavelet representation and decomposition for a general stochastic process*, Athens Conference on Applied Probability and Time Series, vol. II, Lecture Notes in Statist. **115** (1996), Springer-Verlag, New York, 115–129.

[Da] R. Dahlhaus, *Fitting time series models to nonstationary processes*, Ann. Statist. **25** (1997), 1–37.

[Dau] I. Daubechies, *Orthonormal bases of compactly supported wavelets*, Comm. Pure Appl. Math. **41** (1988), 909–996.

[Do] D. L. Donoho, *Denoising via soft thresholding*, IEEE Trans. Inform. Theory **41** (1995), 613–627.

[DJKP] D. L. Donoho, I. M. Johnstone, G. Kerkyacharian, and D. Picard, *Wavelet shrinkage: Symptopia?*, J. Roy. Statist. Soc. Ser. B **57** (1995), 301–337.

[Fla] P. Flandrin, *Temps-fréquence*, Hèrmes, Paris, 1993.

[JoSi] I. M. Johnstone and B. W. Silverman, *Wavelet threshold estimators for data with correlated noise*, J. Roy. Statist. Soc. Ser. B **59** (1997), 319–351.

[KaSh] S. Kawasaki and R. Shibata, *Weak stationarity of a time series with wavelet representation*, Japan J. Indust. Appl. Math. **12** (1995), 37–45.

[Mal] S. G. Mallat, *A theory for multiresolution signal decomposition: The wavelet representation*, IEEE Trans. Pattn. Anal. Mach. Intell. **11** (1989), 674–693.

[MoCh] P. Morettin and Ch. Chang, *Wavelet analysis for stationary processes*, techn. report, Instituto de Matemática e Estatística, São Paulo, Brasil, 1995.

[Na] G. P. Nason, *WaveThresh* 3, techn. report, Univ. Bristol, 1996.

[NaSi] G. P. Nason and B. Silverman, *The stationary wavelet transform and some statistical applications*, Lecture Notes in Statist. **103** (1995), Springer-Verlag, New York, 281–299.

[NaSi2] G. P. Nason and B. W. Silverman, *Some more statistical applications of the stationary wavelet transform*, techn. report, Univ. Bristol, 1996.

[NvS] M. H. Neumann and R. von Sachs, *Wavelet thresholding in anisotropic function classes and application to adaptive estimation of evolutionary spectra*, Ann. Statist. **25** (1997), 38–76.

[Priest] M. B. Priestley, *Spectral analysis and time series*, Academic Press, London, 1981.

[RiVe] O. Rioul and M. Vetterli, *Wavelets and signal processing*, IEEE Signal Process. Magazine **8** (1991), 14–38.

[vS] R. von Sachs, *Wavelet spectrum decomposition of stochastic processes*, 1995, (unpublished manuscript)

[vSNK] R. von Sachs, G. P. Nason, and G. Kroisandt, *Adaptive estimation of the evolutionary wavelet spectrum*, techn. report, Statistics, Stanford Univ., 1997.

[vSS] R. von Sachs and K. Schneider, *Wavelet smoothing of evolutionary spectra by non-linear thresholding*, Appl. Comput. Harmon. Anal. **3** (1996), 268–282.

[Wal] G. G. Walter, *Wavelets and other orthogonal systems with applications*, CRC Press, Boca Raton, Florida, 1994.



FB Mathematik, Universität Kaiserslautern, D-67653 Kaiserslautern, Germany
*E-mail address*: `rvs@mathematik.uni-kl.de`

Department of Mathematics, University of Bristol, University Walk, Bristol BS8 1TW, UK
*E-mail address*: `G.P.Nason@Bristol.ac.uk`

FB Mathematik, Universität Kaiserslautern, D-67653 Kaiserslautern, Germany
*E-mail address*: `kroisandt@mathematik.uni-kl.de`